NONLINEAR OPERATORS AND NONLINEAR
EQUATIONS OF EVOLUTION IN BANACH SPACES

PROCEEDINGS OF SYMPOSIA
IN PURE MATHEMATICS
Volume XVIII, Part 2

NONLINEAR OPERATORS AND NONLINEAR EQUATIONS OF EVOLUTION IN BANACH SPACES

By

FELIX E. BROWDER

AMERICAN MATHEMATICAL SOCIETY
Providence, Rhode Island
1976

Proceedings of the Symposium in Pure Mathematics
of the American Mathematical Society
Held in Chicago, Illinois
April 16–19, 1968

Prepared by the American Mathematical Society
under National Science Foundation Grant GP-8462

AMS 1970 *Subject Classifications*.
Primary 47H05, 47H10, 47H15, 34G05
Secondary 34J05, 34K05, 58C10, 54H25

Library of Congress Cataloging in Publication Data

Browder, Felix E.
  Nonlinear operators and nonlinear equations of evolution in Banach spaces.

  (Proceedings of symposia in pure mathematics; v. 18, pt. 2)
  Paper prepared for the Symposium on Nonlinear Functional Analysis, Chicago, 1968.
  Bibliography: p.
  Includes indexes.
  1. Nonlinear operators. 2. Differential equations, Nonlinear. 3. Functional equations. 4. Banach spaces. I. American Mathematical Society. II. Title. III. Series.
  QA329.8.B76   515'.72   74-34154
  ISBN 0-8218-0244-5

International Standard Book Number 0-8218-0244-5
Copyright © 1976 by the American Mathematical Society

Printed in the United States of America

All rights reserved except those granted to the United States Government.
This book may not be reproduced in any form without the permission of the publishers.

CONTENTS

	Preface	vii
0.	Introduction	1
1.	Contractive mappings	5
2.	Locally Lipschitzian mappings	20
3.	Φ-accretive and Φ-coaccretive mappings	27
4.	Covering space methods	47
5.	Limits of invertible and semi-invertible mappings	63
6.	Fixed point and mapping theory for compact multi-valued mappings	71
7.	Monotone mappings in Banach spaces	79
8.	Nonexpansive mappings in Banach spaces	101
9.	Accretive mappings and nonlinear equations of evolution	121
10.	Existence theorems involving accretive mappings	163
11.	Nonlinear interpolation	176
12.	Generalizations of the topological degree of a mapping	183
13.	Compact perturbations of nonexpansive, monotone, and accretive mappings	220
14.	Nonlinear Fredholm mappings	242
15.	Orientation-preserving and complex analytic mappings	248
16.	Asymptotic fixed point theorems	265
17.	A-proper mappings, approximation methods, and related generalizations of topological degree	273
	Bibliography	285
	Author Index	303
	Subject Index	307

PREFACE

The present volume contains the book-length text of a paper entitled "Nonlinear operators and nonlinear equations of evolution in Banach spaces" composed in its entirety during the calendar year 1968 to be published as part of the Proceedings of the Symposium on Nonlinear Functional Analysis held in connection with the April 1968 meeting in Chicago of the American Mathematical Society. This paper is in fact a detailed treatment in book form of most of the major branches of nonlinear functional analysis as they had developed up to 1968, and no significant alterations or additions have been made since that time except for the correction of errors in detail. Since the manuscript of this work has had a wide circulation in mimeographed form and been referred to in a considerable number of research papers since 1968, an explanation for the rather belated publication of this book is called for.

The observant reader will note that despite the presence of a very lengthy bibliography of papers and books on nonlinear functional analysis which had appeared or existed in preprint form by the end of 1968, the discussion of the text does not give references to the literature for the origin of results presented nor does it indicate that some of the results were developed here for the first time. It was the writer's original intention to add an additional section to the present text developing such references in detail and presenting a historical survey of the development of nonlinear functional analysis in the past decade. It was the difficulties encountered in fulfilling this intention in a completely adequate and precise way together with the pressure of other work which led to the initial delay in publication of this book. The initial delay, and the effect that it had in making some of the historical material which had been prepared out of date in terms of new perspectives and new results on some of the topics of major interest, made it obvious that an additional section surveying the whole field of nonlinear functional analysis and its development would have to contain a good deal of additional material covering new results and new ideas since 1968. As time has passed, it has become clear that this additional section of the present book would be a complete new book in its own right. This is the crucial reason for the publication of this book six years late in 1976 rather than 1969.

It is my firm conviction, which is shared by other workers in this field whose judgement I trust, that despite this delay the publication of the text in its present form is still of great value. This is the case both because of the text's presence as an "underground" part of the literature and the many references to it in the papers since 1968 and because it still is the only treatment in systematic form of the field of nonlinear functional analysis as a whole which deals with the major developments of the 1960's. It is my hope to write the book on the more recent development of nonlinear functional analysis which was the logical outcome of the uncompleted effort involved in the present work and to include the historical survey which was omitted here. The rapid development of this rapidly growing field makes this task far from easy, but one which I regard as of great potential usefulness to the mature development of nonlinear functional analysis. In the meantime, however, the text of the original manuscript is presented here with the hope that it will be of value to workers and students of this field.

FELIX E. BROWDER

NONLINEAR OPERATORS AND NONLINEAR EQUATIONS OF EVOLUTION IN BANACH SPACES

Felix E. Browder

0. Introduction. It is the purpose of the present paper to develop in a reasonably complete and logically coherent way, a significant portion of the results obtained in the most recent development of nonlinear functional analysis in the past several years. For simplicity of presentation (to avoid increasing still further what is already a rather elaborate structure of results and proofs) and because of its key role in applications, we restrict our discussion to the treatment of nonlinear functional equations involving operators or mappings between Banach spaces. The types of problems which we treat concern questions of the following sort:

(a) Given a mapping T from a portion of the Banach space X to the Banach space Y, what is the nature of the set of solution of the equation $T(u) = y$ for a given y in Y (i.e., what is the range of T, what is $T^{-1}(y)$ for fixed y, and how does $T^{-1}(y)$ vary with y)?

(b) Given a mapping f of a portion of the Banach space X into itself (or into 2^X), do there exist fixed points of f (or more generally points x in X such that $x \in f(x)$)?

(c) Given a nonlinear operator T from a portion of X to X (or more generally, a family T_t of such operators indexed by the real number t) do there exist solutions of the equation of evolution

$$du(t)/dt = T(u(t)) \qquad (t \in [0, a])$$

with prescribed initial condition $u(0) = u_0$ (or more generally corresponding solutions of the time dependent equation

$$du(t)/dt = T_t(u(t)) \qquad (t \in [0, a]))?$$

The classical development of nonlinear functional analysis arose contemporaneously with the beginnings of linear functional analysis at about the beginning of the twentieth century in the work of such men as Picard, S. Bernstein, Lyapounov, E. Schmidt, and Lichtenstein and was motivated by the desire to study the existence and properties of boundary value problems for nonlinear partial differential equations. Its most classical tool was the Picard contraction principle (put in its sharpest abstract form by Banach in his Thesis in 1922). Beyond the early development of bifurcation theory by Lyapounov and Schmidt, the second and even more fruitful branch of the classical methods in nonlinear functional analysis was developed in the theory of compact nonlinear mappings in Banach spaces of the late 1920's and early 1930's including Schauder's well-known fixed point theorem and the extension of the Brouwer topological degree by Leray and Schauder in 1934 to mappings in Banach spaces of the form $I + C$ with C compact (as well as interesting related results of Cacciopoli on nonlinear Fredholm mappings).

The central role of compact mappings in this phase of development of nonlinear functional analysis was due in part to the nature of the technical apparatus being developed but also in part to a not always fruitful tendency to see the theory of integral equations as the predestined domain of application of the theory to be developed. Since, however, in fact, the more significant analytical problems lie in the somewhat different domain of boundary value problems for partial differential equations and since the efforts to apply the theory of compact operators (and in particular the Leray-Schauder theory) to the latter problems have given rise to demands for ever more inaccessible (and sometimes, invalid) *a-priori estimates* in these problems, the hope of applying nonlinear functional analysis to problems of this type centers on a general program of creating new theories for significant classes of *noncompact* nonlinear operators. The focus of this study is then on finding such classes which have the opposed characteristics of being narrow enough to have a significant structure of results while also being wide enough to have a significant variety of applications.

To classify the classes of nonlinear operators or mappings that might be considered, one begins with a rough breakdown according to the category of spaces and mappings within which such classes can be defined, e.g.,

(i) topological

(ii) metric

(iii) locally convex topological vector spaces

(iv) metric vector spaces

(v) differentiable manifolds modelled on Banach spaces

(vi) complex analytic manifolds modelled on complex Banach spaces.

We shall consider a variety of mappings defined in Banach spaces but refer them to the broader category as given above within which they can be formulated.

From the point of view of applications to partial differential equations, the most important class of those which are discussed below is that of the *monotone* mappings from a reflexive Banach space X to its dual space X^*.

A mapping T from a subset of the Banach space X (or the locally convex topological vector space X) to the space X^* of continuous linear functionals on X is said to be monotone if (setting (w, u) to be the pairing between the functional w, an element of X^*, and the element u of X), we have $(T(u) - T(v), u - v) \geq 0$ for all u and v in the domain of T. A special case of this definition can be considered in which the mapping T is the Fréchet derivative f' of a real-valued function f of class C^1 on the Banach space X. Then T is monotone in the sense just defined if and only if f is convex. Thus we see that the concept of monotonicity is an extension to general mappings from X to X^* of the characteristic property of the derivatives of convex real functionals on X. It is a familiar fact in the direct method of the calculus of variations, that in the infinite-dimensional case, it is the convexity properties of the functional f (or their perturbations) which are the principal means by which one can prove the existence of critical points of f. The theory of monotone mappings in reflexive Banach spaces exhibits the analogous fact that the monotonicity of a mapping T can be applied to obtain significant information on the character of the range of T. With appropriate modification of hypotheses, such a theory can be created on a general locally convex topological vector space X, and this is the natural category from the formal point of view for the study of monotone mappings from X to X^*.

A second important class of nonlinear, noncompact mappings can be defined within the category of metric spaces, the class of *nonexpansive* mappings U of a subset of a Banach space X into X. Such a mapping U is said to be nonexpansive for a metric space X with distance function d if for all u and v in the domain of U, $d(U(u), U(v)) \leq d(u, v)$. This is a specialization of the more general notion of a Lipschitzian mapping with Lipschitz constant k for which $d(U(u), U(v)) \leq k\, d(u, v)$. For $k < 1$, we have the strict contractions of the Picard contraction principle which asserts that for a complete metric space X, each strict contraction f mapping X into X has an unique fixed point x_0, and if x_1 is an arbitrary point of X, $f^n(x_1)$ converges to x_0 as $n \to \infty$. For the limit class of nonexpansive mappings, the existence of fixed point theorems can be obtained under more restrictive hypotheses involving the uniform convexity of the Banach space X (or slightly more general properties) and the assumption that U maps a bounded closed convex subset G of X into itself. It is of interest to note that under these restrictions, a considerable part of the theory of strict contractions can be mirrored in the theory of nonexpansive mappings.

A third major class of nonlinear mappings consists of the *accretive* mappings from a subset of a Banach space X into X. The most principled (though necessarily vague) definition of this class for a metric vector space X would be that T is accretive if $(-T)$ is the infinitesimal generator of a one-parameter semigroup of nonexpansive self-mappings of X, i.e., roughly if the solutions of the initial value problems

$$du(t)/dt = -T(u(t)), \quad t \geq 0, \quad u(0) = u_0$$

being written in the form $U(t)(u_0) = u(t)$, the mappings $U(t)$ for each fixed $t \geq 0$ are nonexpansive self-mappings of X. For the theory of accretive operators in Banach spaces, the above property is not the definition but a basic theorem under suitable restrictions on the domain or continuity of the mapping T. A closely related property which is used to define the accretiveness property is the following: A mapping T from a subset $D(T)$ of X to X is said to be accretive if for all u and v of $D(T)$,

$$(J(u - v), T(u) - T(v)) \geq 0,$$

where J is a single-valued duality mapping of X into X^*, i.e., a mapping such that for each u in X, $\|J(u)\| = \|u\|$, and $(J(u), u) = \|u\|^2$.

The properties of accretive mappings are obviously closely related as the above remarks indicate with the theory of generation of nonlinear semigroups in §9. There is a strong interplay in addition between the theory of range of accretive mappings on the one side and the fixed point theory of nonexpansive mappings on the other. As a special case, when X is a Hilbert space H, the accretive mappings coincide with the monotone mappings of H into H.

The forms of the definitions for monotone and accretive mappings have obvious resemblances. More general definitions can be used for Φ-accretive mappings where the accretive inequality is replaced by

$$(\Phi(u, v), T(u) - T(v)) \geq 0$$

and related definitions for Φ-accretive and Φ-monotone mappings.

Rather than continuing the classification of mappings by the category of structures within which they are defined, let us turn now to the fourth major theme of the discussion below in terms of the methodology of argument and the methods erected upon these arguments. This fourth theme is that of the *generalized topological degree* as defined for various classes of noncompact nonlinear mappings between Banach spaces.

The topological degree was defined by Leray and Schauder for mappings T from the closure cl (G) of a bounded open set of the Banach space X with values in X and a point y in $X - T(\text{boundary } G)$, where T is of the form $T = I + C$ with C compact. It is an integer-valued function, written $\deg_{\text{LS}} (I + C, G, y)$ below, which has the properties that it is additive on the domain, invariant under suitable homotopies, that it normalized and ± 1 for homeomorphisms $(I + C)$, and that if $\deg_{\text{LS}} (T, G, y) \neq 0$, then there exists x in G such that $T(x) = y$. In the generalized degree theory, we construct similar functions for more general classes of mappings and apply these functions to obtain existence and multiplicity results for solutions of nonlinear functional equations involving mappings in these classes.

In somewhat rough terms, the classes of mappings T which we consider from a subset of a Banach space X to another Banach space Y are those which can be represented in the form $T(u) = S(u, u)$ where S is a mapping of $X \times X$ into Y, with $S(\cdot, v)$ a homeomorphism for each v and $S(u, \cdot)$ compact for each u. We obtain a degree function for an open set G such that T maps cl (G) into Y and for a point y in $Y - T(\text{bdry } (G))$, but a degree function which also depends upon the representation S for T and which we write in variants of the notation deg $([T, S], G, y)$. Special cases of such a definition are obtained when the homeomorphisms $S(\cdot, v)$ are restrained to lie in a given class M of homeomorphisms from X to Y, and if M is a convex class, it is shown that the degree function depends only on T but not on the representation S. Particular examples of such convex classes are the strongly monotone mappings, the strongly accretive mappings, and mappings of the form $I - U$ with U a strict contraction. For mappings represented with respect to such convex classes, we can obtain a degree theory with a degree function $\deg (T, G, y)$ for mappings T which are the uniform limits on bounded sets of mappings T_j which have representations of the form $T_j(u) = S_j(u, u)$. Such limit classes include the monotone mappings, accretive mappings, and mappings of the form $I - U$ with U nonexpansive.

The generalized degree theory can be applied also to the study of nonlinear Fredholm mappings, as well as nonlinear orientation-preserving mappings and complex analytic (holomorphic) mappings in complex Banach spaces. A somewhat different generalization of the degree theory is given for a class of mappings defined by general finite-dimensional approximation methods (such as the methods of Galerkin type) and for which the Leray-Schauder degree function obtained from finite-dimensional uniform approximations of a given mapping is replaced by the limit of the degrees of much weaker finite-dimensional approximations.

Other general methods which are studied in the earlier part of the discussion involve the study of the structure of families of locally Lipschitzian mappings, covering space arguments under delicate hypotheses for uniqueness of solutions of nonlinear functional equations, the study of the ranges and inverse images of limits of invertible and semi-invertible mappings, nonlinear interpolation results for nonexpansive and accretive mappings, compact perturbation theory for various classes of nonlinear mappings (nonexpansive, accretive, and monotone), and asymptotic fixed point theorems.

The great length of the systematic discussion has precluded the inclusion of material on applications to partial differential equations, integral equations, and existence of periodic solutions for general classes of nonlinear initial value problems (for the first of which, we refer to the writer's recent survey paper [**122**]).

The writer should like to express his gratitude to Dr. Chaitan C. Gupta for his invaluable assistance in correcting the manuscript of the present paper, and to Mr. Randell W. Magee for typing the manuscript.

1. Contractive mappings. The simplest and most classical of all the results of nonlinear functional analysis is undoubtedly the Contraction Mapping Theorem of Picard-Banach-Cacciopoli *et al.* In its conventional form, this theorem asserts the following:

Let X be a complete metric space, f a mapping of X into X such that for a fixed constant $k < 1$ and all x and u in X,

$$d(f(x), f(u)) \leq k\, d(x, u)$$

(where $d(\cdot, \cdot)$ denotes the distance function on X). Then f has exactly one fixed point x_0 in X, and for each y_0 in X, the sequence $f^n(y_0)$ converges to x_0 in X as $n \to \infty$ with

$$d(f^n(y_0), x_0) \leq k^{n-1}(1-k)^{-1} d(y_0, f(y_0)).$$

In the present section (as well as some subsequent sections of the present paper) we shall give some extensions, generalizations, and applications of this result.

We note first of all that the Picard theorem as just stated is a special case of the following simple result:

THEOREM (1.1). *Let X_0 be a bounded subset of a complete metric space X, f a continuous mapping of X into X which carries X_0 into X_0 and such that there exists a* *function* $\psi(r)$ *for* $r \geq 0$ *with the following properties:*

(a) *For all* $r > 0$, $0 < \psi(r) < r$. *The function* ψ *is monotone nondecreasing in* r *and is continuous from the right (i.e.,* $\psi(r_j) \to \psi(r')$ *if* $\{r_j\}$ *is a decreasing sequence converging to* r').

(b) *For all* x *and* u *in* X_0,

$$d(f(x), f(u)) \leq \psi(d(x,u)).$$

Then f *has a fixed point* x_0 *in* X, *and for any* y_0 *in* X_0, $f^n(y_0)$ *converges to* x_0 *in* X *with*

$$d(f^n(y_0), x_0) \leq \psi^{n-1}(\text{diam } X_0) \to 0$$

(where ψ^n *denotes the nth iterate of* ψ *and* diam (X_0) *is the diameter of* X_0).

PROOF OF THEOREM (1.1). Let $r_n = \psi^n(\text{diam } X_0)$. We may assume that $r_n > 0$ for all n, and hence

$$r_n = \psi(r_{n-1}) < r_{n-1}, \qquad r_n > 0.$$

Therefore r_n converges to some $r' \geq 0$ as $n \to \infty$. Since ψ is continuous from the right, $\psi(r_n) \to \psi(r')$. Since $\psi(r_n) = r_{n+1}$, it follows that $r' = \psi(r')$. Since $\psi(r') < r'$ by condition (a) for $r' > 0$, we see that $r' = 0$ and $r_n \to 0$ as $n \to \infty$.

On the other hand, let q_n be the diameter of the orbit under f of $y_n = f^n(y_0)$, i.e., $q_n = \sup_{j,k \geq n} d(f^j(y_0), f^k(y_0))$. To prove that $f^n(y_0)$ converges in X, it suffices by the completeness of X to show that $q_n \to 0$ as $n \to \infty$. However,

$$q_n \leq \sup_{j,k \geq n} \psi(d(f^{j-1}(y_0), f^{k-1}(y_0)))$$

and since $\psi(r)$ is monotone nondecreasing in r,

$$\sup_{j,k \geq n} \psi(d(f^{j-1}(y_0), f^{k-1}(y_0))) \leq \psi\left(\sup_{j,k \geq n} d(f^{j-1}(y_0), f^{k-1}(y_0))\right) = \psi(q_{n-1}),$$

i.e., $q_n \leq \psi(q_{n-1})$. Iterating on n, we obtain $q_n \leq \psi^{n-1}(q_1) \leq \psi^{n-1}(\text{diam } X_0) \to 0$. Thus $f^n(y_0)$ converges to some x_0 in X. Finally,

$$d(f^n y_0, x_0) = \lim_{m \to \infty} d(f^n(y_0), f^m(y_0)) \leq q_n \leq \psi^{n-1}(\text{diam } X_0). \qquad \text{q.e.d.}$$

COROLLARY TO THEOREM (1.1). *Let* X *be a complete metric space,* X_0 *a bounded subset of* X, g *a continuous mapping of* X *into* X *which maps* X_0 *into* X_0 *and such that for a given integer* m, $f = g^m$ *satisfies the hypothesis of Theorem* (1.1). *Then* g *has a unique fixed point* x_0 *in the closure of* X_0 *in* X *and for every* y_0 *in* X_0, $g^n(y_0)$ *converges to* x_0 *in* X.

PROOF OF THE COROLLARY. By Theorem (1.1) f has a fixed point x_0 in the closure of X_0 in X and for each y_0 in X_0, $f^k(y_0)$ converges to x_0 in X as $k \to \infty$, i.e., $g^{km}(y_0) \to x_0$ as $r \to \infty$. We assert that the fixed point set of the mapping f on the closure of X_0 (and a fortiori the fixed point set of g on the same set) must consist of the single point x_0. Indeed, suppose x_1 is another fixed point of f in cl (X_0). Let

$r_0 = d(x_0, x_1) > 0$. Since $\psi(r_0) < r_0$, we may choose $\epsilon > 0$ such that $3\epsilon < r_0 - \psi(r_0)$. Since f is continuous, we may choose points y_0 and y_1 in X_0 such that

$$d(f(y_0), x_0) < \epsilon, \quad d(f(y_1), x_1) < \epsilon, \quad \psi(d(y_0, y_1)) < \psi(r_0) + \epsilon.$$

Then we would have

$$r_0 = d(x_0, x_1) \leq d(f(y_0), f(y_1)) + 2\epsilon < \psi(d(y_0, y_1)) + 2\epsilon < \psi(r_0) + 3\epsilon,$$

i.e., $r_0 - \psi(r_0) < 3\epsilon$ which contradicts the choice of ϵ. Thus the fixed point set of g and of f consists of the single point x_0.

Finally, $g^{km+j}(y_0) \to g^j(x_0) = x_0$, for every j with $0 \leq j \leq m$, so that $g^n(y_0) \to x_0$ as $n \to \infty$. q.e.d.

One of the basic applications of the Contraction Mapping Theorem in its original form was for the proof of the local existence theorem for solutions of initial value problems for ordinary differential equations. We now give a sharp result of this type for ordinary differential equations in a Banach space, and then extend it to the case of *nonlinear equations of evolution* in a Banach space, where the latter term will be used in our discussion to refer to differential equations in a Banach space involving operators or mappings which are not everywhere defined or continuous.

THEOREM (1.2). *Let X be a Banach space, x_0 a point of X, R^+ the nonnegative real numbers. Let N be a neighborhood of the point $[0, x_0]$ in $R^+ \times X$ and f a uniformly continuous mapping of N into X. Suppose that there exists a positive function $\beta(t, r)$ for $0 < t < d_0$, $|r| \leq d_1$ with $\beta(t, 0) = 0$ for all $t > 0$, which is continuous in the pair $[t, r]$ on this domain and such that for any solution $u(t)$ on $(0, d_0)$ of the differential equation $du(t)/dt = \beta(t, u(t))$ with $u(t)t^{-1} \to 0$ as $t \to 0$, it follows that $u(t)$ is identically zero on $(0, d_0)$. Suppose finally that for all $[t, u]$ and $[t, v]$ in N,*

(1.1) $$\|f(t, u) - f(t, v)\| \leq \beta(t, \|u - v\|).$$

Then:

(a) *There exists d_2 with $0 < d_2 < d_0$ such that on the interval $[0, d_2]$ we have one and only one solution $x(t)$ of the differential equation*

$$dx(t)/dt = f(t, x(t)) \quad (t \in [0, d_2])$$

with $x(0) = x_0$. Here x denotes a C^1 function from $[0, d_2]$ to X whose values lie in the set X for which f is defined and the derivative is taken in the strong sense.

(b) *If the function $\beta(t, r)$ can be taken to be monotone increasing in r for $r > 0$, then the solution $x(t)$ of part (a) can be obtained as the uniform limit of the successive approximants $x_n(t)$ defined by:*

(1.2) $$x_n(t) = x_0 + \int_0^t f(s, x_{n-1}(s))\, ds \quad (x_0(t) \equiv x_0).$$

Let us remark explicitly that results of the type of Theorem (1.2) are of the type of existence theorems for the infinite-dimensional case involving an assumption of a local Lipschitz condition. Though the inequality (1.1) is much weaker in a concrete analytical sense than the Lipschitz condition in the standard form and the argument for existence a good deal more technically interesting, we shall be concerned in our later discussion in obtaining results on existence and uniqueness with conditions of an essentially weaker type.

The proof of Theorem (1.2) rests upon the use of the following lemma:

LEMMA (1.1). *Let $\beta(t, r)$ be a continuous function for $0 < t < d_0$, $0 \leq r \leq d_1$ with $\beta(t, 0) = 0$ for all t, $\beta(t, r) \geq 0$ for all t and r in its domain. Suppose that for any solution $u(t)$ on the interval $(0, d]$, $d < d_0$, of the differential equation $du/dt = \beta(t, u(t))$ with $u(t)t^{-1} \to 0$ as $t \to 0$, it follows that $u(t)$ is identically zero on $(0, d)$. Let T and M be given positive numbers, $\epsilon > 0$ be given, s a given function with $s(t) \to 0$ as $t \to 0+$.*

Then there exists $\delta > 0$ which depends only upon ϵ, M, s, and T such that if $p(t)$ is any nonnegative solution on the interval $[0, T]$ ($T < d_0$) of the differential inequality for a given function q,

$$p(t) - p(t_1) \leq \int_{t_1}^{t} \beta(r, p(r))\, dr + q(t)$$

for all $t, t_1 > 0$ with $t_1 < t < T$, and if the functions p and q satisfy the following further conditions:

$$|p(t) - p(t_1)| \leq M\, |t - t_1| \qquad (0 \leq t, t_1 \leq T);$$

$$|p(0)| \leq \delta; \qquad t^{-1}p(t) \leq s(t) \quad \text{for } t > 0,$$

$$|q(t)| \leq \delta, \qquad t \in [0, T],$$

then it follows that for all t in $[0, T]$, $|p(t)| \leq \epsilon$.

PROOF OF LEMMA (1.1). Suppose that the conclusion of the lemma is false. Then there exist $M, T > 0$ and two sequences of functions $\{p_n(t)\}$ and $\{q_n(t)\}$ such that $q_n(t)$ converges uniformly to 0 on $[0, T]$ as $n \to \infty$, $p_n(0)$ converges to zero as $n \to \infty$, and we have for all $t_1 < t$ in $[0, T]$, $\sup_{t \in [0, T]} |p_n(t)| \geq \epsilon > 0$, and

$$p_n(t) - p_n(t_1) \leq \int_{t_1}^{t} \beta(r, p_n(r))\, dr + q_n(t),$$

$$|p_n(t) - p_n(t_1)| \leq M\, |t - t_1|; \qquad t^{-1}p_n(t) \leq s(t).$$

By the second of these conditions and Ascoli's theorem, we may pass to an infinite subsequence and assume without loss of generality that $p_n(t)$ converges uniformly on the interval $[0, T]$ to a continuous Lipschitzian function $p(t)$, $p(t) \geq 0$ for all t, which satisfies the conditions:

$$p(t) - p(t_1) \leq \int_{t_1}^{t} \beta(r, p(r))\, dr \qquad (0 < t_1 < t \leq T),$$

$$p(0) = 0, \quad t^{-1}p(t) \leq s(t), \quad t > 0.$$

Furthermore, $\sup_{t \in [0, T]} p(t) \geq \epsilon > 0$.

Let $t_0 > 0$, and consider the minimal solution of the differential equation $dq(t)/dt = \beta(t, q(t))$ with initial value $q(t_0) = p(t_0)$, which exists in a neighborhood of $t = t_0$ by the Peano Existence Theorem. The solution $q(t)$ may be approximated uniformly on its domain of existence by the minimal solutions of the differential equations

$$dq_n(t)/dt = \beta(t, q_n(t)) + 1/n$$

with $q_n(t_0) = q(t_0) = p(t_0)$, and for $t < t_0$, $q_n(t)$ must actually be less than $p(t)$ whenever both are defined, since at the greatest point of equality, less than t_0, the difference of slopes of the curves would have the wrong sign. Hence $q(t) \leq p(t)$ for all $t \leq t_0$ for which $q(t)$ is defined. By the standard continuation properties of solutions of ordinary differential equations involving real-valued functions together with the assumption that $\beta(t, r)$ is continuous for $t > 0$, it follows that $q(t)$ can be continued in t for $t > 0$ up to some point $t_1 > 0$ for which $q(t_1) = 0$, or else that $q(t) \to 0$ as $t \to 0$. In the first case, we continue $q(t)$ for $t < t_1$ by letting $q(t) = 0$ on this domain. In both cases, we obtain a solution $q(t)$ of the differential equation on the open interval $(0, t_0)$ with $q(t) \to 0$ as $t \to 0+$. In the case in which $q(t_1) = 0$ for some $t_1 > 0$, we know already that $q(t)t^{-1} = 0$ on a neighborhood of $t = 0$, and hence by uniqueness of solutions of our given differential equation, $q(t)$ is identically zero on $(0, t_0)$. In the second case, $t^{-1}q(t) \leq t^{-1}p(t) \leq s(t) \to 0$ as $t \to 0+$. Hence $q(t)$ is identically zero on $(0, t_0)$.

It follows that $q(t_0) = p(t_0) = 0$, i.e., p is identically null. This contradicts the construction of the function $p(t)$. q.e.d.

PROOF OF THEOREM (1.2). PROOF OF (a). For each $\epsilon > 0$, we approximate the differential equation for which we desire a solution by the delay-differential equation

$$dx_\epsilon(t)/dt = f(t, x_\epsilon(t - \epsilon)),$$

with the initial condition

$$x_\epsilon(t) = x_0 \quad (t \leq 0).$$

The existence of the solution of this latter equation for any $\epsilon > 0$ follows by a trivial step-by-step argument over integrals of length ϵ, since on each such interval the right-hand term of the delay-differential equation is given by the already constructed values of $x_\epsilon(t)$ on the preceding interval. This solution is obviously unique by the same argument.

We propose to obtain the desired solution $x(t)$ of the original differential equation by taking the limit of $x_\epsilon(t)$ as $\epsilon \to 0$. If we can establish existence, the uniqueness of the solution follows as a very special case of Lemma (1.1). To show that $x_\epsilon(t)$ has a limit as $\epsilon \to 0$, we consider another value of the approximation parameter $\lambda > 0$, and note that the difference $x_\epsilon - x_\lambda$ satisfies the differential equation

$$\frac{dx_\epsilon}{dt}(t) - \frac{dx_\lambda}{dt}(t) = f(t, x_\epsilon(t - \epsilon)) - f(t, x_\lambda(t - \lambda))$$

$$= f(t, x_\epsilon(t)) - f(t, x_\lambda(t)) + R_\epsilon(t) + R_\lambda(t),$$

where $R_\epsilon(t) = f(t, x_\epsilon(t - \epsilon)) - f(t, x_\epsilon(t))$, and $R_\lambda(t)$ is the corresponding difference with ϵ replaced by λ.

The solutions $x_\epsilon(t)$ of the initial value problem for the delay-differential equation above can be constructed on an interval $[0, d]$ with $d > 0$ and independent of $\epsilon > 0$. Indeed, by hypothesis, f is defined and uniformly continuous (and hence uniformly bounded) on some neighborhood N of $[0, x_0]$ in $R^+ \times X$, and we may assume N of the form $[0, d_0] \times B_{d_1}(x_0)$, where $B_{d_1}(x_0)$ is the closed ball of radius d_1 about x_0 in X. For any interval on which $x_\epsilon(t)$ may be constructed, we therefore have

$$\|dx_\epsilon(t)/dt\| \leq M,$$

where M is a bound for the norm of $f(t, x)$ in N. If we take a maximal such interval, $[0, T]$, it follows that either $|t - t_0| \geq d_0$ or $|x - x_0| \geq d_1$ for some point t in $[T, T + \epsilon]$. The first inequality implies that $T \geq t_0 + d_0$, while the second implies that $M(T - \epsilon) \geq d_1$ since it follows that the continuous curve $x_\epsilon(t)$ must emerge from $B_{d_1}(x_0)$ in the preceding subinterval $[T - \epsilon, T]$. In both cases, we obtain a fixed positive lower bound for T which does not depend on the choice of $\epsilon > 0$ for ϵ sufficiently small. For all the solutions thus obtained, we may assume that $\|f(t, x_\epsilon(t))\| \leq M$ since the solutions must lie in the ϵM-neighborhood of N and by shrinking N slightly, we can ensure that f remains bounded uniformly on this neighborhood of N for ϵ sufficiently small. We can also ensure that f is uniformly continuous on the ϵM-neighborhood of N.

Since $\|dx_\epsilon/dt\| \leq M$ for all $\epsilon > 0$ and all t in $[0, T]$, it follows that

$$\|x_\epsilon(t - \epsilon) - x_\epsilon(t)\| \leq \epsilon M.$$

Thus given $\xi > 0$, we can choose $\epsilon(\xi) > 0$ such that for $\epsilon, \lambda < \epsilon(\xi)$, we have

$$\|R_\epsilon(t)\| \leq \xi, \quad \|R_\lambda(t)\| \leq \xi, \quad t \in [0, T].$$

We now consider the nonnegative real-valued function $p_{\epsilon\lambda}(t)$ given by $p_{\epsilon\lambda}(t) = \|x_\epsilon(t) - x_\lambda(t)\|$. Since for $t_1 < t < T$, we have

$$[x_\epsilon(t) - x_\lambda(t)] - [x_\epsilon(t_1) - x_\lambda(t_1)] = \int_{t_1}^t \{[f(s, x_\epsilon(s)) - f(s, x_\lambda(s))] + R_\epsilon(s) + R_\lambda(s)\}\, ds,$$

we apply the triangle inequality and obtain

$$p_{\epsilon\lambda}(t) \leq p_{\epsilon\lambda}(t_1) + \int_{t_1}^t \|f(s, x_\epsilon(s)) - f(s, x_\lambda(s))\|\, ds + q_{\epsilon,\lambda}(t),$$

with

$$q_{\epsilon\lambda}(t) = \int_0^t \{\|R_\epsilon(s)\| + \|R_\lambda(s)\|\}\, ds \leq 2\xi T \quad (0 \leq t \leq T).$$

Applying the assumed bound for f in terms of the given function $\beta(t, r)$, we obtain

$$p_{\epsilon\lambda}(t) - p_{\epsilon\lambda}(t_1) \leq \int_{t_1}^t \beta(s, p_{\epsilon\lambda}(s))\, ds + q_{\epsilon\lambda}(t),$$

$$p_{\epsilon\lambda}(0) = 0, \quad |p_{\epsilon\lambda}(t) - p_{\epsilon\lambda}(t_1)| \leq 2M |t - t_1|,$$

$$t^{-1} p_{\epsilon\lambda}(t) \leq s(t),$$

where
$$s(t) = \operatorname{oscill} \{f(y) \mid \|y - x_0\| \leq Mt\} \to 0 \quad \text{as } t \to 0+.$$

We thus have verified the appropriate conditions for the application of Lemma (1.1) which asserts that since $q_{\epsilon\lambda}(t)$ converges uniformly to 0 as $\epsilon, \lambda \to 0$, it follows that $p_{\epsilon\lambda}(t) \to 0$ uniformly on $[0, T]$. Since X is complete in its metric, it follows that $x_\epsilon(t)$ converges uniformly on $[0, T]$ in X to a continuous function x from $[0, T]$ to X. For each pair t, t_1 in $[0, T]$, and each $\epsilon > 0$, we have

$$x_\epsilon(t) - x_\epsilon(t_1) = \int_{t_1}^{t} f(s, x_\epsilon(s - \epsilon))\, ds,$$

and since $x_\epsilon(s - \epsilon)$ converges uniformly to $x(s)$, we obtain

$$x(t) - x(t_1) = \int_{t_1}^{t} f(s, x(s))\, ds,$$

so that $x(t)$ is indeed a solution of the desired differential equation with the prescribed initial value. q.e.d.

PROOF OF (b). We now assume that the function $\beta(t, r)$ is monotone nondecreasing in r for $r \geq 0$ and each t in $(0, d_0]$. We form the successive approximants x_n ($n \geq 0$) by the recursive definition:

$$x_0(t) = x_0 \qquad (t \in [0, d]),$$

$$x_n(t) = x_0 + \int_0^t f(s, x_{n-1}(s))\, ds \qquad (t \in [0, d], n \geq 1),$$

and we propose to show that these approximants are well defined for t in the interval $[0, d]$, with $d > 0$ and sufficiently small. For a suitable d_2 with $0 < d_2 \leq d$, we shall show that $x_n(t)$ converges uniformly on $[0, d_2]$ as $n \to \infty$ to a solution $x(t)$ of our given initial value problem. Since the existence result of part (a) is not used in the proof, this gives an alternative and simpler existence argument for the case of $\beta(t, r)$ monotone nondecreasing in r.

By hypothesis, there exists a neighborhood $N = [0, d_0] \times B_{d_1}(x_0)$ of $[0, x_0]$ in $R^+ \times X$ on which f is defined and such that for $[t, x]$ in N, $\|f(t, x)\| \leq M$ for a given positive constant M. Suppose that $dM \leq d_1$ and $d \leq d_0$. Then for each $n \geq 0$ and each t in $[0, d]$, the point $[t, x_n(t)]$ lies in the neighborhood N and thereby the $(n + 1)$st approximant $x_{n+1}(t)$ is well defined. Indeed, this is obviously true for $n = 0$. If it is true for a given $n \geq 0$, then for t in $[0, d]$,

$$\|x_{n+1}(t) - x_0\| \leq \int_0^d \|f(s, x_n(s))\|\, ds \leq Md \leq d_1,$$

and the corresponding assertion is valid for x_{n+1}. Hence all the approximants are well defined. Furthermore, we know that for all t in $[0, d]$, $n \geq 0$, $\|dx_n(t)/dt\| \leq M$.

To show that $x_n(t)$ converges uniformly on an interval $[0, d_2]$, it suffices to set $p_{m,n}(t) = \|x_m(t) - x_n(t)\|$ and to show that $p_{m,m}(t) \to 0$ as $m, n \to \infty$, uniformly on

$[0, d_2]$. We remark that, for $0 < t_1 < t \leq d$,

$$x_m(t) - x_n(t) = x_m(t_1) - x_n(t_1) + \int_{t_1}^{t} \frac{d}{ds}[x_m(s) - x_n(s)]\, ds.$$

For each $t > 0$, let

$$s(t) = \text{oscill}\{f(y) \mid \|y - x_0\| \leq Mt\} \qquad (s(t) \to 0 \text{ as } t \to 0+).$$

Then by the above estimate for dx_n/dt, it follows that

$$\left\|\frac{d}{dt}[x_m(t) - x_n(t)]\right\| \leq 2s(t).$$

Thus, if we define a new function $\beta_0(t, r)$ by setting

$$\beta_0(t, r) = \beta(t, r), \quad \text{if } \beta(t, r) \leq 2s(t),$$
$$= 2s(t), \quad \text{otherwise,}$$

we see that

$$\left\|\frac{d}{dt}[x_m(t) - x_n(t)]\right\| \leq \min[2s(t), \beta(t, p_{m-1,n-1}(t))]$$
$$= \beta_0(t, p_{m-1,n-1}(t)).$$

Hence for $0 \leq t_1 < t \leq d$,

$$p_{m,n}(t) - p_{m,n}(t_1) \leq \int_{t_1}^{t} \beta_0(s, p_{m-1,n-1}(s))\, ds.$$

Since $p_{m,n}(0) = 0$ for all m and n, we obtain

$$p_{m,n}(t) \leq \int_0^t \beta_0(s, p_{m-1,n-1}(s))\, ds.$$

For each nonnegative continuous real-valued function u on $[0, d_2]$, where $d_2 = \min(d, 1)$, we define a mapping B which assigns to u the new function Bu on $[0, d_2]$ given by

$$(Bu)(t) = \int_0^t \beta_0(s, u(s))\, ds \qquad (t \in [0, d_2]).$$

In terms of the mapping B, the function $p_{m,n}$ satisfies the inequality $p_{m,n}(t) \leq B(p_{m-1,n-1})(t)$ for all t in $[0, d_2]$. We introduce the pointwise ordering in which $u \leq v$ if and only if $u(t) \leq v(t)$ for all t in $[0, d_2]$. Since β and hence β_0 is monotone nondecreasing in r for fixed t, it follows that $u \leq v$ implies that $Bu \leq Bv$ (where all functions considered are nonnegative). Hence for $m > n$, $p_{m,n} = B^n(p_{m-n,0}) \leq B^n(g_0)$, where g_0 is the constant function $2M$. Since $d_2 \leq 1$ implies that $Bg_0 \leq g_0$, it follows that $B^n(g_0)$ is a decreasing sequence of nonnegative functions which therefore converges pointwise to a nonnegative function $u(t)$. Since $\beta_0(t, r)$ is uniformly bounded, it follows that B maps the whole space of continuous non-negative functions on $[0, d_2]$ into a relatively compact subset of itself. It follows that $B^n g_0$ is relatively compact in the uniform norm and hence converges uniformly to u on $[0, d_2]$ as $n \to \infty$. It follows by the continuity of the mapping B in monotone convergence that $u = \lim B^n g_0 = \lim B^{n+1} g_0 = Bu$, so that for all $t_1 < t$,

$$u(t) - u(t_1) = \int_t^t \beta_0(s, u(s))\, ds \leq \int_{t_1}^t \beta(s, u(s))\, ds.$$

Moreover, we see that $u(t)$ is Lipschitzian in t, $u(0) = 0$, and $t^{-1} u(t) \leq s(t) \to 0$ as $t \to 0+$. Hence applying Lemma (1.1), we see that $u(t)$ is identically zero, i.e., $B^n g_0$ converges uniformly to zero on $[0, d_2]$ as $n \to \infty$.

Finally, we have

$$\|x_m(t) - x_n(t)\| = p_{m,n}(t) \leq B^n(g_0) \to 0$$

as $m, n \to \infty$, $m \geq n$.   q.e.d.

We conclude the present section with an extension of the result of Theorem (1.2) to a more general situation involving a nonlinear perturbation of a general type of nonlinear equation of evolution in a Banach space.

DEFINITION (1.1).   *Let X be a Banach space, and for each t in an interval $[0, T]$ $(0 < T < +\infty)$, suppose that we are given a mapping T_t (in general, nonlinear) with domain $D(T_t)$ in X and with range in X.*

By a strict solution of the equation of evolution

$$du(t)/dt + T_t(u(t)) = f(t)$$

we mean a strongly continuous, once weakly differentiable function u from $[0, T]$ to X such that $u(t)$ lies in $D(T_t)$ for each t in $[0, T]$, and the differential equation is valid.

DEFINITION (1.2).   *By a mild solution of the nonlinear equation of evolution $du/dt + T_t(u(t)) = f(t)$ on $[0, T]$, we mean a strongly continuous function u from $[0, T]$ to X such that there exists a sequence $\{u_n\}$ converging strongly to u on $[0, T]$ as $n \to \infty$, uniformly on $[0, T]$, such that each u_n is the strict solution of an equation of evolution of the form*

$$du_n(t)/dt + T_t(u_n(t)) = f_n(t) \qquad (t \in [0, T]),$$

while f_n converges to f in the sense of the space $L^1([0, T], X)$, i.e.,

$$\int_0^T \|f_n(s) - f(s)\|\, ds \to 0, \qquad n \to \infty.$$

DEFINITION (1.3).   *We shall say that the initial value problem is well posed for the nonlinear equation of evolution above if for each u_0 in X and each f in $L^1([0, T], X)$, there exists one and only one mild solution u on $[0, T]$ of the equation of evolution*

$$du(t)/dt + T_t(u(t)) = f(t),$$

in the sense of Definition (1.2) with $u(0) = u_0$, and if there exists a constant $c > 0$, such that for any pair u_0 and v_0 in X and any pair of inhomogeneous terms f and g in $L^1([0, T], X)$ the corresponding pair of mild solutions u and v satisfies the inequality:

$$\|u(t) - v(t)\| \leq c\left[\|u_0 - v_0\| + \int_0^t \|f(s) - g(s)\|\, ds\right] \qquad (0 \leq t \leq T).$$

Using the above definitions, we can now establish the result corresponding to Theorem (1.2) for the nonlinear perturbations of well posed nonlinear equations of evolution:

THEOREM (1.3). *Let $\{T_t, 0 \leq t \leq T\}$ define a well posed initial value problem in the Banach space X in the sense of Definitions* (1.1), (1.2), *and* (1.3). *Let f be a uniformly continuous mapping from a neighborhood N of a point $[0, x_0]$ in $[0, T] \times X$ into X, and suppose that there exists a function β from $[0, d_0] \times [0, d_1]$ into R^+ such that:*

(a) *$\beta(t, 0) = 0$ for all t in $[0, d_0]$. β is continuous from $[0, d_0] \times [0, d_1]$ to the nonnegative real numbers. $\beta(t, r)$ is monotone nondecreasing in r for each fixed t.*

(b) *Let c be the constant of Definition* (1.3) *appearing in the a-priori bound assumed there. Suppose that p is a nonnegative real-valued function on $[0, d]$ for some $d > 0$ which satisfies the differential equation $dp(t)/dt = c\beta(t, p(t))$ for t in $[0, d]$ with $p(0) = 0$. Then $p(t)$ must be identically zero.*

(c) *For each u and v lying in the ball of radius d_1 about x_0 in X and for each t in $[0, d_0]$,*

$$\|f(t, u) - f(t, v)\| \leq \beta(t, \|u - v\|).$$

Then there exists an interval $[0, d_2]$ with $d_2 > 0$ and on this interval one and only one mild solution u of the equation

$$du(t)/dt + T_t(u(t)) = f(t, u(t)) \qquad (t \in [0, d_2])$$

with $u(0) = x_0$.

PROOF OF THEOREM (1.3). For each f in $L^1([0, T], X)$ and the given x_0 in X, there exists by our assumptions one and only one mild solution on $[0, T]$ of the differential equation

$$du(t)/dt + T_t(u(t)) = f(t), \qquad 0 \leq t \leq T,$$

with $u(0) = x_0$. If f lies in $L^1([0, d_2], X)$ for $d_2 < T$, we extend f to the whole interval $[0, T]$ by setting it equal to zero outside $[0, d_2]$. If we restrict the corresponding mild solution u to the interval $[0, d_2]$, we obtain an element of $C([0, d_2], X)$, the space of continuous functions from $[0, d_2]$ to X with the usual uniform norm. We denote this element by $S_{d_2}(f)$.

For a given d_2 with $0 < d_2 \leq d_0$, let B be the closed ball in $C([0, d_2], X)$ about the constant function with value x_0, given by

$$B = \{u \mid u \in C([0, d_2], X), \|u(t) - x_0\| \leq d_1 \text{ for } 0 \leq t \leq d_2\}.$$

We define a second mapping R_{d_2} where R_{d_2} maps B into $L^1([0, d_2], X)$ by setting

$$R_{d_2}(u)(t) = f(t, u(t)) \qquad (t \in [0, d_2]).$$

An element u of B is a mild solution of the equation

$$du(t)/dt + T_t(u(t)) = f(t), \qquad t \in [0, d_2],$$

with $u(0) = x_0$ if and only if $u = S_{d_2} R_{d_2} u$, i.e., if and only if u is a fixed point of the composite mapping $U = S_{d_2} R_{d_2}$ which carries B into the space $C([0, d_2], X)$ containing B.

To show that such a fixed point exists and is unique, we show first of all that by choosing $d_2 > 0$ sufficiently small, we can ensure that U maps B into B. We note first of all that if $u_0(t) = x_0$ for t in $[0, d_2]$, then u_0 lies in B. Let $v_0 = S_{d_2} R_{d_2} u_0$. Then $v_0(t)$ is a strongly continuous function of t with $v_0(0) = x_0$. Hence there exists an interval $[0, d_2]$ such that for t in this interval, $\|v_0(t) - x_0\| \leq d_1/2$. Let u be any element of B, $w = U(u)$. Then w is the mild solution of the equation

$$dw(t)/dt + T_t(w(t)) = f(t, u(t))$$

with $w(0) = x_0$, and by the bound of Definition (1.3),

$$\|w(t) - v_0(t)\| \leq c \int_0^t \|u(t) - x_0\| \leq c\, d_1 t \leq c\, d_1 d_2, \qquad 0 \leq t \leq d_2.$$

If we choose $cd_2 \leq \frac{1}{2}$, it follows that $\|w(t) - v_0(t)\| \leq d_1/2$ and hence that $\|w(t) - x_0\| \leq d_1$ for all t in $[0, d_2]$. Thus for such a choice of d_2, $U(u)$ lies in B for each element u of B.

Let u and v be two elements of B, w and z given by $w = U(u)$, $z = U(v)$. Considering that w and z are mild solutions of equations with right-hand terms $f(t, u(t))$ and $f(t, v(t))$, respectively, and applying the a-priori bound assumed in Definition (1.3), we obtain the inequality

$$\|w(t) - z(t)\| \leq c \int_0^t \|f(s, u(s)) - f(s, v(s))\| \, ds \qquad (t \in [0, d_2]);$$

while by the inequality relating f to β in the hypothesis of Theorem (1.3), we have

$$\|f(s, u(s)) - f(s, v(s))\| \leq \beta(s, \|u(s) - v(s)\|).$$

Combining these two inequalities, we obtain the basic inequality

$$\|U(u)(t) - U(v)(t)\| \leq c \int_0^t \beta(s, \|u(s) - v(s)\|) \, ds.$$

We now consider the sequence of approximants $u_n = U^n(u_0)$ where u_0 is an arbitrary element of B. To show the existence of a fixed point of U, it suffices to show that $\{u_n\}$ is a Cauchy sequence in the complete metric space B. Let

$$p_{m,n} = \|u_m - u_n\|_B = \max_{t \in [0, d_2]} \|U^m(u_0)(t) - U^n(u_0)(t)\|_X,$$

and let

$$q_{m,n}(t) = \|u_m(t) - u_n(t)\|_X, \qquad t \in [0, d_2].$$

We introduce a third mapping K acting on nonnegative continuous real-valued functions on the interval $[0, d_2]$ by setting

$$(Kq)(t) = \int_0^t c\beta(s, q(s))\, ds.$$

K is a continuous mapping of this space with the uniform norm into a relatively compact subset of itself. Since $\beta(t, r)$ is monotone nondecreasing in r for fixed t, it follows that if we introduce the ordering $q \leq q_1$ for these real-valued functions to mean $q(t) \leq q_1(t)$ for all t in $[0, d_2]$, then $q \leq q_1$ implies that $Kq \leq Kq_1$.

By our basic inequality, we know that

$$q_{m,n}(t) \leq (Kq_{m-1,n-1})(t), \qquad t \in [0, d_2].$$

If we assume $m \geq n$ and iterate the above inequality, we see that

$$q_{m,n}(t) \leq (K^n q_{m-n,0})(t) \leq (K^n q_0)(t)$$

where $q_0(t)$ is the constant $2d_1$.

To complete the proof of Theorem (1.3), it suffices to show that $K^n q_0$ converges to zero, uniformly for t in $[0, d_2]$. We choose d_2 so small that

$$c \int_0^{d_2} \beta(S, 2d_1)\, ds \leq 2d_1.$$

Then $Kg_0 \leq g_0$, and iterating, we obtain $K^n g_0 \leq K^{n-1} g_0$. Thus $K^n g_0(t)$ converges for each t in $[0, d_2]$ to $q(t)$. Since $K^n g_0$ is relatively compact in the space of continuous real-valued functions on $[0, d_2]$, $q(t)$ is continuous in t and $(K^n g_0)(t)$ converges to $q(t)$ uniformly for t in $[0, d_2]$. Since K is continuous as a mapping of the continuous function space, $Kq = q$, i.e.,

$$q(t) = c \int_0^t \beta(s, q(s))\, ds, \qquad t \in [0, d_2].$$

This implies, however, that q is a solution of the differential equation $dq(t)/dt = c\beta(t, q(t))$ on $[0, d_2]$ with $q(0) = 0$. By our assumption (b) on the function β, this implies that $q(t)$ is identically zero. Thus $K^n g_0$ converges to zero in the uniform norm, and the proof of Theorem (1.3) is complete. q.e.d.

As a final application of the arguments of the present section, we introduce another interesting class of problems more general than the differential equations and equations of evolution in a Banach space, namely the *hereditary* or *functional differential equations* or *equations of evolution* in a Banach space.

DEFINITION (1.4). *Let X be a Banach space, T a real number, u a function from $(-\infty, T)$ to X. Then by the past of u at time t, $t < T$, we mean the function $P_t u$ from $(-\infty, 0)$ to X given by $(P_t u)(s) = u(t + s)$ $(s \leq 0)$.*

DEFINITION (1.5). *Let X be a Banach space, $T > 0$ a real number, Y the Banach space $C((-\infty, 0], X)$ of uniformly continuous bounded functions from the nonpositive reals to X. Let $\{T_t : t \in [0, T]\}$ be a family of mappings (in general, nonlinear) with domain and range in X. Let w_0 be a given element of the space*

$Y = C((-\infty, 0], X)$, N a neighborhood of the point $[0, w_0]$ in the product space $[0, T] \times Y_1$ and f a continuous mapping of N into X. Let u be a uniformly strongly continuous, bounded function from $(-\infty, d]$, $d > 0$, into X. Then u is said to be a mild solution of the hereditary differential equation (or equation of evolution)

$$du(t)/dt + T_t(u(t)) = f(t, P_t u) \qquad (t \in [0, d])$$

with initial value $P_0 u = w_0$, if the latter condition holds and u is a mild solution in the sense of Definition (1.2) of the above equation where $f(t, P_t u)$ is considered as a continuous function from $[0, d]$ to X.

THEOREM (1.4). *Consider a hereditary equation of evolution in the Banach space X in the sense of Definition (1.5) and suppose that the nonlinear equation of evolution $du(t)/dt + T_t(u(t)) = f_0(t)$ has a well posed initial value problem in the sense of Definition (1.3). Suppose further that there exists a function $\beta(t, r)$ from $[0, d_0] \times [0, d_1]$ to the nonnegative real numbers such that the following conditions all hold:*

(a) $\beta(t, 0) = 0$ for all t. $\beta(t, r)$ is continuous in $[t, r]$ and is monotone nondecreasing in r for each fixed t.

(b) If c is the constant in the inequality of Definition (1.3) and if p is a solution on an interval $[0, d]$, $d > 0$, of the differential equation $dp(t)/dt = c\beta(t, p(t))$ with $p(0) = 0$, then $p(t)$ is identically zero for t in $[0, d]$.

(c) N is the product of $[0, d_0] \times B_0$ where B_0 is the ball of radius $d_1/2$ about w_0 in Y. For two elements $[t, w]$ and $[t, z]$ of N with the same t-coordinate,

$$\|f(t, w) - f(t, z)\|_X \leq \beta(t, \|w - z\|_Y).$$

Then there exists $d_2 > 0$ sufficiently small such that on the interval $[0, d_2]$ there exists one and only one mild solution u of the hereditary differential equation

$$du(t)/dt + T_t(u(t)) = f(t, P_t u) \qquad (t \in [0, d]),$$

with initial condition $P_0 u = w_0$.

PROOF OF THEOREM (1.4). For $d_2 > 0$, we again define the mapping S_{d_2} analogous to its definition in the proof of Theorem (1.3) by letting $S_{d_2} f$ for f in $L^1([0, d_2], X)$ be the solution v of the initial value problem for the nonlinear equation of evolution

$$dv(t)/dt + T_t(v(t)) = f(t), \qquad t \in [0, d_2],$$

with $v(0) = w_0(0)$. If B is the subset of the Banach space $C((-\infty, d_2), X)$ given by

$$B = \{u \mid u \in C((-\infty, d_2), X), P_0 u = w_0, \|u(t) - w_0(0)\|_X \leq d_1/2, t \in [0, d_2]\},$$

we define a mapping R_{d_2} of B into $L^1([0, d_2], X)$ by setting $R_{d_2}(u)(t) = f(t, P_t u)$. Finally, we define a third mapping U, this time of B into $C((-\infty, d_2), X)$ by setting

$$U(u)(t) = S_{d_2} R_{d_2}(t), \qquad 0 \leq t \leq d_2,$$
$$= w_0(t), \qquad t \leq 0.$$

We note first that if $d_2 > 0$ is sufficiently small, then all the mappings given above are well defined and U maps B into B. S_{d_2} is automatically well defined by hypothesis whenever $d_2 \leq T$. R_{d_2} is well defined if it is true for all u in B that $\|w_0 - P_t u\|_Y \leq d_1$ for $0 \leq t \leq d_2$. However

$$\|w_0 - P_t u\|_Y = \sup_{s \leq 0} \|w_0(s) - u(t+s)\|_X$$

$$\leq \sup_{s \leq 0} \|w_0(s-t) - w_0(s)\|_X + \sup_{-t \leq s \leq 0} \|w_0(s) - w_0(0)\|_X$$

$$+ \sup_{0 \leq s \leq t} \|u(s) - w_0(0)\|_X$$

$$\leq d_1$$

for d_2 sufficiently small by the uniform continuity of $w_0(s)$ in s. Hence R_{d_2} and U are both well defined.

To show that U maps B into B, we begin with the element v_0 of B where $P_0 v_0 = w_0$ and $v_0(t) = w_0(0)$ for $0 \leq t \leq d_2$. If $z_0 = U(v_0)$, then $z_0(t)$ is strongly continuous in t on $[0, d_2]$ and by making d_2 sufficiently small, we may ensure that $\|w_0(0) - z_0(t)\| \leq d_1/4$ for t in $[0, d_2]$. Suppose now that v is any element of B and let $z = U(v)$. Applying the inequality of Definition (1.3) and our assumed inequality relating f and β, it follows that for all t in $[0, d_2]$, we have

$$\|z(t) - z_0(t)\|_X \leq c \int_0^{d_2} \|f(s, P_s v) - f(s, P_s v_0)\|_X \, ds$$

$$\leq c \int_0^{d_2} \beta(s, \|P_s v - P_s v_0\|_Y) \, ds.$$

On the other hand since both v and v_0 coincide with w_0 on the negative reals, we have

$$\|P_s v - P_s v_0\|_Y \leq \max_{0 \leq t \leq d_2} \|v(t) - w_0(0)\|_X \leq d_1/2.$$

Hence

$$\|z(t) - z_0(t)\|_X \leq c \int_0^{d_2} \beta\left(s, \frac{d_1}{2}\right) ds \leq \frac{d_1}{4}$$

if d_2 is sufficiently small. For this choice of d_2, it then follows that for all t in $[0, d_2]$, $\|U(v)(t) - w_0(0)\| \leq d_1/2$ so that $U(v)$ lies in B for each v in B.

To show the existence of the desired solution of the hereditary equation of evolution is equivalent to showing the existence of a fixed point of the mapping U of B into B, and the uniqueness of the solution is equivalent to the uniqueness of the corresponding fixed point. We form the sequence of approximants $u_n = U^n v_0$ ($v_0 \in B$). For any pair of elements u and v of B, it follows from the hypotheses of our theorem and the definition of U that for any t in $[0, d_2]$,

$$\|U(u)(t) - U(v)(t)\|_X \leq c \int_0^t \|f(s, P_s u) - f(s, P_s v)\|_X \, ds$$

$$\leq c \int_0^t \beta(s, \|P_s u - P_s v\|_Y) \, ds,$$

where
$$\|P_s u - P_s v\|_Y = \sup_{r \leq 0} \|u(s+r) - v(v+r)\|_X = \sup_{0 \leq r \leq s} \|u(r) - v(r)\|_X$$
since $u(r) = v(r) = w_0(r)$ for $r \leq 0$. Hence
$$\sup_{0 \leq r \leq t} \|U(u)(t) - U(v)(t)\| \leq c \sup_{0 \leq r \leq t} \int_0^r \beta\left(s, \sup_{0 \leq \lambda \leq s} \|u(\lambda) - v(\lambda)\|_X\right) ds$$
$$\leq c \int_0^t \beta\left(s, \sup_{0 \leq r \leq s} \|u(r) - v(r)\|_X\right) ds.$$

On B, we introduce the seminorms, $d_t(u, v) = \sup_{0 \leq r \leq t} \|u(r) - v(r)\|_X$. Then we have our basic inequality
$$d_t(Uu, Uv) \leq c \int_0^t \beta(s, d_s(u, v)) ds \qquad (t \in [0, d_2]).$$

For each pair of nonnegative integers m and n, $m \geq n$, we set $q_{m,n}(t) = d_t(u_m, u_n)$. We also define the mapping K acting on nonnegative continuous functions on $[0, d_2]$ by $(Kq)(t) = \int_0^t c\beta(s, q(s)) ds$. It follows as before that if $q \leq q_1$, then $Kq \leq Kq_1$ and that for the constant function q_0, $K^n q_0$ converges uniformly to zero as $n \to \infty$ provided that $d_2 > 0$ is sufficiently small. Our basic inequality asserts that
$$q_{m,n} \leq K(q_{m-1,n-1}) \leq \cdots \leq K^n(q_{m-n,0}) \leq K^n(q_0).$$

Thus $q_{m,n}(t) \to 0$, uniformly on $[0, d_2]$, and the approximants u_n thereby converge to a fixed point of U. To show that this fixed point is unique, we let u and v be two fixed points of U, and consider $q(t) = d_t(u, v)$ for $0 \leq t \leq d_2$. It follows from our basic inequality for U that $q \leq K(q)$, and hence by recursion, $q \leq K^n(q) \leq K^n(q_0) \to 0$ as $n \to \infty$. Hence $q = 0$ and therefore $u = v$. q.e.d.

We close this discussion of hereditary differential equations with a remark concerning their relation to the narrower class of differential equations in the ordinary sense. We have considered our differential equations and equations of evolution in a (possibly) infinite-dimensional Banach space X. One of the merits of this degree of generality (aside from the intrinsic necessity of considering problems with an infinite number of degrees of freedom) is that, under very mild restrictions, a hereditary differential equation in the space X of finite or infinite dimension may be transformed into an equivalent differential equation in the ordinary sense in the space $Y = C((-\infty, 0), X)$ which is always of infinite dimension. The transformation which accomplishes this objective is simply that of replacing the original unknown function u from $[-\infty, T]$ to X by a new function w from $[0, T]$ to Y, where $w(t) = P_t(u)$. Suppose that the initial function w_0 is uniformly once differentiable in the strong sense from $(-\infty, 0]$ to X. Then

$$\frac{dw}{dt}(t)(s) = \lim_{h \to 0} [h^{-1}(w(t+h) - w(t))(s)]$$
$$= \lim_{h \to 0} h^{-1}[u(t+s+h) - u(t+s)] = \frac{du}{dt}(t+s) = P_t\left(\frac{du}{dt}\right)(s).$$

Our hereditary differential equation $du(t)/dt = f(t, P_t(u))$ then becomes

$$dw(t)/dt = g(t, w(t)), \quad 0 \leq t \leq T,$$

where $g(t, w(t)) = P_t h$, $h(s) = f(s, w(s))$, with the initial condition $w(0) = w_0$.

2. **Locally Lipschitzian mappings.** It is our purpose in the present section to establish some general results on the continuity under suitable perturbations of the open mapping property at a point, i.e., the property that the image of the point lies in the interior of the image under a given mapping f of an open set containing the original point. The case of primary interest for our later applications is that of the class of locally Lipschitzian mappings of open subsets of a Banach space, or more generally in complete metric spaces.

Results of the type in which we are interested can be considered as an oblique generalization of the inverse function theorem or the implicit function theorem. We begin our discussion with two variants of this theorem.

DEFINITION (2.1). *Let X and Y be Banach spaces, L a bounded linear mapping from X to Y, L^* its adjoint map from Y^* to X^*. We set*

$$\alpha(L) = c_0, \quad c_0 = \text{infimum of constants } c \geq 0 \text{ such that}$$
$$\|y^*\| \leq c \|L^*y^*\| \quad \text{for all } y^* \text{ in } Y^*.$$

$$\alpha(L) = +\infty, \quad \text{if no such constants } c \text{ exist.}$$

LEMMA (2.1). *Let X and Y be Banach spaces, L a bounded linear mapping from X to Y. Then:*

(a) *The range of L, $R(L)$, is all of Y if and only if $\alpha(L) < \infty$.*

(b) *If $R(L) = Y$, then for any $\epsilon > 0$ and any y in Y, there exists x in X with $L(x) = y$ such that $\|x\| \leq (\alpha(L) + \epsilon) \|y\|$. If X is reflexive, we may actually take $\epsilon = 0$.*

(c) *Let $\beta(L)$ be the infimum of all constants c such that for each y in Y, we may find x in X such that $L(x) = y$ and $\|x\| \leq c \|y\|$. Then $\alpha(L) = \beta(L)$.*

PROOF OF LEMMA (2.1). By the well-known closed range theorem of Banach [183], the range of L is all of Y if and only if L^* is injective and L^* has a bounded inverse on its range. This latter property is equivalent to the existence of a constant $c > 0$ such that

$$\|y^*\| \leq c \|Ly^*\|, \quad y^* \in Y^*,$$

and this in turn is equivalent to the condition that $\alpha(L) < \infty$. Hence the conclusion of (a) is valid.

Let $Z = R(L^*)$, where $R(L) = Y$. By the closed range theorem, $R(L^*)$ is closed in the weak* topology on X^* and consists exactly of the annihilator of $N(L)$, where $N(L)$ is the null space of L. Hence $Z = (X/N(L))^*$, in the sense that if p is the projection mapping of X onto $X/N(L)$, then p^* is an isometric injection of $(X/N(L))^*$ into X^* which identifies the space $(X/N(L))^*$ with Z.

The mapping L induces a mapping L_1 of $X/N(L)$ into Y such that $L = L_1 p$. If $R(L) = Y$, it follows that $R(L_1)$ is equal to Y, and since L_1 is an injection, it follows that L_1 is a bicontinuous mapping of $X/N(L)$ onto Y by the open mapping theorem. Hence there exists a constant $c_0 = \|L_1^{-1}\|$ such that for each y in Y, there exists an element u of $X/N(L)$ such that $L_1 u = y$ and $\|u\| \leq c_0 \|y\|$. Obviously, we cannot replace c_0 by a smaller constant in this last inequality. By the definition of the norm in a quotient space, for each $\epsilon > 0$ there exists an element x in X such that $p(x) = u$ and $\|x\| \leq (1 + \epsilon) \|u\|$, and it follows that for each $\epsilon > 0$, we can find x in X such that $L(x) = y$ and $\|x\| \leq (c_0 + \epsilon) \|y\|$. Moreover, if c_0 is replaced by a smaller constant, this assumption can no longer be valid since $u = p(x)$ for any such x must satisfy the conditions $\|u\| \leq \|x\|$, $L_1 u = y$. Hence, $c_0 = \beta(L)$.

However, $\|L_1^{-1}\| = \|(L_1^{-1})^*\| = \|(L_1^*)^{-1}\|$, since $(L_1^{-1})^* = (L_1^*)^{-1}$. On the other hand, if we identify $(X/N(L))^*$ with Z by the mapping p^*, L_1^* is identical with L^* taken as a mapping of Y^* onto $Z = R(L^*)$. Hence $\|(L_1^*)^{-1}\| = \alpha(L)$, and therefore $\alpha(L) = \beta(L)$ proving (b) and (c). q.e.d.

THEOREM (2.1). *Let X and Y be Banach spaces, L and L_0 two bounded linear mappings of X into Y such that $R(L) = Y$, while $\|L - L_0\| \, \alpha(L) < 1$. Then $R(L_0) = Y$.*

PROOF OF THEOREM (2.1). It suffices by Lemma (2.1) to prove that $\alpha(L_0) < \infty$. By the definition of $\alpha(L)$, $\|y^*\| \leq \alpha(L) \|L^* y^*\|$ for all y^* in Y^*. Since

$$\|L_0^* y^*\| \geq \|L^* y^*\| - \|L^* y^* - L_0^* y^*\| \geq [\alpha(L)]^{-1} \|y^*\| - \|L^* - L_0^*\| \cdot \|y^*\|$$

while $\|L^* - L_0^*\| = \|L - L_0\|$, we see that

$$\|L_0^* y^*\| \geq \frac{1 - \alpha(L) \|L - L_0\|}{\alpha(L)} \|y^*\|$$

for all y^* in Y^*. Thus

$$\alpha(L_0) \leq \alpha(L)[1 - \alpha(L) \|L - L_0\|]^{-1} < \infty. \qquad \text{q.e.d.}$$

DEFINITION (2.2). *Let X and Y be Banach spaces, f a continuous mapping of an open subset G of X into Y, x_0 a point of G. Then f is said to be G-differentiable at x_0 if there exists a bounded linear mapping $L = f'_{x_0}$ of X into Y such that for each h in X,*

$$\frac{f(x_0 + th) - f(x_0)}{t} \to L(h) = f'_{x_0}(h)$$

as $t \to 0$.

This is a variant on the usual definition of the Gateaux derivative in which we impose the condition that the derivative $f'_x(h)$ should be a bounded linear function of h. The convergence in Definition (2.2) is strong convergence in the Banach space Y for the usual definition of Gateaux derivative, and if we consider weak convergence, we denote f'_x as the weak G-derivative.

THEOREM (2.2). *Let X and Y be Banach spaces, f a continuous mapping of an open subset G of X into Y. Suppose that there exists $\lambda > 0$ such that for a given point x_0 of G, the open ball $B_\lambda(x_0)$ of radius λ about x_0 is contained in G and on that ball,*

f is weakly G-differentiable. Suppose further that for x in $B_\lambda(x_0)$, $\|f'_x - f'_{x_0}\| \alpha_0 \le r < 1$, where we suppose that f'_{x_0} maps X onto Y and that $\alpha_0 = \alpha(f'_{x_0})$.

Then the mapping f is open at x_0, and more precisely, $f(B_\lambda(x_0))$ contains the open ball about $f(x_0)$ of radius λ_1 where $\lambda_1 = \lambda \alpha_0^{-1}(1 - \alpha_0 r)$.

PROOF OF THEOREM (2.2). Let $d_0 = \sup \|f'_x - f'_{x_0}\|, x \in B_\lambda(x_0)$. Then $d_0 \alpha_0 \le r$. Suppose y lies in $B_{\lambda_1}(y_0)$, $y_0 = f(x_0)$. Then there exists $\alpha > \alpha_0$ such that $\|y - y_0\| (1 - \alpha d_0)^{-1} \alpha < \lambda$. We shall show that such an element y lies in $f(B_\lambda(x_0))$.

To find x so that $f(x) = y$, we obtain x as the limit of a sequence of approximants $\{x_n\}$ chosen so that the following recursive hypotheses hold: x_0 is the given point. If $\{x_0, \ldots, x_n\}$ have been constructed and if $y_j = f(x_j)$, $h_j = x_j - x_{j-1}$, then

$$\|h_j\| \le (d_0 \alpha)^{j-1} \alpha \|y - y_0\| \qquad (1 \le j \le n)$$

and

$$\|y - y_j\| \le (d_0 \alpha)^j \|y - y_0\| \qquad (0 \le j \le n).$$

To construct the new approximant x_{n+1}, we set $x_{n+1} = x_n + h_{n+1}$ and seek to determine h_{n+1} so that the above estimates hold for the $(n+1)$st step. To do this, we need the following simple lemma:

LEMMA (2.2). *Let X and Y be Banach spaces, G an open subset of X, f a continuous function from G to Y which is weakly G-differentiable from some ball $B_\lambda(x_0)$ in G to Y. Then for each x in $B_\lambda(x_0)$ and each h in X such that $x + h$ lies in $B_\lambda(x_0)$, we have*

$$f(x + h) = f(x) + \int_0^1 f'_{x+th}(h) \, dt,$$

where (for simplicity) the integral is taken in the weak sense.

PROOF OF LEMMA (2.2). Let y^* be an element of Y^*, and let $g(t) = (f(x + th), y^*)$, where we use (y, y^*) to denote the pairing between the element y of Y and the functional y^*. By the definition of weak G-derivative, g is differentiable for each t in $[0, 1]$ and $g'(t) = (f'_{x+th}(h), y^*)$. Since $|g'(t)| \le \|f'_{x+th}\| \cdot \|h\| \cdot \|y^*\|$, g' is uniformly bounded for each t in $[0, 1]$. Hence, by the mean value theorem, g is Lipschitzian and absolutely continuous with respect to Lebesgue measure on $[0, 1]$. Thus

$$g(1) - g(0) = \int_0^1 g'(t) \, dt,$$

i.e.,

$$(f(x + h), y^*) = (f(x), y^*) + \int_0^1 (f'_{x+th}(h), y^*) \, dt \qquad (y^* \in Y^*),$$

which is equivalent to the assertion of the lemma. q.e.d.

PROOF OF THEOREM (2.2) COMPLETED. By Lemma (2.2),

$$f(x_{n+1}) = f(x_n + h_{n+1}) = f(x_n) + \int_0^1 f'_{x_n + th_{n+1}}(h_{n+1}) \, dt.$$

We rewrite the right side in the form

$$f(x_n) + f'_{x_0}(h_{n+1}) + \int_0^1 [f'_{x_n+th_{n+1}} - f'_{x_0}](h_{n+1})\, dt.$$

Since $f(x_n) = y_n$, if we set $f'_{x_0}(h_{n+1}) = y - y_n$, we may obtain h_{n+1} in X satisfying the inequality $\|h_{n+1}\| \leq \alpha \|y - y_n\| \leq (d_0\alpha)^n \alpha \|y - y_0\|$ by the recursive hypothesis. By the recursive hypothesis, once more,

$$\|x_n - x_0\| \leq \sum_{j=1}^n \|h_j\| \leq \alpha \|y - y_0\| \sum_{j=0}^{n-1} (d_0\alpha)^j < (1 - d_0\alpha)^{-1} \alpha \|y - y_0\|,$$

and

$$\|x_{n+1} - x_0\| \leq \sum_{j=1}^{n+1} \|h_j\| < (1 - d_0\alpha)^{-1} \alpha \|y - y_0\|.$$

By hypothesis, $(1 - d_0\alpha)^{-1}\alpha \|y - y_0\| < \lambda$ so that both x_n and x_{n+1} lie in $B_\lambda(x_0)$ and the validity of the integral formula is assured. For each t in $[0, 1]$, $x_n + th_{n+1}$ also lies in $B_\lambda(x_0)$. Hence $\|f'_{x_n+th_{n+1}} - f'_{x_0}\| \leq d_0$. Thus $f(x_{n+1}) = y_n + (y - y_n) + R_{n+1}$, with

$$R_{n+1} = \int_0^1 [f'_{x_n+th_{n+1}} - f'_{x_n}](h_{n+1})\, dt.$$

If we set $y_{n+1} = f(x_{n+1})$, it follows that $\|y - y_{n+1}\| = \|R_{n+1}\| \leq d_0 \|h_{n+1}\| \leq (d_0\alpha)^{n+1} \|y - y_0\|$. Thus all the bounds of the recursive hypothesis have been verified for the $(n + 1)$st step.

Since for $n > m$,

$$\|x_n - x_m\| \leq \sum_{j=m+1}^n \|h_j\| \to 0 \qquad (m \to \infty),$$

it follows that x_n converges in X to an element x of X. Since $\|x_n - x_0\| \leq (1 - d_0\alpha)^{-1} \alpha \|y - y_0\| < \lambda$, it follows that x lies in the open ball of radius λ about x_0. Since $\|f(x_n) - y\| = \|y_n - y\| \to 0$, it follows that $f(x) = y$. q.e.d.

DEFINITION (2.3). *Let X and Y be metric spaces, with distance functions both denoted by d. Let f be a mapping from X into Y. Then f is said to be Lipschitzian if for a constant c,*

$$d(f(x), f(u)) \leq c\, d(x, u)$$

for all x and u in X.

The Lipschitz norm of f on X is defined to be the least constant c in this last inequality, written $\|f\|_{\mathrm{Lip}}$.

f is said to be locally Lipschitzian if each point u_0 of X has an open neighborhood N such that the restriction of f to N is a Lipschitzian map of N into Y.

DEFINITION (2.4). *Let X be a metric space, Y a normed linear space, $\{f_\lambda, \lambda \in [0,1]\}$ a family of Lipschitzian mappings of X into Y. Then this family is said to be a continuous curve in the space of Lipschitzian maps if for each λ_0 in $[0, 1]$, $\|f_\lambda - f_{\lambda_0}\|_{\mathrm{Lip}} \to 0$ as $\lambda \to \lambda_0$. (Here $f_\lambda - f_{\lambda_0}$ denotes the mapping whose value at x in X is given by $f_\lambda(x) - f_{\lambda_0}(x)$.)*

DEFINITION (2.5). *Let X be a metric space, Y a normed linear space, $\{f_\lambda, \lambda \in [0, 1]\}$ a family of locally Lipschitzian mappings of X into Y. Then this family is said to be a continuous curve in the space of locally Lipschitzian mappings if each point u_0 of X has a neighborhood N such that, considered as Lipschitzian maps of N into Y, the family $\{f_\lambda\}$ is a continuous curve.*

THEOREM (2.3). *Let X be a metric space, Y a Banach space, T a mapping of X into 2^Y, the space of subsets of Y. Let $\{f_\lambda : \lambda \in [0, 1]\}$ be a continuous family of locally Lipschitzian mappings of X into Y in the sense of Definition (2.5), and suppose that $T + f_0$ is an open mapping of X into Y, i.e., $(T + f_0)(W)$ is open in Y for each open subset W of X. Suppose further that the following condition holds:*

For each point x_0 of X, there exists an open neighborhood N of x_0 in X and for each λ in $[0, 1]$, a constant $c_\lambda > 0$ such that for all x and u in N, y in $(T + f_\lambda)(x)$ and w in $(T + f_\lambda)(u)$, we have $\|w - y\| \geq c_\lambda d(x, u)$. Then for each λ in $[0, 1]$, $(T + f_\lambda)$ is an open mapping of X into Y.

THEOREM (2.4). *Suppose that the hypotheses of Theorem (2.3) hold and that in addition $(T + f_1)$ is a closed mapping, i.e., maps closed sets in X into closed sets in Y. Then $R(T + f_1) = Y$.*

PROOF OF THEOREM (2.3). It suffices to prove that $T + f_\lambda$ is open on a neighborhood N of each point x_0 of X. Hence, we may replace X by N and assume that $\{f_\lambda : \lambda \in [0, 1]\}$ is a continuous family of Lipschitzian mappings in the sense of Definition (2.4).

We remark next that the constants c_λ of the inequality assumed in the hypothesis of Theorem (2.3) may be chosen independent of λ in $[0, 1]$. Indeed, suppose that λ_0 lies in $[0, 1]$. Then for λ sufficiently close to λ_0, for any w_λ in $(T + f_\lambda)(u)$, y_λ in $(T + f_\lambda)(x)$, there exist points w in $(T + f_{\lambda_0})(u)$, y in $(T + f_{\lambda_0})(x)$ such that

$$\|(w - w_\lambda) - (y - y_\lambda)\| < c_{\lambda_0}/[2d(x, u)]^{-1}.$$

Since $\|w - y\| \geq c_{\lambda_0} d(x, u)$, it follows that for λ in some interval about λ_0,

$$\|w_\lambda - y_\lambda\| \geq (c_{\lambda_0}/2) d(x, u).$$

Since the compact set $[0, 1]$ can be covered by a finite number of these intervals, we can take the least constant for this finite family of intervals and assume without loss of generality that for all λ in $[0, 1]$, all x and u in X, and each w_λ in $(T + f_\lambda)(u)$, y_λ in $(T + f_\lambda)(x)$, we have $\|w_\lambda - y_\lambda\| \geq c\, d(u, x)$.

We remark that it suffices to show that there exists $d_0 > 0$ such that if $(T + f_\lambda)$ is an open mapping for a given parameter value λ and if ξ is another parameter value with $|\xi - \lambda| < d_0$, then $(T + f_\xi)$ is an open mapping of X into Y. Indeed, if this were the case, we could pass from the openness of $T + f_0$ to the corresponding property for $(T + f_\lambda)$ for any λ in $[0, 1]$ in a finite number of steps of length close to d_0.

By hypothesis, $\|f_\lambda - f_\xi\|_{\text{Lip}}$ is a continuous function of λ and ξ on $[0, 1]$ which vanishes for $\lambda = \xi$. Hence, we can choose $d_0 > 0$ so small that for $|\lambda - \xi| < d_0$, we have $\|f_\lambda - f_\xi\|_{\text{Lip}} < c/2$. Suppose that u_0 is a point of X, $w_0 \in (T + f_\xi)(u_0)$. We wish to show that there exists an open neighborhood of w_0 which lies in $(T + f_\xi)(X)$. (It suffices to restrict ourselves to the image of X, since otherwise we can replace the metric space X by any open subset of X for which the same argument applies.)

Since $w_0 \in (T + f_\xi)(u_0)$, there exists z_0 in $T(u_0)$ such that $w_0 = z_0 + f_\xi(u_0)$. Let $w_1 = z_0 + f_\lambda(u_0)$. Then w_1 lies in $(T + f_\lambda)(X)$, and since we are assuming that $(T + f_\lambda)(X)$ is open in Y, there exists a ball $B_d(w_1)$ which is contained in $(T + f_\lambda)(X)$.

We normalize the family $\{f_\lambda : \lambda \in [0, 1]\}$ by setting $f_\lambda(p_0) = q_0$ for two points p_0 in X and q_0 in Y. After this normalization which does not affect the continuity of the families in the Lipschitzian sense, it follows that for all u in X, $\|f_\lambda u - f_\xi u\| \leq c_{\lambda,\xi} d(u, p_0)$, with $c_{\lambda,\xi} \to 0$ as $\lambda - \xi \to 0$. This normalization process consists simply of adding the constant element $q_0 - f_\lambda(p_0)$ to each function f_λ. It alters $T + f_\lambda$ by the same additive constant for each fixed λ, and hence does not affect the property that $(T + f_\lambda)(X)$ should be open. In particular, we may carry through this normalization with $p_0 = u_0$, with u_0 as given above. Hence we may assume without loss of generality that $f_\lambda(u_0) = f_\xi(u_0)$ for each ξ in $[0, 1]$.

By the inequality assumed in the hypothesis and refined in the discussion above, the mapping $S_\lambda = (T + f_\lambda)^{-1}$ is a single-valued Lipschitzian mapping from $R(T + f_\lambda)$ into X with Lipschitz constant $\leq c^{-1}$. For each w in $B_d(w_1)$, which is contained in $R(T + f_\lambda)$, the element $u_\lambda = S_\lambda(w)$ is well defined and by the definition of S_λ, there exists y_λ in $T(u_\lambda)$ such that $w = y_\lambda + f_\lambda(u_\lambda)$.

We wish to show that for ξ sufficiently close to λ and for each point y sufficiently close to w_1, where $w_1 \in (T + f_\xi)(u_0)$ for each ξ in $[0, 1]$ by our normalization, the point y must lie in $(T + f_\xi)(X)$. For this purpose, we shall look for an element w of $B_d(w_1)$ such that y lies in $(T + f_\xi)(S_\lambda w)$, or more specifically, we shall seek to make $y = y_\lambda + f_\xi S_\lambda w \in (T + f_\xi)(S_\lambda w)$ with y_λ as above. This ansatz is equivalent by the above equation involving y_λ and w to setting $y_\lambda = w - f_\lambda S_\lambda w$ in the equation $y = y_\lambda + f_\xi S_\lambda w$, i.e., $y = w + [f_\xi - f_\lambda] S_\lambda w$. If we put $S = [f_\lambda - f_\xi] S_\lambda$, we wish to solve the equation $(I - S)w = y$.

Since S_λ is Lipschitzian with Lipschitz constant $\leq c^{-1}$, if we choose $d_0 > 0$ so small that $\|f_\xi - f_\lambda\|_{\text{Lip}} < c/2$, then S will be Lipschitzian with Lipschitz constant $\leq c^{-1} c/2 = \frac{1}{2}$. Suppose now that $\|y - w_1\| < d/4$. We shall solve the equation $w - Sw = y$ by iterating the contraction mapping U_y where $U_y(w) = Sw + y$. Since U_y for any y is Lipschitzian with constant < 1, it follows that we can apply the Contraction Mapping Principle to obtain a solution of $U_y w = w$, i.e., $w - Sw = y$, as the limit of $U_y^n(w_1)$ with $n \to \infty$ provided that all the iterates lie in the set $B_d(w_1)$ on which U_y is defined. We note that $U_y(w_1) = y$ since $S(w_1) = y$ since $S(w_1) = 0$, and

$$\|U_y^n(w_1) - U_y^{n-1}(w_1)\| \leq \tfrac{1}{2} \|U_y^{n-1}(w_1) - U_y^{n-2}(w_1)\|$$
$$\leq 2^{-(n-1)} \|U_y(w_1) - w_1\|$$
$$\leq 2^{-(n-1)} \|y - w_1\|.$$

Thus
$$\|U_y^n(w_1) - w_1\| \leq \sum_{j=0}^{n-1} 2^{-j} \|y - w_1\| \leq 2\|y - w_1\| \leq \frac{d}{2},$$

as long as all the iterates $U_y^j(w_1)$ for $j \leq n$ lies in $B_d(w_1)$. By recursion, it follows from this estimate that all the iterates $U_y^n(w_1)$ do lie in $B_d(w_1)$ and indeed in a closed sub-ball which is compact. Hence, the approximants converge to a point w of the open ball $B_d(w_1)$ for which $w - Sw = y$, and it follows that all points y in $B_{d/4}(w_1)$ do indeed lie in $(T + f_\xi)(X)$ for $|\xi - \lambda| < d_0$. q.e.d.

PROOF OF THEOREM (2.4). By Theorem (2.3), $(T + f\lambda)(X)$ is an open subset of Y for each λ in $[0, 1]$. By the hypothesis of Theorem (2.4), $(T + f_\lambda)(X)$ is a closed subset of Y. Hence it is both open and closed in the connected space Y, and being nonempty, $(T + f_1)(X) = Y$. q.e.d.

As a simple consequence of Theorem (2.4), we have the following more specialized result:

THEOREM (2.5). *Let X be a metric space, Y a Banach space, T a mapping of X into 2^Y. Suppose that $\{f_\lambda : \lambda \in [0, 1]\}$ is a continuous curve of locally Lipschitzian mappings of X into Y in the sense of Definition (2.5) and there exists a neighborhood N of each point u_0 of X and for each λ, a constant c_λ (depending upon N) such that for each x and u in N, each y in $(T + f_\lambda)(x)$, and each w in $(T + f_\lambda)(u)$, we have*
$$\|w - y\| \geq c_\lambda d(u, x).$$

Suppose further that $(T + f_0)$ is an open mapping of X into Y and that $T + f_1$ satisfies one of the two following conditions:

(a) *$(T + f_1)$ is proper, i.e., for each compact subset K of Y, $(T + f_1)^{-1}(K)$ is a compact subset of X.*

(b) *X is complete, the graph of $(T + f_1)$ is closed and there exist a compact mapping C of X into Y and a continuous strictly increasing function $c(r)$ for $r \geq 0$ with $c(0) = 0$ such that for all x and u in X, each y in $(T + f_1)(x)$, and each w in $(T + f_1)(u)$,*
$$\|y - w\| \geq c(d(x, u)) - \|C(x) - C(u)\|.$$

Then the range of $(T + f_1)$ is all of the space Y.

PROOF OF THEOREM (2.5). By Theorems (2.3) and (2.4), it suffices to show that $(T + f_1)$ is a closed mapping or even just that $(T + f_1)(x)$ is closed in Y. Suppose then that $\{w_n\}$ is a sequence in Y converging to an element w of Y, where for each n, $w_n \in (T + f_1)(u_n)$. In case (a) in which $(T + f_1)$ is proper, we choose for our compact set K, the union of the elements of the sequence $\{w_n\}$ together with its limit point w. Then the sequence $\{u_n\}$ lies in the compact set $(T + f_1)^{-1}(K)$ and by passing to an infinite subsequence, we may assume that $u_n \to u$ in X as $n \to \infty$. If a given u_n appears in the sequence infinitely often, then $w = w_n$ lies in $(T + f_1)(X)$. If u is distinct from all the elements u_n, we may find a neighborhood N of u such that for x in N, x_1 in N, $(T + f_1)(x) \cap (T + f_1)(x_1) = \varnothing$ if $x \neq x_1$, and we can assume without loss of generality that all the elements u_n lie in N. Then $(T + f_1)(u)$ has a point in common with K but not with $(T + f_1)(u_n)$ for any

n. In this case, this point in $K \cap (T + f_1)(u)$ must be the point w, and w lies in $(T + f_1)(X)$. Finally, suppose that $u_n \to u_{n_0}$ for some n_0 while u_n differs from u_{n_0} for $n > N$. Then passing to a subsequence, we can assume that we have eliminated u_{n_0} from the sequence, and we return to the second case.

In all three cases, w lies in $(T + f_1)(X)$. Since w is an arbitrary limit point of $(T + f_1)(X)$, it follows that $(T + f_1)(X)$ is closed in Y and the conclusion of Theorem (2.5) follows in case (a).

In case (b), we shall show that $(T + f_1)$ is proper under the hypotheses of case (b). To show that a mapping is proper, it suffices to let K be the union of the elements of a sequence $\{w_n\}$ of Y which converges to a point w of Y together with its limit point w. Let $\{u_n\}$ be an infinite sequence in $(T + f_1)^{-1}(K)$, where we may assume without loss of generality that $w_n \in (T + f_1)(u_n)$. Passing to an infinite subsequence, we may assume that $C(u_n)$ converges in Y. Hence for $n > m$,

$$c(d(u_n, u_m)) \leq \|w_m - w_n\| + \|C(x_n) - C(x_m)\| \to 0$$

as $n, m \to \infty$. Since c is a strictly increasing continuous function with $c(0) = 0$, its inverse function is continuous and strictly increasing. It follows that $d(u_n, u_m) \to 0$ as $n, m \to \infty$, and hence that u_n converges to a point u of X since X, by the assumptions of case (b), is a complete metric space. Hence $[u_n, w_n] \to [u, w]$, and since the graph of $(T + f_1)$ is closed in $X \times Y$ by hypothesis, $w \in (T + f_1)(u)$.

q.e.d.

3. **Φ-accretive and Φ-coaccretive nonlinear mappings.** The two classes of noncompact nonlinear mappings in Banach spaces with which we are most intimately concerned in the later sections of the present paper are the class of monotone mappings T from a Banach space X to its conjugate space X^* and the class of accretive mappings A from a Banach space X to X. We recall that if X is a Banach space, X^* its conjugate space, and if (w, u) denotes the pairing between the functional w lying in X^* and the element u of X, then a mapping T from X to 2^{X^*} is said to be monotone if for each pair x and u of X, $w \in T(u)$, $y \in T(x)$, we have $(w - y, u - x) \geq 0$, i.e., for single-valued mappings T, $(T(u) - T(x), u - x) \geq 0$ for all u and x for which $T(u)$ and $T(u)$ are defined, i.e., x and u in $D(T)$, the domain of T.

The class of accretive mappings A from X to 2^X is defined in a rather similar way if one employs the concept of a duality mapping J from X to X^*. By such a (single-valued) mapping, we mean a mapping J from X to X^* such that for each u in X, $J(u)$ is an element of X^* which satisfies the two following conditions:

$$(J(u), u) = \|Ju\| \cdot \|u\|; \qquad \|Ju\| = \|u\|.$$

The existence of at least one such mapping follows trivially from the Hahn-Banach theorem. If X^* is strictly convex (or rotund in the terminology of Day [164]), then J is unique and is continuous from the strong topology of X to the weak* topology of X^*.

Let A be a mapping from X to 2^X for a Banach space X, J a duality mapping from X to X^*. Then A is said to be accretive if for each x and u in X, w in $A(u)$, y in $A(x)$, $(J(u - x), w - y) \geq 0$. The definition seems to depend upon the choice of J, but as we shall show in §9, in a variety of cases, it is actually independent of the choice of J.

It is our purpose in the present section to develop the theory of a wider class of operators between Banach spaces that includes the theory of accretive operators as a special case and part of the theory of monotone operators, at least under the additional hypothesis that the mappings involved are locally Lipschitzian. This class is the class of Φ-accretive mappings which we now proceed to define:

DEFINITION (3.1). *Let X be a Banach space. Then by a Φ-system on X, we mean a covering $\{N_\alpha\}$ of X by open subsets together with a family $\{\Phi_\alpha\}$ indexed on the same set such that for each α, Φ_α is a mapping of $N_\alpha \times N_\alpha$ into X^* which satisfies the following conditions:*

(a) *For each x and u in N_α, $\|\Phi_\alpha(x, u)\| \leq \|x - u\|$.*

(b) *There exists a constant $c > 0$ such that for all α and all pairs x and u in N_α $(\Phi_\alpha(x, u), x - u) \geq c \|x - u\|^2$.*

DEFINITION (3.2). *Let X be a Banach space, and suppose that we are given a Φ-system on X. Then a mapping A of X into 2^X is said to be Φ-accretive with respect to the given Φ-system if for each α, each pair x and u in N_α, each w in $A(u)$, and each y in $A(x)$, $(\Phi_\alpha(x, u), y - w) \geq 0$. In the case in which A is a single-valued mapping of the subset $D(A)$ of X into X, the condition for A to be Φ-accretive becomes: $(\Phi_\alpha(x, u), A(x) - A(u)) \geq 0$.*

A mapping A is accretive (without modifiers) in the sense described above if and only if A is Φ-accretive with respect to the Φ-system consisting of the single neighborhood X and the corresponding function $\Phi_\alpha(x, u) = J(x - u)$, with J a given duality mapping of X into X^*. To point up the special properties following from the fact of the existence of a global Φ_α, we introduce the following subsidiary definition:

DEFINITION (3.3). *Let X be a Banach space. By a global Φ-system on an open subset G of X, we mean a mapping Φ of $G \times G$ into X^* which satisfies the two conditions for the mappings in a Φ-system, namely, $\|\Phi(x, u)\| \leq \|x - u\|$, and $(\Phi(x, u), x - u) \geq c_0 \|x - u\|^2$ for a fixed constant c_0 and all x and u in G.*

We say that a mapping A of X into X is globally Φ-accretive on G if for all x and u in G, $(\Phi(x, u), A(x) - A(u)) \geq 0$, or, for the case of multivalued mappings, if for all $w \in A(u)$, $y \in A(x)$, $(\Phi(x, u), y - w) \geq 0$.

By the preceding remark, each accretive mapping A of a Banach space X into itself, or more particularly every monotone mapping (single-valued or multi-valued) from a Hilbert space H into itself, is globally Φ-accretive with respect to an appropriate Φ-function $\Phi(x, u) = J(x - u)$.

The inverse mappings of globally Φ-accretive mappings form another interesting class of nonlinear maps, the Φ-coaccretive mappings:

DEFINITION (3.4). *Let X be a Banach space with a global Φ-system on X given by the mapping Φ of $X \times X$ into X^*. Then a mapping T from X into 2^X is said to be (globally) Φ-coaccretive if for each x and u in X, y in $T(x)$, w in $T(u)$, we have*

$(\Phi(y, w), x - u) \geq 0$, i.e., for single-valued mappings T, $(\Phi(Tx, Tu), x - u) \geq 0$ for all x and u in $D(T)$.

It is obvious by inspection that a mapping A of X into 2^X is globally Φ-accretive with respect to a given global Φ-system on X if and only if $T = A^{-1}$ is globally Φ-coaccretive with respect to the same Φ-system, where

$$A^{-1}(x) = \{u \mid u \in X, x \in A(u)\}.$$

As a special case of Definition (3.4), we have the following definition which we give in full for reference in other sections of this paper:

DEFINITION (3.5). *Let X be a Banach space, T a mapping of X into 2^X. Then T is said to be coaccretive if and only if for each pair x and u of X and for each y in $T(x)$, w in $T(u)$, we have $(J(y - w), x - u) \geq 0$.*

The definition seems to depend upon the choice of the duality mapping J, but, as in the case of accretive mappings, it is often independent of J. This fact, indeed, follows from the case of accretive mappings since T is coaccretive if and only if $A = T^{-1}$ is accretive.

Two definitions which are of special importance in the theory of accretive and monotone mappings are the following:

DEFINITION (3.6). *Let X be a Banach space, A a mapping of X into 2^X which is Φ-accretive with respect to a given Φ-system. Then*

(a) *A is said to be maximal Φ-accretive with respect to the given Φ-system if the graph of A is not properly included in the graph of A_1, for another mapping A_1 of X into 2^X which is also Φ-accretive with respect to the given Φ-system.*

(b) *A is said to be hypermaximal Φ-accretive if $R(A + I)$, the range of $A + I$, is the whole space X.*

Similarly, we have

DEFINITION (3.7). *Let X be a Banach space with a global Φ-system, T a mapping of X into 2^X which is Φ-coaccretive. Then:*

(a) *T is maximal Φ-coaccretive if T is properly contained in no other mapping T_1 from X to 2^X which is Φ-coaccretive with respect to the same Φ-system.*

(b) *T is hypermaximal Φ-coaccretive if $R(I + T) = X$.*

As special cases of the above definitions which we shall frequently refer to in the following sections, we have

DEFINITION (3.8). *Let X be a Banach space, A and T mappings of X into 2^X. Then:*

(a) *A is maximal accretive if A is accretive and maximal in the set of accretive mappings from X to 2^X, ordered by inclusion of graphs.*

(b) *A is hypermaximal accretive if A is accretive and $R(A + I) = X$.*

(c) *T is maximal coaccretive if T is coaccretive and maximal in the set of coaccretive mappings from X to 2^X, ordered by inclusion of graphs.*

(d) *T is hypermaximal coaccretive if T is coaccretive and $R(I + T) = X$.*

We now begin the discussion of existence theorems involving Φ-accretive and Φ-coaccretive mappings.

THEOREM (3.1). *Let X be a Banach space with a given Φ-system, G an open subset of X, A a locally Lipschitzian mapping of G into X with A Φ-accretive with respect to the given Φ-system on G, i.e., $(\Phi_\alpha(x, u), A(x) - A(u)) \geq 0$ for all x and u in $N_\alpha \cap G$.*

Then $(A + I)$ is an open mapping of G into X.

COROLLARY TO THEOREM (3.1). *Let X be a Banach space with a Φ-system, A a locally Lipschitzian Φ-accretive mapping of X into X. Then $(A + I)$ is a local homeomorphism of X into X (i.e., each point x_0 of X has a neighborhood N in X which is mapped by A homeomorphically onto a neighborhood N' of $A(x_0)$ in X).*

PROOF OF THEOREM (3.1). For each $\lambda \geq 1$, consider the mapping $(A + \lambda I)$. Since

$$A + \lambda I = \lambda(I + \lambda^{-1}A) = \lambda(I + \xi A) \qquad (\xi = \lambda^{-1}),$$

it suffices to prove that for each ξ in $[0, 1]$, the mapping $g_\xi = I + \xi A$ is an open mapping of G into X. Since $g_0 = I$ is obviously an open mapping, it follows from Theorem (2.3) that since the family $\{g_\xi : \xi \in [0, 1]\}$ is obviously a continuous curve of locally Lipschitzian mappings from G to X, that g_ξ is an open mapping for each ξ in $[0, 1]$ if we can establish that for each α and ξ, there exists a constant $c_{\alpha, \xi} > 0$ with $\|g_\xi(x) - g_\xi(u)\| \geq c_{\alpha, \xi} \|x - u\|$ for all x and u in $G \cap N_\alpha$.

To prove this last assertion on the validity of the inequality, we consider

$$(\Phi_\alpha(x, u), g_\xi(x) - g_\xi(u)) = (\Phi_\alpha(x, u), x - u) + \xi(\Phi_\alpha(x, u), A(x) - A(u))$$
$$\geq (\Phi_\alpha(x, u), x - u)$$

by the Φ-accretiveness of A. Since $(\Phi_\alpha(x, u), x - u) \geq c_0 \|x - u\|^2$ with $c_0 > 0$, it follows that

$$c_0 \|x - u\|^2 \leq (\Phi_\alpha(x, u), g_\xi(x) - g_\xi(u))$$
$$\leq \|\Phi_\alpha(x, u)\| \cdot \|g_\xi(x) - g_\xi(u)\|$$
$$\leq \|x - u\| \cdot \|g_\xi(x) - g_\xi(u)\|.$$

Hence

$$\|g_\xi(x) - g_\xi(u)\| \geq c_0 \|x - u\| \qquad (x, u \in N_\alpha \cap g). \qquad \text{q.e.d.}$$

PROOF OF THE COROLLARY TO THEOREM (3.1). By Theorem (3.1), $A + I$ is an open mapping. Since $A + I$ is continuous, in order to prove that $A + I$ is a local homeomorphism, it suffices to show that $(A + I)$ is locally one-to-one. Suppose x and u lie in N_α. Then

$$\|(A + I)x - (A + I)u\| \cdot \|x - u\| \geq (\Phi_\alpha(x, u), (A + I)x - (A + I)u)$$
$$\geq (\Phi_\alpha(x, u), x - u) \geq c_0 \|x - u\|^2.$$

Hence $(A + I)x = (A + I)u$ implies that $x = u$. q.e.d.

THEOREM (3.2). *Let X be a Banach space, A a locally Lipschitzian mapping of X into X such that A is Φ-accretive with respect to a global Φ-system on X.*

Then $(A + I)$ is a homeomorphism of X onto X and $(A + I)^{-1}$ is a Lipschitzian mapping of X into X with Lipschitz constant $\leq c_0^{-1}$.

In particular, A is hypermaximal Φ-accretive on X.

PROOF OF THEOREM (3.2). By Theorem (3.1), $(A + I)$ is an open mapping of X into X. Hence $(A + I)(X)$ is an open subset of X. If we can show that there exists a constant $c > 0$ such that

$$\|(A + I)x - (A + I)(u)\| \geq c \|x - u\|,$$

it follows from Theorem (2.5) of §2 that $(A + I)(X)$ is closed in X, $(A + I)(X) = X$, and, trivially from the inequality, that $A + I$ is one-to-one. These facts together imply that $(A + I)$ is a homeomorphism. Since the inequality implies that $(A + I)^{-1}$ exists and satisfies a Lipschitz condition with constant c^{-1}, it suffices to show that the inequality holds with $c = c_0$.

However, we may consider for x, u in X

$$(\Phi(x, u), (A + I)(x) - (A + I)(u)) = (\Phi(x, u), A(x) - A(u)) + (\Phi(x, u), x - u)$$
$$\geq (\Phi(x, u), x - u) \geq c_0 \|x - u\|^2.$$

From this chain of inequalities, it follows that

$$c_0 \|x - u\|^2 \leq \|\Phi(x, u)\| \cdot \|(A + I)(x) - (A + I)u\|$$
$$\leq \|x - u\| \cdot \|(A + I)x - (A + I)u\|$$

and the desired inequality follows. q.e.d.

THEOREM (3.3). *Let X be a Banach space with a global Φ-system, T a locally Lipschitzian mapping of X into X which is Φ-coaccretive with respect to this global Φ-system.*

Then for each $\lambda > 0$, $(T + \lambda I)$ is a homeomorphism of X onto X with a Lipschitzian inverse.

PROOF OF THEOREM (3.3). Fix $\lambda_0 > 0$, and consider the half-infinite interval $[\lambda_0, \infty)$. For λ in this interval $(T + \lambda I) = \lambda(I + \lambda^{-1}T) = \lambda(I + \xi T)$ for $\xi = \lambda^{-1}$. To prove that $(T + \lambda I)$ is a homeomorphism of X onto X with a Lipschitzian inverse, it suffices to prove the same fact for $h_\xi = I + \xi T$ with ξ running through the interval $[0, \xi_0]$ where $\xi_0 = \lambda_0^{-1}$. Since $h_0 = I$ is an open mapping and the family $\{h_\xi : \xi \in [0, \xi_0]\}$ is a continuous curve of locally Lipschitzian mappings of X into X, it follows from Theorem (2.3) that it suffices to prove the inequality $\|h_\xi(x) - h_\xi(u)\| \geq c_\xi \|x - u\|$ for some constant $c_\xi > 0$ and all x and u in X. Indeed, it then follows from Theorem (2.3) that each h_ξ is an open mapping, while from the inequality itself, it follows that h_ξ is one-to-one and is a closed mapping. Thus $h_\xi(X)$, being open and closed in X, must equal X, and h_ξ^{-1} would satisfy a Lipschitz condition on X.

The desired inequality follows from the following calculation: For x and u in

X, we have

$$(\Phi(Tx, Tu), h_\xi(x) - h_\xi(u)) = (\Phi(Tx, Tu), x - u) + \xi(\Phi(Tx, Tu), Tx - Tu)$$
$$\geq \xi(\Phi(Tx, Tu), Tx - Tu) \geq \xi c_0 \|Tx - Tu\|^2$$

by the Φ-coaccretiveness of T. For $\xi = 0$, the inequality is trivial. For $\xi \neq 0$, we obtain

$$\xi c_0 \|Tx - Tu\|^2 \leq \|\Phi(Tx, Tu)\| \cdot \|h_\xi(x) - h_\xi(u)\| \leq \|Tx - Tu\| \cdot \|h_\xi(x) - h_\xi(u)\|$$

so that it follows that

$$\xi c_0 \|Tx - Tu\| \leq \|h_\xi(x) - h_\xi(u)\|.$$

On the other hand,

$$\|x - u\| \leq \|h_\xi(x) - h_\xi(u)\| + \|\xi T(x) - \xi T(u)\|$$
$$\leq (1 + c_0^{-1}) \|h_\xi(x) - h_\xi(u)\|. \qquad \text{q.e.d.}$$

THEOREM (3.4). *Let X be a Banach space, G an open subset of X, and suppose a given Φ-system on X. Suppose that A is a locally Lipschitzian mapping of G into X such that for each neighborhood N_α of the Φ-system, there exists a constant $c_\alpha > 0$ such that*

$$(\Phi_\alpha(x, u), A(x) - A(u)) \geq c_\alpha \|x - u\|^2$$

for all x and u in N_α.

Then A is an open mapping and a local homeomorphism of G into X.

PROOF OF THEOREM (3.4). For each x and u in N_α, the assumed inequality implies that

$$c_\alpha \|x - u\|^2 \leq \|Ax - Au\| \cdot \|\Phi_\alpha(x, u)\| \leq \|x - u\| \cdot \|Ax - Au\|,$$

so that $c_\alpha \|x - u\| \leq \|Ax - Au\|$. Hence, A is one-to-one on N_α and has a continuous inverse from $A(N_\alpha)$ to N_α. To show that A is a local homeomorphism, it suffices to show that A is an open mapping.

To show that A is an open mapping, we can restrict our attention to a neighborhood N_α, replace G by N_α, and assume that A is globally Φ-accretive on G. Then $A = (A - c_\alpha I) + c_\alpha I$, and $(A - c_\alpha I)$ is also Φ-accretive on G since

$$(\Phi_\alpha(x, u), (A - c_\alpha I)(x) - (A - c_\alpha I)(u))$$
$$= (\Phi_\alpha(x, u), A(x) - A(u)) - c_\alpha(\Phi_\alpha(x, u), x - u)$$
$$\geq c_\alpha \|x - u\|^2 - c_\alpha \|\Phi_\alpha(x, u)\| \cdot \|x - u\| \geq 0.$$

Hence $A = (A - c_\alpha I) + c_\alpha I$, $c_\alpha > 0$, is open by Theorem (3.1). q.e.d.

THEOREM (3.5). *Let X be a Banach space with a global Φ-system, A a mapping of X into 2^X which is hypermaximal Φ-accretive with respect to this Φ-system, A_0 a locally Lipschitzian mapping of X into X which is Φ-accretive with respect to this Φ-system. Then $(A + A_0 + I)$ has all of X as its range and has a single-valued Lipschitzian inverse, i.e., $(A + A_0)$ is hypermaximal Φ-accretive with respect to the given Φ-system.*

COROLLARY TO THEOREM (3.5). *Let X be a Banach space with a global Φ-system, A a mapping of X into 2^X which is hypermaximal Φ-accretive, A_0 a locally Lipschitzian mapping of X into X such that for all x and u of X and a fixed constant $c > 0$,*

$$(\Phi(x, u), A_0(x) - A_0(u)) \geq c \|x - u\|^2.$$

Then $(A + A_0)$ has all of X as its range.

PROOF OF THEOREM (3.5). We consider the family of mappings

$$A_\lambda = A + \lambda A_0 + I, \quad 0 \leq \lambda \leq 1.$$

For $\lambda = 0$, A_λ becomes $A + I$ which is an open mapping. Since λA_0 is a continuous curve of locally Lipschitzian mappings of X, we may apply Theorem (2.3) of §2 and conclude that in order to prove that A_λ is open for every λ in $[0, 1]$, it suffices to show that there exists a constant c_λ such that for each y in $A_\lambda(x)$, w in $A_\lambda(u)$, $\|w - y\| \geq c_\lambda \|u - x\|$. However, $w = u + \lambda A_0(u) + z$ for some z in $A(u)$, while $y = x + \lambda A_0(x) + v$ for some v in $A(x)$. Hence

$$(\Phi(x, u), y - w) = (\Phi(x, u), x - u + \lambda(A_0(x) - A_0(u)) + v - z)$$
$$\geq (\Phi(x, u), x - u) + \lambda(\Phi(x, u), A_0(x) - A_0(u))$$
$$\geq (\Phi(x, u), x - u) \geq c_0 \|x - u\|^2.$$

Hence

$$c_0 \|x - u\|^2 \leq \|\Phi(x, u)\| \cdot \|w - y\| \leq \|x - u\| \cdot \|w - y\|,$$

so that $c_0 \|x - u\| \leq \|w - y\|$. Thus each A_λ is an open mapping for λ in $[0, 1]$, and in particular $A_1(X)$ is an open subset of X.

On the other hand, the inequality proved above implies that A_λ^{-1} is a single-valued Lipschitzian mapping, whose domain consists of $A_\lambda(X)$. We assert that $A_\lambda(X)$ is closed in X. Indeed, suppose that w_n lies in $A_\lambda(X)$ for a given λ and all n, and suppose that $w_n \to w$ in X. If $x_n = A_\lambda^{-1}(w_n)$, then $w_n = x_n + \lambda A_0(x_n) + y_n$ for some y_n in $A(x_n)$. Since A_0, being locally Lipschitzian, is a continuous mapping from X to X, it follows that since

$$\|x_n - x_m\| \leq c_0^{-1} \|w_n - w_m\| \to 0 \quad (m, n \to \infty),$$

$x_n \to x$ and $A_0(x_n) \to A_0(x)$ for some element x of X. Hence

$$y_n = w_n - x_n - \lambda A_0(x_n) \to w - x - \lambda A_0(x),$$

while $y_n + x_n \to w - \lambda A_0(x)$. Since $y_n + x_n \in (A + I)(x_n)$ and $(A + I)^{-1}$ is defined on all of X and is continuous, it follows that $(A + I)^{-1}(w - \lambda A_0(x)) = x$, i.e., $w - \lambda A_0(x) - x \in A(x)$, or finally $w \in (A + \lambda A_0 + I)(x)$.

Thus $A + I$ has a range which is both open and closed in X, and hence coincides with X. q.e.d.

PROOF OF THE COROLLARY TO THEOREM (3.5). If we replace A_0 by $c^{-1}A_0$ and A by $c^{-1}A$, $c > 0$, then the mapping

$$A + A_0 = (A + [A_0 - cI] + cI) = c(c^{-1}A + c^{-1}[A_0 - cI] + I)$$

will have all of X as its range by Theorem (3.5) provided that $A_0 - cI$ is Φ-accretive and $c^{-1}A$ is hypermaximal Φ-accretive. The first of these facts is true since

$$(\Phi(x, u), [A_0 - cI](x) - [A_0 - cI](u)) = (\Phi(x, u), A_0x - A_0u) - (\Phi(x, u), cx - cu)$$
$$\geq c \|x - u\|^2 - c \|x - u\|^2 \geq 0$$

for all x and u in X. Hence it suffices to show that for each $c > 0$, A is hypermaximal Φ-accretive implies that $c^{-1}A$ is also hypermaximal Φ-accretive.

However, the family of mappings $c^{-1}A + I - c^{-1}(A + cI)$ will have all of X as their range if and only if this is true for $A + cI$, $c > 0$. For the latter, we note that if w lies in $(A + cI)$, y in $(A + cI)(x)$, then

$$(\Phi(x, u), y - w) \geq c(\Phi(x, u), x - u) \geq cc_0 \|x - u\|^2.$$

Hence

$$cc_0 \|x - u\|^2 \leq \|\Phi(x, u)\| \cdot \|y - w\| \leq \|x - u\| \cdot \|y - w\|$$

and therefore $cc_0 \|x - u\| \leq \|y - w\|$. If we let c run over an interval $[k_0, 1]$ or $[1, k_1]$ with $0 < k_0 < 1 < k_1$, then it follows from Theorem (2.5) that for each $c > 0$, $(A + cI)$ maps X onto X.   q.e.d.

THEOREM (3.6). *Let X be a Banach space with a given global Φ-system where $\Phi(x, u) = \psi(x - u)$, for a mapping ψ of X into X^* such that $\|\psi(x)\| \leq \|x\|$ for all x in X and $(\psi(x), x) \geq c_0 \|x\|^2$, while $\psi(tv) = c_t(v)\psi(v)$ with $c_t(v)$ a positive constant for $t > 0$, $v \in X$. Let A be a Φ-accretive mapping of X into X which is hemicontinuous (i.e., continuous from each line segment in X to the weak topology of X).*

Then A is maximal Φ-accretive.

PROOF OF THEOREM (3.6). To show that A is maximal Φ-accretive, we take a point $[u_0, w_0]$ in $X \times X$ which lies in the graph of a supposed Φ-accretive extension A_1 of A as a mapping from X to 2^X and show that $w_0 = A(u_0)$. By the assumption, $(\Phi(x, u_0), A(x) - w_0) \geq 0$ $(x \in X)$, i.e., $(\psi(x - u_0), A(x) - w_0) \geq 0$.

We take an arbitrary element v of X, and let $x_t = u_0 + tv$, $t > 0$. Setting x_t for x in the above inequality, we obtain $c_t(v)(\psi(v), A(x_t) - w_0) \geq 0$. Cancelling the positive factor $c_t(v)$, we see that $(\psi(v), A(x_t) - w_0) \geq 0$. Since $x_t \to u_0$ as $t \to 0+$ and since A is continuous on the line segment joining u_0 to $u_0 + v$ to the weak topology of X, we see that $(\psi(v), A(u_0) - w_0) \geq 0$. Letting $v = w_0 - A(u_0)$ we obtain $-(\psi(v), v) \geq 0$, $(\psi(v), v) \geq c_0 \|v\|^2$ so that $\|v\|^2 \leq 0$, and $v = 0$. Hence $w_0 = A(u_0)$.   q.e.d.

THEOREM (3.7). *Let X be a Banach space with a global Φ-system given by a function Φ such that the map $x \to \Phi(x, u)$ for a fixed u is continuous from the strong topology of X to the weak* topology of X^*. Let A be a mapping from X to 2^X which is hypermaximal Φ-accretive.*

Then A is maximal Φ-accretive.

PROOF OF THEOREM (3.7). Suppose that for a pair $[u_0, w_0]$ in $X \times X$, we have $(\Phi(x, u_0), w - w_0) \geq 0$ for all x in X, w in $A(x)$. Then

$$(\Phi(x, u_0), w + x - w_0 - u_0) = (\Phi(x, u_0), w - w_0) + (\Phi(x, u_0), x - u_0)$$
$$\geq (\Phi(x, u_0), w - w_0) + c_0 \|x - u_0\|^2 \geq 0.$$

By hypothesis $(A + I)$ has all of X as its range, and since A is Φ-accretive, $(A + I)^{-1}$ is single valued and Lipschitzian. For each $t > 0$ and an arbitrary v in X, we let $x_t = (A + I)^{-1}(w_0 + u_0 + tv)$, so that for some w_t in $A(x_t)$, we have $x_t + w_t = w_0 + u_0 + tv$.

If we introduce this pair $[x_t, w_t]$ in the graph of A into the inequality

$$(\Phi(x_t, u_0), w_t + x_t - w_0 - u_0) \geq 0,$$

we obtain $t(\Phi(x_t, u_0), v) \geq 0$. If we cancel the positive constant t, we find that $(\Phi(x_t, u_0), v) \geq 0$. We let $t \to 0+$ and note that since $w_0 + u_0 + tv \to w_0 + u_0$ and since $(A + I)^{-1}$ is continuous, x_t converges strongly to $(A + I)^{-1}(w_0 + u_0)$. Hence $\Phi(x_t, u_0)$ converges weakly to $\Phi(x, u_0)$ with $x = (A + I)^{-1}(w_0 + u_0)$, so that $(\Phi(x, u_0), v) \geq 0$ $(v \in X)$.

We now let $v = u_0 - x$. Then

$$0 \leq (\Phi(x, u_0), v) = -(\Phi(x, u_0), x - u_0) \leq -c_0 \|x - u_0\|^2 \leq 0,$$

so that $x = u_0$. Hence $u_0 = (A + I)^{-1}(w_0 + u_0)$, i.e., $w_0 + u_0 \in u_0 + A(u_0)$, so that $w_0 \in A(u_0)$.   q.e.d.

THEOREM (3.8). *Let X be a Banach space with a global Φ-system given by a function $\Phi(x, u)$ which is continuous in each variable separately. Let A be a mapping of X into 2^X which is maximal Φ-accretive. Let $\{x_n\}$ be a sequence in X such that x_n converges strongly to x_0 in X, $w_n \in A(x_n)$ for each n, and suppose that w_n converges weakly to w_0 in X. Then $w_0 \in A(x_0)$.*

PROOF OF THEOREM (3.8). Let x be any element of X, $w \in A(x)$. By the Φ-accretiveness of A, $(\Phi(x_n, x), w_n - w) \geq 0$. As $n \to \infty$, $x_n \to x_0$ and hence $\Phi(x_n, x) \to \Phi(x_0, x)$ strongly in X^*. Similarly, $w_n - w$ converges weakly to $w_0 - w$. Hence $(\Phi(x_n, x), w_n - w) \to (\Phi(x_0, x), w_0 - w)$ and therefore $(\Phi(x_0, x), w_0 - w) \geq 0$ for all x in X, w in $A(x)$. Since A is maximal Φ-accretive, $w_0 \in A(x_0)$.   q.e.d.

In connection with the extension of Theorems (3.7) and (3.8) to the class of monotone mappings from a Banach space X to its conjugate space X^*, it is convenient to introduce a wide-ranging generalization of the concept of Φ-system and of Φ-accretive mapping, namely the following:

DEFINITION (3.9). *Let X and Y be locally convex topological vector spaces over the reals (both X and Y Hausdorff). Let (y', y) denote the pairing between y in Y and y' in Y', the space of continuous linear functionals on Y.*

If G is an open subset of X, a global Φ-system in the general sense for the pair $[X, Y]$ over G consists of a mapping Φ, not necessarily continuous, of $G \times G$ into Y'.

By a Φ-system in general for the pair $[X, Y]$, we mean a covering N_α of X and for each α, a Φ-system for N_α in the sense of the preceding paragraph, i.e., a mapping Φ_α of $N_\alpha \times N_\alpha$ into Y'.

DEFINITION (3.10). *Let X and Y be a pair of locally convex topological vector spaces, and suppose a given Φ-system for the pair $[X, Y]$. Then a mapping A of X into 2^Y is said to be Φ-monotone if for each α, all pairs x and u in N_α, y in $A(x)$, w in $A(u)$, we have $(\Phi_\alpha(x, u), y - w) \geq 0$.*

In particular, if A is a single-valued mapping of a subset $D(A)$ of X into Y, then A is Φ-monotone if for each pair x, u in $D(A) \cap N_\alpha$, $(\Phi_\alpha(x, u), A(x) - A(u)) \geq 0$.

As a special case of the above definitions, we have the definition of monotonicity in the ordinary sense for mappings of X into X':

DEFINITION (3.11). *Let X be a topological vector space which is Hausdorff and locally convex, X' its dual space. Then a mapping A of X into $2^{X'}$ is said to be monotone if for all x, u in X, y in $A(x)$, w in $A(u)$, we have $(y - w, x - u) \geq 0$. In particular, if A is a single-valued mapping of a subset $D(A)$ of X into X', A is monotone if*

$$(A(x) - A(u), x - u) \geq 0 \qquad (x, u \in D(A)).$$

The definition of monotone mapping in Definition (3.11) becomes a special case of the general definition of Φ-monotone mapping by setting $Y = X'$ and defining the mapping Φ of $X \times X$ into Y' by setting $\Phi(x, u) = x - u$.

DEFINITION (3.12). *Let X and Y be two locally convex separated topological vector spaces, with a given Φ-system for the pair $[X, Y]$. Suppose that A is a Φ-monotone mapping of X into 2^Y. Then*

(a) A is said to be maximal Φ-monotone if the graph of A is not properly contained in the graph of another Φ-monotone mapping A_1 from X to 2^Y.

(b) A is said to be hypermaximal Φ-monotone if there exists a Φ-monotone mapping A_2 from X to Y such that $R(A + A_2) = Y$, and $(A + A_2)^{-1}$ is a single-valued continuous mapping of Y into X.

In this definition, we use the familiar convention that the sum $A + A_2$ of two possibly multivalued mappings is given by:

$$(A + A_2)(u) = A(u) + A_2(u), \quad \text{if } A(u) \text{ and } A_2(u) \text{ are both nonempty,}$$
$$= \varnothing, \qquad \text{otherwise.}$$

THEOREM (3.9). *Let X and Y be two locally convex, separated, topological vector spaces with a given Φ-system for the pair $[X, Y]$ given by a global mapping Φ of $X \times X$ into Y' such that for each fixed u in X, the mapping $x \to \Phi(x, u)$ is continuous from the topology of X to the weak* topology on Y' (i.e., the weakest topology on Y' such that the functions (y', y) are continuous on Y' for each y in Y). Suppose further that $\Phi(x, u) = 0$ implies that $x = u$.*

Let A be a hypermaximal Φ-monotone mapping of X into 2^Y.

Then: A is maximal Φ-monotone from X to 2^Y.

PROOF OF THEOREM (3.9). Suppose that u_0 lies in X, w_0 lies in Y, and that for all x in X, w in $A(x)$, we have $(\Phi(x, u_0), w - w_0) \geq 0$. It suffices to show that $w_0 \in A(u_0)$.

By hypothesis, A is hypermaximal Φ-monotone with respect to the given Φ-system, i.e., there exists a Φ-monotone mapping A_2 of X into Y such that $(A + A_2)$ has all of Y as its range and such that $(A + A_2)^{-1}$ is a single-valued continuous mapping of X into Y. Thus for all x in X, $(\Phi(x, u_0), A_2(x) - A_2(u_0)) \geq 0$, and adding this to the preceding inequality, we obtain

(i) $\qquad (\Phi(x, u_0), w + A_2(x) - (w_0 + A_2(u_0))) \geq 0$.

Let v be a fixed but arbitrary element of Y, $t > 0$. We set

$$w_t = (w_0 + A_2(u_0)) + tv,$$

and set $x_t = (A + A_2)^{-1}(w_t)$. We substitute $x = x_t$ in the inequality (i), and note that since $w_t \in (A + A_2)(x_t)$, there exists y_t in $A(x_t)$ such that $y_t + A_2(x_t) = w_t$. Hence, we get

$$0 \leq (\Phi(x_t, u_0), w_t - (w_0 + A_2(u_0))) = t(\Phi(x_t, u_0), v),$$

and as a consequence, $(\Phi(x_t, u_0), v) \geq 0$.

As $t \to 0+$, w_t converges in Y to $w_0 + A_2(u_0)$. Since $(A + A_2)^{-1}$ is continuous from Y to X by our hypothesis, it follows that

$$x_t = (A + A_2)^{-1}(w_t) \to (A + A_2)^{-1}(w_0 + A_2(u_0)) = x_0$$

as $t \to 0+$. Since $\Phi(x_t, u_0)$ is continuous in the first variable to the weak* topology on Y', it follows that $(\Phi(x_t, u_0), v) \to (\Phi(x_0, u_0), v)$ $(t \to 0+)$. Hence $(\Phi(x_0, u_0), v) \geq 0$ $(v \in Y)$.

Since we may replace v by $(-v)$ in the last inequality without altering the left-hand term in the pairing, we have $(\Phi(x_0, u_0), v) = 0$ for all v in Y, i.e., $\Phi(x_0, u_0) = 0$. By hypothesis, this is possible only if $u_0 = x_0$, so that $u_0 = (A + A_2)^{-1}(w_0 + A_2(u_0))$. Hence $w_0 + A_2(u_0) \in A_2(u_0) + A(u_0)$, and we conclude that w_0 lies in $A(u_0)$. q.e.d.

THEOREM (3.10). *Let X and Y be a pair of Banach spaces with a global Φ-system for the pair $[X, Y]$ given by the global Φ-function $\Phi : X \times X \to Y'$ with Φ continuous in each of its arguments separately. Let A be a maximal Φ-monotone mapping of X into 2^Y, and let $G(A)$, the graph of A, be given as usual by $G(A) = \{[x, y] \mid x \in X, y \in A(x)\}$ considered as a subset of $X \times Y$.*

Suppose that $\{[x_n, y_n]\}$ is a sequence in $G(A)$ with x_n converging strongly to x in X, y_n converging weakly to y in Y. Then $y \in A(x)$, i.e., $[x, y]$ lies in $G(A)$.

PROOF OF THEOREM (3.10). For each $[u, w]$ in $G(A)$ and each integer n, it follows from the Φ-monotonicity of A that $(\Phi(u, x_n), w - y_n) \geq 0$. By hypothesis, x_n converges strongly to x in X. By hypothesis $\Phi(u, x_n)$ is strongly continuous in x_n on X, so that $\Phi(u, x_n) \to \Phi(u, x)$ strongly in Y. By assumption, $y_n \to y$ weakly in Y, so that $w - y_n$ converges weakly to $w - y$. Hence $(\Phi(u, x_n), w - y_n) \to (\Phi(u, x), w - y)$, and therefore

$$(\Phi(u, x), w - y) \geq 0 \qquad ([u, w] \in G(A)).$$

Since A is maximal Φ-monotone, it follows that $[x, y] \in G(A)$. q.e.d.

THEOREM (3.11). *Let X and Y be two Banach spaces with a global Φ-system for the pair $[X, Y]$ given by the function $\Phi: X \times X \to Y^*$ such that Φ is continuous in each variable separately. Let A be a maximal Φ-monotone mapping of X into 2^Y. Then:*

(a) *For each x in X, $A(x)$ is a weakly closed convex subset of Y.*

(b) *Suppose further that Y is reflexive and that A is locally bounded on an open subset G of X (i.e., each point x_0 of G has a neighborhood N such that $A(N)$ is bounded in Y). Then considered as a mapping from X to 2^Y with X given its strong topology and Y its weak topology, A is an upper-semicontinuous set-valued mapping on G.*

(c) *Under the hypotheses of (b), the effective domain of A,*

$$D(A) = \{u \mid A(u) \text{ is nonempty}\},$$

is a closed subset of G.

Before beginning the proof of Theorem (3.11), we recall the definition of the concept of upper-semicontinuous set-valued mapping employed in part (b) of the theorem.

DEFINITION (3.13). *Let S and S_1 be two topological spaces, A a mapping of S into 2^{S_1}. Then A is said to be an upper-semicontinuous set-valued mapping of S if for each point s_0 of S and each open neighborhood V of $A(s_0)$ in S_1, there exists a neighborhood U of s_0 in S (with U depending upon V) such that $A(U) \subset V$.*

REMARK. We use the convention throughout that for any mapping A of S into 2^{S_1} and for any subset S' of S,

$$A(S') = \bigcup_{s \in S'} A(s).$$

PROOF OF THEOREM (3.11). PROOF OF (a). Since A is maximal Φ-monotone, the pair $[x, w]$ lies in $G(A)$ if and only if $(\Phi(x, u), w - y) \geq 0$ for all elements $[u, y]$ in $G(A)$. Thus

$$A(x) = \bigcap_{[u,v] \in G(A)} \{w \mid w \in Y, (\Phi(x, u), w - y) \geq 0\}.$$

Each of the sets in the intersection is flat, and hence $A(x)$, being the intersection of weakly closed convex subsets of Y, is itself a weakly closed convex subset of Y.
q.e.d.

PROOF OF (b). Since the property of A of being an upper-semicontinuous set-valued mapping is a local property on the neighborhoods of the various points of G and since A is locally bounded on G, we may assume without loss of generality that A is bounded on G, i.e., $A(G)$ is a bounded subset of Y.

Let x_0 be a point of G, V a weak open neighborhood of $A(x_0)$. Since Y is reflexive and $A(G)$ is bounded, there exists a weakly compact subset K of Y such that $A(G) \subset K$. Suppose that A is not an upper-semicontinuous mapping at x_0. Then there exists a sequence $\{x_n\}$ in G converging strongly to x_0 such that for each n, we may find $w_n \in A(x_n) - V \subset K - V$. Since K is weakly sequentially compact, we may pass to an infinite subsequence and assume that w_n converges weakly

to some element w of E. Since V is weakly open in Y, w must lie in $K - V$. By Theorem (3.10), however, w lies in $A(x_0) \subset V$. This is a contradiction, proving the assertion (b).
\hfill q.e.d.

PROOF OF (c). Suppose that $A(x_0)$ is empty, and suppose that we can find a sequence $x_n \to x_0$ strongly in G such that $A(x_n)$ is nonempty for each n. We may find a weakly compact set K in Y such that $A(x_n) \subset K$ for all n. If we choose w_n in $A(x_n)$ for each n, we may pass to an infinite subsequence and assume without loss of generality that w_n converges weakly to some w in Y as $n \to \infty$. By Theorem (3.10), however, w lies in $A(x_0)$ contradicting the assumption that $A(x_0)$ is empty.
\hfill q.e.d.

THEOREM (3.12). *Let X and Y be two Banach spaces with a global Φ-system for the pair $[X, Y]$ given by a function $\Phi: X \times X \to Y^*$ such that for all x and u in X, $\Phi(x, u) = \psi(x - u)$ where $\psi: X \to Y^*$ is a function satisfying the conditions: (1) $\psi(X)$ is dense in Y^*, (2) for each v in X and $t > 0$, $\psi(tv) = c_t(c)\psi(v)$ for some constant $c_t(v) > 0$. Then*

(a) *Let G be an open subset of X, A a Φ-monotone mapping of G into 2^Y such that for each x in G, $A(x)$ is a bounded closed convex subset of Y which is nonempty. Suppose that on each closed line segment in G, A is an upper-semicontinuous set-valued function from that line segment to 2^Y with the weak topology on Y. Let u_0 be a point of G, w_0 an element of Y and suppose that $(\Phi(x, u_0), y - w_0) \geq 0$ for all x in G, y in $A(x)$.*

Then w_0 lies in $A(x_0)$.

(b) *Let A be a Φ-monotone mapping of X into Y such that for each x in X, $A(x)$ is a nonempty bounded closed convex subset of Y. Suppose that A is upper-semicontinuous from each line segment in X to 2^Y with the weak topology on Y. Then A is maximal Φ-monotone from X to 2^Y.*

PROOF OF THEOREM (3.12). PROOF OF (a). Suppose that w_0 does not lie in $A(u_0)$. Since $A(u_0)$ is closed and convex, there exists an element y^* of Y^* such that

$$(y^*, w_0) > \sup_{w \in A(u_0)} (y^*, w).$$

Since $A(u_0)$ is bounded, there is a ball about y^* in Y^* such that for all z^* in this ball, the same inequality holds with y^* replaced by z^*. Since $\psi(X)$ is dense in Y^*, we may choose z^* of the form $\psi(v)$ for some v in X, i.e.,

$$(\psi(v), w_0) > (\psi(v), w) \qquad (w \in A(u_0)).$$

Since G is open, there exists $t_0 > 0$ such that for each t in the interval $(0, t_0]$, $x_t = u_0 + tv$ lies in G. For each such x_t, $A(x_t)$ is nonempty by hypothesis, and we choose an element w_t in $A(x_t)$. If V is the weak neighborhood of the set $A(u_0)$ in Y given by

$$V = \{w \mid w \in Y, (\psi(v), w - w_0) < 0\},$$

it follows from the assumed upper-semicontinuity of A on each line segment in G

to the weak topology of Y that for a suitable $t_1 > 0$ and t in the interval $(0, t_1)$, w_t must lie in V.

However, by the assumed inequality above involving u_0 and w_0, we have $(\psi(x_t - u_0), w_t - w_0) \geq 0$. We note that by the choice of x_t,

$$\psi(x_t - u_0) = \psi(tv) = c_t(v)\psi(v) \qquad (c_t(v) > 0 \text{ for } t > 0).$$

Hence $c_t(v)(\psi(v), w_t - w_0) \geq 0$. Cancelling the positive factor $c_t(v)$, we find that for each t in $(0, t_1)$, $0 > (\psi(v), w_t - w_0) \geq 0$, which is a contradiction proving assertion (a).   q.e.d.

PROOF OF (b). Assertion (b) is simply a translation of the assertion of part (a) for the special case in which $G = X$.   q.e.d.

THEOREM (3.13). *Let X and Y be two Banach spaces with a global Φ-system for the pair $[X, Y]$ given by a mapping $\Phi: X \times X \to Y^*$, where $\Phi(x, u) = \psi(x - u)$ for each x and u in X and $\psi: X \to Y^*$ is a function satisfying the following conditions:*

(1) *ψ is locally uniformly continuous.*
(2) *For v in X and $t > 0$, $\psi(tv) = t\psi(v)$.*
(3) *If B_1 is the unit ball in X, $\psi(B_1)$ is an absorbing set in Y.*

Let G be an open subset of X, A a Φ-monotone mapping of G into 2^Y such that $A(x)$ is nonempty for each x in G. Suppose that A is locally bounded on each line segment in G.

Then A is locally bounded on G.

PROOF OF THEOREM (3.13). We suppose that the conclusion of the theorem is false. Then there exist a sequence $\{u_n\}$ in G converging strongly to u_0 in G and a corresponding sequence $\{w_n\}$ in Y with $\|w_n\| \to \infty$ as $n \to \infty$ such that for each n, w_n lies in $A(u_n)$. Let $p_n = \|w_n\|$, $r_n = \|u_n - u_0\|$. We consider a fixed element v of X with $\|v\| \leq 1$ and an arbitrary sequence t_n with $t_n > 0$ for each n and $t_n \to 0$ as $n \to \infty$.

Let $x_n = u_0 + t_n v$. Since G is open and ψ is assumed to be locally uniformly continuous, we may find a neighborhood N of 0 in X on which ψ is uniformly continuous. Thus there exists a function $\beta(r)$ with $\beta(r) \to 0$ as $r \to 0$ such that for all x and u in N, $\|\psi(x) - \psi(u)\| \leq \beta(\|x - u\|)$. We may assume without loss of generality that the sequences $\{x_n - u_0\}$ and $\{x_n - u_n\}$ both lie in N. Since A is locally bounded on the segment $[u_0, u_0 + t_0 v]$ in G, we may find a constant $M > 0$ such that for any $y_n \in A(x_n)$, $\|y_n\| \leq M$. We choose such elements y_n for each n, whose existence follows from our hypothesis that $A(x)$ is nonempty for each x in G.

Since the mapping A is Φ-monotone with $\Phi(x, u) = \psi(x - u)$, we have $(\psi(x_n - u_n), y_n - w_n) \geq 0$, i.e., $(\psi(x_n - u_n), w_n) \leq (\psi(x_n - u_n), y_n)$. For the right-hand term, we have

$$(\psi(x_n - u_n), y_n) = (\psi(x_n - u_0), y_n) + (\psi(x_n - u_n) - \psi(x_n - u_0), y_n).$$

By the definition of $\beta(r)$, we have

$$\|\psi(x_n - u_n) - \psi(x_n - u_0)\| \leq \beta(\|u_n - u_0\|) = \beta(r_n).$$

Moreover, $(\psi(x_n - u_0), y_n) = t_n(\psi(v), y_n) \le M_1 t_n$ for a constant M_1 independent of n. Hence $(\psi(x_n - u_n), w_n) \le \beta(r_n) + M_1 t_n$ $(n \ge 1)$.

Finally,

$$t_n(\psi(v), w_n) = (\psi(t_n v), w_n) = (\psi(x_n - u_n), w_n) + (\psi(t_n v) - \psi(x_n - u_n), w_n)$$
$$\le \beta(r_n) + M_1 t_n + \|w_n\| \beta(\|t_n v - x_n + u_n\|)$$
$$\le \beta(r_n) + M_1 t_n + p_n \beta(r_n).$$

By condition (3) on the function ψ, $\psi(B_1)$ is an absorbing set in Y. Hence for y in Y, $y = \psi(v)$ for some v and

$$t_n(w_n, y) \le \beta(r_n) + M_1 t_n + p_n \beta(r_n).$$

We choose t_n so that $\varepsilon_n = t_n^{-1} \beta(r_n) \to 0$. Then for each y in Y, $(w_n, y) \le M(y) + \varepsilon_n \|w_n\|$. If $\varepsilon_n \|w_n\|$ is bounded, $\|w_n\|$ is bounded by the uniform boundedness principle. If not, $z_n = \varepsilon_n^{-1} \|w_n\|^{-1} w_n$ is bounded by the same principle, while $\|z_n\| = \varepsilon_n^{-1}$.   q.e.d.

COROLLARY TO THEOREM (3.13). *The condition (3) in the hypothesis of Theorem (3.13) on ψ may be replaced by the stronger hypothesis:*
(3)' *ψ maps X onto Y^*.*

PROOF OF THE COROLLARY. If B_n is the ball of radius n, we see that $\psi(B_n) = n\psi(B_1)$. Hence if $\psi(X) = Y^*$, it follows that $\psi(B_1)$ is a set in Y^* containing 0 such that its multiples absorb Y^*.   q.e.d.

THEOREM (3.14). *Let X and Y be a pair of Banach spaces with a global Φ-system for the pair $[X, Y]$ given by a function $\Phi: X \times X \to Y^*$ of the form $\Phi(x, u) = \psi(x - u)$, where ψ satisfies the conditions (1), (2), and (3) of Theorem (3.13). Suppose that Y is reflexive, and that for an open subset G of X, we are given a Φ-monotone mapping A of G into 2^Y satisfying the following conditions:*
(i) *$A(x)$ for each x in G is a nonempty closed convex subset of Y.*
(ii) *For each line segment in G, A is an upper-semicontinuous set-valued mapping from that segment to 2^Y, with the weak topology on Y.*

Then A is an upper-semicontinuous mapping from G to 2^Y with respect to the weak topology on Y.

PROOF OF THEOREM (3.14). By Theorem (3.13), A is a locally bounded mapping from G to 2^Y. By Theorem (3.11)(b) and Theorem (3.12)(b), A is therefore upper-semicontinuous from G to 2^Y with the weak topology on Y.   q.e.d.

THEOREM (3.15). *Let X and Y be a pair of Banach spaces with a global Φ-system for the pair $[X, Y]$ given by a mapping $\Phi: X \times X \to Y^*$ with $\Phi(x, u) = \psi(x - u)$ for a function $\psi: X \to Y^*$ which satisfies the conditions (1), (2), and (3) of Theorem (3.13). Let G be an open subset of X, A a maximal Φ-monotone mapping of*

X into 2^Y such that for each x in G, $A(x)$ is nonempty. Suppose further that A is locally bounded on each line segment in G.

Then A is upper-semicontinuous from G to 2^Y, with the weak topology on Y.

PROOF OF THEOREM (3.15). By Theorem (3.13), A is locally bounded on G. The conclusion then follows from Theorem (3.11)(b). q.e.d.

We now consider the specialization of the preceding general theory for Φ-monotone mappings to the special cases of accretive and monotone mappings.

THEOREM (3.16). *Let X be a Banach space with a strictly convex conjugate space X^*, A an accretive mapping from X to 2^X. Then*

(a) *If A is hypermaximal accretive, it is maximal accretive.*

(b) *Suppose that X^* is uniformly convex and A is maximal accretive. Then the following conclusions are valid:*

(I) *If for a sequence $\{x_n\}$ in x with $w_n \in A(x_n)$, we have x_n converges strongly to x_0 in X, w_n converges weakly to w_0 in X, then w_0 lies in $A(x_0)$.*

(II) *If for an open subset G of X, $A(x)$ is nonempty for x in G and A is locally bounded on each line segment in G, then A is locally bounded in G and is upper-semicontinuous as a set-valued mapping from G to 2^X, with the weak topology on the image space.*

(c) *Suppose that X^* is uniformly convex and A is an accretive mapping of X into 2^X. Then*

(III) *Suppose that for an open subset G of X, $A(x)$ is nonempty for x in G and A is locally bounded on each line segment in G. Then A is locally bounded in G.*

(IV) *Suppose that for each x in G, $A(x)$ is nonempty, closed and convex while A is upper-semicontinuous and locally bounded from each line segment in G to 2^X, with the weak topology on the image space. Then A is upper-semicontinuous as a set-valued mapping from G to 2^X with the weak topology on the image space.*

(V) *Suppose that for each x in X, $A(x)$ is nonempty, closed, and convex and that for each line segment in X, A is a locally bounded upper-semicontinuous mapping from the line segment to 2^X, with the weak topology on the image space. Then A is maximal accretive.*

The proof of Theorem (3.16) is based upon the following simple result:

PROPOSITION (3.1). *Let X be a Banach space with a strictly convex conjugate space X^*. Then there exists exactly one duality mapping J of X into X^* with the normalization $\|Ju\| = \|u\|$, $u \in X$. This mapping J is continuous from the strong topology of X to the weak* topology of X^*. If in addition X^* is uniformly convex, then J is continuous from the strong topology of X to the strong topology of X^*, and indeed uniformly strongly continuous on each bounded subset of X.*

PROOF OF PROPOSITION (3.1). We recall that by a (single-valued) duality mapping J of X into X^* with the given normalization, we mean any mapping J which satisfies the two conditions:

$$(J(u), u) = \|Ju\| \cdot \|u\|, \qquad \|Ju\| = \|u\|,$$

for all u in X. For each u in X, there exists an element x^* in X^* by the Hahn-Banach theorem such that $\|x^*\| = \|u\|$, $(x^*, u) = \|u\|^2$. Hence, a duality mapping J exists for any Banach space though it need not be continuous. Suppose, however, that X^* is strictly convex, (i.e., any convex nonempty subset of a sphere in X^* contains exactly one point). For any given u in X, the set of possible values w for $J(u)$ is contained in the set

$$C_u = \{w \mid w \in X^*, \|w\| \leq \|u\|, (w, u) = \|u\|^2\}$$

and the set C_u is certainly convex, being the intersection of a ball and a flat in X^*. For any w in C_u, moreover, $\|u\|^2 = (w, u) \leq \|w\| \cdot \|u\|$, so that (except for the trivial case in which $u = 0$), $\|u\| \leq \|w\| \leq \|u\|$. Hence, C_u is contained in the sphere of radius $\|u\|$ about 0, and hence consists of a single point by the strict convexity of X^*. Thus J is unique.

Suppose once more that X^* is strictly convex and that we are given a sequence $\{x_n\}$ in X converging strongly in X to x_0. We wish to show that $J(x_n)$ converges in the weak* topology to $J(x_0)$. To establish this assertion, it suffices to show that for each u_0 in X, $(J(x_n), u_0) \to (J(x_0), u_0)$. Since the sequence $\{x_n\}$ and the points x_0 and u_0 are all contained in a separable subspace X_0 of X, we can assume without loss of generality that X itself is separable. If the latter is the case, however, then the weak* topology restricted to a bounded subset of X^* is metrizable. Since $\{J(x_n)\}$ is contained in a bounded subset of X^*, and since each closed ball in X^* is weak* closed and indeed weak* compact, it follows that by passing to an infinite subsequence, we may asume that $J(x_n)$ converges in the weak* topology to an element w_0 of X^*. In particular,

$$(J(x_n), x_n) \to (w_0, c_0), \qquad \|w_0\| \leq \underline{\lim} \|J(x_n)\| = \underline{\lim} \|x_n\| = \|x_0\|,$$

while $(J(x_n), x_n) = \|x_n\|^2 \to \|x_0\|^2$. By the remarks of the preceding paragraph, it follows that $w_0 \in C_{x_0}$, i.e., $w_0 = J(x_0)$. However, we have shown that every weak* convergent infinite subsequence of the weak* relatively compact sequence $\{J(x_n)\}$ has $J(x_0)$ as its weak* limit. Hence $J(x_n)$ converges weak* to $J(x_0)$ and J is continuous from the strong topology of X to the weak* topology of X^*.

Suppose finally that X^* is uniformly convex. Then for x_n converging strongly to x_0 in X, $J(x_n)$ converges weakly to $J(x_0)$ in X^* since X^* is reflexive while $\|J(x_n)\| = \|x_n\| \to \|x_0\| = \|J(x_0)\|$. By the uniform convexity of X^*, it follows that $J(x_n)$ converges strongly in X^* to $J(x_0)$ and J is continuous in the strong topology. To complete the proof of Proposition (3.1), we show that J is uniformly continuous in the strong topology on bounded subsets of X.

To prove that J is uniformly continuous in the norm topologies on bounded subsets of X, it suffices to consider only the ball B_M of radius $M > 0$ about the origin in X. Let $\epsilon > 0$ be given. We must show that there exists $\xi > 0$, depending upon ϵ and M, such that for any pair u and v in B_M with $\|u - v\| < \xi$, then $\|Ju - Jv\| < \epsilon$.

For the sake of normalizing the notation, we assume that $\|v\| \leq \|u\|$. We note first that if $\|u\| < \epsilon/2$, then

$$\|Ju - Jv\| \le \|Ju\| + \|Jv\| = \|u\| + \|v\| \le 2\|u\| < \epsilon$$

and there is nothing to prove in this case. Hence, we need only consider pairs with $\|u\| \ge \epsilon/2$. Suppose we are given such a pair u and v, and set $u_1 = \|u\|^{-1}Mu$, $v_1 = \|u\|^{-1}Mv$. Then

$$\|u_1 - v_1\| = M\|u\|^{-1}\|u - v\| \le 2M\epsilon^{-1}\xi$$

and

$$\|Ju - Jv\| = \|u\|M^{-1}\|Ju_1 - Jv_1\| \le \|Ju_1 - Jv_1\|.$$

If we can find $\delta > 0$ such that for $\|u_1 - v_1\| < \delta$, we have $\|Ju_1 - Jv_1\| < \epsilon$, then by choosing $\xi = \epsilon\delta(2M)^{-1}$, it follows that for $\|u - v\| < \xi$, $\|Ju - Jv\| < \epsilon$. If we replace our original u and v by u_1 and v_1, respectively, it follows that it suffices to consider only the case in which $\|u\| = M$.

Suppose then that $\|u\| = M$. We note first that $(Ju - Jv, u - v) \le 2M\xi$. On the other hand,

$$(Ju - Jv, u - v) = (Ju, u) + (Jv, v) - (Ju, v) - (Jv, u)$$
$$= [\|u\|^2 + \|v\|^2 - 2\|u\| \cdot \|v\|]$$
$$+ [\|Ju\| \cdot \|v\| - (Ju, v)] + [\|Jv\| \cdot \|u\| - (Jv, u)]$$

where each of the terms in square brackets is nonnegative. Hence

$$\|v\| \cdot \|u\| - (Jv, u) = \|Jv\| \cdot \|u\| - (Jv, u) \le 2M\xi.$$

Let $w = 2^{-1}(Ju + Jv)$. Then $\|Ju\| = \|u\| = M$, $\|Jv\| = \|v\| \le \|u\| \le M$, while

$$M\|w\| \ge (w, u) = 2^{-1}(Ju + Jv, u) = 2^{-1}\|u\|^2 + 2^{-1}(Jv, u)$$
$$\ge \|u\|^2 + \tfrac{1}{2}\|u\|[\|v\| - \|u\|] - M\xi \ge M^2 + \tfrac{1}{2}M\xi - M\xi$$

since $|\|v\| - \|u\|| \le \|u - v\| < \xi$. Hence $\|w\| \ge M - \tfrac{3}{2}\xi$. By the uniform convexity of the Banach space X^*, however, there exists $\lambda > 0$ such that if for two elements y and y_1 in X^* with $\|y\| \le M$, $\|y_1\| \le M$, we have $\|y + y_1\| \ge 2M - 2\lambda$, then $\|y - y_1\| < \epsilon$. If we set $\xi = 2\lambda/3$, it follows that if $\|u - v\| < \xi$, then for $y = Ju$, $y_1 = Jv$, $\|Ju - Jv\| = \|y - y_1\| < \epsilon$.   q.e.d.

REMARK. The result on the uniform continuity of J on bounded sets for uniformly convex X^* is due originally to Smulian in 1941, who proved moreover that this condition is necessary as well as sufficient. In his original formulation, the result is that the norm on X is continuously Fréchet differentiable on the unit sphere in X with uniformly continuous derivative if and only if X^* is uniformly convex. Our duality mapping J is precisely the Fréchet derivative in this case of the real-valued function $g(u) = \tfrac{1}{2}\|u\|^2$. Smulian's result belongs to the general family of theorems concerning the relation between smoothness properties of the space X and rotundity properties of the space X^*.

PROOF OF THEOREM (3.16). If we apply Proposition (3.1), we can derive the various assertions of Theorem (3.16) from the preceding theorems of this section.

Thus assertion (a) follows from Theorem (3.9) if we note that for strictly convex X^*, the Φ-function $\Phi(x, u) = J(x - u)$ is continuous from the strong topology of X to the weak* topology of $Y' = X^*$. The assertion of (b)(I) is a special case of Theorem (3.10) if we note that for X^* uniformly convex, the function $\Phi(x, u)$ is continuous in x and u on X to the strong topology of X^*. Assertion (b)(II) is a special case of Theorem (3.11)(b). Assertion (c)(III) is a special case of Theorem (3.13) if we apply the result of Proposition (3.1) on the uniform continuity of J on bounded sets in X. Assertions (c)(IV) and (c)(V) follow from Theorem (3.14) and Theorem (3.12), respectively. q.e.d.

To complete the discussion of this section, we turn finally to the consideration of related results for the class of monotone mappings from a Banach space X to its conjugate space X^*, and in this case results of a sharper type can be obtained.

THEOREM (3.17). *Let X be a Banach space, X^* its conjugate space, A a monotone mapping from X to 2^{X^*} such that for each x in an open subset G of X, $A(x)$ is nonempty. Then A is locally bounded on G.*

PROOF OF THEOREM (3.17). A monotone mapping is Φ-monotone with respect to the mapping Φ of $X \times X$ into $X^{**} = Y^*$, with $Y = X^*$, given by $\Phi(x, u) = x - u$, where we identify X in the canonical way with a subset of X^{**}. Since Φ is obviously uniformly continuous, we may apply all the preceding results of this section. In particular, we know by Theorem (3.13) that in order to prove that A is locally bounded on G, it suffices to prove that A is locally bounded on each line segment in G. Hence, it suffices to prove that for any finite-dimensional subspace F of X, A is locally bounded on $G \cap F$. Let j_F be the injection map of this subspace F into X, j_F^* the dual mapping which projects X^* onto F^*.

By the uniform boundedness theorem, to prove that A is locally bounded, it suffices to show that for each subspace F_1 of finite dimension in X that $j_{F_1}^* A$ is locally bounded on $F \cap G$. If we enlarge F to a bigger finite-dimensional space containing F_1, it suffices to show that $j_F^* A$ is locally bounded on $F \cap G$. On the other hand, if $A_F = j_F^* A : F \to 2^{F^*}$, then it follows from the monotonicity of A that A_F is a monotone mapping of F into 2^{F^*}. Indeed, if u and v are elements of F, then $w_F \in A_F(u)$ and $y_F \in A_F(v)$ if and only if $w_F = j_F^*(w)$ for some w in $A(u)$, $y_F = j_F^*(y)$ for some y in $A(v)$. Hence

$$(w_F - y_F, u - v) = (j_F^*(w - y), u - v) = (w - y, j_F(u - v))$$
$$= (w - y, u - v) \geq 0.$$

We see therefore that if we replace A by A_F, it suffices to consider only the case in which the Banach space X is of finite dimension.

Suppose finally that X is of finite dimension, and that for a sequence $\{u_n\}$ in G with $u_n \to u_0 \in G$, we have $w_n \in A(u_n)$ with $\|w_n\| \to +\infty$. Let x be any point of G, y any point of $A(x)$, where the latter is nonempty by hypothesis. Then $(w_n - y, u_n - x) \geq 0$. Set $y_n = p_n^{-1} w_n$ with $p_n = \|w_n\|$. Then $\|y_n\| = 1$, and passing to an infinite subsequence, we may assume that $y_n \to y_0$ in the finite-dimensional space X^* with $\|y_0\| = 1$. In terms of y_n, our inequality becomes $p_n(y_n - p_n^{-1}y, u_n - x) \geq 0$. Cancelling the positive factor p_n, we obtain $(y_n - p_n^{-1}y, u_n - x) \geq 0$. Letting $n \to \infty$, we see that $(y_0, u_0 - x) \geq 0$. If v is an arbitrary element of X, $x_t = u_0 - tv$ lies in G for $t > 0$ sufficiently small. Replacing x by x_t in the last inequality, we see that $t(y_0, v) \geq 0$ i.e., $(y_0, v) = 0$ for all v in X. Hence $y_0 = 0$, which contradicts the fact that $\|y_0\| = 1$. This contradiction is based upon the assumption that A is not locally bounded on G. Hence the proof of Theorem (3.17) is complete. q.e.d.

Applying corresponding specializations and sharpenings of the other theorems the present section, we obtain the following results for the case of monotone of mappings:

THEOREM (3.18). *Let X be a Banach space, A a monotone mapping of X into 2^{X^*} where X^* is the conjugate space of X. Then*

(a) *If A is maximal monotone from X to 2^{X^*}, then for any sequence $\{x_n\}$ in X which converges strongly to x_0 in X with a corresponding sequence $\{w_n\}$ in X^* with $w_n \in A(x_n)$ for each n, if w_n converges to w_0 in the weak* topology on X^*, then $w_0 \in A(x_0)$.*

(b) *If A is maximal monotone from X to 2^{X^*} and for any pair of sequences $\{x_n\}$ in X and $\{w_n\}$ in X^* with $w_n \in A(x_n)$ for each n, if x_n converges weakly to x_0 in X and w_n converges strongly to w_0 in X^*, then $w_0 \in A(x_0)$.*

(c) *If A is maximal monotone from X to 2^{X^*} and $A(x)$ is nonempty for each x in the open set G of X, then A is upper-semicontinuous as a set-valued function from G to 2^{X^*}, with X^* given the weak* topology.*

(d) *Let G be an open subset of X such that for each point x in G, $A(x)$ is a nonempty weak* closed convex subset of X^*. Suppose that on each line segment of G. A is an upper-semicontinuous set-valued function from that line segment to 2^{X^*}, with X^* given its weak* topology. Then for any u_0 in G, w_0 in X^* such that $(w - w_0, u - u_0) \geq 0$ for each u in G and each w in $A(u)$, we must have $w_0 \in A(u_0)$.*

(e) *Suppose that for each x in X, $A(x)$ is a nonempty weak* closed convex subset of X^* and that for each line segment in X, A is an upper-semicontinuous set-valued mapping from the line segment to 2^{X^*}, with X^* given its weak* topology. Then A is maximal monotone.*

PROOF OF THEOREM (3.18). PROOF OF (a). For each $[x, y]$ in $G(A)$, $(w_n - y, x_n - x) \geq 0$. As $n \to \infty$, $w_n - y$ converges weak* to $w_0 - y$, while $x_n - x$ converges strongly to $x_0 - x$. Hence $(w_n - y, x_n - x) \to (w_0 - y, x_0 - x)$ so that $(w_0 - y, x_0 - x) \geq 0$ for all $[x, y]$ in $G(A)$. Since A is maximal monotone, $[x_0, w_0] \in G(A)$. q.e.d.

PROOF OF (b). For each $[x, y]$ in $G(A)$, $(w_n - y, x_n - x) \geq 0$. As $n \to \infty$, $w_n - y$ converges strongly to $w_0 - y$ in X^*, while $x_n - x$ converges weakly to $x_0 - x$. Hence $(w_n - y, x_n - x) \to (w_0 - y, x_0 - x)$. Thus for all $[x, y]$ in $G(A)$

$$(w_0 - y, x_0 - x) \geq 0,$$

and by the maximal monotonicity of A, $[x_0, w_0] \in G(A)$. q.e.d.

PROOF OF (c). By Theorem (3.17), A is locally bounded on G. If we wish to show that A is upper-semicontinuous from G to 2^{X^*} it suffices to do this on a neighborhood N of each given point x_0 of G. Hence, we may assume without loss of generality that $A(G)$ is a bounded subset of X^*.

Suppose that A is not upper-semicontinuous in the appropriate sense at x_0. Then there exist sequences $\{x_n\}$ in G and $\{w_n\}$ in X^* as well as a weak* neighborhood V of $A(x_0)$ in X^* such that x_n converges to x_0 strongly in X, w_n lies in $A(x_n)$ for each n, and all the w_n lie in $A(G) - V$. Since $A(G)$ is bounded, it is contained in a weak* compact subset K of X^*. If for each integer m, we define $R_m = \bigcup_{n \geq m} \{w_n\}$, then by the weak* compactness of K, there exists a point w_0 of $K - V$ which lies in the weak* closure of each of the sets R_m.

Let $[x, y]$ be a given point of $G(A)$. For each n, since A is monotone, we have $(w_n - y, x_n - x) \geq 0$. Hence

$$(w_n - y, x_0 - x) = (w_n - y, x_n - x) + (w_n - y, x_0 - x_n)$$
$$\geq (w_n - y, x_0 - x_n) \geq -2M \|x_0 - x_n\|$$

since $\|w_n\| + \|y\| \leq M$, depending upon y, for all integers n. Hence if $n \geq m$ with m sufficiently large so that $\|x_0 - x_n\| \leq \epsilon(2M)^{-1}$, we have $(w_n - y, x_0 - x) \geq \epsilon$ for a given $\epsilon > 0$. Since this is true for all w_n in R_m and since the function $(w - y, x_0 - x)$ is continuous in the weak* topology on w for y, x_0, and x fixed, it follows that the same inequality holds with w_n replaced by the element w_0 of the weak* closure of R_m, i.e., $(w_0 - y, x_0 - x) \geq -\epsilon$.

Since $\epsilon > 0$ is arbitrary in the above inequality, if follows that

$$(w_0 - y, x_0 - x) \geq 0 \qquad ([x, y] \in G(A)).$$

Since A is maximal monotone from X to 2^{X^*}, it follows that w_0 lies in $A(x_0)$, which contradicts the fact that w_0 lies outside of the neighborhood V of $A(x_0)$. This contradiction proves that A is upper-semicontinuous as a set-valued function from G to 2^{X^*} for X^* with the weak* topology. q.e.d.

PROOF OF (d). Suppose that $(w - w_0, u - u_0) \geq 0$ for u_0 in G, all u in G, and w in $A(u)$. Suppose that w_0 does not lie in $A(u_0)$. Since $A(u_0)$ is a weak* closed convex subset of X^*, if w_0 lies in the complement of $A(u_0)$, there exists an element v of X such that $(w_0, v) > \sup_{w \in A(u_0)} (w, v)$.

We set $x_t = u_0 + tv$, $t > 0$, for the given element v of X. For t sufficiently small, x_t lies in G. Since $A(x_t)$ is nonempty by hypothesis, there exists w_t in $A(x_t)$. Let V be the weak* neighborhood of $A(u_0)$ given by $V = \{w \mid w \in X^*, (w, v) < (w_0, v)\}$. Since A is upper-semicontinuous from the segment $[u_0, u_0 + t_0v]$ to 2^{X^*} with the weak* topology on X^*, it follows that for $t > 0$ sufficiently small, w_t must lie in V. However, $0 \leq (w_t - w_0, x_t - u_0) = t(w_t - w_0, v) < 0$ since w_t lies in V. This is a contradiction, which proves that the conclusion of (d) is valid. q.e.d.

PROOF OF (e). This is simply the special case of assertion (d) for $G = X$. q.e.d.

4. Covering space methods. In the discussion of Φ-accretive mappings in §3, we established that mappings of a suitable type were local homeomorphisms of the Banach space X into itself, i.e., the mapping f involved mapped a neighborhood N of any prescribed point of X homeomorphically on an open subset of X. It is of great interest in this situation, as well as in a number of others which we shall consider below, to determine conditions under which such local homeomorphisms are actually global homeomorphisms. Such conditions will be derived in the present section by methods which connect up the theory of local homeomorphisms with the related but not identical theory of covering mappings.

We begin by making the definitions of the key terms in our discussion precise:

DEFINITION (4.1). *Let S_0 and S_1 be two topological spaces, f a continuous mapping of S_0 into S_1. Then f is said to be a local homeomorphism of S_0 into S_1 if for each point s_0 of S_0, there exists an open neighborhood N of s_0 in S_0 which is mapped by f homeomorphically onto an open neighborhood N_1 of $f(s_0)$ in S_1.*

DEFINITION (4.2). *Let S_0 and S_1 be two topological spaces, f a continuous mapping of S_0 into s_1. Then f is said to be a covering mapping of S_0 onto S_1 if $f(S_0) = S_1$ and if for each point s_1 of S_1, there exists an open neighborhood N_1 of s_1 in S_1 such that $f^{-1}(N_1)$ is the union of a pairwise disjoint family of open subsets N_α of S_0, each of which is mapped homeomorphically by f onto N_1.*

We note that if f is a covering mapping and if $f(s_0) = s_1$, then the particular N_α which contains s_0 is certainly a neighborhood of s_0 in S_0 mapped homeomorphically by f onto N_1. Thus each covering mapping is trivially a local homeomorphism. It is not true in general, however, that each local homeomorphism is a covering mapping, even if $f(S_0) = S_1$. A simple example to the contrary is provided by the mapping of the complex plane into itself given by the holomorphic function $f(z) = \int_0^z \exp(\zeta^2) d\zeta$ which is a local homeomorphism since $f'(z)$ is never null. However, f cannot be a covering mapping of the complex plane onto itself, for while the image of f is the whole complex plane, if f were a covering mapping, it would be a homeomorphism. It is easy to see that the only holomorphic homeomorphisms of the complex plane on itself are linear mappings.

We begin with one of the simplest sufficient conditions for a local homeomorphism to be a covering mapping.

THEOREM (4.1). *Let S_0 and S_1 be two topological spaces with S_0 normal, S_1 connected and having a countable base of neighborhoods at each point. Let f be a local homeomorphism of S_0 into S_1, and suppose that f is a closed mapping of S_0 into S_1, i.e., f maps closed sets of S_0 on closed sets of S_1.*

Then f is a finite covering mapping of S_0 on S_1, i.e., the inverse image under f of each point of S_1 is a finite set.

PROOF OF THEOREM (4.1). We note first that it suffices to prove that for each point s_1 of $S_1, f^{-1}(s_1)$ is a finite set in S_0. Indeed suppose that this is the case, and let $\{t_1, \ldots, t_r\} = f^{-1}(s_1)$. For each t_j, there exists a neighborhood M_j in S_0 mapped homeomorphically by f onto an open set $N_j = f(M_j)$ in S_1. Let $N' = \bigcap_{j=1}^r N_j$. Then N' is an open neighborhood of s_1 and $f^{-1}(N') \cap M_j$ is an open subset of M_j mapped homeomorphically onto N' by f. Since we can take the neighborhoods M_j pairwise disjoint, the covering space condition would be satisfied by N' if we knew that $f^{-1}(N')$ were contained in $\bigcup_{j=1}^{r} M_j$. This is not obviously the case, however, but if we replace N' by any smaller neighborhood N'' of s_1 in N', the covering space condition will be satisfied for N'' by the same argument if N'' can be taken so that $f^{-1}(N'') \subset \bigcup_{j=1}^{r} M_j$. However, let $F = S_0 - \bigcup_{j=1}^{r} M_j$. Then F is a closed subset of S_0 and since f is assumed to be a closed mapping, $f(F)$ must be a closed subst of S_1 which does not contain the point s_1 since $f^{-1}(s_1)$ is contained in the union of the finite family $\{M_1, \ldots, M_r\}$. Hence $N'' = N' \cap (S_1 - f(F))$ is an open subset of S_1 which is contained in N' and contains s_1, with the additional property that

$$f^{-1}(N'') \subset \bigcup_{j=1}^{r} M_j.$$

Hence $f^{-1}(N'') \cap M_j$ is an open subset of M_j mapped homeomorphically by f onto N'', and it would follow from the fact that $f^{-1}(s_1)$ is finite for each s_1 in S_1 that f is indeed a covering mapping.

It suffices to prove that $f^{-1}(s_1)$ is finite for each s_1 in S_1. Suppose then that for a given s_1, $f^{-1}(s_1)$ is an infinite set. Since f is a local homeomorphism, each point t of $f^{-1}(s_1)$ has a neighborhood in S_0 which includes no other points of $f^{-1}(s_1)$. Hence $f^{-1}(s_1)$ is a discrete subset of S_0. Since this set is infinite, we may write it as $f^{-1}(s_1) = Q_1 \cup Q_2$, where Q_1 and Q_2 are disjoint subsets of $f^{-1}(s_1)$ with Q_1 a countable infinite set, $Q_1 = \bigcup_{j=1}^{\infty} \{t_j\}$. Since Q_1 and Q_2 are both closed subsets of S_0, $f^{-1}(s_1)$ being discrete in S_0, and since S_0 is assumed to be a normal space, we can find disjoint open subsets G_1 and G_2 of S_0 such that $Q_1 \subset G_1$, $Q_2 \subset G_2$.

We now proceed to define by recursion a family of open neighborhoods $\{M_j\}$ in S_0 with M_j a neighborhood of t_j for each t_j in Q_1, having the following properties:

(1) $M_j \subset G_1$. For $j = k$, cl $(M_j) \cap$ cl $(M_k) = \varnothing$ (where cl (M) for a subset M of a topological space S denotes its closure).

(2) f maps M_j homeomorphically onto an open neighborhood N_j of s_1. For each j, cl $(N_{j+1}) \subset N_j$. The family $\{N_j\}$ forms a base of neighborhoods for the topology of S_1 at s_1.

For the purposes of this recursive construction, we define M_1 by noting that $\{t_1\}$ and $f^{-1}(s_1) - \{t_1\}$ are disjoint closed subsets of the normal space S_0 and hence we can find a neighborhood M_1' of t_1 in S_0 such that cl (M_1') does not contain any other points of $f^{-1}(s_1)$. By taking M_1 smaller than M_1', we may assure that f maps M_1 homeomorphically onto an open neighborhood N_1 of s_1 in S_1 and that $M_1 \subset G_1$.

Let $\{U_j\}$ be a countable base of neighborhoods of S_1 at s_1. To show that a given sequence $\{N_j\}$ of neighborhoods of s_1 is a base at s_1, it suffices to ensure that for each j, $N_j \subset U_j$.

Suppose we have constructed $\{M_1, \ldots, M_m\}$ for some integer $m \geq 1$ such that for $j \leq m$, $k \leq m$, cl (M_j) and cl (M_k) are disjoint for $j \neq k$, each M_j lies in G_1, f maps M_j homeomorphically onto a neighborhood N_j of s_1 with $N_j \subset U_j$, and cl $(N_j) \subset N_{j-1}$. We seek to extend the construction to $j = m + 1$. We note first that the two sets

$$\{t_{m+1}\} \quad \text{and} \quad \bigcup_{j=1}^{m} \text{cl}\,(M_j) \cup (f^{-1}(s_1) - \{t_{m+1}\})$$

are closed disjoint subsets of the normal space S_0. Hence, we can find a neighborhood M'_{m+1} of t_{m+1} in S_0 such that cl (M'_{m+1}) is disjoint from cl (M_j) for $j \leq m$ and contains no other points of $f^{-1}(s_1)$ than t_{m+1}. We may assume by shrinking M'_{m+1} that we can obtain an open neighborhood M_{m+1} of t_{m+1} which is contained in $M'_{m+1} \cap G$, such that f maps M_{m+1} onto an open subset N_{m+1} of S_1 containing s_1, and by shrinking M_{m+1} even further, that cl $(N_{m+1}) \subset N_m \cap U_{m+1}$. This open set M_{m+1} satisfies the conditions of the recursive construction for the $(m+1)$st step and hence the construction can be carried through for all the integers $j \geq 1$.

For each j, we may choose a point x_j in $M_j \cap f^{-1}(N_j) -$ cl (N_{j+1}). Then $f(x_j) \neq s_1$, and $f(x_j) \in N_j$. Moreover, $f(x_j)$ converges to s_1 in S_1. Let

$$F_0 = \bigcup_{j=1}^{\infty} \{x_j\}.$$

We assert that F_0 is closed in S_0. Suppose indeed that y_0 lies in cl (F_0) but not in F_0. Then each neighborhood V of y_0 must contain points x_j with arbitrarily large j. It follows that if V_1 is any neighborhood of $f(y_0)$, V_1 must intersect N_j in a nonempty set. Since f is both an open and closed mapping, f maps S_0 onto S_1. Since S_0 is normal and hence Hausdorff by assumption and f is a local homeomorphism, S_1 is a Hausdorff space. If $f(y_0)$ differs from s_1, there must exist neighborhoods N_j of s_1 and V_1 of $f(y_0)$ which have an empty intersection. Hence $f(y_0) = s_1$, so that y_0 lies in $f^{-1}(s_1)$. This is impossible however since by our construction of the sets M_j, each point t of $f^{-1}(s_1)$ has a neighborhood in S_0 which contains at most one of the points $\{x_j\}$. Hence F_0 is closed in S_0.

Since f is a closed mapping, $f(F_0)$ must be closed in S_1. However, s_1 lies in cl $(f(F_0))$ but not in $f(F_0)$. This is a contradiction to the assumption that $f^{-1}(s_1)$ is infinite.

Thus f maps S_0 onto S_1 and satisfies the covering space condition, and is thereby a covering mapping. q.e.d.

There are a number of variants of Theorem (4.1), some of which we give in the next theorem. We remark that in [57], the writer has established that the conclusion of Theorem (4.1) is valid without the assumption that S_1 has a countable base at each point and without the assumption that S_0 is normal, but with additional conditions of a local character that S_1 is locally pathwise connected and locally simply connected.

THEOREM (4.2). *Let S_0 and S_1 be topological spaces with S_1 regular and connected, f a local homeomorphism of S_0 into S_1. Suppose that S_1 has a countable base of neighborhoods at each point (or more generally satisfies the condition that a subset F of S_1 is closed in S_1 if and only if its intersection with each compact subset K of S_1 is compact).*

Then f is a finite covering mapping of S_0 onto S_1 if and only if f is a proper mapping, i.e., for each compact subset K of S_1, $f^{-1}(K)$ is compact in S_0.

PROOF OF THEOREM (4.2). We begin with several remarks on the hypotheses.
(1) If S_1 has a countable base of neighborhoods at each point, then S_1 does

satisfy the condition that a subset F of S_1 is closed if and only if its intersection with each compact K in S_1 is compact. Obviously, if F is closed, its intersection with the compact K is closed in K and hence compact. Suppose on the other hand that F has the property that its intersection with every compact K is compact. Let x_0 be a point of the closure of F. Since S_1 has a countable base of neighborhoods at x_0, we may choose an infinite sequence from F converging to x_0. If $\{x_j\}$ is this infinite sequence, then the set $K = \bigcup_{j=1}^{\infty} \{x_j\} \cup \{x_0\}$ is a compact subset of S_1. Since $K \cap F$ is compact and hence closed, x_0 lies in $K \cap F$ since it is a limit point of $K \cap F$. Hence F includes all its limit points and is closed.

(2) Suppose that S_1 satisfies the condition that a set F is closed if and only if $F \cap K$ is compact for each compact K in S_1. Then each proper mapping f of S_0 into S_1 is a closed mapping. Indeed, let F_0 be a closed subset of S_0 and set $F = f(F_0)$. If K is any compact subset of S_1, $f^{-1}(K)$ is compact since f is proper. Hence $f^{-1}(K) \cap F_0$ is compact. Moreover, $f(f^{-1}(K) \cap F_0) = K \cap F$, so that $K \cap F$ is compact. Hence F is a closed subset of S_1, and f is a closed mapping.

We now turn to the direct proof of Theorem (4.2). Suppose first that f is a finite covering mapping. Let K be a compact subset of S_1. For each point s_1 of K, we can find a neighborhood N of s_1 such that $f^{-1}(N)$ is the union of a finite family of disjoint open sets $\{M_1, \ldots, M_r\}$ each of which is mapped homeomorphically onto N by f. For each such neighborhood N, we can find a smaller neighborhood N' whose closure is contained in N. Using the compactness of K, we can cover K by a finite family of these neighborhoods N'. For each N', $K \cap \mathrm{cl}\,(N')$ is a compact subset of S_1 and $f^{-1}(K \cap \mathrm{cl}\,(N'))$ is the union of a finite family of homeomorphs of $K \cap \mathrm{cl}\,(N')$ and hence is compact. Since $f^{-1}(K)$ is the union of a finite family of $f^{-1}(K \cap \mathrm{cl}\,(N'))$, each compact, $f^{-1}(K)$ is compact and f is a proper mapping.

Suppose on the other hand that f is a local homeomorphism which is a proper mapping. Then by the discussion under (2) above, f is a closed mapping. Since a one-point set $\{s_1\}$ is compact, $f^{-1}(s_1)$ is a compact discrete subset of S_0 for each s_1 in S_1. Hence $f^{-1}(s_1)$ is a finite set $\{t_1, \ldots, t_r\}$. For each t_j, we can find a neighborhood M_j mapped homeomorphically by f onto a neighborhood N_j of s_1. If we set $N' = \bigcap_j N_j$, then N' is a neighborhood of s_1 in S_1. Let

$$N'' = N' \cap \left(S_1 - f\left(S_0 - \bigcup_{j=1}^{r} M_j\right)\right).$$

Since each M_j is open, $S_0 - \bigcup_j M_j$ is closed in S_0. Hence its image under f is closed in S_1. Hence N'' is open in S_1. On the other hand,

$$f^{-1}(N'') = \bigcup_{j=1}^{r} [M_j \cap f^{-1}(N'')]$$

where $M_j \cap f^{-1}(N'')$ is mapped homeomorphically onto N'' by f. Since the M_j could be taken pairwise disjoint, it follows that $f^{-1}(N'')$ is the union of a finite family of open subsets of S_0 each of which is mapped by f homeomorphically onto N''.

Since f is both an open and a closed mapping, $f(S_0)$ is both open and closed in the connected space S_1. Hence f maps S_0 onto S_1, and f is a covering mapping of S_0 onto S_1. Since $f^{-1}(s_1)$ is a finite set for each s_1 in S_1, f is a finite covering mapping. q.e.d.

THEOREM (4.3). *Let S_0 and S_1 be two regular topological spaces with S_0 completely normal, S_1 connected and having a countable base of neighborhoods at each of its points. Suppose further that S_1 has a base of pathwise connected, simply connected open sets. Let f be a local homeomorphism of S_0 into S_1. Then f is a covering mapping of S_0 onto S_1 if and only if f satisfies the following condition:*

(c) *For each point s_1 of S_1, there exists a neighborhood N_1 of s_1 in S_1 such that each component C of $f^{-1}(N_1)$ has the property that f is a closed mapping of C into N_1.*

PROOF OF THEOREM (4.3). Let us note first that every covering mapping f must satisfy the condition (c) above. By the covering mapping condition, each point s_1 of S_1 has a neighborhood N such that $f^{-1}(N)$ is the union of a disjoint family of open sets M_α, each mapped homeomorphically by f onto N. If this is true for a given neighborhood N of s_1, it is also true for any smaller neighborhood. Hence, we may take N as a connected open set. Then each M_α is connected and consists exactly of one component of $f^{-1}(N)$ and the mapping f being a homeomorphism of M_α on N is certainly a closed mapping of M_α into N.

For the converse, we must show that condition (c) implies that f is a covering mapping. To do this, we note first that if we consider a neighborhood N_2 of s_1 which is contained in the neighborhood N_1 involved in condition (c), then condition (c) holds with N_1 replaced by N_2. Indeed, let C_2 be any component of $f^{-1}(N_2)$. Then C_2 is contained in a component C_1 of $f^{-1}(N_1)$. If F_2 is a closed subset of C_2, then $F_2 = C_2 \cap F_1$ where F_1 is the closed subset of C_1 obtained by taking the closure of F_2 in C_1. Moreover, $F_2 = F_1 \cap f^{-1}(N_2)$ since any point of $f^{-1}(N_2)$ lying in the closure of C_2 in S_0 must lie in C_2 itself. Hence, $f(F_2) = N_2 \cap f(F_1)$. (Indeed, it is obvious that $f(F_2) \subset N_2 \cap f(F_1)$ since $F_2 = f^{-1}(N_2) \cap F_1$. On the other hand, for any points of $f(F_1)$ which lie in N_2, the point x of F_1 such that $f(x) = s$ must lie in $F_1 \cap f^{-1}(N_2) = F_2$.) Since F_1 is closed in C_1, it follows by condition (c) applied to N_1 that $f(F_1)$ is closed in N_1. Hence $f(F_2) = f(F_1) \cap N_2$ is closed in N_2 and condition (c) holds for N_2.

We choose N_2 to be a connected, simply connected neighborhood of s_1 contained in N_1 and let C_2 be any component of $f^{-1}(N_2)$. Then f is a local homeomorphism of C_2 into N_2 and f is a closed mapping of C_2 into N_2. Since N_2 satisfies the first axiom of countability and C_2 is normal (being an open subset of a completely normal space S_0), we can apply Theorem (4.1) and conclude that f is a covering mapping of C_2 onto N_2. Since N_2 is simply connected and C_2 is connected and locally pathwise connected, and hence pathwise connected, such a covering mapping must be a homeomorphism. Hence, we have verified that $f^{-1}(N_2)$ is the union of a pairwise disjoint family of open sets, the components of the open set $f^{-1}(N_2)$ in the locally connected space S_0, each of which is mapped homeomorphically onto N_2 by f. Hence f is a covering mapping if we can show that f maps S_0 onto S_1.

Obviously, $f(S_0)$ is open in S_1 since f is an open mapping. We assert in addition that $f(S_0)$ is closed in S_1. Indeed, suppose that s_1 lies in cl $(f(S_0))$ and choose N, as in the preceding paragraph. Then $f^{-1}(N_2)$ is nonempty, and each component C_2 of $f^{-1}(N_2)$ is mapped by f onto N_2. In particular, s_1 lies in $f(S_0)$ and hence $f(S_0)$ is closed in S_1. Thus $f(S_0) = S_1$ since $f(S_0)$ is both open and closed in the connected space S_1. q.e.d.

REMARK. All the separation axioms assumed for S_0 and S_1 in the three preceding theorems will hold if both S_0 and S_1 are metrizable spaces, and in particular if they are given to us as metric spaces.

If we wish to strengthen the force and applicability of results like Theorem (4.1), we need to replace the condition that f is a closed mapping by another condition which, under favorable circumstances, can be verified by examining the behavior of the mapping f in the neighborhood of each point s_0 of S_0. We devote the remainder of the discussion of this section to deriving such a condition and using it to obtain stronger existence theorems concerning Φ-accretive mappings than those derived in the first part of §3.

To begin this discussion, we introduce the following definition:

DEFINITION (4.3). *Let S_0 and S_1 be two topological spaces, f a local homeomorphism of S_0 into S_1. If h is a continuous mapping of another topological space S into S_1, then a mapping g of S into S_0 is said to be a lifting of h against f if $h = fg$.*

THEOREM (4.4). *Let S_0 and S_1 be two topological spaces, f a local homeomorphism of S_0 into S_1. Let S be a connected topological space, h a continuous mapping of S into S_1, x_0 a point in S. Suppose that we have two liftings g and g_1 of h against f such that $g(x_0) = g_1(x_0)$.*

Then g and g_1 coincide.

PROOF OF THEOREM (4.4). Let C be the subset of all the points x in S such that $g(x) = g_1(x)$. C is closed since g and g_1 are continuous. By hypothesis x_0 lies in C, so that C is nonempty. It suffices to show that C is open in S, since then C will be a nonempty open and closed subset of the connected space S and hence must coincide with S.

Suppose x_1 lies in C, so that $g(x_1) = g_1(x_1) = s_0$. Since f is a local homeomorphism, there exists a neighborhood N of s_0 which is mapped by f homeomorphically onto an open subset N_1 of S_1 containing $s_1 = f(s_0)$. Since $h = fg$, it follows that $h(x_1) = s_1$. Since g, g_1, and h are all continuous, there exists a neighborhood N' of x_1 in S such that for x in N', $h(x)$ lies in N_1 while both $g(x)$ and $g_1(x)$ lie in N. Since $h(x) = f(g(x)) = f(g_1(x))$ and f is injective on N, it follows that for x in N', $g(x) = g_1(x)$. Hence $N' \subset C$, and C is open in S. q.e.d.

THEOREM (4.5). *Let S_0 and S_1 be two metrizable topological spaces, f a local homeomorphism of S_0 into S_1, s_0 a given point of S_0. Let S be the unit interval $[0, 1]$, and let the continuous mapping $h : [0, 1] \to S_1$ represent a path in S_1 such that $f(s_0) = h(0)$. Suppose that the following condition is satisfied:*

(1) For any t with $0 < t < 1$, and for any continuous mapping $g_t : [0, t) \to S_0$

such that $g_t(0) = s_0$ and $f(g_t(s)) = h(s)$ for $0 \le s < t$, the image of g_t is relatively compact in S_0.

Then h has a lifting $g:[0, 1] \to S_0$ such that $g(0) = s_0$.

PROOF OF THEOREM (4.5). We note first that there exists $t_0 > 0$ such that h has a lifting g_{t_0} on the interval $(0, t_0)$ with $g_t(0) = s_0$. Indeed, s_0 has a neighborhood N_0 mapped homeomorphically by f on a neighborhood N_1 of $h(0) = f(x_0)$. Since h is continuous, there exists $t_0 > 0$ such that for $0 \le s < t_0$, $h(s)$ lies in N_1. Hence for s in $(0, t_0)$, we can define $g_{t_0}(s)$ to be the unique element of N_0 such that $f(g_{t_0}(s)) = h(s)$, i.e., $g_{t_0} = (f|_{N_0})^{-1}h$.

Since two liftings of h on the same interval $[0, t)$ with $g(0) = s_0$ must coincide, we can find a maximal lifting g_t of h against f with $g_t(0) = s_0$ on an interval $[0, t)$ with $0 < t \le 1$. We shall show that $t = 1$ and that g_t can be extended to a lifting g of h on $[0, 1]$.

By the hypothesis, condition (1) holds for the mapping g_t. Hence $g_t([0, t))$ is relatively compact in S_0. We choose a sequence $\{t_j\}$ in $[0, t)$ with $t_j \to t$, and by passing to a suitable infinite subsequence, we can assume that $g_t(t_j)$ converges in S_0 to an element y_0. Since f is a local homeomorphism, there exists a neighborhood N_0 of y_0 which is mapped by f homeomorphically onto a neighborhood N_1 of $f(y_0)$. Since $g_t(t_j) \to y_0$ while $f(g_t(t_j)) = h(t_j) \to h(t)$, it follows from the continuity of f that $f(y_0) = h(t)$. If j is sufficiently large, $g_t(t_j)$ lies in N_0 and $h(t_j)$ lies in N_1. Hence we can define a lifting k of h against f on the interval $[t_j, t]$ by setting $k(s) = (f|_{N_0})^{-1}h(s)$. For $s = t_j$, $k(t_j) = g_t(t_j)$. Hence by Theorem (4.4), g_t and k must coincide on the interval $[t_j, t)$. The lifting k can be extended to the interval $(t - \epsilon, t + \epsilon)$ for some $\epsilon > 0$ since in that interval $h(t)$ lies in N_1. We can therefore extend the lifting g_t to the interval $[0, t + \epsilon)$ by setting $g_t(s) = k(s)$ for s in $[t, t + \epsilon)$. All this is true if t is less than 1 and ϵ is small and less than $1 - t$, but the existence of such an extension for g_t contradicts the maximality of the lifting g_t. Hence $t = 1$, and g_t can be extended to a lifting g of h on $[0, 1]$ with $g(0) = s_0$. q.e.d.

THEOREM (4.6). *Let S_0 and S_1 be two topological spaces, f a local homeomorphism of S_0 into S_1. Let $S = [0, 1] \times [0, 1]$ and suppose that we are given two continuous mappings $H:S \to S_1$, $g:[0, 1] \to S_0$ such that for all r in $[0, 1]$, $f(g(r)) = H(r, 0)$. For each r in $[0, 1]$, let $h^{(r)}:[0, 1] \to S_1$ be the continuous curve in S_1 given by $h^{(r)}(t) = H(r, t)$, $0 \le t \le 1$. Suppose that for each r in $[0, 1]$, $h^{(r)}$ can be lifted against f to a continuous curve $g^{(r)}:[0, 1] \to S_0$ such that $g^{(r)}(0) = g(r)$.*

Then the mapping $H:S \to S_1$ can be lifted to a continuous mapping $G:S \to S_0$ against f such that for each r in $[0, 1]$, $G(r, 0) = g(r)$.

PROOF OF THEOREM (4.6). For each $[r, t]$ in S, we set

$$G(r, t) = g^{(r)}(t).$$

Then it is clear that $G(r, 0) = g(r)$ and that $f(G(r, t)) = f(g^{(r)}(t)) = h^{(r)}(t) = H(r, t)$. Hence it suffices to show that G as thus defined is a continuous mapping of S into S_0.

For each r_0 in $[0, 1]$, there exists a neighborhood $N(r_0)$ of $g(r_0)$ in S_0 mapped homeomorphically by f on a neighborhood $N_1(r_0)$ in S_0 of $f(g(r_0)) = H(r_0, 0)$. There exists $\epsilon > 0$ such that for $|r - r_0| \leq \epsilon, 0 \leq t < \epsilon, H(r, t)$ lies in $N_1(r_0)$. For r in $[r_0 - \epsilon, r_0 + \epsilon]$, the lifting $g^{(r)}(t)$ of $h^{(r)}(t)$ with $t < \epsilon$ is given by $(f|_{N(r_0)})^{-1}(H(r, t))$. Hence for such r and t, $G(r, t) = (f|_{N(r_0)})^{-1}(H(r, t))$ is continuous in the variable $[r, t]$. Therefore, there exists a subset of S of the form $[0, 1] \times [0, t_0)$ on which G is continuous.

We now let t_0 be the largest value in $[0, 1]$ such that G is continuous on $[0, 1] \times [0, t_0)$ and we shall show that
(1) G is continuous on $[0, 1] \times [0, t_0]$.
(2) $t_0 = 1$.
It will then follow that G is continuous on the whole of S.

Let r_0 be a point in $[0, 1]$, and consider $G(r_0, t_0) = s_0$. There exists a neighborhood N_0 of this point in S_0 which is mapped homeomorphically by f on a neighborhood N_1 of $f(G(r_0, t_0)) = H(r_0, t_0)$. Since H is continuous, there exists $\epsilon > 0$ such that for all $[r, t]$ in S with $|r - r_0| < \epsilon, |t - t_0| < \epsilon, H(r, t)$ lies in N_1. We define a mapping k of the subset

$$S' = \{[r, t] \mid [r, t] \in S, |r - r_0| < \epsilon, |t - t_0| < \epsilon\}$$

into S_0 by setting $k(r, t) = (f|_{N_0})^{-1}(H(r, t))$. Then k is a continuous mapping of S' into S_0 and to show that G is continuous on $[0, 1] \times [0, t_0]$, it suffices to show that k and G coincide on S'. For $r = r_0$, the curve $k^{(r_0)}:[t_0 - \epsilon t_0] \to S_0$ is a lifting of $h^{(r_0)}$ on the interval $[t_0 - \epsilon, t_0]$ with $k^{(r_0)}(t_0) = g^{(k_0)}(t_0)$. By Theorem (4.4), it follows that $k^{(r_0)}$ and $g^{(r_0)}$ coincide on the interval $[t_0 - \epsilon, t_0]$, i.e.,

$$G(r_0, t) = k(r_0, t) \qquad (t_0 - \epsilon < t \leq t_0).$$

Moreover, $G(r_0, t) = k(r_0, t)$ for all points $[r_0, t]$ in S for which both are defined. Similarly, for each t in $[t_0 - \epsilon, t_0 + \epsilon]$ with $t \leq 1$, we have two continuous curves

$$k^{(t)}(r) = k(r, t), \qquad G^{(t)}(r) = G(r, t),$$

both of which are liftings against f of the curve $H^{(t)}:[r_0 - \epsilon, r_0 + \epsilon] \to S_1$ given by $H^{(t)}(r) = H(r, t)$, with $k^{(t)}(r_0) = k(r_0, t) = G(r_0, t) = G^{(t)}(r_0)$. Hence these liftings coincide by Theorem (4.4), i.e., G and k coincide on S' and G is continuous on $[0, 1] \times [0, t_0]$.

Suppose that $t_0 < 1$. Then by the preceding argument we can extend G to a continuous mapping from $[0, 1] \times [0, t_0 + \epsilon)$ for some $\epsilon > 0$. This contradicts the maximality of t_0. Hence $t_0 = 1$, and G is continuous on S. q.e.d.

THEOREM (4.7). *Let S_0 and S_1 be two topological spaces, f a local homeomorphism of S_0 into S_1. Suppose that each point of S_1 has an open connected, simply connected neighborhood in S_1 and that S_1 is connected and locally pathwise connected.*

Then f is a covering mapping of S_0 onto S_1 if and only if every continuous curve $h:[0, 1] \to S_1$ has a lifting g against f with $g(0) = s_0$ for a given s_0 such that $f(s_0) = h(0)$.

PROOF OF THEOREM (4.7). The existence of liftings against covering mappings f is a consequence of the Covering Homotopy Theorem for covering spaces (Steenrod [431]). We shall prove the converse assertion that the existence of liftings implies that f is a covering mapping.

Let s_1 be a point of S_1. By hypothesis, s_1 has an open neighborhood N_1 in S_1 with N_1 connected and simply connected. Since S_1 is locally pathwise connected and N_1 is open and connected, N_1 is pathwise connected. Since S_0 is locally homeomorphic to S_1, S_0 is locally pathwise connected and hence locally connected. If C is any component of the open set $f^{-1}(N_1)$, it follows that C is an open subset of S_0. We shall show that f maps each such C homeomorphically onto N_1.

We note first that $f(C) = N_1$. Indeed, $f(C)$ is a nonempty subset of N_1. Let s' be a point in $f(C)$ and for each s_1 in N_1, take a continuous curve C_{s_1} connecting s' to s_1. (Such a curve exists since N_1 is pathwise connected.) By hypothesis, we can lift this curve to a curve $g:[0, 1] \to S_0$ such that $g(0) = s_0$ where $f(s_0) = s'$. Then $f(g(1)) = s_1$ and s_1 lies in $f(C)$ since the whole curve g beginning at s_0 in C must lie in C. Hence $f(C) = N_1$.

Since f is an open mapping, in order to show that f is a homeomorphism of C onto N_1, it suffices to show that f is one-to-one on C. Suppose not. Then there must exist two points s_0 and s_0' in C such that $f(s_0) = f(s_0')$ with s_0 distinct from s_0'. Since C is a connected open set in a locally pathwise connected space, C is pathwise connected. Hence there exists a continuous mapping $g:[0, 1] \to S_0$ such that $g(0) = s_0$, $g(1) = s_0'$. The curve $h:[0, 1] \to S_1$ given by $h(r) = f(g(r))$, $r \in [0, 1]$ is a closed curve in N_1 beginning and ending at $s_1 = f(s_0) = f(s_0')$. Since N_1 is simply connected, this curve can be contracted to a point over N_1, i.e., there exists a continuous mapping $H:[0, 1] \times [0, 1] \to S_1$ such that

$$H(r, 0) = h(r), \qquad 0 \leq r \leq 1,$$

$$H(0, t) = s_1 = H(1, t), \qquad 0 \leq t \leq 1,$$

$$H(r, 1) = s_1, \qquad 0 \leq r \leq 1.$$

Since every continuous curve in S_1 can be lifted against f with an arbitrary initial value, it follows from Theorem (4.6) that the mapping H of $[0, 1] \times [0, 1]$ into S_1 can be lifted to a continuous mapping $G:[0, 1] \times [0, 1] \to S_0$ such that $G(r, 0) = g(r)$. Since $f(G(0, t)) = s_1 = f(G(1, t))$ for all t in $[0, 1]$, $G(0, t)$ and $G(1, t)$ are continuous functions of t in $[0, 1]$ lying in the discrete set $f^{-1}(s_1)$. Hence $G(0, t) = G(0, 0) = g(0) = s_0$, $G(1, t) = G(1, 0) = g(1) = s_0'$ for all t in $[0, 1]$. However, $f(G(r, 1)) = H(r, 1) = s_1$ for all r in $[0, 1]$ implies that $G(r, 1)$ varies continuously in r for $0 \leq r \leq 1$ and has its values in $f^{-1}(s_1)$ which is a discrete set. Hence $G(0, 1) = G(1, 1)$, i.e., $s_0 = s_0'$, contradicting the assumption that the two points are distinct. Hence f is a one-to-one mapping of C onto N_1.

Finally, we assert that f is a covering mapping of S_0 onto S_1. We have verified the covering condition and must merely show that $f(S_0) = S_1$. Since $f(S_0)$ is a nonempty open subset of S_1 and S_1 is connected, it suffices to show that $f(S_0)$ is closed in S_1. Let s_1 be a point of cl $(f(S_0))$ and choose a neighborhood N_1 of s_1 as in the preceding discussion. Then $f^{-1}(N_1)$ is nonempty, and for each component C of $f^{-1}(N_1)$, f maps C onto N_1 so that s_1 lies in $f(S_0)$. Hence $f(S_0)$ is closed in S_1 and therefore coincides with S_1. q.e.d.

We now derive a finer variant of Theorem (4.7) in metric spaces under metric hypotheses upon the mapping f. We employ the following definitions:

DEFINITION (4.4). *Let X be a metric space, $h:[0, 1] \to X$ a continuous curve in X. Then h is said to be rectifiable if there exists a constant $M > 0$ such that for any subdivision of $[0, 1]$ of the form*

$$0 = t_0 < t_1 < t_2 < \cdots < t_m = 1$$

for arbitrary m, we have

$$\sum_{j=1}^{m} d(h(t_j), h(t_{j-1})) \leq M$$

where d is the distance function on X.

The least such constant M is called the *length* of the curve. If the curve is not rectifiable, it is said to have infinite length.

DEFINITION (4.5). *Let X be a metric space, $h:[0, 1] \to X$ a half-open curve in X. Then the curve is said to be rectifiable if there exists a constant $M > 0$ such that for any sequence of points*

$$0 = t_0 < t_1 < t_2 < \cdots < t_m < 1$$

with m arbitrary, we have

$$\sum_{j=1}^{m} d(h(t_j), h(t_{j-1})) \leq M.$$

The length of the curve is defined to be the least such constant M, and the curve is said to have infinite length if it is not rectifiable.

PROPOSITION (4.1). *Let X be a complete metric space, $h_0:[0, 1) \to X$ a continuous half-open curve in X. Then if the curve h_0 is rectifiable, it can be extended to a continuous curve $h:[0, 1] \to X$ of the same length.*

PROOF OF PROPOSITION (4.1). Let M_0 be the length of the curve h_0. Then for a given $\epsilon > 0$, there exists a sequence on $[0, 1)$ of the form

$$0 = t_0 < t_1 < \cdots < t_m < 1$$

such that

$$\sum_{j=1}^{m} d(h_0(t_j), h_0(t_{j-1})) > M_0 - \epsilon.$$

Let t be a parameter value with $t_m < t < 1$. Then we may extend the sequence by setting $t_{m+1} = t$, and we see that

$$\sum_{j=1}^{m} d(h_0(t_j), h_0(t_{j-1})) + d(h_0(t), h_0(t_m)) \leq M_0.$$

Hence for $t > t_m$, $d(h_0(t), h_0(t_m)) < \epsilon$. Suppose that $t_m < s < 1$. Then we have similarly $d(h_0(s), h_0(t_m)) < \epsilon$, and as a consequence $d(h_0(s), h_0(t)) < 2\epsilon$ for $s, t > t_m$.

Thus $d(h_0(t), h_0(s)) \to 0$ $(s, t \to 1)$, and since the metric space X is assumed to be complete, there exists an element x_0 of X such that $h_0(t) \to x_0$ as $t \to 1-$. We set

$$h(t) = h_0(t), \quad t < 1,$$
$$= x_0, \quad t = 1.$$

Then h is the desired extension of h_0 to a continuous mapping of $[0, 1]$ into X and it follows immediately that h is a rectifiable curve with length equal to the length of h_0. q.e.d.

DEFINITION (4.6). *Let X be a metric space. Then X is said to be rectifiably pathwise connected if each pair of points of X can be joined by a continuous rectifiable curve $h: [0, 1] \to X$.*

DEFINITION (4.7). *Let X be a metric space. Then X is said to be rectifiably simply connected if given a continuous closed curve $h: [0, 1] \to X$ with $h(0) = h(1) = x_0$, there exists a continuous homotopy $H: [0, 1] \times [0, 1] \to X$ of h to the constant mapping into x_0 which satisfies the following conditions:*

(1) $H(s, 0) = h(s), s \in [0, 1]$.
(2) $H(0, t) = H(1, t) = H(s, 1) = x_0$ *for all s, t in $[0, 1]$.*
(3) *For each s in $[0, 1]$, the curve $H^{(s)}: [0, 1] \to X$ given by*

$$H^{(s)}(t) = H(s, t),$$

is a rectifiable curve in X.

PROPOSITION (4.2). *Let X be a Banach space, U an open subset of X. Then:*
(a) *If U is connected, it is rectifiably pathwise connected.*
(b) *If U is convex, U is rectifiably simply connected.*

PROOF OF PROPOSITION (4.2). PROOF OF (a). Since X is locally pathwise connected, each connected open subset U of X is pathwise connected. Let x_0 and x_1 be two points of U, and let $h_1: [0, 1] \to U$ be a continuous path with $h_1(0) = x_0$ and $h_1(1) = x_1$. For each parameter value t in $[0, 1]$, there exists a ball of radius $d(t)$ about $h_1(t)$ which is contained in the open set U. Since the image of h_1 is compact in U, we may assume that $d(t)$ is a constant d_0 for all t. Since h_1 is uniformly continuous, we may find a sequence of parameter values $0 = s_0 < s_1 < \cdots < s_r = 1$ such that $d(h(s_j), h(s_{j-1})) < d_0$. Hence the line segment joining $h(s_{j-1})$ to $h(s_j)$ is completely contained in U. We replace the curve h_1 by the polygonal line consisting of these line segments. This polygonal line is a continuous rectifiable curve joining x_0 to x_1 in U. q.e.d.

PROOF OF (b). Let $h: [0, 1] \to U$ be a continuous curve. We define the homotopy H by taking a fixed point x_0 of U and setting

$$H(s, t) = (1 - t)h(s) + tx_0.$$

The verification of the three conditions of Definition (4.7) follows by an obvious argument. q.e.d.

THEOREM (4.8). *Let X and Y be metric spaces, f a local homeomorphism of X into Y. Suppose that X is a complete metric space which is locally pathwise connected, that Y is connected, and that each point y_0 of Y has an open neighborhood N in Y which is rectifiably pathwise connected and rectifiably simply connected. Suppose further that if $h:[0, 1) \to Y$ is a continuous rectifiable half-open curve, then for any lifting g of h against f, $g:[0, 1) \to X$, then g is also rectifiable.*

Then f is a covering mapping of X onto Y. In particular, if X is connected and Y is simply connected, then f is a homeomorphism of X onto Y.

PROOF OF THEOREM (4.8). Let y_0 be a point of Y, N an open neighborhood of y_0 in Y which is rectifiably pathwise connected and rectifiably simply connected. Since $f^{-1}(N)$ is an open subset of X and X is locally pathwise connected, each component C of $f^{-1}(N)$ is pathwise connected. We consider one such component C, and we show that for each such C, f maps C homeomorphically onto N. Since we may assume that C is nonempty if $f^{-1}(N)$ is nonempty, we may find a point in $f(C)$ in N. Let y_1 be such a point and y any other point of N. We shall show that:

(1) y lies in $f(C)$.
(2) $f^{-1}(y)$ consists of a single point in C.

Since N is rectifiably pathwise connected, there exists a continuous rectifiable curve in N, $h:[0, 1] \to W$ such that $h(0) = y_1$, $h(1) = y$. We consider liftings $g_t:[0, t) \to X$ of h_t, the restriction of h to the interval $[0, t)$, $0 < t \leq 1$. Such liftings exist for t sufficiently small with $g_t(0) = x_1$, where x_1 is a given point in C such that $f(x_1) = y_1$. By Theorem (4.4) any two such liftings must coincide on their common domain of definition. For any such lifting g_t, it follows by the hypothesis of the theorem that $g_t:[0, t) \to X$ is a rectifiable curve. By Proposition (4.1), since X is a complete metric space, $g_t(s)$ converges to a point x_2 in X as $s \to t-$. Since $f(g_t(s)) \to h(t)$, it follows that x_2 lies in $f^{-1}(h(t))$ and hence in $f^{-1}(N)$. Since C is closed in $f^{-1}(N)$, x_2 lies in C. Hence the image of g_t is relatively compact in C. Applying Theorem (4.5), we see that there exists a lifting g of h against f on the whole interval $[0, 1]$, i.e., $g:[0, 1] \to X$ such that $fg = h$. In particular, $f(g(1)) = h(1) = y$, while $g(1)$ must lie in the same component C of $f^{-1}(N)$ as x_1. Hence y lies in $f(C)$ and f maps C onto N.

Thus we have proved step (1) as listed above, and it follows in particular that f maps X onto Y. Indeed, $f(X)$ is open in Y since f is an open mapping of X into Y. Let y_0 be a point of cl$(f(X))$ and consider the neighborhood N treated above. Since N intersects $f(X)$, if follows that $f^{-1}(N)$ is nonempty and hence by the preceding argument, $N \subset f(X)$. In particular, $f(X)$ contains its own closure and must be closed in Y. Since Y is connected, the open and closed subset $f(X)$ must coincide with Y. Hence $f(X) = Y$.

To complete our argument, we must show that f is a homeomorphism of C onto N, and since f is a local homeomorphism of C onto N as above, it suffices to prove that f is a one-to-one mapping of C into N. Suppose that we have two points x_0 and x_1 in C such that $y = f(x_0) = f(x_1)$. Since C is pathwise connected, we may find a curve $g:[0, 1] \to C$ such that $g(0) = x_0$, $g(1) = x_1$. Let $h = fg:[0, 1] \to Y$.

Then h is a closed curve in N, and since N is rectifiably simply connected, there exists a homotopy H of h to the constant mapping into y such that H satisfies the conditions of Definition (4.7). For each s in $[0, 1]$, the curve $H^{(s)}:[0, 1] \to Y$ given by $H^{(s)}(t) = H(s, t)$ is rectifiable and starts at $h(s)$. By Theorem (4.5) and our assumption on liftings of rectifiable curves against f, each curve $H^{(s)}$ can be lifted to a curve in X starting at $g(s)$.

If we apply Theorem (4.6), it follows that the mapping H of $[0, 1] \times [0, 1] \to Y$ can be lifted to a continuous mapping $G:[0, 1] \times [0, 1] \to X$ such that $G(x, 0) = g(s)$. Since $f(G(0, t)) = f(G(1, t)) = f(G(s, 1)) = y$, it follows as in the proof of Theorem (4.7) that $G(1, t) = G(0, t)$ for all t and that $G(0, t) = g(0)$, $G(1, t) = g(1)$ for all t in $[0, 1]$. Hence $g(0) = x_0 = g(1) = x_1$, and f is an injective mapping of C into N.

Thus we have verified all the conditions for f to be a covering mapping.   q.e.d.

THEOREM (4.9). *Let X be a complete metric space, Y a metric space, f a local homeomorphism of X into Y. Suppose that X is locally pathwise connected, Y is connected, and that each point y_0 of Y has a neighborhood N in Y such that N is rectifiably pathwise connected and rectifiably simply connected. Suppose that there exists a function $c: R^+ \to R^+$ ($R^+ = $ the nonnegative real numbers) with c non-increasing in r, continuous, and with $\int^{+\infty} c(r)\, dr = +\infty$ such that if $d(u) = d(u, u_0)$ for a fixed u_0 in X and all u in X, then each point x_0 in X has a neighborhood N in X on which the inequality*

$$d(f(u), f(v)) \geq c(\max(d(u), d(v))) \cdot d(u, v)$$

holds for all u and v in N.

Then f is a covering mapping of X onto Y.

PROOF OF THEOREM (4.9). By Theorem (4.8), it suffices to show that if $g:[0, 1) \to X$ has as its projection the rectifiable curve $h:[0, 1) \to Y$, then the continuous curve g is also rectifiable. Consider the function $d(g(t))$, t in $[0, 1)$. If $d(g(t))$ is bounded on $[0, 1)$, then since g is bounded away from 0 on each bounded interval of $[0, \infty)$, it follows that there exists $c_0 > 0$ such that for each t, $d(f(u), f(v)) \geq c_0\, d(u, v)$ for all u and v in some neighborhood N_t of $g(t)$.

Consider a sequence $0 = t_0 < t_1 < \cdots < t_m < 1$, and the corresponding sum $\sum_{j=1}^{m} d(g(t_j), g(t_{j-1}))$. Since the sum is not decreased by introducing more points into the sequence of points $\{t_j\}$, we may assume that we can intersperse new points in the interval $[0, t_m]$ such that for any j, t_{j-1} and t_j lie in a neighborhood N_t as above. In this case, it follows that $d(h(t_j), h(t_{j-1})) \geq c_0\, d(g(t_j), g(t_{j-1}))$. Summing on j, we have

$$\sum_{j=1}^{m} d(g(t_j), g(t_{j-1})) \leq c_0^{-1} \text{ length } (h) < \infty$$

and we have shown that g is rectifiable in this case.

In the other case, $d(g(t))$ is unbounded on $[0, 1)$. For each positive integer n, we let $[s_n, t_n]$ be the unique subinterval of $[0, 1)$ which satisfies the following conditions:

(i) $d(g(t)) < n$ for $t < s_n$.

(ii) $n \leq d(g(t)) \leq n + 1$, $t \in [s_n, t_n]$.

(iii) For some $\epsilon_n > 0$, $d(g(t)) > n + 1$ for $t_n < t < t_n + \epsilon_n$.

The intervals $[s_n, t_n]$ are well defined for $n \geq N$ and are pairwise disjoint. For each $n \geq N$, $t_n \leq s_{n+1}$. Consider any finite subset of the subdivision of the interval $[0, 1)$ by the collection of points s_n and t_n. Then for $n \geq N$,

$$d(g(t_n), g(s_n)) \geq d(g(t_n), u_0) - d(g(s_n), u_0) = d(g(t_n)) - d(g(s_n)) \geq (n+1) - n = 1.$$

On the other hand, on each interval $[s_n, t_n]$ we may interpolate points

$$s_n = t_{n,0} < t_{n,1} < \cdots < t_{n,r_n} = t_n$$

such that for each k with $0 \leq k \leq r_n - 1$, the pair of successive points $g(t_{n,k})$ and $g(t_{n,k+1})$ lie in a neighborhood N on which we have the inequality

$$d(f(u), f(v)) \geq c(\max(d(u), d(v))) \cdot d(u, v).$$

Applying this inequality to each such successive pair of points, we obtain

$$d(h(t_{n,k}), h(t_{n,k+1})) \geq c(\max(d(g(t_{n,k})), d(g(t_{n,k+1})))) \, d(g(t_{n,k}), g(t_{n,k+1})).$$

Since $d(g(t))$ lies between n and $(n+1)$ on the interval $[s_n, t_n]$, this inequality becomes

$$d(h(t_{n,k}), h(t_{n,k+1})) \geq c(n+1) \, d(g(t_{n,k}), g(t_{n,k+1})).$$

If we sum this inequality over k, we find that

$$\sum_{k=0}^{r_n-1} d(h(t_{n,k}), h(t_{n,k+1})) \geq c(n+1) \sum_{k=0}^{r_n-1} d(g(t_{n,k}), g(t_{n,k+1}))$$

$$\geq c(n+1) \, d(g(t_n), g(s_n)) \geq c(n+1).$$

We now sum the last inequality over n for $n \geq N$, and obtain

$$\sum_{n \geq N} \sum_{k=0}^{r_n-1} d(h(t_{n,k}), h(t_{n,k+1})) \geq \sum_{n \geq N} c(n+1).$$

The left-hand double-sum is bounded by the length of the curve h since the intervals $[s_n, t_n]$ can overlap only at their endpoints at most, and by construction, the length of the curve h is finite. Hence

$$\infty > \sum_{n \geq N} c(n+1) \geq \int_{N+1}^{\infty} c(r) \, dr = +\infty$$

which is a contradiction proving that the assumed case of $d(g(t))$ being unbounded is impossible. Hence the proof of our theorem is complete.   q.e.d.

As a simple application of Theorem (4.9), we have the following result which we shall apply to a number of important classes of nonlinear mappings in Banach spaces:

THEOREM (4.10).   *Let X and Y be Banach spaces, f a local homeomorphism of X into Y. Suppose that for $r \geq 0$, there exists a continuous nonincreasing function*

$c(r) > 0$ such that $\int^{+\infty} c(r)\,dr = +\infty$ such that the following condition holds: For each x_0 in X_0, there is an open neighborhood N in X such that for all u and v in N,

$$\|f(u) - f(v)\| \geq c(\max\|u\|, \|v\|)\|u - v\|.$$

Then f is a homeomorphism of X onto Y.

COROLLARY TO THEOREM (4.10). *Let X and Y be Banach spaces, f an open mapping of X into Y. Suppose that there exists a constant $c_0 > 0$ such that each point x_0 of X has a neighborhood N in X with the inequality $\|f(u) - f(v)\| \geq c_0\|u - v\|$ holding for all u and v in N.*
Then f is a one-to-one mapping of X onto Y with a continuous inverse.

PROOF OF THEOREM (4.10). This is a special case of Theorem (4.9) if we note that by Proposition (4.2), Y is rectifiably locally connected and rectifiably simply connected. The corollary is obviously a specialization of the theorem with $c(r) = c_0 > 0$ for all $r > 0$. q.e.d.

We now apply Theorem (4.10) to strengthen the theory of Φ-accretive mappings constructed in §3.

THEOREM (4.11). *Let X be a Banach space with a given Φ-system in the sense of Definition (3.1) (not necessarily global) and let A be a locally Lipschitzian Φ-accretive mapping of X into X. Then*

(a) *For every $c_0 > 0$, $(A + c_0 I)$ is a one-to-one bicontinuous mapping of X onto X.*

(b) *Suppose that there exists a continuous nonincreasing function $c(r)$ for $r \geq 0$ with $c(r) > 0$ for all r and $\int^{+\infty} c(r)\,dr = +\infty$ such that each point x_0 in X has a neighborhood N in X such that for u and v in $N \cap N_\alpha$*

$$(\Phi_\alpha(u, v), A(u) - A(v)) \geq c(\max(\|u\|, \|v\|))\|u - v\|^2.$$

Then A is a one-to-one-bicontinuous mapping of X onto X.

PROOF OF THEOREM (4.11). PROOF OF (a). By Theorem (3.1), $A + cI$ is a local homeomorphism of X into X. Suppose that u and v lie in N_α. Then

$$\|u - v\| \cdot \|(A + cI)(u) - (A + cI)(v)\| \geq (\Phi_\alpha(u, v), (A + cI)(u) - (A + cI)(v))$$
$$= (\Phi_\alpha(u, v), Au - Av) + c(\Phi_\alpha(u, v), u - v)$$
$$\geq c(\Phi_\alpha(u, v), u - v)$$
$$\geq cc_0\|u - v\|^2.$$

Cancelling $\|u - v\|$, we obtain $cc_0\|u - v\| \leq \|(A + cI)u - (A + cI)v\|$. Hence the conclusion of part (a) follows from the corollary to Theorem (4.10). q.e.d.

PROOF OF (b). We may assume that $N \subset N_\alpha$ for some α, since each point x_0 of X is contained in $N \cap N_\alpha$ for our original choices of N. By Theorem (3.4), A is a local homeomorphism of X into X. On the other hand, for u and v in N,

$$\|u - v\| \cdot \|Au - Av\| \geq (\Phi_\alpha(u, v), Au - Av) \geq c(\max(\|u\|, \|v\|))\|u - v\|^2$$

from which it follows that $\|Au - Av\| \geq c(\max(\|u\|, \|v\|)) \|u - v\|$. The conclusion of part (b) then follows from Theorem (4.10). q.e.d.

5. Limits of invertible and semi-invertible mappings. As we shall establish in the following sections (and have already shown for locally-Lipschitzian Φ-accretive mappings), nonlinear mappings of various significant classes, monotone, accretive, Φ-accretive, and others, can be obtained as the uniform limits on bounded sets of mappings with continuous inverses defined on the whole of the range space. In the present section, it is our purpose to study the general question of the properties of mappings which are approximable by mappings having useful properties for their inverses.

The key definitions which we employ are the following:

DEFINITION (5.1). *Let S_0 and S_1 be two topological spaces, T a mapping of S_0 into 2^{S_1}. Then T is said to be invertible if the mapping T^{-1} of S_1 into 2^{S_0} given by*

$$T^{-1}(v) = \{u \mid u \in S_0, v \in T(u)\} \qquad (v \in S_1)$$

is a single-valued continuous mapping of S_1 into S_0.

DEFINITION (5.2). *Let S_0 and S_1 be two topological spaces, T a mapping of S_0 into 2^{S_1}. Then T is said to be semi-invertible if $R(T) = S_1$ (where $R(T)$ denotes the range of T), and if the mapping T^{-1} satisfies the two following conditions:*

(a) *For each v in S_1, $T^{-1}(v)$ is a connected subset of S_0.*

(b) *T^{-1} is an upper-semicontinuous set-valued function from S_1 to 2^{S_0}.*

It follows from these two definitions that an invertible mapping is always semi-invertible and that a semi-invertible mapping is invertible if and only if T^{-1} is single-valued.

We shall consider mappings of S_0 into 2^{S_1} which are approximable in a suitable sense by invertible and semi-invertible mappings. To define the sense of this approximation, we introduce several further definitions.

DEFINITION (5.3). *Let S_0 be a set, Y a uniform space, T a mapping of S_0 into 2^Y, $\{T_\xi : \xi \in \Omega\}$ a directed set of mappings of S_0 into 2^Y. Then T_ξ is said to converge to T uniformly on S_0 if for each ϵ in the uniformity, there exists an index $\xi_0(\epsilon)$ in Ω such that for $\xi > \xi_0(\xi)$, and each s in S_0, $T_\xi(s)$ lies in the ϵ-neighborhood of $T(s)$ and $T(s)$ lies in the ϵ-neighborhood of $T_\xi(s)$.*

In the usual terms, what is required in Definition (5.3) is the uniform convergence of $T_\xi(s)$ to $T(s)$ in the Hausdorff uniformity on 2^Y (i.e., the obvious generalization of the Hausdorff metric to the space of subsets of a uniform space). Obviously, for single-valued mappings T and T_ξ, Definition (5.3) coincides with the usual definition of uniform convergence on S_0.

DEFINITION (5.4). *Let S be a topological space. Then by a bounding system on S, we mean a family $\{F_\alpha\}$ of closed subsets of S with the property that for each F_α in the family, there exists F_β in the family such that F_α is contained in the interior of F_β.*

A subset R of S is said to be bounded with respect to the bounding system if there exists a set F_α in the system which contains R.

An important role in the later argument is played by the following simple result from point set topology:

PROPOSITION (5.1). *Let S_0 be a connected topological space, S a topological space, W an upper-semicontinuous set-valued mapping from S_0 to 2^S such that for each s_0 in S_0, $W(s_0)$ is a connected set. Then $W(S_0)$ is connected.*

PROOF OF PROPOSITION (5.1). Suppose that $W(S_0)$ is not connected. Then there exist two disjoint nonempty open subsets G_1 and G_2 of $W(S_0)$ such that $W(S_0) = G_1 \cup G_2$. For each s_0 in S_0, $W(s_0)$ is connected. Hence it must lie entirely in one or the other of these two sets, i.e., $W^{-1}(G_1)$ and $W^{-1}(G_2)$ are disjoint subsets of S_0. Obviously, $S_0 = W^{-1}(G_1) \cup W^{-1}(G_2)$.

Since G_1 is open in $W(S_0)$, there exists an open set U_1 in S such that $G_1 = U_1 \cap W(S_0)$. For a given s_0 in $W^{-1}(G_1)$, U_1 is an open neighborhood of $W(s_0)$. Hence, by the upper-semicontinuity of W, there exists a neighborhood N of s_0 in S_0 such that for t in N, $W(t) \subset U_1$. For such t, $W(t) \subset U_1 \cap W(S_0) = G_1$, i.e., $N \subset G_1$. Hence $W^{-1}(G_1)$ is open in S_0. Similarly, $W^{-1}(G_2)$ is open in S_0.

Finally, S_0 is connected while S_0 is the union of the two disjoint nonempty open subsets $W^{-1}(G_1)$ and $W^{-1}(G_2)$. This is a contradiction which proves the proposition.   q.e.d.

Our basic result concerning the range of mappings which are approximable by semi-invertible mappings is the following:

THEOREM (5.1). *Let S be a topological space with a bounding system of subsets $\{F_\alpha\}$, Y a connected, locally connected uniform space. Let T be a mapping of S into 2^Y and suppose that there exists a directed set $\{T_\xi : \xi \in \Omega\}$ of semi-invertible mappings from S to 2^Y converging uniformly to T in the sense of Definition (5.3) on each F_α. Suppose further that for each y_0 in Y, there exists a neighborhood N of y_0 in Y such that $T^{-1}(N)$ is bounded with respect to the given bounding system.*

Then:

(a) *$R(T)$, the range of T, is dense in Y.*

(b) *If, in addition, $T(F_\alpha)$ is closed in Y for each α, then $R(T) = Y$.*

To simplify the presentation of the proof of Theorem (5.1), we shall break up the development of the proof into a number of subsidiary results. The first is the following localization of Theorem (5.1):

THEOREM (5.2). *Let S be a topological space, Y a uniform space, y_0 a point of Y at which Y has a base of connected neighborhoods. Let F_0 and F_1 be closed subsets of S with F_0 contained in the interior of F_1, and suppose that we are given a mapping T of F_1 into 2^Y such that there exists a directed set $\{T_\xi : \xi \in \Omega\}$ of semi-invertible mappings from S to 2^Y converging uniformly to T on F_1 in the sense of Definition (5.3). Suppose further that there exists a neighborhood N of y_0 in Y such that $T^{-1}(N) \subset F_0$. Then if y_0 lies in the closure of $T(F_0)$, it follows that y_0 lies in the interior of $\mathrm{cl}\,(T(F_1))$.*

PROOF OF THEOREM (5.1) FROM THEOREM (5.2). Under the hypotheses of Theorem (5.1), for each point y_0 of $\mathrm{cl}\,(R(T))$, there exist a pair of closed subsets F_0 and F_1 of the bounding system and a neighborhood N of y_0 such that the hypotheses of Theorem (5.2) hold. (Indeed, we need only take N to be a neighborhood of y_0 such that $T^{-1}(N)$ is bounded with respect to the given bounding system, F_0 an element of the bounding system which contains $T^{-1}(N)$, and F_1 another element of the bounding system which contains F_0 in its interior. Since y_0 lies in the closure of $T(S)$ while for all s in $S - F_0$, $T(s)$ does not intersect N, it follows that y_0 lies in the closure of $T(F_0)$.) It follows from Theorem (5.2) that cl $(R(T))$ is a subset of the interior cl $(R(T))$. Hence cl $(R(T))$ is both open and closed in Y. Since Y is connected and cl $(R(T))$ is nonempty, it follows that $Y =$ cl $(R(T))$, and the proof of (a) is complete.

For the proof of the assertion of (b), we note that if cl $(T(F_1))$ is closed, it follows that y_0 lies in $T(F_1)$. Hence if $T(F_a)$ is closed in Y for all F_a in the bounding system, it follows that $Y = $ cl $(R(T)) = R(T)$. q.e.d.

We now turn to the proof of Theorem (5.2).

PROOF OF THEOREM (5.2). Since Y has a base of connected neighborhoods at y_0, we may choose a connected neighborhood N_0 of y_0 contained in N and such that for some element ϵ of the uniformity, the ϵ-neighborhood of N_0 is contained in N. We may assume that ϵ is chosen so small that the ϵ-neighborhood of y_0 is contained in N_0. Since y_0 lies in the closure of $T(F_0)$, we may find a point s_0 of F_0 such that for some point y_1 in the γ-neighborhood of y_0, $y_1 \in T(s_0)$, where $\gamma^2 < \epsilon$. Since T_ξ converges uniformly to T on F_1 in the sense of Definition (5.3), we may choose ξ_0 in Ω such that for $\xi > \xi_0$, there exists a point y_ξ in the ϵ-neighborhood of y_0 lying in $T(s_0)$. Hence for such ξ, $T_\xi(F_0)$ has a nonempty intersection with N_0.

By hypothesis, T_ξ is semi-invertible so that T_ξ^{-1} is an upper-semicontinuous set-valued mapping from Y to 2^S with the property that for each y in Y, $T_\xi^{-1}(y)$ is a connected subset of S. Since N_0 is connected, it follows that by Proposition (5.1), $T_\xi^{-1}(N_0)$ is a nonempty connected subset of S. We now choose ξ_0 in Ω so that in addition for $\xi > \xi_0$, and for each s in F_1, $T_\xi(s)$ lies in the ϵ-neighborhood of $T(s)$. We assert that for each such ξ, $T_\xi^{-1}(N_0)$ is a subset of F_1. Indeed, suppose not. Then since $T_\xi^{-1}(N_0)$ is connected and intersects F, it must contain a point of the boundary of F_1 in S. Since F_0 lies in the interior of F_1 in S, such a point s_ξ must lie in $F_1 - F_0$. Let y_ξ be a point of $T_\xi(s_\xi)$ in N_0. By the choice of ξ, there must be a point of $T(s_\xi)$ in the ϵ-neighborhood of y_ξ. If Z_ξ is such a point, then z_ξ lies in the ϵ-neighborhood of N_0, and thus must lie in N by the assumption that the ϵ-neighborhood of N_0 is contained in N. In particular, it follows that the point s_ξ of $F_1 - F_0$ lies in $T^{-1}(N)$, contradicting the fact that $T^{-1}(N)$ is contained in F_0. This contradiction shows that for each $\xi > \xi_0$, $T_\xi^{-1}(N_0)$ is contained in F_1.

Let y be any point of N_0. We shall show that y lies in the closure of $T(F_1)$. Let γ be any uniformity on Y. For each $\xi > \xi_0$, we may find a point s_ξ in F_1 such that y lies in $T_\xi(s_\xi)$. For $\xi > \xi_\gamma$, there exists a point s_ξ in $T(s_\xi)$ lying in the γ-neighborhood of y. Since the γ-neighborhoods of y form a base for the topology of Y at y, y lies in the closure of $T(F_1)$. q.e.d.

We now consider some applications of Theorem (5.1) to Φ-accretive and Φ-coaccretive mappings. (Sharper applications to the study of monotone and accretive mappings are given in later sections.)

THEOREM (5.3). *Let X be a Banach space with a given Φ-system, A a hypermaximal Φ-accretive mapping of X into 2^X with respect to the given Φ-system. Suppose that each point x_0 of X has a neighborhood N in X such that $A^{-1}(N)$ is bounded in X. Then $R(A) = X$.*

PROOF OF THEOREM (5.3). By the corollary to Theorem (3.5), for each real number $\xi > 0$, $A_\xi = A + \xi I$ is an invertible mapping of X onto X and as ξ tends to 0, A_ξ converges to A in the sense of Definition (5.3) uniformly on each bounded subset of X. Hence we may apply Theorem (5.1) with the bounding system given by the closed balls about the origin in X, and we obtain the conclusions of Theorem (5.3). q.e.d.

THEOREM (5.4). *Let X be a Banach space with a given Φ-system, A a locally Lipschitzian mapping of X into X which is Φ-accretive with respect to the given Φ-system. Suppose that for each point x_0 of X, there exists a neighborhood N of x_0 such that $A^{-1}(N)$ is bounded. Then:*

(a) *$R(A)$ is dense in X.*

(b) *If for each of a family of closed balls B_r about the origin in X with r tending to infinity, $A(B_r)$ is closed in X, then $R(A) = X$.*

PROOF OF THEOREM (5.4). By Theorem (4.11)(a), A is hypermaximal Φ-accretive. Hence the conclusions of the present theorem follow from those of Theorem (5.3). q.e.d.

THEOREM (5.5). *Let X be a Banach space with a given global Φ-system, A a mapping of X into 2^X which is hypermaximal Φ-accretive, A_0 a locally Lipschitzian mapping of X into X which is Φ-accretive. Suppose that for each point x_0 in X, there exists a neighborhood N of x_0 such that $(A + A_0)^{-1}(N)$ is bounded. Then*

(a) *The range of $(A + A_0)$ is dense in X.*

(b) *If for a family of closed balls $\{B_r\}$ about the origin in X with r tending to infinity, $(A + A_0)(B_r)$ is closed in X, then the range of $(A + A_0)$ is the whole of X.*

PROOF OF THEOREM (5.5). By Theorem (3.5), $(A + A_0)$ is hypermaximal Φ-accretive. Hence the conclusions of the present theorem follow from those of Theorem (5.3). q.e.d.

THEOREM (5.6). *Let X be a Banach space with a global Φ-system, T a locally Lipschitzian mapping of X into X which is Φ-coaccretive with respect to the given Φ-system. Suppose that for each point x_0 of X, there exists a neighborhood N of x_0 in X such that $T^{-1}(N)$ is bounded.*

Then:

(a) *$R(T)$ is dense in X.*

(b) *If for a sequence $\{B_r\}$ of closed balls about the origin in X with r tending to infinity, $T(B_r)$ is closed in X, then $R(T) = X$.*

PROOF OF THEOREM (5.6). By Theorem (3.3), for each $\delta > 0$, the mapping $T_\delta = T + \delta I$ is a mapping of X onto X with a single-valued Lipschitzian inverse.

Since T_δ converges to T uniformly on each bounded subset of X, the conclusions of Theorem (5.6) follow immediately from those of Theorem (5.1).   q.e.d.

For the remainder of this section, we turn to another type of property of mappings approximable by invertible mappings, namely the structure of $T^{-1}(y)$ for a given point y of Y.

We begin with the relatively elementary case of monotone mappings from a Banach space X to its conjugate space X^*, both for its intrinsic interest and as a motivation for corresponding though more complex results in the theory of Φ-accretive mappings.

PROPOSITION (5.2). *Let X and Y be two Banach spaces with a given global Φ-system for the pair $[X, Y]$ given by the mapping $\Phi: X \times X \to Y^*$. Let $\{T_1, \ldots, T_n\}$ be a finite family of mappings of X into 2^Y, each of which is maximal Φ-monotone for the given Φ-system. Let y be a given point of Y.*

Then y lies in $T(x)$ for a given x in X and T the mapping of X into 2^Y given by $T = T_1 + T_2 + \cdots + T_n$ (i.e., $T(x) = \sum_{j=1}^{n} T_j(x)$ whenever all the summands are nonempty, $T(x) = \varnothing$ otherwise), if and only if the following condition is satisfied:

There exists a family $(y_1, y_2, \ldots, y_n)$ in Y such that $y = y_1 + y_2 + \cdots + y_n$ while for each $2n$-tuple $\{u_n, \ldots, u_n; w_1, \ldots, w_n\}$ with each u_j in X, each w_j in Y, and w_j in $T_j(u_j)$, we have

$$\sum_{j=1}^{n} (\Phi(x, u_j), y_j - w_j) \geq 0.$$

PROOF OF PROPOSITION (5.2). If y lies in $T(x)$, then $y = y_1 + y_2 + \cdots + y_n$ with $y_j \in T_j(x)$. It follows by the Φ-monotonicity of each T_j that for w_j in $T_j(u_j)$, we have $(\Phi(x, u_j), y_j - w_j) \geq 0$. Summing on j, we obtain the necessity of the given condition above.

Suppose now that we wish to prove the sufficiency of the condition. We carry through the proof by induction on n. For $n = 1$, the condition is equivalent to the maximal Φ-monotonicity of the single mapping T_1. Suppose the proposition is true for $(n-1)$, and that we are given the above condition for a pair x in X, y in Y, with $y = y_1 + y_2 + \cdots + y_n$ as above. Since T_n is maximal Φ-monotone, if y_n is not an element of $T_n(x_n)$, then there exists a pair $[u_n, w_n]$ in the graph of T_n such that $(\Phi(x, u_n), y_n - w_n) < 0$. Taking arbitrary elements $[u_j, w_j]$ in $G(T_j)$ for $1 \leq j \leq (n-1)$, it follows that

$$\sum_{j=1}^{n-1} (\Phi(x, u_j), x_j - w_j) > 0.$$

If we apply the inductive hypothesis, it follows that $y_j \in T_j(x)$ for $j \leq n - 1$. Hence we may set $u_j = x$ and $w_j = y_j$ for each j in the preceding inequality. We obtain

$$0 = \sum_{j=1}^{n-1} (\Phi(x, x), y_j - y_j) > 0,$$

which is a contradiction. Hence y_n lies in $T_n(x_n)$. Since the sequence may be reordered to make any term the last, it follows that $y_j \in T_j(x)$ for $1 \leq j \leq n$. Thus the inductive step is complete, and the proposition follows.   q.e.d.

THEOREM (5.7). *Let X be a Banach space, $\{T_1, \ldots, T_n\}$ a family of maximal monotone mappings of X into 2^{X^*}. Let $T = T_1 + T_2 + \cdots + T_n$.*
  Then for each y in X^*, $T^{-1}(y)$ is a convex subset of X.

PROOF OF THEOREM (5.7). Let X_n^* be the product of n copies of the space X^*. For each element $\{y_1, \ldots, y_n\}$ in X_n^* such that $y = y_1 + \cdots + y_n$, and for each pair of n-tuples $u = \{u_1, \ldots, u_n\}$ in X and $\{w_1, \ldots, w_n\}$ in X^* such that $w_j \in T_j(u_j)$ for all j with $1 \leq j \leq n$, we let

$$T_{y,u,w} = \left\{ x \,\Big|\, x \in X, \sum_{j=1}^n (x - u_j, y_j - w_j) \geq 0 \right\}.$$

By Proposition (5.2), $T^{-1}(y)$ consists of the union over $\{y_1, \ldots, y_n\}$ of the intersections over u and w of the sets $T_{y,u,w}$.

The inequality defining the set $T_{y,u,w}$ may be written in the following form:

$$\left(x, y - \sum_{j=1}^n w_j\right) - \sum_{j=1}^n (u_j, v_j) + \sum_{j=1}^n (u_j, w_j) \geq 0.$$

The set of ordered $(n+1)$-tuples $[x, y_1, \ldots, y_n]$ in $X \times X_n^*$ which satisfy this inequality for a given pair of n-tuples u and w is obviously convex. Hence, so is its intersection over u and w. Hence, so is its intersection with the hyperplane $y_1 + \cdots + y_n = y$. Let P be the projection mapping of $X \times X_n^*$ on its first component. P maps the preceding intersection onto $T^{-1}(y)$. Since P is linear, $T^{-1}(y)$ is a convex subset of X.   q.e.d.

COROLLARY TO THEOREM (5.7). *If T is maximal monotone, then $T^{-1}(y)$ is a closed convex subset of X, and hence weakly closed in X.*

Similar results are given for accretive mappings in the detailed discussion of the latter below.

In the general context of Theorems (5.1) and (5.2), we have the following simple result on the structure of $T^{-1}(y_0)$ for a given point y_0 of Y:

THEOREM (5.8). *Let S be a topological space, Y a metric space, y_0 a point of Y at which Y has a base of connected neighborhoods. Let F_0 and F_1 be closed subsets of S with F_0 contained in the interior of F_1, and suppose that we are given a mapping T of F_1 into 2^Y such that there exists a directed set $\{T_\xi : \xi \in \Omega\}$ of semi-invertible mappings of S into 2^Y converging uniformly to T on F_1. Suppose that there exists a neighborhood N of y_0 in Y such that $T^{-1}(N) \subset F_0$. Suppose further that T is a proper mapping (i.e., that for each compact subset K of Y, $T^{-1}(K)$ is compact in S), and maps compact subsets of S into compact subsets of Y.*
  Then $T^{-1}(y_0)$ is a connected compact subset of S.

PROOF OF THEOREM (5.8). Since T is proper, $T^{-1}(y_0)$ is a compact subset of S. Suppose that $T^{-1}(y_0)$ is not connected. Then it can be decomposed into the union of two disjoint nonempty compact subsets C_1 and C_2, and we can find two disjoint open subsets U_1 and U_2 of S which contain C_1 and C_2 respectively.

Since T is a proper mapping, $T(F_1 - (U_1 \cup U_2))$ is a closed subset of Y which does not contain y_0. Indeed, suppose that $\{y_k\}$ is a convergent sequence from $T(F_1 - (U_1 \cup U_2))$ with limit y in Y. Let K be the compact subset of Y consisting of the elements of the sequence $\{y_k\}$ together with the limit y. Since T is proper, $T^{-1}(K)$ is compact in S. Hence its intersection with the closed subset $F_1 - (U_1 \cup U_2)$ is compact in S. Hence $T(T^{-1}(K) - (U_1 \cup U_2))$ is a compact subset of S, which includes all the elements $\{y_k\}$. Hence it includes y.

Let N_0 be a neighborhood of y_0 which does not intersect $T(F_1 - (U_1 \cup U_2))$. Then if we assume that $N_0 \subset N$, it follows that $T^{-1}(N_0)$ is a subset of F_1 and hence a subset of $U_1 \cup U_2$. We assume that there exists a ball B_0 about y_0 for some $\epsilon > 0$ contained in N_0. Then for $\xi > \xi_0$, we may assume that $T_\xi(s)$ is contained in the $\epsilon/2$ neighborhood of $T(s)$ for each s in $F_1 - (U_1 \cup U_2)$ and hence does not intersect the $(\epsilon/2)$-ball about y_0. Hence for such ξ, $T_\xi^{-1}(y_0)$ is a subset of $U_1 \cup U_2$, and indeed $T_\xi^{-1}(B_{\epsilon/2}(y_0))$ is a subset of $U_1 \cup U_2$.

We choose a neighborhood N_2 of y_0 in Y such that N_2 is connected and is contained in the ball of radius $(\epsilon/2)$ about y_0. Then by Proposition (5.1), $T_\xi^{-1}(N_2)$ is a connected subset of $U_1 \cup U_2$. If s_1 is a point of $T^{-1}(y_0) \cap U_1$, then for $\xi > \xi_1$, $T_\xi(s_1)$ intersects N_2, i.e., s_1 lies in $T_\xi^{-1}(N_2)$. Similarly, if s_2 is a point of $T^{-1}(y_0) \cap U_2$, then for $\xi > \xi_2$, $T_\xi^{-1}(N_2)$ contains the point of s_2. Hence, finally, for a suitable ξ_3 in Ω and all $\xi > \xi_3$, $T_\xi^{-1}(N_2)$ contains points of both of the disjoint open sets U_1 and U_2. This contradicts the connectedness of $T_\xi^{-1}(N_2)$, and this contradiction proves the theorem. q.e.d.

For sharper results, we must sharpen our hypotheses on the space Y and upon the approximating mappings T_ξ.

THEOREM (5.9). *Let S be a metrizable space, Y a Banach space, y_0 a point of Y. Let T be a proper continuous mapping of S into Y, and suppose that for a given bounding system $\{F_\alpha\}$ on S, there exists a sequence $\{T_j\}$ of invertible continuous mappings of S into Y which converges to T uniformly on each bounded set F_α. Suppose that for a given point y_0 of $\text{cl}(R(T))$, there exists a neighborhood N of y_0 such that $T^{-1}(N)$ is bounded.*

Then $T^{-1}(y_0)$ is the intersection of a decreasing sequence of compact absolute retracts and hence has trivial fundamental group and trivial Čech cohomology.

To carry through the proof of Theorem (5.9), we shall need the following auxiliary result:

PROPOSITION (5.3). *Let S be a metrizable space, M a compact subset of S, $\{R_k\}$ a sequence of subsets of X, each of which is a compact absolute retract. Suppose that the following conditions both hold:*

(i) *M is contained in each R_k.*

(ii) *For each neighborhood V of M in X, there exists an infinite subsequence $\{R_{k_j}\}$ of $\{R_k\}$ such that each R_{k_j} is contained in V.*

Then M is an R_δ, i.e., the intersection of a descending sequence of compact absolute retracts.

PROOF OF PROPOSITION (5.3). We may suppose a metric given on S. For each k, since R_k is an absolute retract (in the category of metrizable spaces), there exists a continuous retraction ζ_k of S on R_k. We begin by a recursive construction of an infinite subsequence $\{R_{j(k)}\}$ of the given sequence $\{R_k\}$, such that if for each $m < k$, we set

$$\beta_m^k = \zeta_{j(m)}\zeta_{j(m-1)} \cdots \zeta_{j(k-1)},$$

then for each x in $R_{j(k)}$, we have $d(x, \beta_m^k(x)) < k^{-1}$ (where d is the distance function on S).

We begin this recursive construction by setting $j(1) = 1$. Then $R_{j(1)} = R_1$. Suppose that $j(k)$ has been selected as a strictly increasing function of k for $k \leq m$. Then for each $m < (n+1)$, the mapping $\beta_m^{n+1} = \zeta_{j(m)}\zeta_{j(m-1)} \cdots \zeta_{j(n)}$ is a well-defined continuous mapping of X into X with $\beta_m^{n+1}(x) = x$ on M. It follows that there is a neighborhood V of M such that $d(x, \beta_m^{n+1}(x)) < (n+1)^{-1}$ for all x in V and each $m \leq n$, and we may assume that V is contained in the $(1/n)$-neighborhood of M. We choose $j(n+1)$ to be the least integer k after $j(n)$ such that $R_{j(n+1)} \subset V$, which is possible by hypothesis (ii).

Hence, the recursive construction may be extended to all integers k, and we note that in addition $R_{j(k)}$ is contained in the $(k-1)^{-1}$-neighborhood of M. Thus the intersection of the sequence $\{R_{j(k)}\}$ coincides with M.

For each positive integer n, we now consider the subset Q_n of the new compact metric space X, where X is the Cartesian product of $R_{j(k)}$, $k \geq 1$, and

$$Q_n = \{x \mid x \in X, x_j = \beta_j^n(x_n) \text{ for } j < n\}.$$

Obviously, Q_n decreases with n, while for each n, Q_n is homeomorphic to the Cartesian product of $R_{j(k)}$, $k \geq n$. Since the Cartesian product of a sequence of compact absolute retracts is also a compact absolute retract, Q_n is a compact absolute retract for each n. Let M_0 be the subset of Q_n given by $x_j = x \in M$ $j \geq 1$, i.e., M_0 is the diagonal set of the Cartesian product of the subset M of $R_{j(k)}$ for each k. Then M_0 is a subset of the intersection $\bigcap_n Q_n$, and M_0 is homeomorphic to M. However, if x lies in $\bigcap_n Q_n$, then for each j and n with $j < n$,

$$x_j = \beta_j^n(x_n), \qquad d(x_n, \beta_j^n(x_n)) < n^{-1}.$$

Hence x_n converges to x_j as $n \to \infty$ for each j, i.e., $x_j = x_m$ for each j and m. Thus x lies on the diagonal of the Cartesian product, and since $\bigcap_k R_{j(k)} = M$, $x_j \equiv x$ lies in M for each j. Hence $\bigcap_n Q_n = M_0$ is an R_δ. q.e.d.

PROOF OF THEOREM (5.9). Let $K = T^{-1}(y_0)$. Then K is a bounded compact set, and T_j converges to T uniformly on K as $j \to \infty$. Hence there exists a sequence $\epsilon_j \to 0$ as $j \to \infty$ such that $T_j(K)$ is a subset of the ball $B_{\epsilon_j}(y_0)$ in Y. Let K_j be the convex closure of $T_j(K)$ in Y. Then K_j is a compact subset of $B_{\epsilon_j}(y_0)$ and since K_j is convex, K_j is a compact absolute retract.

Since T_j is a homeomorphism, $T_j^{-1}(K_j) = R_j$ is also a compact absolute retract for each j. By construction K is a subset of R_j for each j. Let V be a neighborhood of K in S. We assert that there exists j_V such that for $j \geq j_V$, R_j must be a subset of V. Indeed, suppose not. Then there would exist a sequence $\{x_{j(k)}\}$ with $x_{j(k)} \in R_{j(k)} - V$ and $j(k) \to \infty$ as $k \to \infty$. Since $x_{j(k)}$ lies in $R_{j(k)}$, it follows that

$$d(T_{j(k)}(x_{j(k)}), y_0) \leq \epsilon_{j(k)} \to 0 \qquad (k \to \infty).$$

Under the assumptions of our theorem, $\{x_{j(k)}\}$ lies in a bounded set. Hence

$$d(T(x_{j(k)}), T_{j(k)}(x_{j(k)})) \to 0 \qquad (k \to \infty),$$

i.e., $d(T(x_{j(k)}), y_0) \to 0$. Since T is assumed to be proper, it follows that by passing to a subsequence (which we identify with the original subsequence), we may assume that $x_{j(k)}$ converges to some element x of S. For x, moreover, it follows that $T(x) = y_0$, contradicting the fact that x does not lie in V. Hence we obtain a contradiction showing that $R_j \subset V$ for $j \geq j_V$.

Thus the assumptions of Proposition (5.3) are satisfied, and $K = T^{-1}(y_0)$ is an intersection of a descending sequence of compact absolute retracts. q.e.d.

We may apply Theorem (5.9) to the classes of Φ-accretive and Φ-coaccretive mappings.

THEOREM (5.10). *Let X be a Banach space with a given Φ-system, A a locally Lipschitzian mapping of X into X which is Φ-accretive with respect to the given Φ-system. Suppose that A is proper and that for each point x_0 of X, there exists a neighborhood N of x_0 such that $A^{-1}(N)$ is bounded. Then $R(A) = X$, and for each point x_0, $A^{-1}(x_0)$ is a compact intersection of compact absolute retracts.*

PROOF OF THEOREM (5.10). Since A is proper, it is a closed mapping and it follows from Theorem (5.4) that $R(A) = X$. We apply Theorem (5.9) with $T = A$, $T_j = A + j^{-1}I$, and use Theorem (4.11) and the proof of Theorem (5.3) above to verify that the hypotheses of Theorem (5.9) apply in this case. q.e.d.

THEOREM (5.11). *Let X be a Banach space with a global Φ-system, T a locally Lipschitzian mapping of X into X which is Φ-coaccretive with respect to the given Φ-system. Suppose that T is proper and that for each point x_0 of X, there exists a neighborhood N of x_0 in X such that $T^{-1}(N)$ is bounded.*

Then $R(T) = X$, and for each x_0 in X, $T^{-1}(x_0)$ is a compact R_δ.

PROOF OF THEOREM (5.11). This follows from Theorem (5.9) combined with the proof of Theorem (5.6). q.e.d.

6. **Fixed point and mapping theory for compact multi-valued mappings.** The discussion and results of the previous sections have been based upon arguments of a metric and point-set topological character. At this point of our discussion, we begin the application of methods based upon the most elementary of the basic results of algebraic topology, the Brouwer fixed point theorem. We use the latter in the following standard form: *A continuous self-mapping f of a finite-dimensional*

simplex σ into itself must have at least one fixed point. We recall that by a simplex, one means the convex span of a finite set of linearly independent points in a Euclidean space, and that the Brouwer theorem is essentially equivalent to the simple fact that the homology groups of the sphere are not all trivial for positive dimension.

Our basic technique of argument is to relate the generalizations of the Brouwer theorem to infinite dimensions (like the theorem of Schauder-Tychonoff and its generalizations which we consider below) involving fixed point theorems for mappings of a subset K of a locally convex topological vector space X into X, to other theorems concerning the properties of mappings of K into X', the space of continuous linear functionals on X. We obtain thereby an extremely simple and useful interplay between fixed point and mapping theorems based upon the separation theory for convex subsets of locally convex topological vector spaces (i.e., essentially upon the Hahn-Banach theorem) as combined with the Brouwer fixed point theorem.

The first and simplest example of such a mapping theorem is the following:

THEOREM (6.1). *Let X be a locally convex topological vector space, K a compact convex subset of X, f a continuous mapping of K into X'. (We denote the pairing between X' and X by (w, x) for w in X' and x in X.)*

Then there exists an element x_0 of K such that for all x in K,

$$(6.1) \qquad (f(x_0), x - x_0) \geq 0.$$

PROOF OF THEOREM (6.1). Suppose that the assertion of the theorem were false for a given mapping f. Then for each x_0 in K, there would exist an element $y = y(x_0)$ in K such that $(f(x_0), y - x_0) < 0$. For a given y in K, we consider the subset N_y of K of those x_0 for which y serves as above, i.e., $N_y = \{x \mid x \in K, (f(x), y - x) < 0\}$. By our preceding remark, it follows that the family $\{N_y : y \in K\}$ is a covering of the compact set K.

For each fixed y in K, we consider the real-valued function g_y on K given by

$$g_y(x) = (f(x), y - x) \qquad (x \in K).$$

We assert that g_y is continuous on K. Indeed, let x and u be two points of K. Then

$$g_y(x) - g_y(u) = (f(x) - f(u), y - u) + (f(x), u - x).$$

We hold x fixed in K, and shall show that for u in some neighborhood of x in K, we may make $|g_y(x) - g_y(u)| < \epsilon$ for a prescribed constant $\epsilon > 0$. Since the set $y - K$ is compact and hence bounded in the topological vector space X and since convergence in X' is uniform convergence on bounded subsets of X, it follows from the continuity of the mapping f from K to X' that we can choose a neighborhood U_1 of x in K such that for u in U_1 $|(f(x) - f(u), y - u)| < \epsilon/2$. Obviously, we may choose a neighborhood U_2 of x in K such that for u in U_2, $|(f(x), u - x)| < \epsilon/2$. Hence for u in $U_1 \cap U_2$, $|g_y(x) - g_y(u)| < \epsilon$, and g_y is continuous on K.

Since g_y is continuous and since $N_y = g_y^{-1}((-\infty, 0))$, it follows that each N_y is an open subset of K, i.e., the family $\{N_y : y \in K\}$ is an open covering of K. Hence, since K is compact, we may choose a finite subcovering $\{N_{y_1}, \ldots, N_{y_r}\}$ of K and a partition of unity subordinated to this covering, i.e., a family $\{\alpha_1, \ldots, \alpha_r\}$ of continuous real-valued functions on K such that:

$$0 \leq \alpha_j(x) \leq 1 \quad (x \in K, 1 \leq j \leq r);$$

$\alpha_{j_r}(x) = 0$ outside some closed subset of N_{y_j} $\quad (1 \leq j \leq r);$

$$\sum_{j=1}^{r} \alpha_j(x) = 1 \quad (x \in K).$$

Using this partition of unity, we define a mapping p of K into X by setting $p(x) = \sum_{j=1}^{r} \alpha_j(x) y_j$. By the definition of the α_j, p is a continuous mapping of K into X. Since each y_j is an element of the convex set K and $p(x)$ is a convex linear combination of the finite set of points $\{y_1, \ldots, y_r\}$, $p(x)$ lies in K for each x in K. More precisely, $p(x)$ lies in the finite-dimensional convex set K_0 spanned by $\{y_1, \ldots, y_r\}$ and p is a continuous self-mapping of K_0. Hence there exists a point x_1 in K_0 by the Brouwer theorem such that $p(x_1) = x_1$.

For each x in K, we note that

$$(f(x), p(x) - x) = \sum_{j=1}^{r} \alpha_j(x)(f(x), y_j - x) < 0$$

since each nonvanishing summand must be negative and at least one summand does not vanish for each x in K. However, it follows that $0 = (f(x_1), p(x_1) - x_1) < 0$, which is a contradiction proving the theorem. q.e.d.

An extension of Theorem (6.1) is the following result (Theorem (6.2)) based upon the following definition:

DEFINITION (6.1). *Let X be a locally convex topological vector space, X' its conjugate space, M a subset of $X \times X'$. Then M is said to be a monotone set if for each pair $[x, y]$ and $[u, w]$ in M, $(y - w, x - u) \geq 0$ (i.e., M is the graph of a monotone mapping from X to $2^{X'}$).*

THEOREM (6.2). *Let X be a locally convex topological vector space, K a compact convex subset of X, f a continuous mapping of K into X', the space of continuous linear functionals on X. Let M be a monotone subset of $K \times X'$ (i.e., a monotone subset of $X \times X'$ such that for all $[x, y]$ in M, x lies in K).*

Then there exists an element x_0 of K such that for all $[u, w]$ in M,

(6.2) $$(f(x_0) - w, x_0 - u) \geq 0.$$

PROOF OF THEOREM (6.2). Suppose that the assertion of the theorem were false for a given mapping f and a given monotone set M. Then for each x_0 in K, there would exist an element $[u, w]$ of M such that $(f(x_0) - w, x_0 - u) < 0$. For each pair $[u, w]$ of M, we set

$$N_{u,w} = \{x \mid \in K, (f(x) - w, x - u) < 0\}.$$

By our preceding remark, the family $\{N_{u,w}:[u,w] \in M\}$ is a covering of the compact set K. Since $(f(x) - w, x - u) = -g_u(x) + (w, u - x)$, where g_u is the function defined in the proof of Theorem (6.1), it follows from that proof that the function $h_{u,w}$ given by $h_{u,w}(x) = (f(x) - w, x - u)$ for a fixed $[u, w]$ in M is continuous on K. Hence each $N_{u,w}$ is an open subset of K.

By the compactness of K, we may choose a finite subcovering $\{N_{u_1,w_1}, \ldots, N_{u_r,w_r}\}$ of K and a partition of unity $\{\beta_1, \ldots, B_r\}$ subordinated to this covering. We now form two mappings p and q by setting

$$p(x) = \sum_{j=1}^{r} \beta_j(x) u_j, \quad q(x) = \sum_{j=1}^{r} \beta_j(x) w_j.$$

Since p is a continuous mapping of the convex span of the finite subset $\{u_1, \ldots, u_r\}$ of K into itself, it has a fixed point x_1 in K by the Brouwer fixed point theorem, i.e., $x_1 - p(x_1) = 0$.

On the other hand, we consider

$$\lambda(x) = (f(x) - q(x), x - p(x)) \qquad (x \in K).$$

We may write $\lambda(x)$ in the form

$$\lambda(x) = \sum_{j,k=1}^{r} \beta_j(x) \beta_k(x)(f(x) - w_j, x - u_k)$$

and decompose the double sum in two pieces $\lambda(x) = \lambda_1(x) + \lambda_2(x)$, where

$$\lambda_1(x) = \sum_{j=1}^{r} [\beta_j(x)]^2 (f(x) - w_j, x - u_j),$$

and

$$\lambda_2(x) = \sum_{j \neq k} \beta_j(x) \beta_k(x)(f(x) - w_j, x - u_k).$$

For the first sum, we have $\lambda_1(x) < 0$ since whenever $\beta_j(x) \neq 0$, the corresponding factor $(f(x) - w_j, x - u_j) < 0$ since x must lie in N_{u_j,w_j}, while at least one $\beta_j(x)$ is nonnull for each x in K. For the second sum, we note that

$$\lambda_2(x) = \sum_{1 \leq j < k \leq r} \beta_j(x) \beta_k(x)[(f(x) - w_j, x - u_k) + (f(x) - w_k, x - u_j)].$$

For the term in square brackets, we obtain

$(f(x) - w_j, x - u_k) + (f(x) - w_k, x - u_j)$
$= (f(x) - w_j, x - u_j) + (f(x) - w_k, x - u_k) + (w_k - w_j, u_j - u_k) \leq 0$

whenever $\beta_j(x)\beta_k(x) \neq 0$ since then x lies in $N_{u_j,w_j} \cap N_{u_k,w_k}$ and we have

$$(f(x) - w_j, x - u_j) < 0, \qquad (f(x) - w_k, x - u_k) < 0,$$

while

$$(w_k - w_j, u_j - u_k) = -(w_k - w_j, u_k - u_j) \leq 0$$

by the monotonicity of the set M. Hence $\lambda(x) = \lambda_1(x) + \lambda_2(x) < 0$ $(x \in K)$. Finally, $0 = (f(x_1) - q(x_1), x_1 - p(x_1)) = \lambda(x_1) < 0$, which yields a contradiction proving the theorem. q.e.d.

THEOREM (6.3). *Let X be a locally convex topological vector space, K a compact convex subset of X, T an upper-semicontinuous mapping of K into 2^X such that for each x in K, $T(x)$ is a nonempty closed convex subset of X. For each x in K, let $W_K(x)$ be the closure in X of the set $\{y \mid y \in X,\text{ there exists } \lambda > 0 \text{ such that } x + \lambda y \in K\}$. Suppose that for each x in K, there exists w in $W_K(x)$, y in $T(x)$, and $\xi \geq 0$ such that $y - x = \xi w$.*

Then there exists x_0 in K such that $x_0 \in T(x_0)$ (i.e., x_0 is a fixed point of T in the usual sense for multi-valued mappings T).

COROLLARY 1 TO THEOREM (6.3). *Let K be a compact convex subset of the locally convex space X, T an upper-semicontinuous mapping of K into 2^X such that for each x in K, $T(x)$ is a nonempty closed convex subset of X. Suppose that for each x in K, there exists u in K, and $\xi \geq 0$ such that $x + \xi(u - x) \in T(x)$. Then T has a fixed point in K.*

COROLLARY 2 TO THEOREM (6.3). *Let K be a compact convex subset of the locally convex topological vector space X. Let T be an upper-semicontinuous mapping of K into 2^X such that for each x in K, $T(x)$ is a closed convex subset of X whose intersection with K is nonempty. Then T has a fixed point in K.*

PROOF OF COROLLARY 1 TO THEOREM (6.3). For each x in K, $W_K(x)$ includes all points w of the form $w = u - x$, where u is any element of K. q.e.d.

PROOF OF COROLLARY 2 TO THEOREM (6.3). If $T(x) \cap K$ includes a point u for a given x in K, then $x + (u - x)$ lies in $T(x)$. q.e.d.

PROOF OF THEOREM (6.3). Suppose that the conclusion of Theorem (6.3) is false for a given mapping T. Then for each x in K, 0 does not lie in the closed convex subset $x - T(x)$ of X. Hence there exists a linear functional v in X' such that $(v, x - y) > \delta > 0$ $(y \in T(x))$, with v and δ depending upon x in K.

For each v in X', we let

$$N_{v,\delta} = \{x \mid x \in K, (v, x - y) > \delta, y \in T(x)\}.$$

By our preceding remarks, the family $\{N_v : v \in X'\}$ is a covering of the compact set K. On the other hand, for $0 < \gamma < \delta$, $N_{v,\delta}$ contains N_v in its interior. Hence Int $(N_{v,\gamma})$ forms an open covering of K.

By the compactness of K, we may extract a finite subcovering $\{N_{v_1}, \ldots, N_{v_s}\}$ of K and a corresponding partition of unity $\{\psi_1, \ldots, \psi_s\}$ subordinated to this covering. Using this partition of unity, we define a continuous mapping q of K into X' by setting $q(x) = \sum_{j=1}^{s} \psi_j(x) v_j$. We note that for each x in K, y in $T(x)$,

$$(q(x), x - y) = \sum_{j=1}^{s} \psi_j(x)(v_j, x - y) > 0$$

since for any of the nonempty set of indices j such that $\psi_j(x) \neq 0$, x lies in N_{v_j} and $(v_j, x - y) > 0$ for all y in $T(x)$.

Since q is a continuous mapping of K into X', we may apply Theorem (6.1) to the mapping q and thereby obtain a point x_0 in K such that for all x in K, $(q(x_0), x - x_0) \geq 0$. Since $(q(x_0), x - x_0)$ is a continuous function of x in X, it follows that for any w in $W_K(x_0)$, $(q(x_0), w) \geq 0$ (since each neighborhood N of w in X includes a point w' such that for some $\lambda > 0$, $x_0 + \lambda w'$ lies in K, so that $(q(x_0), \lambda w') = \lambda(q(x_0), w') \geq 0$, and $(q(x_0), w') \geq 0$).

On the other hand, we may find a point y in $T(x_0)$, an element w in $W_K(x_0)$ and $\xi > 0$ such that $y = x_0 + \xi w$. Hence

$$0 < (q(x_0), x_0 - y) = -\xi(q(x_0), w).$$

For this w, we obtain $(q(x_0), w) < 0$, which is a contradiction proving the theorem.
q.e.d.

THEOREM (6.4). *Let X and Y be locally convex topological vector spaces, K a compact convex subset of X, K_1 a compact convex subset of Y. Let T and S be two mappings of K into 2^{K_1} such that the following conditions hold:*

(i) *T is an upper-semicontinuous mapping of K into 2^{K_1} such that for each x in K, $T(x)$ is a nonempty closed convex subset of K_1.*

(ii) *For each x in K, $S(x)$ is an open subset of K_1. For each y in K_1, $S^{-1}(y)$ is a nonempty convex subset of K.*

Then there exists x_0 in K such that $T(x_0) \cap S(x_0) \neq \emptyset$.

PROOF OF THEOREM (6.4). For each y in K_1, $S^{-1}(y)$ is nonempty. Hence the family $\{S(x) : x \in K\}$ is an open covering of the compact set K_1 and we can extract a finite subcovering $\{S(x_1), \ldots, S(x_r)\}$. Let $\{\alpha_1, \ldots, \alpha_r\}$ be a partition of unity on K_1 subordinated to this covering. We form the mapping s of K_1 into K by setting

$$s(y) = \sum_{j=1}^{r} \alpha_j(y) x_j.$$

Since the points $\{x_1, \ldots, x_r\}$ all lie in K, $s(y)$ being a convex linear combination of points of the convex set K must itself lie in K. Since the α_j are all continuous, s is a continuous mapping of K_1 into K. Moreover, for each y in K_1, $\alpha_j(y)$ is nonnull only if y lies in $S(x_j)$, i.e., only if $x_j \in S^{-1}(y)$. Since $S^{-1}(y)$ is convex for each y in K_1, it follows that $s(y)$ lies in $S^{-1}(y)$ for each y, i.e., $y \in S(s(y))$ $(y \in K_1)$.

We now define a mapping R of K_1 into 2^{K_1} by setting $R(y) = T(s(y))$. It follows from the properties of T that for each y in K_1, $R(y)$ is a nonempty closed convex subset of K_1. Since s is continuous and T is upper-semicontinuous, it follows that R is upper-semicontinuous from K_1 to 2^{K_1}. Hence, we may apply Corollary 2 to Theorem (6.3), and we conclude that there exists a point y_0 in K_1 such that $y_0 \in R(y_0)$, i.e., $y_0 \in T(s(y_0))$. Since y_0 also lies in $S(s(y_0))$, it follows that for $x_0 = s(y_0)$, $T(x_0) \cap S(x_0) \neq \emptyset$.
q.e.d.

We shall now apply Theorem (6.4) to obtain an extension of Theorem (6.2) to the case where the mapping f of that theorem is multi-valued. To carry through this extension, we shall use the following auxiliary result:

PROPOSITION (6.1). *Let K be a convex subset of the locally convex topological vector space X, K_1 a compact convex subset of X'. Suppose that for each w in K_1, there exists u in K such that $(w, u) < 0$.*

Then there exists u_0 in K such that $(w, u_0) < 0$ for all w in K_1.

PROOF OF PROPOSITION (6.1). Suppose the assertion were not true. Then for each u in K, there exists w in K_1 such that $(w, u) \geq 0$. For each u in K, we set $T(u) = \{w \mid w \in K_1, (w, u) \geq 0\}$. For each u in K, $T(u)$ is a closed convex subset of K_1 which is nonempty by our preceding remark. We assert that T is an upper-semicontinuous mapping of K into 2^{K_1}. Indeed, suppose not. Then there would exist u_0 in K and a neighborhood V of $T(u_0)$ in K_1 such that for each neighborhood U of u_0 in K, $T(U) \cap (K_1 - V) \neq \varnothing$. Since K_1 is compact, it follows that there exists a point w in $K_1 - V$ such that w lies in the closure of $T(U)$ for each neighborhood U of u_0 in K. For a given $\epsilon > 0$, choose the neighborhood U of u_0 so small that for u in U, and all y in K_1, $|(y, u - u_0)| < \epsilon$. Choose a neighborhood V' of w so small that for w' in V', $|(w - w', x)| < \epsilon$ for all x in K. We may choose such a point w' in $V' \cap T(U)$. Hence $(w', u) \geq 0$ for some u in U. Thus

$$(w, u_0) = (w, u_0 - u) + (w - w', u) + (w', u) > -2\epsilon.$$

Since $\epsilon > 0$ is arbitrary, it follows that $(w, u_0) \geq 0$, and w lies in $T(u_0)$ which contradicts the fact that w does not lie in the neighborhood V of $T(u_0)$. Hence T is upper-semicontinuous.

We define another mapping S of K into 2^{K_1} by setting

$$S(u) = \{w \mid w \in K_1, (w, u) < 0\} \qquad (u \in K).$$

For each u in K, $S(u)$ is an open subset of K_1. For each w in K_1, $S^{-1}(w)$ is convex and it is nonempty by the hypothesis of the proposition.

We now apply Theorem (6.4) to this pair of mappings T and S to conclude that there exist u_0 in K, w_0 in K_1 such that $w_0 \in T(u_0) \cap S(u_0)$, i.e., $(w, u_0) \geq 0$, $(w, u_0) < 0$, which is a contradiction proving the proposition.   q.e.d.

For convenience, we prove the following further auxiliary result:

PROPOSITION (6.2). *Let K and K_1 be two compact topological spaces, T a mapping of K into 2^{K_1} with $T(x)$ closed for each x in K. Then a necessary and sufficient condition that T be an upper-semicontinuous mapping from K to 2^{K_1} is that the graph of T be a closed subset of $K \times K_1$ (where $G(T)$, the graph of T, is defined as usual by $G(T) = \{[u, w] \mid u \in K, w \in K_1, w \in T(u)\}$).*

PROOF OF PROPOSITION (6.2). Suppose first that T is upper-semicontinuous and that $[u, w]$ is a point of $K \times K_1$ outside of $G(T)$. Then w is not a point of the closed set $T(u)$. Since K_1 is regular, there exist neighborhoods V_1 of w and V_2 of

$T(u)$ which are disjoint. Since T is upper-semicontinuous, there exists a neighborhood U of u in K such that for x in U, $T(x) \subset V_2$. Hence $U \times V_1$ is a neighborhood of $[u, w]$ in $K \times K_1$ which does not intersect $G(T)$, i.e., $G(T)$ is closed in $K \times K_1$.

Suppose conversely that $G(T)$ is closed in $K \times K_1$ and hence compact. Let u be a point in K, V an open neighborhood of $T(u)$. Suppose that T were not upper-semicontinuous at u with respect to V. Then for each neighborhood U of u in K, we would have

$$G(T) \cap (\text{cl}\,(U) \times (K_1 - V)) \neq \varnothing.$$

This is a family of compact sets having the finite intersection property. Hence

$$G(T) \cap \bigcap_U (\text{cl}\,(U) \times (K_1 - V)) \neq \varnothing.$$

For a point $[x, y]$ of this intersection, $x = u$ and $y \in T(u) - V$, which contradicts the assumption that $T(u) \subset V$. Hence T is upper-semicontinuous from K to 2^{K_1}.

q.e.d.

THEOREM (6.5). *Let X be a locally convex topological vector space, K a compact convex subset of X, K_1 a compact convex subset of X'. Let T be an upper-semicontinuous mapping of K into 2^{K_1} such that for all x in K, $T(x)$ is a nonempty closed convex subset of K_1. Let M be a monotone subset of $K \times X'$.*

Then there exist u_0 in K, w_0 in $T(u_0)$ such that for all $[u, w]$ in M,

(6.3) $$(w_0 - w, u_0 - u) \geq 0.$$

PROOF OF THEOREM (6.5). For each finite subset $F = \{[u_1, w_1], \ldots, [u_r, w_r]\}$ of M and for each $\epsilon > 0$, let

$$H_{\epsilon, F} = \{[u, w] \mid [u, w] \in G(T), (w - w_j, u - u_j) \geq -\epsilon, 1 \leq j \leq r\}.$$

Each $H_{\epsilon, F}$ is compact and the intersection of the $H_{\epsilon, F}$ for all finite subsets F of M and all $\epsilon > 0$ is exactly the set of $[u_0, w_0]$ in $G(T)$ which satisfy the inequality (6.3) above for all elements of M. Hence since the intersection of a finite number of the $H_{\epsilon, F}$ is itself one of this family of sets, it suffices to prove that for each $\epsilon > 0$, and each finite subset F of M, $H_{\epsilon, F}$ is nonempty.

We define a mapping S of K into 2^{K_1} by setting

$$S(u) = \{w \mid w \in K_1, (w - w_j, u - u_j) > -\epsilon, 1 \leq j \leq r\}$$

for each u in K. We note that for each u in K, $S(u)$ is an open subset of K_1. For each w in K_1,

$$S^{-1}(w) = \{u \mid u \in K, (w - w_j, u - u_j) > -\epsilon\}$$

is the intersection of half-spaces and hence is convex. Since $S^{-1}(w)$ contains the set

$$Z_w = \{u \mid u \in K, (w - y, u - x) \geq 0 \text{ for all } [x, y] \text{ in } M\}$$

to show that $S^{-1}(w)$ is nonempty for each w in K_I, it suffices to show that the set Z_w is nonempty. However, Z_w corresponds to the same problem for the single-valued function $T(x) = w$ with a single constant value. Hence Z_w is nonempty by Theorem (6.2).

Finally, we apply Theorem (6.4) to the pair of mappings T and S to obtain an element u_0 such that $T(u_0)$ and $S(u_0)$ intersect. For the pair $[u_0, w_0]$, however, it follows that $[u_0, w_0]$ lies in $H_{\epsilon,F}$ and the latter set is nonempty. Hence the proof of the theorem is complete. q.e.d.

7. **Monotone mappings in Banach spaces.** We turn in the present section to the study of one of the principal classes of nonlinear noncompact mappings in Banach spaces encountered in applications, the theory of monotone mappings from a reflexive Banach space X to its conjugate space X^*.

DEFINITION (7.1). *Let X be a Banach space, G a closed convex subset of X, T a mapping of G into 2^{X^*}. Then*

(a) *T is said to be monotone from G to 2^{X^*} if for each x and u in G, each y in $T(x)$, and each w in $T(u)$, $(y - w, x - u) \geq 0$.*

(b) *T is said to be maximal monotone from G to 2^{X^*} if T is monotone and if, in addition, there exists no mapping T_1 from G to 2^{X^*} which is monotone, such that $G(T) \subset G(T_1)$, and T_1 is different from T.*

(c) *For each such T, $D(T)$, the effective domain of T, and $G(T)$, the graph of T, are the subsets of G and $G \times X^*$, respectively, defined as*

$$D(T) = \{u \mid u \in G, T(u) \text{ is nonempty}\},$$

$$G(T) = \{[u, w] \mid [u, w] \in G \times X^*, w \in T(u)\}.$$

DEFINITION (7.2). *Let X be a Banach space, G a closed convex subset of X, T a mapping of G into 2^{X^*}. Then T is said to be finitely continuous if the following conditions all hold:*

(a) *$D(T) = G$, and for each x in G, $T(x)$ is a nonempty closed convex set.*

(b) *For each finite subset $f = \{x_1, \ldots, x_r\}$ of G, if G_f is the convex closure of f, then T is an upper-semicontinuous mapping of G_f into 2^{X^*}, with X^* given its weak* topology.*

DEFINITION (7.3). *Let X be a Banach space, G a closed convex subset of X, T a mapping of G into 2^{X^*}, w_0 an element of X^*. Then T is said to be coercive on G with respect to w_0 if for some $R > 0$ and for all x with $\|x\| > R$, $(T(x), x) - (w_0, x) > 0$.*

DEFINITION (7.4). *A mapping T of G into 2^{X^*} is said to be bounded if for each $R > 0$, there exists $k(R) > 0$ such that for each u with $\|u\| \leq R$, there exists w in $T(u)$ with $\|w\| \leq k(R)$.*

In terms of this set of definitions, we can state our basic result on monotone mappings, from which many other results will be derived later in the section:

THEOREM (7.1). *Let X be a reflexive Banach space, G a closed convex subset of X, T and T_0 two monotone mappings of G into 2^{X^*}. We assume for simplicity that $0 \in G$ and that $0 \in T(0)$. Suppose that T_0 is finitely continuous, bounded, and coercive with respect to a given element w_0 of X^*.*

Then there exists a function g independent of w_0 and an element $[u_0, z_0]$ in $G(T_0)$ with $\|u_0\| \leq g(\|w_0\|)$, $\|z_0\| \leq g(\|w_0\|)$ such that for all $[x, y]$ in $G(T)$,

(7.1) $$(y + z_0 - w_0, x - u_0) \geq 0.$$

For the single-valued case for the mappings T and T_0, the inequality (7.1) becomes the slightly more transparent inequality

(7.2) $\quad (T(x) + T_0(u_0) - w_0, x - u_0) \geq 0 \qquad (x \in D(T))$,

and is interpreted as indicating that u_0 is a "weak solution" of the variational inequality

(7.3) $\quad (T(u_0) + T_0(u_0) - w_0, x - u_0) \geq 0 \qquad (x \in G)$.

The latter inequality is itself a generalization of the simpler functional equation $w_0 = (T + T_0)(u_0)$ for the single-valued case, or $w_0 \in (T + T_0)(u_0)$ for the multi-valued case, a generalization corresponding to the restriction that the solution must lie in the given convex set G. We indicate below in detail in the discussion of the application of this result to variational problems how the introduction of variational inequalities in place of functional equations is analogous to the consideration of variational problems on convex sets in X rather than on X itself.

We shall preface the proof of Theorem (7.1) with the following auxiliary result:

PROPOSITION (7.1). *Let C be a closed bounded convex subset in a reflexive Banach space Y, C_1 a convex subset of the conjugate space Y^*. Suppose that for each w in C_1, there exists an element u of C such that $(w, u) \geq 0$. Then there exists an element x_0 of C such that $(w, x_0) \geq 0$ for all w in C_1.*

PROOF OF PROPOSITION (7.1). Suppose the conclusion of the proposition were false. Then for each x in C, there would exist $w = w(x)$ in C_1 such that $(w, x) < 0$. For each fixed w in C_1, let $N_w = \{x \mid x \in C, (w, x) < 0\}$. Then N_w is an open set in the weak topology on C, and each point x of C is contained in at least one N_w. By our hypothesis that Y is reflexive and C is bounded, closed, and convex, it follows that C is weakly compact. Hence there exists a finite subcovering of C by sets $\{N_{w_1}, N_{w_2}, \ldots, N_{w_r}\}$. Let $\{\alpha_1, \ldots, \alpha_r\}$ be a partition of unity in the weak topology on C, subordinated to this covering. Then each α_j is a weakly continuous mapping from C to the interval $[0, 1]$ with support in N_{w_j} and such that $\sum_{j=1}^{r} \alpha_j(x) = 1$ for each x in C.

We form the mapping p of C into C_1 given by

$$p(x) = \sum_{j=1}^{r} \alpha_j(x) w_j.$$

Then p is continuous from the weak topology on C to the strong topology on the finite-dimensional subset of C_1 which is the convex closure of the finite set of points $w_1, \ldots, w_r$. For each x in C, we know moreover that

$$(p(x), x) = \sum_{j=1}^{r} \alpha_j(x)(w_j, x) < 0$$

since $\alpha_j(x)$ vanishes unless x lies in N_{w_j}, if x lies in N_{w_j}, $(w_j, x) < 0$, and at least one $\alpha_j(x)$ is positive for each x in C.

We now form a mapping T from C_1 to 2^C by setting

$$T(w) = \{x \mid x \in C, (w, x) \geq 0\} \qquad (x \in C_1).$$

By hypothesis, $T(w)$ is nonempty for each w in C_1. By its definition, $T(w)$ is a weakly closed convex subset of C for each w in C_1. We assert moreover that T is upper-semicontinuous as a set-valued mapping from C_1 in its strong topology to 2^C, with C given its weak topology. Indeed, suppose that $w_j \to w_0$ in C_1 in the strong topology and that for a given weak neighborhood V of $T(w_0)$, $T(w_j) \cap (C - V)$ is nonempty for each j. If we choose a point y_j in $T(w_j) \cap (C - V)$ for each j and pass to an infinite subsequence, it follows from the fact that $C - V$ is weakly compact and hence weakly sequentially compact that we may assume y_j converges weakly to some y_0 in $C - V$. Hence $(w_j, y_j) \to (w_0, y_0) \geq 0$, i.e., $y_0 \in T(w_0) \cap (C - V)$. Since $T(w_0) \subset V$, this contradiction proves that T is upper-semicontinuous from the strong topology on C_1 to the weak topology of C.

We now define the composite mapping S from C to 2^C by $S(x) = T(p(x))$. Since p is single-valued and continuous from the weak topology on C to the strong topology on C_1 while T is upper-semicontinuous from the strong topology on C_1 to 2^C, with C given its weak topology, it follows that S is upper-semicontinuous as a set-valued function from C to 2^C, with C as both domain and range space given its weak topology. For every x in C, $S(x) = T(p(x))$ is a weakly-closed convex subset of C. Hence, we may apply Corollary 2 to Theorem (6.3) of §6 and we obtain the existence of a fixed point x_0 in C of the mapping S, i.e., $x_0 \in S(x_0)$. For such an x_0, we have $x_0 \in T(p(x_0))$, from which it follows that $(p(x_0), x_0) \geq 0$. Since $(p(x), x) < 0$ for all x in C, this is a contradiction proving the proposition. q.e.d.

Another useful proposition which simplifies the notation of the proof of Theorem (7.1) and is essential to its later generalizations is the following:

PROPOSITION (7.2). *Let X be a reflexive Banach space, A a bounded subset of X, x_0 a point in the weak closure of A. Then there exists an infinite sequence $\{x_k\}$ in A converging weakly to x_0 in X.*

PROOF OF PROPOSITION (7.2). We prove first that there exists a countable subset A_0 of A such that x_0 lies in the weak closure of A_0. For each integer n, let B^n be the product of n copies of the unit ball in X^*. Then B^n is compact in the product of the weak topologies. For each element $[w_1, \ldots, w_n]$ of B^n and each positive integer m, we may find an element x of A such that

$$|(w_j, x - x_0)| < m^{-1} \qquad (1 \leq j \leq n).$$

Moreover, this inequality holds on a weak neighborhood of the point $[w_1, \ldots, w_n]$ in B^n for the same choice of x. Since B^n is compact, we may find a finite set $F_{n,m}$ of points x such that for each $[w_1, \ldots, w_n]$ at least one point x of $F_{n,m}$ serves to verify the above inequality. Set $A_0 = \bigcup_{n,m} F_{n,m}$, the union being taken over all pairs of positive integers n and m. Then A_0 is countable, and it follows by the above prescription that x_0 lies in the weak closure of A_0.

Let X_0 be the separable closed subspace of X spanned by the points of A_0. Then x_0 lies in the closure of $A \cap X_0$, and X_0 is reflexive since it is a closed subspace of a reflexive space. To show that x_0 is the weak limit of a sequence from A_0 in X, it suffices to do the same in X_0. Since X_0 is separable and reflexive, however, the weak topology on X_0 is metrizable and each limit point of A_0 in the weak topology on X_0 is the weak limit of an infinite sequence from A_0. q.e.d.

PROOF OF THEOREM (7.1). We begin the proof of Theorem (7.1) by introducing the notation and definitions for an appropriate approximation argument using finite-dimensional subsets of G.

Let f be a finite subset of G. We let G_f be the convex closure of f, which is a finite-dimensional subset of G and compact in the topology induced by the topology of G. Let F_f be the finite-dimensional subspace of X spanned by f, j_f the injection mapping of F_f in X, and j_f^* the projection mapping of X^* onto F_f^* dual to j_f.

We shall always assume that 0 is an element of each finite subset f which we consider, so that 0 is always a point of G_f.

PROPOSITION (7.3). *Let X be a Banach space, G a closed convex subset of X, T and T_0 two mappings of G into 2^{X^*} such that T is monotone, T_0 is finitely continuous, bounded, and coercive with respect to a given element w_0 of X^* on G. Suppose that $0 \in G$ and that $0 \in T(0)$. For each finite subset f of G with $0 \in f$, let G_f, F_f, F_f^*, j_f, and j_f^* be defined as above. We define the mappings T_f and $T_{0,f}$ of G_f into 2^{F^*} by setting*

$$T_f(u) = j_f^*(T(u)), \qquad T_{0,f}(u) = j_f^*(T_0(u)) \qquad (u \in G_f).$$

Then there exist constants R_1 and $R_2 > 0$ independent of the choice of the finite subset f such that for any f, there exists at least one solution $[u_f, w_f]$ in $G(T_{0,f})$ of the inequality system:

$(7.1)_f$ $(y_f + w_f - j_f^*(w_0), x_f - u_f) \geq 0$ *(all $[x_f, y_f]$ in $G(T_f)$ with $\|u_f\| \leq R_1$*

and $w_f = j_f^*(z_f)$, for an element z_f in $T_0(u_f)$ with $\|z_f\| \leq R_2$).

PROOF OF PROPOSITION (7.3). By hypothesis, there exists a constant $R > 0$ such that for $\|u\| > R$, $u \in G$, $w \in T_0(u)$, we have $(w, u) - (w_0, u) > 0$. We choose $R_1 = R$.

Let R_0 be a constant with $R_0 > R$. By hypothesis, T_0 is bounded so that there exists $k(R_0) > 0$ such that for each u in G with $\|u\| \leq R_0$, there exists w in $T_0(u)$ with $\|w\| \leq k(R_0)$. Let $B_{k(R_0)}$ be the closed ball in X^* with center at 0 and radius $k(R_0)$. Since each G_f is bounded, we may find a constant $k_f \geq k(R_0)$ such that for each u in G_f, there exists w in $T_0(u)$ such that $\|w\| \leq k_f$. We now define a new mapping $T'_{0,f}$ of G_f into 2^{F_f} by setting

$$\begin{aligned}T'_{0,f}(u) &= j_f^*(T_0(u) \cap B_{k(R_0)}), & \|u\| &< R_0 \\ &= j_f^*(T_0(u) \cap B_{k_f}), & \|u\| &\geq R_0\end{aligned} \qquad (u \in G_f).$$

I.e., $T'_{0,f}(u) = j_f^*(T'_0(u))$, where

$$T'_0(u) = T_0(u) \cap B_{k(R_0)}, \quad \|u\| < R_0,$$
$$= T_0(u) \cap B_{k_f}, \quad \|u\| \geq R_0.$$

We verify from the properties of T_0 and the choices of the constants $k(R_0)$ and k_f that for each u in G_f, $T'_0(u)$ is a nonempty closed bounded subset of X^* and hence weakly compact. Moreover, T'_0 is an upper-semicontinuous set-valued mapping from G_f to 2^{X^*} with X^* given its weak topology, as follows immediately from the corresponding property for T_0. Since j_f^* is continuous from the weak topology of X^* to F_f^*, it follows that for each u in G_f, $T'_{0,f}(u)$ is a nonempty convex subset of F_f^* which is compact and hence closed. Moreover, the mapping $T'_{0,f}$ is upper-semicontinuous from G_f to 2. Since $G(T'_{0,f}) \subset G(T_{0,f})$, it suffices to prove Proposition (7.3) for $T'_{0,f}$.

We now apply Theorem (6.5) of §6 with the mapping T of that theorem replaced by $(-T'_{0,f})$, X by F_f, X' by F_f^*, and with the monotone set M given as $G(T_f) - j_f^*(w_0)$. By our construction above, each $T'_{0,f}(u)$ is a closed convex subset of $j_f^*(B_{k_f})$ which is a compact convex subset in F_f^*. The conclusion of that theorem tells us that there exists a point $[u_f, w_f]$ in $G(T'_{0,f})$ with $w_f = j_f^*(z_f)$, $z_f \in T'_0(u_f)$ such that for all $[x_f, y_f]$ in $G(T_f)$, we have $(y_f + w_f - j_f^*(w_0), x_f - u_f) \geq 0$.

In particular, we can choose $x_f = 0$, $y_f = 0$ since $[0, 0] \in G(T)$. We obtain $0 \leq (j_f^*(z_f - w_0), -u_f) = -(z_f - w_0, u_f)$, i.e., $(z_f - w_0, u_f) \geq 0$. Since z_f lies in $T'_0(u_f) \cap T_0(u_f)$, it follows that $\|u_f\| \leq R < R_0$. Hence $\|z_f\| \leq k(R_0)$. Since $R = R_1$ and if we set $R_2 = k(R_0)$, our desired inequalities on the norms of u_f and z_f are fulfilled. q.e.d.

PROOF OF THEOREM (7.1) CONTINUED. For each finite subset f_0 of G with 0 in f_0, we may form the subset W_{f_0} of $G \times X^*$ given by

$$W_{f_0} = \bigcup_{f_0 \subset f} \{[u_f, z_f]\}$$

where $[u_f, z_f]$ is an element of $G \times X^*$ which satisfies the conclusions of Proposition (7.3) chosen for each f. Since X and X^* are reflexive so that closed balls are weakly compact, it follows from the fact that the family of sets W_{f_0} satisfies the finite intersection property that we can find a point $[u_0, v_0]$ of $G \times X^*$ which lies in the weak closure of W_{f_0} for each f_0.

Let $[x, y]$ be a given point of $G(T)$ and choose the finite subset f_0 of G to consist of the three point set $\{0, u_0, x\}$. For each f containing f_0, we have $[x, j_f^*(y)]$ in $G(T_f)$, and we obtain

$$0 \leq (j_f^*(y + z_f - w_0), x - u_f) = (y + z_f - w_0, x - u_f).$$

On the other hand, for any $[u, w]$ in $G(T_0)$, it follows from the monotonicity of T_0 that $(w - z_f, u - u_f) \geq 0$. Adding the last inequality to the preceding one, we obtain

$$0 \leq (y + z_f - w_0, x - u_f) + (w - z_f, u - u_f)$$
$$= (y, x - u_f) + (z_f, x - u) + (w, u - u_f) - (w_0, x - u_f).$$

Consider the function g on $(B_{R_1} \times B_{R_2}) \cap (G \times X^*)$ given by

$$g(\xi, \zeta) = (y, x - \xi) + (\zeta, x - u) + (w, u - \xi) - (w_0, x - \xi).$$

Since g is obviously continuous in the weak topology and is nonnegative for $[\xi, \zeta]$ in W_{f_0}, it must be nonnegative on the weak closure of W_{f_0}. In particular, $g(u_0, v_0) \geq 0$, i.e.,

$$0 \leq (y, x - u_0) + (v_0, x - u) + (w, u - u_0) - (w_0, x - u_0)$$
$$= (y + v_0 - w_0, x - u_0) + (w - v_0, u - u_0).$$

Hence the inequality

(7.4) $\qquad (y + v_0 - w_0, x - u_0) + (w - v_0, u - u_0) \geq 0$

holds for all $[x, y]$ in $G(T)$ and all $[u, w]$ in $G(T_0)$.

We may assume without loss of generality that the monotone mapping T from G to 2^{X^*} is actually maximal monotone from G to 2^{X^*} since by Zorn's lemma, each monotone mapping T is extendable to a maximal monotone mapping T_1 from G to 2^{X^*} and if we replace T by T_1, the requirements of the conclusion of Theorem (7.1) become more stringent.

We assert now that the inequality (7.4) holding for all $[x, y]$ in $G(T)$ and all $[u, w]$ in $G(T_0)$ implies the two following systems of inequalities:

(7.5) $\qquad (y + v_0 - w_0, x - u_0) \geq 0 \qquad ([x, y] \in G(T))$,

and

(7.6) $\qquad (w - v_0, u - u_0) \geq 0 \qquad ([u, w] \in G(T_0))$.

Indeed, suppose first that for some $[u, w]$ in $G(T_0)$, $(w - w_0, u - u_0) < 0$. Then by the inequality system (7.4), it follows that for all $[x, y]$ in $G(T)$, $(y + v_0 - w_0, x - u_0) > 0$. Since T is assumed to be maximal monotone from G to 2^{X^*}, it follows from the last inequality that $(w_0 - v_0) \in T(u_0)$, i.e., $[u_0, w_0 - v_0] \in G(T)$. Replacing $[x, y]$ by this pair, we obtain

$$0 = ((w_0 - v_0) + v_0 - w_0, u_0 - u_0) > 0,$$

which is a contradiction. Hence $(w - v_0, u - u_0) \geq 0$ $([u, w] \in G(T_0))$, and the inequality system (7.6) has been established.

Suppose, on the other hand, that for a given $[x, y]$ in $G(T)$, $(y + v_0 - w_0, x - u_0) < 0$. Then choosing an element of the form $[u_0, v]$ in $G(T_0)$, we obtain

$$0 \leq (y + v_0 - w_0, x - u_0) + (v - v_0, u_0 - u_0) < 0,$$

which is a contradiction. Hence

$$(y + v_0 - w_0, x - u_0) \geq 0 \qquad ([x, y] \in G(T)),$$

i.e., the inequality system (7.5) has been established.

To conclude the proof of Theorem (7.1), we must pass from the pair of inequality systems (7.5) and (7.6) to the inequality system (7.1). We do so as follows: Let

x be any element of G, and for $0 < t < 1$, let
$$u_t = (1-t)u_0 + tx = u_0 + t(x - u_0).$$
Since u_0 and x lies in G and G is convex, each u_t lies in G. Let w_t be a point of $T_0(u_t)$ and substitute the pair $[u_t, w_t]$ of $G(T_0)$ for $[u, w]$ in the inequality (7.6). We obtain $t(w_t - v_0, x - u_0) \geq 0$, i.e.,
$$(w_t - v_0, x - u_0) \geq 0.$$
Since T_0 is bounded, we may assume that there exists $R_3 > 0$ such that $\|w_t\| \leq R_3$ for all $t > 0$. For a subsequence $w_j = w_{t_j}$ with $t_j \to 0$, we may assume that $w_j \to w'$ as $j \to \infty$ in the weak topology on X^*. Hence $(w' - v_0, x - u_0) \geq 0$. Since T_0 is upper-semicontinuous as a set-valued function from G to 2^{X^*}, with X^* given the weak topology, w' lies in $T_0(u_0)$. Hence for each point $x - u_0$ in the convex subset $G - u_0$ of X, we can find an element $w' - v_0$ in the weakly compact convex set $(T(u_0) \cap B_{R_2}) - v_0$ in X^* such that $(w' - v_0, x - u_0) \geq 0$.

We now apply Proposition (7.1) with $Y = X^*$, $Y^* = X$, and the convex sets described above. By the result of that proposition, there exists an element z_0 of $T_0(u_0)$ such that for all x in G, $(z_0 - v_0, x - u_0) \geq 0$, i.e.,
$$(z_0, x - u_0) \geq (v_0, x - u_0) \qquad (x \in G).$$

Returning to the inequality system (7.5), we find that for each $[x, y]$ in $G(T)$,
$$(y + z_0 - w_0, x - u_0) \geq (y + v_0 - w_0, x - u_0) \geq 0,$$
and the inequality system (7.1) is valid for the element $[u_0, z_0]$ in $G(T_0)$. q.e.d.

THEOREM (7.2). *Let X be a reflexive Banach space, G a closed convex subset of X, T and T_0 monotone mappings of G into 2^{X^*} with T_0 finitely continuous, bounded, and coercive with respect to a given element w_0 of X^*. Suppose that T is maximal monotone from G to 2^{X^*} and $0 \in D(T)$.*

Then there exist u_0 in G, y_0 in $T(u_0)$ and z_0 in $T_0(u_0)$ such that $w_0 = y_0 + z_0$, i.e., $w_0 \in (T + T_0)(u_0)$.

PROOF OF THEOREM (7.2). By Theorem (7.1), there exists an element $[u_0, z_0]$ in $G(T_0)$ such that for all $[x, y]$ in $G(T)$
$$(y - (w_0 - z_0), x - u_0) \geq 0.$$
Since T is maximal monotone from G to 2^{X^*}, it follows that $[u_0, w_0 - z_0] \in G(T)$, i.e., $y_0 = w_0 - z_0$ satisfies the conditions of the present theorem. q.e.d.

THEOREM (7.3). *Let X be a reflexive Banach space, G a closed convex subset of X, T a maximal monotone mapping of G into 2^{X^*} which is coercive on G with respect to a given w_0 in X^* and with $0 \in D(T)$.*
Then there exists u_0 in G such that $w_0 \in T(u_0)$.

PROOF OF THEOREM (7.3). Let J be the mapping of X into 2^{X^*} given by
$$J(u) = \{w \mid w \in X^*, \|w\| = \|u\|, (w, u) = \|u\|^2\}.$$

J is the (multi-valued) duality mapping of X into 2^{X^*}. By the Hahn-Banach theorem, $J(u)$ is nonempty for each u in X. For each u, $J(u)$ may also be characterized as the set $\{w \mid \|w\| \leq \|u\|, (w, u) = \|u\|^2\}$ and hence is convex and weakly closed in X^*. Obviously, $J(u)$ is bounded for each u. Moreover, J is upper-semicontinuous from X to 2^{X^*}, with the weak topology on X^*. Indeed, suppose that u_j converges strongly to u_0 in X and that we have a sequence $\{w_j\}$ in X^* with $w_j \in J(u_j)$ and w_j lying outside of a weak neighborhood V of $J(u_0)$. Passing to an infinite subsequence we may assume that w_j converges weakly to an element w_0 of X^*. Then we have

$$\|w_0\| \leq \lim_j \|w_j\| = \lim_j \|u_j\| = \|u_0\|,$$

and

$$(w_0, u_0) = \lim_j (w_j, u_j) = \lim_j \|u_j\|^2 = \|u_0\|.$$

Hence w_0 is an element of $J(u_0)$, contradicting the fact that w_0 must lie in the exterior of V.

For a sequence $\epsilon_j \to 0$, we may consider the sequence of mappings $T + \epsilon_j J \cdot \epsilon J$ is coercive with respect to any w_0 in X^* on each closed convex subset G of X, for indeed

$$(\epsilon J(u) - w_0, u) = \epsilon \|u\|^2 - \|w_0\| \cdot \|u\| > 0$$

if $\|u\| > \epsilon^{-1} \|w_0\|$. Hence by Theorem (7.2), we may find an element u_j in G for and each j, corresponding elements z_j in $J(u_j)$ and w_j in $T(u_j)$ such that $w_0 = \epsilon_j z_j + w_j$.

Since T is coercive with respect to w_0, there exists a constant $R > 0$ such that for $\|u\| > R$ and all w in $T(u)(u - w_0, u) > 0$. On the other hand,

$$(w_j - w_0, u_j) = -\epsilon_j (z_j, u_j) \leq 0.$$

Hence, $\|u_j\| \leq R$ and since $\|z_j\| = \|u_j\|$, it follows that

$$\|w_0 - w_j\| = \epsilon_j \|z_j\| \leq \epsilon_j R \to 0 \qquad (j \to +\infty).$$

Since X is reflexive and the sequence $\{u_j\}$ is bounded, we may pass to an infinite subsequence and assume that u_j converges weakly to some element u_0 of G as $j \to \infty$. Since T is monotone, we know that for each $[x, y]$ in $G(T)$ and all j, $(y - w_j, x - u_j) \geq 0$. Since $(y - w_j)$ converges strongly to $(y - w_0)$ while $(x - u_j)$ converges weakly to $(x - u_0)$, it follows that $(y - w_j, x - u_j) \to (y - w_0, x - u_0)$. Thus for all $[x, y]$ in $G(T)$, $(y - w_0, x - u_0) \geq 0$. Since T is maximal monotone from G to 2^{X^*}, we have $w_0 \in T(u_0)$. q.e.d.

THEOREM (7.4). *Let X be a reflexive Banach space, G a closed convex subset of X, T and T_0 two monotone mappings of G into 2^{X^*} with T maximal monotone from G to 2^{X^*} and T_0 a finitely continuous mapping from G to 2^{X^*}. Let G_0 be a bounded convex subset of G such that $T_0(G_0)$ is bounded in X^*.*

Then $(T + T_0)(G_0)$ is closed in X^*.

PROOF OF THEOREM (7.4). We consider a sequence $\{u_j\}$ in G_0 and a corresponding sequence $\{w_j\}$ in X^* where $w_j \in (T + T_0)(u_j)$ and w_j converges strongly to w_0 in X^*. Since $w_j = v_j + z_j$ with $v_j \in T(u_j)$ and $z_j \in T_0(u_j)$ and since $T_0(G_0)$ is bounded, we have $\|z_j\| \leq M$. Passing to an infinite subsequence, we may assume without loss of generality that u_j converges weakly to some u_0 in G and that z_j converges weakly to some z_0 in X^*. It suffices to show that $w_0 \in (T + T_0)(u_0)$.

For each element $[x, y]$ in $G(T)$, we have

$$(y, u_j - x) \leq (v_j, u_j - x) = (v_j, u_j - u_0) + (v_j, u_0 - x).$$

We know that for any element z_0' in $T_0(u_0)$, $(z_0' - z_j, u_0 - u_j) \geq 0$. Since w_j converges strongly to w_0 and $(u_j - u_0)$ converges weakly to 0, $(w_j, u_j - u_0) \to 0$. Hence

$$(v_j, u_j - u_0) = (w_j, u_j - u_0) + (z_0' - z_j, u_j - u_0) - (z_0', u_j - u_0)$$

which implies that
$$\overline{\lim} \, (v_j, u_j - u_0) \leq 0.$$

On the other hand,

$$(v_j, u_0 - x) = (w_j - z_j, u_0 - x) \to (w_0 - z_0, u_0 - x),$$

and $(y, u_j - x) \to (y, u_0 - x)$. Combining these relations, we obtain

$$(y, u_0 - x) \leq (w_0 - z_0, u_0 - x)$$

for all $[x, y]$ in $G(T)$. Since T is maximal monotone from G to 2^{X^*}, it follows that $w_0 - z_0 \in T(u_0)$.

Let $w_1 = w_0 - z_0$. Then for each j,

$$(z_j, u_j - u_0) \leq (v_j - w_1, u_j - u_0) + (z_j, u_j - u_0)$$
$$\leq (w_j - w_1, u_j - u_0) \to 0,$$

so that $\overline{\lim} \, (z_j, u_j - u_0) \leq 0$. Hence for any $[u, w]$ in $G(T_0)$, we obtain

$$(w, u_j - u) \leq (z_j, u_j - u) = (z_j, u_j - u_0) + (z_j, u_0 - u).$$

Since

$$(w, u_j - u) \to (w, u_0 - u), \qquad (z_j, u_0 - u) \to (z_0, u_0 - u),$$

we find that $(w, u_0 - u) \leq (z_0, u_0 - u)$.

Let v be a given element of G, and let $u_t = u_0 + t(v - u_0)$ for $0 < t \leq 1$, $w_t \in T_0(u_t)$. From the last inequality, we find that $t(w_t, u_0 - v) \leq t(z_0, u_0 - v)$. If we cancel the positive factor t, we obtain $(w_t, u_0 - v) \leq (z_0, u_0 - v)$. If we consider a sequence $t_j \to 0$ and $w_j = w_{t_j}$, then by choosing a suitable weakly convergent subsequence, we may ensure that there exists an element w_2 of $T_0(u_0)$ such that $(w_0, u_0 - v) \leq (z_0, u_0 - v)$. From our construction, w_0 depends on the element v of G. Applying Proposition (7.2), however, it follows that there exists w_2 in $T_0(u_0)$ such that the inequality $(w_2, u_0 - v) \leq (z_0, u_0 - v)$ holds for all v in G.

Since $w_0 = w_1 + z_0$ with w_1 in $T(u_0)$, we know that for all $[x, y]$ in $G(T)$

$$(y - w_0 + z_0, x - u_0) \geq 0.$$

From the result of the preceding paragraph, we know that $(z_0, x - u_0) \leq (w_2, x - u_0)$. Hence $(y + w_2 - w_0, x - u_0) \geq 0$, and since T is maximal monotone from G to 2^{X^*}, it follows that $w_0 - w_2$ lies in $T(u_0)$. Since $w_2 \in T_0(u_0)$, we see that $w_0 \in T(u_0) + T_0(u_0)$. q.e.d.

THEOREM (7.5). *Let X be a reflexive Banach space, G a closed convex subset of X, T a maximal monotone mapping from G to 2^{X^*}, T_0 a finitely continuous, bounded monotone mapping from G to 2^{X^*}. Suppose that $[0, 0] \in G(T)$ and that $(T + T_0)$ is coercive on G with respect to a given element w_0 of X^*.*

Then there exists an element u_0 of G such that $w_0 \in (T + T_0)(u_0)$.

PROOF OF THEOREM (7.5). By the assumption of coercivity of $(T + T_0)$ with respect to w_0, there exists $R > 0$ such that for all u with $\|u\| > R$ in G and all w in $(T + T_0)(u)$, $(w - w_0, u) > 0$. By the assumption of boundedness of T_0, if R_1 is a fixed constant ($R_1 > R$), there exists a constant $K(R_1) > 0$ such that for every u of G with $\|u\| \leq R_1$, there exists an element w of $T_0(u)$ with $\|w\| \leq K(R_1)$. We now may decrease the set of values of T_0 for all u with $\|u\| < R_1$ by setting $T_0'(u) = T_0(u) \cap B_{K(R_1)}$. The resulting mapping is still finitely continuous and monotone, while for $\|u\| < R_1$, the image set $T_0'(u)$ is a subset of the ball $B_{K(R_1)}$. Since a proof of our theorem for T_0' implies a proof for the original T_0, we may assume that for our original mapping T_0, $T_0(B_R)$ is a bounded set in X^*.

Let J be the (multi-valued) duality mapping described in the proof of Theorem (7.3) and let $T_\epsilon = T_0 + \epsilon J$, $\epsilon > 0$. For each $\epsilon > 0$, T_ϵ is finitely continuous, bounded, and coercive with respect to any element w_0 of X^*. Hence we may apply Theorem (7.2) to obtain an element u_ϵ of G for each $\epsilon > 0$ such that $w_0 \in (T + T_0 + \epsilon J)(u_\epsilon)$, i.e.,

$$\epsilon^{-1}(w_0 - v_\epsilon) = J(u_\epsilon); \quad v_\epsilon \in (T + T_0)(u_\epsilon).$$

Hence $(v_\epsilon - w_0, u_\epsilon) = -\epsilon \|u_\epsilon\|^2 \leq 0$, which implies that $\|u_\epsilon\| \leq R$, and $T_0(u_\epsilon)$ is contained in $T_0(B_R)$. Moreover, $\|w_0 - v_\epsilon\| = \epsilon \|u_\epsilon\| \leq \epsilon R \to 0$, i.e., w_0 lies in the strong closure of $(T + T_0)(B_R \cap G)$. Applying Theorem (7.4), we see that w_0 lies in $(T + T_0)(B_R \cap G)$. q.e.d.

THEOREM (7.6). *Let X be a reflexive Banach space which is either separable or has an equivalent locally uniformly convex norm. Let G be a closed convex subset of X, T and T_0 two monotone mappings of G into 2^{X^*} with T maximal monotone from G to 2^{X^*} and T_0 finitely continuous and bounded. Suppose that for each point w_0 of X^*, there exists a neighborhood N of w_0 in X^* such that $(T + T_0)^{-1}(N)$ is bounded in G.*

Then $(T + T_0)(G) = X^*$.

PROOF OF THEOREM (7.6). We shall derive the conclusion of Theorem (7.6) by an application of Theorem (5.1) of §5 by approximating $(T + T_0)$ by invertible mappings from G to 2^{X^*}. We note first that if X is reflexive and separable, it was shown by Kadec [262] that X has an equivalent locally uniformly convex norm. Hence, we can assume this latter property in any case. (We recall that a Banach space X is said to be locally uniformly convex if for each u_0 with $\|u_0\| = 1$ and each $\epsilon > 0$, there exists $\alpha > 0$ such that if u is any point with $\|u\| \leq 1$ for which $\|u + u_0\| \geq 2(1 - \alpha)$, then $\|u - u_0\| < \epsilon$.) It follows immediately from the definition of local uniform convexity that such a space is strictly convex. If J is the multi-valued duality mapping, $J: X \to 2^{X^*}$, then it follows that J^{-1} is single-valued. More precisely, we know that for a bounded sequence $\{u_j\}$ and an element u with $z_j \in J(U_j)$, $z \in J(u)$, $(z_j - z, u_j - u) \to 0$ then u_j converges strongly to u in X.

(Indeed, suppose that $(z_j - z, u_j - u) \to 0$. Since

$$(z_j - z, u_j - u) = (\|u_j\| - \|u\|)^2 + [\|z\| \cdot \|u_j\| - (z, u_j)] + [\|z_j\| \cdot \|u\| - (Jz_j, u)]$$

with each of the three terms nonnegative, it follows that $\|u_j\| \to \|u\|$, $(z, u_j) \to \|u\|^2$. Hence $(u + u_j, z) = (u, z) + (u_j, z) \to 2\|u\|^2$ which implies that $\varliminf \|u + u_j\| \geq 2\|u\|$. By the local uniform convexity, u_j converges strongly to u in X.)

We approximate the mapping $(T + T_0)$ of G into 2^{X^*} by the mappings $T_\epsilon = T + T_0 + \epsilon J$. The approximation is uniform on bounded subsets of G in the sense of §5 for multi-valued mappings. Each T_ϵ maps G onto X^* by Theorem (7.2) since ϵJ is coercive with respect to any point w_0 of X^* for each fixed $\epsilon > 0$. Finally, each T_ϵ is invertible as we show by the following argument. Suppose that for a given $\epsilon > 0$, $w_j \in T_\epsilon(u_j)$ with w_j converging strongly to w_0 in X^*, and let u_0 be an element of G such that w_0 lies in $T_\epsilon(u_0)$. Then there exist elements v_j in $(T + T_0)(u_j)$, v_0 in $(T + T_0)(u_0)$, z_j in $J(u_j)$ and z_0 in $J(u_0)$ such that

$$w_j = v_j + \epsilon z_j, \qquad w_0 = v_0 + \epsilon z_0.$$

Then

$$(w_j - w_0, u_j - u_0) = (v_j - v_0, u_j - u_0) + \epsilon(z_j - z_0, u_j - u_0)$$
$$\geq \epsilon(z_j - z_0, u_j - u_0)$$

while since $w_j - w_0$ converges to 0 strongly in X^* and $(u_j - u_0)$ is bounded, it follows that $(w_j - w_0, u_j - u_0) \to 0$. Hence $(z_j - z_0, u_j - u_0) \to 0$, and it follows from our remarks above that u_j converges strongly to u.

Thus all the conditions of Theorem (5.1) are satisfied including the condition that $(T + T_0)(G_0)$ is closed for each bounded convex set G_0. Our conclusion follows from that theorem. q.e.d.

We now extend the preceding results to the more general case in which the mapping T_0 is no longer monotone but satisfies the more general condition of being pseudo-monotone in the sense of the following definition:

DEFINITION (7.5). *Let X be a Banach space, G a closed convex subset of X, T_0 a mapping of G into 2^{X^*}. Then T is said to be pseudo-monotone from G to 2^{X^*} if it satisfies the following condition:*

(p): *For any sequence $\{u_j\}$ in G converging weakly to an element u_0 of G and for* any sequence $\{w_j\}$ in X^* with $w_j \in T_0(u_j)$ for each $j \geq 1$, for which

$$\limsup (w_j, u_j - u_0) \leq 0,$$

for each v in G there exists an element w_0 of $T_0(u_0)$, such that

$$(w_0, u_0 - v) \leq \liminf (w_j, u_j - v).$$

To orient this definition with respect with monotonicity, we note the following fact:

PROPOSITION (7.4). *Let X be a reflexive Banach space, G a closed convex subset of X, T_0 a monotone mapping of G into 2^{X^*} such that for each x in G, $T_0(x)$ is a closed convex nonempty subset of X^*. Suppose further that T_0 is upper-semicontinuous as a set-valued function from each line segment in G to the weak topology on X^*.*

Then T_0 is pseudo-monotone from G to 2^{X^*}.

PROOF OF PROPOSITION (7.4). Let $\{u_j\}$ be a sequence in G converging weakly to the element u_0 of G, $\{w_j\}$ a sequence in X^* with $w_j \in T_0(u_j)$ for each $j \geq 1$, and such that $\limsup (w_j, u_j - u_0) \leq 0$. Let w_0 be any element of $T_0(u_0)$, where the latter set is nonempty by hypothesis. By the monotonicity of T_0, for each $j \geq 1$, $(w_0, u_j - u_0) \leq (w_j, u_j - u_0)$, while

$$(w_0, u_j - u_0) \to 0 \qquad (j \to \infty).$$

Hence $(w_j, u_j - u_0) \to 0$.

Let v be any element of G, w_1 an element of $T_0(v)$. Since T_0 is monotone, it follows that for each $j \geq 1$,

$$(w_1, u_j - v) \leq (w_j, u_j - v) = (w_j, u_j - u_0) + (w_j, u_0 - v).$$

Hence,

$$\liminf (w_j, u_j - v) = \liminf (w_j, u_0 - v),$$

while since $(w_1, u_j - v) \to (w_1, u_0 - v)$, we have $(w_1, u_0 - v) \leq \liminf (w_j, u_0 - v)$. For a given z in G, we replace v in this last inequality by

$$v_t = u_0 + t(z - u_0) \qquad (0 < t \leq 1).$$

We obtain for $w_t \in T_0(v_t)$,

$$t(w_t, u_0 - z) \leq \liminf t(w_j, u_0 - z),$$

and, after cancelling the positive constant t,

$$(w_t, u_0 - z) \leq \liminf (w_j, u_0 - z).$$

Since T_0 is upper-semicontinuous from line segments in G to 2^{X^*} with the weak topology on X^*, we may find an infinite sequence of parameter values $t_j \to 0$ and a corresponding sequence $w_j = w_{t_j}$ such that w_j converges weakly in X^* to an element w_0 of $T_0(u_0)$. Thus

$$(w_0, u_0 - z) \leq \liminf (w_j, u_0 - z)$$

for the given z of G. q.e.d.

THEOREM (7.7). *Let X be a reflexive Banach space, G a closed convex subset of X, T and T_0 two mappings of G into 2^{X^*}. Suppose that T is maximal monotone from G to 2^{X^*} and that T_0 is a finitely continuous pseudo-monotone mapping of G into 2^{X^*}. Let G_0 be a bounded closed convex subset of G such that $T_0(G_0)$ is bounded. Then $(T + T_0)(G_0)$ is closed in X^*.*

PROOF OF THEOREM (7.7). Let w_0 be a limit point of $(T + T_0)(G_0)$. We must show that w_0 lies in $(T + T_0)(G_0)$. By assumption, there exist a sequence $\{u_j\}$ in G_0 and a sequence $\{w_j\}$ in X^* with $w_j \in (T + T_0)(u_j)$ for each j such that w_j converges strongly to w_0 in X^*. By its definition, $w_j = v_j + z_j$ with $v_j \in T(u_j)$ and $z_j \in T_0(u_j)$. Since $T_0(G_0)$ is bounded, there exists a constant R such that for all j, $\|z_j\| \leq R$.

Since X and X^* are reflexive, G_0 is weakly sequentially compact, as is each closed ball in X^*. By passing to an infinite subsequence, we may assume without loss of generality that u_j converges weakly to some element u_0 of G_0 and that z_j converges weakly to an element z_0 of X^*.

Let $[x, y]$ be any element of $G(T)$. Then

$$(z_j, u_j - u_0) = (z_j, u_j - x) - (z_j, u_0 - x).$$

By the weak convergence of z_j to z_0, $(z_j, u_0 - x) \to (z_0, u_0 - x)$. On the other hand,

$$(z_j, u_j - x) \leq (z_j + (v_j - y), u_j - x) = (w_j - y, u_j - x) \to (w_0 - y, u_0 - x).$$

Hence

$$\limsup_{j \to \infty} (z_j, u_j - u_0) \leq (w_0 - z_0 - y, u_0 - x)$$

for all $[x, y]$ in $G(T)$.

We assert that

$$\inf_{[x,y] \in G(T)} (w_0 - z_0 - y, u_0 - x) \leq 0.$$

Indeed, suppose the contrary. Then there must exist $\beta > 0$ such that for all $[x, y]$ in $G(T)$, $(w_0 - z_0 - y, u_0 - x) \geq \beta > 0$.

Since T is maximal monotone from G to 2^{X^*}, it follows that $[u_0, w_0 - z_0] \in G(T)$. Hence we may replace $[x, y]$ in the inequality by this element and we obtain $0 = ([w_0 - z_0] - [w_0 - z_0], u_0 - u_0) \geq \beta > 0$. This is a contradiction, proving that the asserted infimum is nonpositive.

Therefore, we have $\limsup (z_j, u_j - u_0) \leq 0$. Since T_0 is pseudo-monotone, it follows that for each v in G, there exists an element $z(v)$ of $T_0(u_0)$ such that $(z(v), u_0 - v) \leq \liminf (z_j, u_j - v)$. On the other hand, $(z_j, u_j - v) = (z_j, u_j - u_0) + (z_j, u_0 - v)$, so that

$$\liminf (z_j, u_j - v) \leq \limsup (z_j, u_j - u_0) + (z_0, u_0 - v) \leq (z_0, u_0 - v).$$

Thus $(z(v), u_0 - v) \leq (z_0, u_0 - v)$. Finally, if we apply Proposition (7.1), we find that there exists a single element z of $T_0(u_0)$ such that for all v in G, $(z, u_0 - v) \leq (z_0, u_0 - v)$.

For each $[x, y]$ in $G(T)$,

$$\limsup (z_j, u_j - u_0) \leq (w_0 - z_0 - y, u_0 - x).$$

Since T_0 is pseudo-monotone, we know that

$$0 = (z, u_0 - u_0) \leq \liminf (z_j, u_j - u_0) \leq (w_0 - z_0 - y, u_0 - x).$$

It follows that

$$(w_0 - z - y, u_0 - x) = (w_0 - z_0 - y, u_0 - x) + (z_0 - z, u_0 - x) \geq 0$$

for each $[x, y]$ in $G(T)$ since both terms of the sum are nonnegative. Hence by the maximal monotonicity of T from G to 2^{X^*}, we know that

$$w_0 \in (T + T_0)(u_0). \quad \text{q.e.d.}$$

THEOREM (7.8). *Let X be a reflexive Banach space, G a closed convex subset of X, T and T_0 two mappings of G into 2^{X^*} with T monotone, $[0, 0]$ in the graph of T, and T_0 a finitely continuous pseudo-monotone mapping of G into 2^{X^*} such that for each bounded subset B of G, $T_0(B)$ is bounded in X^*. Suppose that T_0 is coercive on G with respect to the given element w_0 of X^*.*

Then there exists an element $[u_0, z_0]$ of $G(T_0)$ such that for all $[x, y]$ in $G(T)$ $(y + z_0 - w_0, x - u_0) \geq 0$.

COROLLARY TO THEOREM (7.8). *If in addition T is maximal monotone from G to 2^{X^*}, then u_0 satisfies the equation $w_0 \in (T + T_0)(u_0)$.*

PROOF OF THEOREM (7.8). We may assume without loss of generality that T is indeed maximal monotone from G to 2^{X^*}. Indeed, if we consider the family of monotone extensions T_1 of T from G to 2^{X^*}, each linearly ordered subfamily with respect to inclusion certainly has an upper bound in the family. Applying Zorn's lemma, we see that T has a maximal monotone extension T_1 from G to 2^{X^*}. If we consider the corresponding problem with T replaced by T_1, any solution $[u_0, z_0]$ for the new problem will certainly be a solution for the original case. Hence we may assume that T itself is maximal.

If T is maximal monotone, however, the inequality $(y + z_0 - w_0, x - u_0) \geq 0$ for all $[x, y]$ in $G(T)$ implies that $(w_0 - z_0)$ lies in $T(u_0)$ and is in fact equivalent to this last assertion by an obvious argument. Since z_0 is an otherwise unrestricted element in $T_0(u_0)$, this last condition: $(w_0 - z_0) \in T(u_0)$, is equivalent to the fact that $w_0 \in (T + T_0)(u_0)$. Hence, the corollary stated above does follow from the result of the theorem.

To show that w_0 lies in $(T + T_0)(u_0)$, we assert that by Theorem (7.7) it suffices to show that for each $\epsilon > 0$, w_0 lies in $(T + T_0 + \epsilon J)(u_0)$. Indeed, suppose that for each $\epsilon > 0$, there exists u_ϵ in G such that $w_0 \in (T + T_0 + \epsilon J)(u_\epsilon)$. Then there exist elements v_ϵ in $T(u_\epsilon)$, z_ϵ in $T_0(u_\epsilon)$, and q_ϵ in $J(u_\epsilon)$ such that $w_0 = v_\epsilon + z_\epsilon + \epsilon q_\epsilon$.

By the assumption that T_0 is coercive with respect to w_0, we know that there exists $R > 0$ such that for all u with $\|u\| > R$ and all z in $T_0(u)$, we have $(z - w_0, u) > 0$. Hence if $\|u_\epsilon\| > R$,

$$0 = (z_\epsilon - w_0, u_\epsilon) + (v_\epsilon, u_\epsilon) + \epsilon(q_\epsilon, u_\epsilon),$$

where $(v_\epsilon, u_\epsilon) = (v_\epsilon - 0, u_\epsilon - 0) \geq 0$ since T is monotone and $[0, 0]$ lies in $G(T)$, while $(q_\epsilon, u_\epsilon) = \|u_\epsilon\|^2 \geq 0$. Since $(z_\epsilon - w_0, u_\epsilon) > 0$ for $\|u_\epsilon\| > R$, it follows for that case that $0 > 0$, and this contradiction proves that for each $\epsilon > 0$, $\|u_\epsilon\| \leq R$. Hence $\|q_\epsilon\| = \|u_\epsilon\| \leq R$, while by the hypothesis that $T_0(B)$ is bounded for B bounded there exists a constant $R_0 > 0$, independent of ϵ for $\epsilon > 0$, such that $\|z_\epsilon\| \leq R_0$. On the other hand

$$\|w_0 - (v_\epsilon + z_\epsilon)\| = \epsilon \|q_\epsilon\| \leq \epsilon R \to 0 \qquad (\epsilon \to 0).$$

Hence w_0 is a limit point of the sequence $\{v_{\epsilon_j} + z_{\epsilon_j}\}$, $(\epsilon_j \to 0)$ with $\|z_{\epsilon_j}\| \leq R_0$, and by Theorem (7.7), w_0 must lie in $(T + T_0)(G \cap B_R)$ since each element $v_\epsilon + z_\epsilon$ lies in that set. Hence, we may replace the consideration of the question as to whether w_0 lies in the range of $(T + T_0)$ by the corresponding question for $(T + T_0 + \epsilon J)$ for a given $\epsilon > 0$.

We now apply the following auxiliary result:

PROPOSITION (7.5). *Let X be a strictly convex reflexive Banach space with a strictly convex conjugate space X^*, G a closed convex subset of X, T a maximal monotone mapping of G into 2^{X^*} with $[0, 0] \in G(T)$, J the duality mapping of X into 2^{X^*}. Then for each $\epsilon > 0$, $(T + \epsilon J)$ is maximal monotone from G to 2^{X^*} and has all of X^* as its range. Moreover, $(T + \epsilon J)$ is one-to-one and has an inverse mapping R which is continuous from the strong topology of X^* to the weak topology on G.*

PROOF OF PROPOSITION (7.5). For each u in X, $J(u)$ may be characterized by

$$J(u) = \{w \mid w \in X^*, \|w\| \leq \|u\|, (w, u) = \|u\|^2\}.$$

Hence $J(u)$ is a convex subset of a sphere in X^*, and if X^* is strictly convex, then $J(u)$ is a single point for each u in X and J is a single-valued mapping. If X is strictly convex, in addition, then J^{-1} which is the duality mapping from X^* to $2^X = 2^{X^{**}}$ is also single-valued.

We assert that if $\{u_j\}$ is a sequence in X for which $(Ju_j - Ju, u_j - u) \to 0$ for a given element u of X, then u_j converges weakly to u in X. Indeed, we calculate easily that

$$(Ju_j - Ju, u_j - u) = [\|u_j\| - \|u\|]^2 + [\|Ju_j\| \cdot \|u\| - (Ju_j, u)]$$
$$+ [\|Ju\| \cdot \|u_j\| - (Ju, u_j)].$$

Since each of the three terms in the sum is nonnegative, it follows that

$$\|u_j\| \to \|u\|; \qquad (Ju_j, u) \to \|u\|^2; \qquad (Ju, u_j) \to \|u\|^2.$$

Since $\{u_j\}$ is a bounded sequence in a reflexive space X, to show that u_j converges weakly, it suffices to choose a weakly convergent subsequence (which for simplicity of notation we identify with the original sequence) such that u_j converges weakly to u_0 for some u_0 in X and to show that $u_0 = u$. However, $(Ju, u_j) \to (Ju, u_0)$ and $\|u_0\| \leq \lim \|u_j\| \leq \|u\| = \|Ju\|$. Hence $u_0 = J^{-1}(J(u)) = u$, and our assertion is proved.

The fact that $(T + \epsilon J)$ has all of X^* as its range follows from Theorem (7.2) since for each $\epsilon > 0$, ϵJ is coercive from G to 2^{X^*} with respect to any element w of X^*. The fact that $(T + \epsilon J)^{-1} = R$ is single-valued follows from the fact if $w \in (T + \epsilon J)(u_1) \cap (T + \epsilon J)(u_2)$, there exist elements v_1 and v_2 of $T(u_1)$ and $T(u_2)$, respectively, such that $w = v_1 + \epsilon J(u_1) = v_2 + \epsilon J(u_2)$. Hence

$$(v_1 - v_2, u_1 - u_2) + \epsilon(J(u_1) - J(u_2), u_1 - u_2) = 0.$$

Since both terms are nonnegative, it follows that $(J(u_1) - J(u_2), u_1 - u_2) = 0$. By the trivial case of the continuity argument above, $u_1 = u_2$ and R is single-valued.

To show that R is continuous from the strong topology on X^* to the weak topology on G, consider a strongly convergent sequence $\{w_j\}$ in X^* with limit w, and for each j, let $u_j = R(w_j)$. Then there exists v_j in $T(u_j)$ such that $w_j = v_j + \epsilon J(u_j)$. Let $u = R(w)$, and v in $T(u)$ such that $w = v + \epsilon J(u)$. Then

$$(w_j - w, u_j - u) = (v_j - v, u_j - u) + \epsilon(J(u_j) - J(u), u_j - u) \to 0$$

as $j \to \infty$ since $w_j - w$ converges to zero strongly in X^* while the sequence $\{u_j\}$ is bounded in G since

$$\|w_j\| \cdot \|u_j\| \geq (w_j, u_j) = (v_j, u_j) + \epsilon(Ju_j, u_j) \geq \epsilon \|u_j\|^2,$$

so that $\|u_j\| \leq \epsilon^{-1} \|w_j\| \leq R$ for some constant R. By our remarks above, it follows that u_j converges weakly to u in G, i.e., $R(w_j)$ converges weakly to $R(w)$ in G.

Finally, we wish to show that $(T + \epsilon J)$ is maximal monotone from G to 2^{X^*}. Suppose then that for some element $[u_0, w_0]$ of $G \times X^*$ and all elements $[x, y]$ of $G(T)$, we have $(y + \epsilon J(x) - w_0, x - u_0) \geq 0$. For each $t > 0$, we set $w_t = w_0 + tz$ for an arbitrary element z of X^* and set $x_t = R(w_t)$. Then $w_t = y_t + \epsilon J(x_t)$ for some y_t in $T(x_t)$. If we replace $[x, y]$ by $[x_t, y_t]$, we obtain

$$0 \leq (w_t - w_0, x_t - x_0) = t(z, R(w_t) - x_0).$$

Cancelling the positive factor t, we get $(z, R(w_t) - x_0) \geq 0$. Letting $t \to 0+$ and using the fact that R is continuous from the strong topology in X^* to the weak topology in G, we see that $(z, R(w_0) - x_0) \geq 0$ for all z in X^*. Hence $x_0 = R(w_0)$ and $w_0 \in (T + \epsilon J)(u_0)$. Hence $(T + \epsilon J)$ is maximal monotone. q.e.d.

PROOF OF THEOREM (7.8) COMPLETED. By a result of Asplund [10], each reflexive Banach space X has an equivalent norm in which both X and X^* are strictly convex. We may assume without loss of generality that this is our given norm. We wish to consider the range of $(T + \epsilon J + T_0)$ for a given $\epsilon > 0$, and by Proposition (7.5), $(T + \epsilon J)$ is a maximal monotone mapping from G to 2^{X^*} and the range of $(T + \epsilon J)$ is the whole of X^*. Hence if we replace T by $(T + \epsilon J)$ in the original assertion of our theorem, we may assume without loss of generality that our given maximal monotone mapping T from G to 2^{X^*} has all of X^* in its range.

We shall carry through the proof of Theorem (7.8) by using Proposition (7.3) as in the proof of Theorem (7.1) but with an altered limiting argument on the approximations obtained from finite subsets of G. Let f be a finite subset of G containing 0, G_f its convex closure in G. By Proposition (7.3), there exist constants R_1 and R_2 independent of the subset f such that for each f, there exist an element u_f in G_f and an element z_f of $T_0(u_f)$ such that the following conditions hold:

(1) For each element $[x, y]$ in $G(T)$ with x in G_f, we have the inequality

$(7.7)_f$ $$(y + z_f - w_0, x - u_f) \geq 0.$$

(2) We have the bounds

$$\|u_f\| \leq R_1, \quad \|z_f\| \leq R_2.$$

For each finite subset f_0 of G containing 0, we construct the following subset of $G \times X^*$:

$$W_{f_0} = \bigcup_{f \geq f_0} \{[u_f, z_f]\}$$

where $[u_f, z_f]$ is a solution of conditions (1) and (2) above for the given f and is chosen for f independent of f_0. The subsets W_{f_0} of $G \times X^*$ are uniformly bounded and satisfy the finite intersection property. It follows from the reflexivity of X that there exists an element $[u_0, z_0]$ of $G \times X^*$ such that $[u_0, z_0]$ lies in the weak closure of W_{f_0} for each finite subset f_0 of G containing 0.

By our remarks at the beginning of this final part of the proof of Theorem (7.8), we may assume that the mapping T has all of X^* in its range. Hence, we can find an element x_0 of G such that $(w_0 - z_0)$ lies in $T(x_0)$. Let $[x, y]$ be an arbitrary element of $G(T)$. We choose the finite subset f_0 of G to consist of the three points $\{0, x_0, x\}$. Since $[u_0, z_0]$ lies in the weak closure of W_{f_0} for this choice of f_0, we may find a sequence $\{[u_j, z_j]\}$ in W_{f_0} by Proposition (7.2) such that $[u_j, z_j]$ converges weakly to $[u_0, z_0]$ in $G \times X^*$, i.e., u_j converges weakly to u_0 in G and z_j converges weakly to z_0 in X^*. Since x_0 lies in f_0 and hence in each f which contains f_0, we know that by the inequality $(7.7)_f$ above,

$$((w_0 - z_0) + z_j - w_0, x_0 - u_j) \geq 0,$$

i.e., $(z_j, u_j - x_0) \leq (z_0, u_j - x_0)$. Hence $(z_j, u_j - u_0) = (z_j, u_j - x_0) + (z_j, x_0 - u_0)$ implies that

$$\limsup (z_j, u_j - u_0) \leq \limsup [(z_j, x_0 - u_0) + (z_0, u_j - x_0)] = 0$$

since $(z_j, x_0 - u_0) + (z_0, u_j - x_0) \to (z_0, x_0 - u_0) + (z_0, u_0 - x_0) = 0$. If we apply the fact that T_0 is pseudo-monotone, we conclude that for each element v of G, there exists an element z_v of $T_0(u_0)$ such that

$$(z_v, u_0 - v) \leq \liminf (z_j, u_j - v).$$

In particular, taking $v = u_0$, we obtain $0 \leq \liminf (z_j, u_j - u_0)$, so that $(z_j, u_j - u_0) \to 0$ $(j \to \infty)$.

On the other hand, for each v in G,

$$(z_j, u_j - v) = (z_j, u_j - u_0) + (z_j, u_0 - v)$$

and hence

$$\lim (z_j, u_j - v) = (z_0, u_0 - v).$$

By the preceding paragraph, there exists z_v in $T_0(u_0)$ such that

$$(z_v, u_0 - v) \leq \lim (z_j, u_j - v) = (z_0, u_0 - v).$$

If we apply Proposition (7.1) with respect to the convex sets $T_0(u_0) - z_0$ and $u_0 - G$, we see that there exists an element z of $T_0(u_0)$ such that for all v in G, we have the inequality

$$(z, u_0 - v) \leq (z_0, u_0 - v) \qquad (v \in G).$$

By our construction of f_0 and the inequality $(7.7)_f$, we know that for the given element $[x, y]$ of $G(T)$ and every $j \geq 1$, $(y + z_j - w_0, x - u_j) \geq 0$. Hence $(z_j, u_j - x) \leq (y - w_0, x - u_j)$, and since

$$(z_j, u_j - x) \to (z_0, u_0 - x); \qquad (y - w_0, x - u_j) \to (y - w_0, x - u_0)$$

we obtain $(z_0, u_0 - x) \leq (y - w_0, x - u_0)$; thus $(z, u_0 - x) \leq (y - w_0, x - u_0)$, i.e.,

$$(y + z - w_0, x - u_0) \geq 0$$

for a fixed z in $T_0(u_0)$ and all $[x, y]$ in $G(T)$. Since T is maximal monotone from G to 2^{X^*}, it follows that $(w_0 - z) \in T(u_0)$, i.e., $w_0 \in z + T(u_0) \subset (T + T_0)(u_0)$. q.e.d.

The preceding theorems of this section make it clear that it is of importance for the application of these results to be able to determine if a given monotone mapping T from G into 2^{X^*} is maximal monotone from G to 2^{X^*}.

A simple basic result in this direction is the following:

THEOREM (7.9). *Let X be a reflexive Banach space, G a closed convex subset of X, T a monotone mapping of G into 2^{X^*}. Then a necessary and sufficient condition that T be maximal monotone is that there exist an equivalent norm on X such that for the corresponding duality mapping J, $(T + J)$ has all of X^* as its range.*

PROOF OF THEOREM (7.9). Since any duality mapping J satisfies the conditions on T_0 in Theorem (7.2), it follows that if T is maximal monotone from G to 2^{X^*}, then by Theorem (7.2) the range of $(T + J)$ is indeed the whole of X^*.

To prove the converse, we choose a norm on X equivalent to the given norm such that in the new norm both X and X^* are strictly convex. For the corresponding duality mapping J, it follows that J and J^{-1} are single-valued, and by the argument for Proposition (7.5), for any monotone mapping T, $(T + J)^{-1}$ exists and is continuous from the strong topology of X^* to the weak topology of G. By the same argument, if the range of $T + J$ is all of X^*, then $T + J$ is a maximal monotone mapping of G into 2^{X^*}.

We assert that it follows that T itself is maximal monotone from G to 2^{X^*}. Indeed suppose $[u_0, w_0]$ is an element of $G \times X^*$ such that for all $[x, y]$ in $G(T)$, $(w_0 - y, u_0 - x) \geq 0$. Since J is monotone, for any y_0 in $J(u_0)$, we have for each z in $J(x)$, $(y_0 - z, u_0 - x) \geq 0$. Adding these two inequalities, we obtain

$$(y_0 + w_0 - (y + z), u_0 - x) \geq 0$$

for every element $[x, y + z]$ in $G(T + J)$. Since $(T + J)$ is already known to be maximal monotone, it follows that $(y_0 + w_0) \in (T + J)(u_0)$, i.e., $w_0 + J(u_0) \in J(u_0) + T(u_0)$ since J is single-valued. Hence w_0 lies in $T(u_0)$. q.e.d.

As an example of a nontrivial class of monotone mappings from a convex set G in X to 2^{X^*}, we have the one given in the following definition:

DEFINITION (7.6). *Let G be a convex subset in a Banach space X, g a function from G to $R^1 \cup \{+\infty\}$. We normalize g by assuming that $g(0) = 0$, where 0 is a point of G. Suppose that g is convex in the usual sense and lower-semicontinuous (in the sense of single-valued, extended real-valued functions). Then ∂g, the subgradient mapping of g, is the mapping of G into 2^{X^*} defined as follows: For x in G, w lies in $(\partial g)(x)$ if and only if for all y in G, $g(y) \geq g(x) + (w, y - x)$.*

PROPOSITION (7.6). *Let X be a Banach space, G a closed convex subset of X. Then:*

(a) For each lower-semicontinuous convex, extended real-valued function g, ∂g is a monotone mapping of G into 2^{X^*}.

(b) If $G = X$ and $g(x) = \frac{1}{2} \|x\|^2$, then $\partial g = J$, the duality mapping of X into 2^{X^*}.

(c) If $G = X$, and $g(x) = \Phi(\|x\|)$ where $\Phi'(0) = \Phi(0) = 0$ and Φ is a C^1 convex function on $R^+ = \{r \mid r \in R^1, r \geq 0\}$, then $\partial g = J_\varphi$, the generalized duality mapping of X into 2^{X^*} with gauge function $\psi = \Phi'$ given by

$$J(u) = \{w \mid w \in X^*, \|w\| = \psi(\|u\|), (w, u) = \|w\| \cdot \|u\|\}.$$

(d) If G is a closed convex set in G with $0 \in G$ and g is the indicator function of G, i.e.,

$$g(x) = +\infty, \quad u \notin G,$$
$$= 0, \quad u \in G,$$

then $\partial g = S_G$, where S_G is the support map of G into 2^{X^*} given by

$$S_G(u) = \{w \mid w \in X^*, (w, u - x) \geq 0 \text{ for all } x \text{ in } G\}.$$

PROOF OF PROPOSITION (7.6). PROOF OF (a). Suppose that $w_1 \in (\partial g)(x_1)$ and $w_2 \in (\partial g)(x_2)$. Then $g(x_2) \geq g(x_1) + (w_1, x_2 - x_1)$ and $g(x_1) \geq g(x_2) + (w_2, x_1 - x_2)$. Adding the two inequalities, we obtain

$$g(x_1) + g(x_2) \geq [g(x_1) + g(x_2)] - (w_1 - w_2, x_1 - x_2)$$

i.e., $(w_1 - w_2, x_1 - x_2) \geq 0$. Thus (∂g) is monotone. q.e.d.

PROOF OF (b). The assertion of (b) is a special case of the assertion of (c).

PROOF OF (c). Let $\Phi(r)$ be a C^1 convex function from R^+ to R with $\Phi'(0) = \Phi(0) = 0$, $\psi = \Phi'$, and let $g(x) = \Phi(\|x\|)$. We shall show that $\partial g = J_\psi$.

Suppose first that $w \in (\partial g)(u)$. Then for all v in X, $\Phi(\|v\|) \geq \Phi(\|u\|) + (w, v - u)$. Since g is an even function on X and convex with $g(0) = 0$, it follows that $g(x) \geq 0$ for all x. If we set $v = (1 + \epsilon)u$ for $\epsilon > 0$ and small in the above inequality, we obtain

$$\epsilon(w, u) \leq \Phi((1 + \epsilon)\|u\|) - \Phi(\|u\|).$$

Dividing by ϵ and letting $\epsilon \to 0+$, we obtain

$$(w, u) \leq \Phi'(\|u\|)\|u\| = \psi(\|u\|)\|u\|.$$

If we set $v = (1 - \epsilon)u$, with $\epsilon > 0$, we obtain

$$\epsilon(w, u) \geq \Phi(\|u\|) - \Phi((1 - \epsilon)\|u\|).$$

Letting $\epsilon \to 0+$, we find that

$$(w, u) \geq \Phi'(\|u\|)\|u\| = \psi(\|u\|)\|u\|.$$

Hence $(w, u) = \psi(\|u\|)\|u\|$. On the other hand, if for any x in X with $x \neq 0$, we set $v = u + \epsilon x$ with $\epsilon > 0$, we find that $\epsilon(w, x) \leq \Phi(\|u + \epsilon x\|) - \Phi(\|x\|)$. Φ is increasing since by convexity $\Phi'(r)$ is nondecreasing and by hypothesis $\Phi'(0) = 0$. Hence since $\|u + \epsilon x\| \leq \|u\| + \epsilon \|x\|$,

$$\Phi(\|x + \epsilon x\|) \leq \Phi(\|u\| + \epsilon \|x\|) = \Phi(\|u\|) + \int_0^\epsilon \Phi'(\|u\| + s\|x\|)\|x\|\, ds.$$

Dividing by ϵ and letting $\epsilon \to 0+$, we find that

$$(w, x) \leq \lim_{\epsilon \to 0+} \epsilon^{-1} \int_0^\epsilon \Phi'(\|u\| + s\|x\|)\, ds = \Phi'(\|u\|)\|x\|.$$

Thus $\|w\| \leq \Phi'(\|u\|) = \psi(\|u\|)$.

Because

$$\psi(\|u\|)\|u\| \leq (w, u) - \|w\| \cdot \|u\| \leq \psi(\|u\|)\|u\|,$$

it follows that $\|w\| = \psi(\|u\|)$. Therefore $w \in J_\psi(u)$.

To prove the converse, suppose that w lies in $J_\psi(u)$ so that $(w, u) = \|u\|\psi(\|u\|)$ and $\|w\| = \psi(\|u\|)$. We wish to show that for each v in X,

$$\alpha(v) = \Phi(\|v\|) - \Phi(\|u\|) - (w, v - u) \geq 0.$$

We know that if we set $\|v\| = s$, $\|u\| = r$, we have

$$\alpha(v) = \Phi(s) - \Phi(r) + \Phi'(r)r - (w, v)$$
$$\geq \Phi(s) - \Phi(r) + \Phi'(r)r - \Phi'(r)s.$$

We remark that since Φ is convex, $\Phi'(r)$ is nondecreasing in r. Hence

$$\Phi(s) - \Phi(r) + \Phi'(r)[r - s] = \int_r^s \Phi'(t)\, dt + \Phi'(r)[r - s] \geq 0$$

if $s \geq r$, since then $\int_r^s \Phi'(t)\, dt \geq \Phi'(r)[s-r]$, as well as if $r \geq s$, since then $\int_r^s \Phi'(t)\, dt \geq -[r-s]\Phi'(r)$. Thus $\alpha(v) \geq 0$ for all cases, and it follows that w lies in $(\partial g)(u)$. q.e.d.

PROOF OF (d). This is a formal translation of the definitions of ∂g and S_G. q.e.d.

THEOREM (7.10). *Let X be a reflexive Banach space, G a closed convex subset of X, g an extended real-valued convex function on G which is lower-semicontinuous with $g(0) = 0$. Then ∂g is a maximal monotone mapping from G to 2^{X^*}.*

PROOF OF THEOREM (7.10). By Theorem (7.9), it suffices to show that $\partial g + J$ has all of X^* as its range. We may assume without loss of generality that X^* is strictly convex. If we set $g_0(x) = \frac{1}{2}\|x\|^2$, we know that $J = \partial g$ and that J is a demi-continuous mapping from X to X^* by the strict convexity of X^*. Hence g_0 is weakly Gateaux differentiable at every point and $g_0'(x) = J(x)$. Hence $\partial(g + g_0) = \partial g + \partial g_0 = \partial g + J$, and therefore in order to show that $\partial g + J$ maps onto X^*, it suffices to prove this for $\partial(g + g_0)$. We note that for a convex function g, 0 lies in $\partial g(u)$ if and only if g has a minimum at u. Hence w will lie in $\partial(g + g_0)(u)$ if and only if $g(x) + g_0(x) - (w, x)$ has a minimum at the point u. Since $g(x) - (w, x)$ is also a proper, lower-semicontinuous convex function, it suffices to show under the hypotheses of the theorem that $g(x) + g_0(x)$ has a minimum at some point u of G.

Since $g + g_0$ is convex and lower-semicontinuous, it is weakly lower-semicontinuous. It follows that $g + g_0$ will assume a minimum at some point of G provided that $g(x) + g_0(x) \to +\infty$ as $\|x\| \to +\infty$. Suppose that this latter condition did not hold. Then we could find a sequence of points $\{x_k\}$ in G such that $\|x_k\| t_k = \to \infty$ while $g(x_k) + g_0(x_k) \leq M$ for some constant M independent of k. It follows that

$$g(x_k) \leq M - \tfrac{1}{2} t_k^2.$$

Let $u_k = t_k^{-1} x_k$. Then u_k lies in the unit ball in G, and since $g(0) = 0$,

$$g(u_k) \leq g_k^{-1} g(x_k) \leq t_k^{-1} M - \tfrac{1}{2} t_k \to (-\infty) \quad (k \to \infty).$$

On the other hand, the intersection of G with the unit ball is weakly compact since X is reflexive and g is lower-semicontinuous and convex and hence weakly lower-semicontinuous on G. Hence g assumes its minimum on the intersection of the unit ball with G. Since $g(u)$ is never $(-\infty)$, this contradicts the existence of a sequence $\{u_k\}$ with $g(u_k) \to -\infty$. This contradiction proves the theorem. q.e.d.

Using the results for monotone operators on convex sets or the result of Theorem (7.10) for the maximal monotonicity of subgradients, one can easily derive results from the preceding theorems for the solvability of nonlinear variational inequalities, i.e., solutions u_0 of systems of inequalities of the form

$$(T(u_0) - w_0, v - u_0) \geq g(u_0) - g(v) \quad (v \in G)$$

where g is a proper lower-semicontinuous convex function and T is a monotone or pseudo-monotone mapping from G to 2^{X^*}. Indeed, the inequality given is exactly equivalent to the equation: $w_0 \in (\partial g + T)(u_0)$.

We note that our normalizing hypothesis that $[0, 0] \in G(T)$ in the theorems of this section can easily be removed at the expense of a slight complication of hypotheses for coercivity if one replaces the operator T by the new maximal monotone mapping T_1 with $T_1(u) = T(u - u_0) - v_0$, for $[u_0, v_0]$ in $G(T)$. Then $[0, 0]$ lies in $G(T_1)$, and the corresponding results for T can be derived by a direct application of the normalized theorems to T_1.

The results we have derived above are of a particularly simple form and have interesting consequences when the convex set G which is involved is bounded. As an example, we have the following:

THEOREM (7.11). *Let X be a reflexive Banach space, G a closed bounded convex subset of X, T_0 a mapping of G into 2^{X^*} which is finitely continuous, pseudo-monotone, and such that $T_0(G)$ is bounded. Suppose that for each point x of the boundary of G, $T_0(x)$ and $(-S_G(x))$ have no nonzero elements in common.*

Then there exists an element u_0 of G such that $0 \in T_0(u_0)$.

COROLLARY TO THEOREM (7.11). *Let G be the ball of radius R about the origin in X. If for each y in $T_0(x)$ for x on the boundary of G, λy does not lie in $-J(x)$ for any $\lambda > 0$, then T has a zero in G.*

PROOF OF THEOREM (7.11). We apply Theorem (7.8) with the monotone mapping T being given by the single-valued mapping $T(x) = 0$ for each x in G. The coercivity conditions in the hypothesis of that theorem are vacuous since G is bounded. Hence there exists a solution for $w_0 = 0$, i.e., there exists $[u_0, z_0]$ in $G(T_0)$ such that

$$(z_0, x - u_0) \geq 0 \quad (x \in G)$$

or we may rewrite this in the form

$$(-z_0, u_0 - x) \geq 0 \quad (x \in G).$$

This last inequality tells us that $z_0 \in -S_G(u_0)$, and by the hypothesis, $T_0(u_0) \cap (-S_G(u_0))$ can have only 0 in common. Hence $z_0 = 0$, and $0 \in T(u_0)$.   q.e.d.

PROOF OF THE COROLLARY. We note that for $G = B_R$, $S_G(x) = \{w \mid w = \lambda v, \lambda > 0, v \in J(x)\}$.   q.e.d.

For explicit reference, we state the existence form for nonlinear variational inequalities that follows by combining Theorem (7.10) with Theorem (7.8). (More general results are possible from the use of that theorem by a different technique: applying Theorem (7.8) in the space $Y = X \times R^1$ on the convex set G_1 which is the super-graph of the convex function g. In this framework, we have the further freedom of arbitrarily specifying the monotone mapping T, which is not tied to the convex function.)

THEOREM (7.12). *Let X be a reflexive Banach space, G a closed convex subset of X, g a proper extended real-valued lower-semicontinuous convex function on G with*

$0 \in G$ and g having its minimum at 0. Let T_0 be a finitely continuous pseudo-monotone mapping of G into 2^{X^*} such that for each bounded set B of G, $T_0(B)$ is bounded. Suppose that T_0 is coercive on G with respect to a given element w_0 of X^*.

Then there exists an element u_0 of G and an element z_0 of $T_0(u_0)$ such that for all v in G,

$$(z_0 - w_0, v - u_0) \geq g(u_0) - g(v).$$

PROOF OF THEOREM (7.12). By Theorem (7.10), g is a maximal monotone mapping of G into 2^{X^*}. Since g has its minimum at 0, $0 \in (\partial g)(0)$. The desired inequality is equivalent to the statement that $(w_0 - z_0) \in (\partial g)(u_0)$, i.e., $w_0 \in (\partial g + T_0)(u_0)$. The existence of such an element u_0 is a consequence of Theorem (7.8) with the given T_0 and $T = \partial g$. q.e.d.

8. Nonexpansive mappings in Banach spaces. In the present section, we shall study a class of mappings of Banach spaces whose theory has very important links with the theory of monotone and accretive mappings.

DEFINITION (8.1). Let X be a metric space, Y a subset of X, f a mapping of Y into X. Then f is said to be *nonexpansive* if for each pair of elements y_0 and y_1 of Y,

$$d(f(y_0), f(y_1)) \leq d(y_0, y_1).$$

In other words, nonexpansive mappings are Lipschitzian mappings with Lipschitz constant 1.

We shall be concerned in our discussion with the fixed point theory of nonexpansive mappings which we can regard in a heuristic fashion as a possible generalization of the Picard-Banach theory of fixed points of strict contractions. The limitations of this heuristic viewpoint are exhibited rather sharply by the fact that the class of nonexpansive mappings includes all isometries of metric spaces and therefore that there can be no nontrivial fixed point results for this broader class of mappings without sharply delimited additional hypotheses (at least in the category of metric spaces).

THEOREM (8.1). *Let G be a closed bounded convex subset of a Banach space X, U a nonexpansive mapping of G into G. Then there exists a sequence $\{u_k\}$ in G such that $u_k - U(u_k) \to 0$.*

PROOF OF THEOREM (8.1). We let $\lambda_k = (1 - k^{-1})$ and for a given point x_0 of G, we define the mapping U_k of G into G by setting

$$U_k(x) = (1 - \lambda_k)x_0 + \lambda_k U(x).$$

By the nonexpansiveness of U, we see that

$$\|U_k(x) - U_k(u)\| = \lambda_k \|U(x) - U(u)\| \leq \lambda_k \|x - u\|$$

for all x and u in G, i.e., U_k is a strict contraction with constant $\lambda_k < 1$ for each k. The mapping U_k does carry G into G since for each x in G, $U_k(x)$ is a convex linear combination of the points x_0 and $U(x)$ of G and G is a convex set. Hence U_k has a unique fixed point u_k in G, $U_k(u_k) = u_k$. If we calculate $u_k - U(u_k)$, we obtain

$$u_k - U(u_k) = [u_k - U_k(u_k)] + [U_k(u_k) - U(u_k)]$$
$$= (1 - \lambda_k)[x_0 - U(u_k)]$$

so that

$$\|u_k - U(u_k)\| \leq (1 - \lambda_k) \operatorname{diam}(G) = k^{-1} \operatorname{diam}(G) \to 0$$

as $k \to \infty$. q.e.d.

THEOREM (8.2). *Let X be a strictly convex Banach space, G a convex set in X, U a nonexpansive mapping of G into X.*

Then $F(U)$, the fixed point set of U, is a convex subset of G.

PROOF OF THEOREM (8.2). Let x_0 and x_1 be points of $F(U)$ and for $0 < \lambda < 1$, let $x_\lambda = (1 - \lambda)x_0 + \lambda x_1$ be the corresponding convex linear combination of x_0 and x_1. By assumption $U(x_0) = x_0$, $U(x_1) = x_1$. Since U is nonexpansive,

$$\|U(x_\lambda) - x_0\| = \|U(x_\lambda) - U(x_0)\| \leq \|x_\lambda - x_0\| = \lambda \|x_1 - x_0\|,$$

and

$$\|U(x_\lambda) - x_1\| = \|U(x_\lambda) - U(x_1)\| \leq \|x_\lambda - x_1\| = (1 - \lambda) \|x_1 - x_0\|.$$

Hence,

$$\|x_1 - x_0\| \leq \|x_1 - U(x_\lambda)\| + \|U(x_\lambda) - x_0\| \leq \|x_0 - x_1\|.$$

Since X is strictly convex, it follows that

$$x_1 - U(x_\lambda) = c[U(x_\lambda) - x_0], \quad c > 0.$$

Taking the norms of the two sides of this equation, we see that $c\lambda = (1 - \lambda)$. Hence

$$x_1 - U(x_\lambda) = (1 - \lambda)\lambda^{-1}[x_1 - x_0 - (x_1 - U(x_\lambda))],$$

so that

$$(1 + (1 - \lambda)\lambda^{-1})(x_1 - U(x_\lambda)) = (1 - \lambda)\lambda^{-1}(x_1 - x_0),$$

i.e., $x_1 - U(x_\lambda) = (1 - \lambda)(x_1 - x_0) = x_1 - x_\lambda$. It follows immediately that $U(x_\lambda) = x_\lambda$. q.e.d.

THEOREM (8.3). *Let X be a uniformly convex Banach space, G a bounded convex subset of X, U a nonexpansive mapping of G into X. Then given $\epsilon > 0$, there exists $\xi(\epsilon) > 0$ such that for any pair of points x_0 and x_1 in G with*

$$\|U(x_0) - x_0\| \leq \xi(\epsilon), \quad \|U(x_1) - x_1\| \leq \xi(\epsilon),$$

it follows that, for each point x on the segment joining x_0 to x_1, $\|U(x) - x\| \leq \epsilon$.

PROOF OF THEOREM (8.3). Any point x of the segment joining x_0 to x_1 may be written in the form $x = (1 - \lambda)x_0 + \lambda x_1$ for some λ with $0 \leq \lambda \leq 1$. We note first that if $\|x_1 - x_0\| < \epsilon/3$, then for every x in the segment $\|x - x_0\| < \epsilon/3$. It then follows that

$$\|U(x) - x\| \leq \|U(x) - U(x_0)\| + \|U(x_0) - x_0\| + \|x_0 - x\|$$
$$\leq 2 \|x - x_0\| + \xi(\epsilon) < \epsilon$$

if $\xi(\epsilon) < \epsilon/3$. Hence we need only consider pairs of points with $\|x_1 - x_0\| \geq \epsilon/3$. If we set d_0 to be the diameter of the set G, then for $\lambda < \epsilon/3d_0$, we have $\|x - x_0\| = \lambda \|x_1 - x_0\| < \epsilon/3$, and by the same argument, $\|U(x) - x\| < \epsilon$. Hence we need only consider $\lambda \geq \epsilon/3d_0$. Moreover, if $(1 - \lambda) < \epsilon/3d_0$, we have

$$\|x - x_1\| = (1 - \lambda) \|x_1 - x_0\| < \epsilon/3,$$

and still another application of the same argument assures us that $\|U(x) - x\| < \epsilon$ for $\xi(\epsilon) < \epsilon/3$. Hence we may assume without loss of generality that λ runs through the interval $[\epsilon/3d_0, 1 - \epsilon/3d_0]$ and that $\|x_1 - x_0\| \geq \epsilon/3$.

Let $y = U(x)$. Then

$$\|y - x_0\| \leq \|U(x) - U(x_0)\| + \|U(x_0) - x_0\| \leq \xi(\epsilon) + \lambda \|x_1 - x_0\|$$

and

$$\|y - x_1\| \leq \|U(x) - U(x_1)\| + \|U(x_1) - x_1\| \leq \xi(\epsilon) + (1 - \lambda) \|x_1 - x_0\|.$$

Let

$$z_0 = \lambda^{-1} \|x_1 - x_0\|^{-1} (y - x_0), \qquad z_1 = (1 - \lambda)^{-1} \|x_1 - x_0\|^{-1} (x_1 - y).$$

Then

$$\|z_0\| \leq 1 + \xi(\epsilon) 9 d_0 \epsilon^{-2}, \qquad \|z_1\| \leq 1 + \xi(\epsilon) 9 d_0 \epsilon^{-2},$$

while $\|\lambda z_0 + (1 - \lambda) z_1\| = \|x_1 - x_0\| \cdot \|x_1 - x_0\|^{-1} = 1$ for λ in the interval $[\epsilon/3d_0, 1 - \epsilon/3d_0]$. Hence by the uniform convexity of X, if $\xi(\epsilon)$ is sufficiently small and positive, it follows that $\|z_1 - z_0\| < \epsilon/d_0$, i.e.,

$$\|y - x\| = \|(1 - \lambda)(y - x_0) - \lambda(x_1 - y)\| \leq \|x_1 - x_0\| \lambda(1 - \lambda) \|z_1 - z_0\| < \epsilon.$$

Since $y = U(x)$, it follows that for this choice of $\xi(\epsilon)$, $\|U(x) - x\| < \epsilon$. q.e.d.

THEOREM (8.4). *Let X be a uniformly convex Banach space, G a bounded closed convex subset of X, U a nonexpansive mapping of G into X. Then*

(a) *If $\{u_j\}$ is a weakly convergent sequence in G with weak limit u_0 and if $(I - U)(u_j)$ converges strongly to an element w in X, then $(I - U)(u_0) = w$.*

(b) *$(I - U)(G)$ is a closed subset of X.*

PROOF OF THEOREM (8.4). PROOF OF (a). If we replace the mapping U by the mapping U_w with $U_w(u) = U(u) + w$ for the given w in the hypothesis of part (a), the new mapping is also nonexpansive while $(I - U_w)(u_j) = (I - U) \times (u_j) - w$ converges to 0. If $(I - U_w)(u_0) = 0$, it follows that $(I - U)(u_0) = w$. Hence, we may assume without loss of generality that $w = 0$.

Since $\|(I - U)(u_j)\| = \epsilon_j \to 0$ as $j \to \infty$, we may thin out the sequence to make the convergence faster, and we do this in such a way that for each j, $\epsilon_j \leq \xi(\epsilon_{j-1}) < \epsilon_{j-1}$, where $\xi(\epsilon)$ for any $\epsilon > 0$ is the constant described in the conclusion of Theorem (8.3). With this condition imposed upon the constants ϵ_j, we assert that for any point u in the convex closure of the set $\bigcup_{j \geq k} \{u_j\}$, $\|u - U(u)\| \leq \epsilon_{k-1}$.

To verify this last assertion, we note that each point in this convex closure is a strong limit point of points in the convex closures of the finite sets

$$u_k, u_{k+1}, \ldots, u_s \quad (s \geq k)$$

and that it suffices to prove the desired inequality in convex cl $(u_k, \ldots, u_s)$ for each $s \geq k$.

For each fixed s, we prove this assertion by induction on decreasing k for $k \leq s$. For $k = s$, the inequality is valid. Suppose it is true for a given $k \leq s$, $k \geq 1$, and consider a point u in the convex closure of $\{u_{k-1}, u_k, \ldots, u_s\}$. Then u lies on a segment joining u_{k-1} with a point v in the convex closure of $\{u_k, \ldots, u_s\}$. By the inductive hypothesis, $\|U(v) - v\| \leq \epsilon_{k-1} \leq \xi(\epsilon_{k-2})$. Moreover

$$\|U(u_{k-1}) - u_{k-1}\| \leq \epsilon_{k-1} \leq \xi(\epsilon_{k-2}).$$

Applying Theorem (8.3), it follows that $\|U(u) - u\| \leq \epsilon_{k-2}$, and the inductive step is proved.

Hence we have proved that for each point x in the convex closure of the set $\bigcup_{j \geq k} \{u_j\}$ we have $\|x - U(x)\| \leq \epsilon_{k-1}$. However, this convex closure is weakly closed and contains the weak limit u_0 of the sequence $\{u_j\}$. Hence $\|u_0 - U(u_0)\| \leq \epsilon_j$ $(j \geq 1)$, i.e., $(I - U)(u_0) = 0$. q.e.d.

PROOF OF (b). Suppose that w lies in the closure of $(I - U)(G)$. Then there exists a sequence $\{u_j\}$ in G such that $(I - U)(u_j)$ converges to w as $j \to \infty$. Since G is a bounded weakly closed set in the reflexive space X, it is weakly sequentially compact. Hence we may assume that u_j converges weakly to some u_0 in G. By the conclusion of part (a), $(I - U)(u_0) = w$, i.e., $(I - U)(G)$ is closed in X. q.e.d.

THEOREM (8.5). *Let X be a uniformly convex Banach space, G a bounded closed convex subset of X, U a nonexpansive mapping of G into G. Then U has a fixed point in G.*

PROOF OF THEOREM (8.5). By Theorem (8.1), there exists a sequence $\{u_k\}$ in G such that $(I - U)(u_k) \to 0$. Hence 0 lies in the closure of $(I - U)(G)$. By Theorem (8.4), $(I - U)(G)$ is closed in X. Hence there exists u_0 in G such that $(I - U)(u_0) = 0$, i.e., $U(u_0) = u_0$. q.e.d.

THEOREM (8.6). *Let X be a uniformly convex Banach space, G a closed bounded convex subset of X, G_0 an open subset of X containing the ϵ-neighborhood of G for some $\epsilon > 0$. Suppose that U is a nonexpansive mapping of G_0 into X which carries the boundary of G into G.*

Then U has a fixed point in G.

PROOF OF THEOREM (8.6). For each positive constant d with $0 < d < \epsilon$, let $G_d = \{x \mid d(x, G) \leq d\}$, where $d(x, G)$ denotes the distance of x from the closed set G. Each G_d is a closed bounded convex subset of X contained in G_0. We shall construct a mapping V of G_0 into X by setting $V_\lambda(x) = \lambda U(x) + (1 - \lambda)(x)$. Since V_λ is the convex linear combination of nonexpansive mappings, it is nonexpansive from G_0 to X. We shall show that for each $d < \epsilon$ and for $\lambda > 0$, sufficiently small, V_λ maps G_d into G_d and has the same fixed points as the mapping U.

Let x be a point of G_d. If x lies outside of G, then $\|x - u\| = d(x, G)$ for some point u of the boundary of G, and $d(x, G) \leq d$. Since $U(u)$ lies in G by hypothesis, it follows that

$$d(U(x), G) \leq \|U(x) - U(u)\| \leq \|x - u\| \leq d.$$

Hence $U(x)$ lies in the convex set G_d and thus $V_\lambda(x)$, being a convex linear combination of the points $U(x)$ and x of G_d, is itself an element of G_d. For x in G, we see that $\|V_\lambda(x) - x\| = \lambda \|U(x) - x\| \leq 3\lambda \operatorname{diam}(G)$ since for any point u of the boundary of G,

$$\|U(x) - x\| \leq \|U(x) - U(u)\| + \|U(u) - u\| + \|u - x\|$$
$$\leq 2\|x - u\| + \|U(u) - u\| \leq 3 \operatorname{diam}(G).$$

Hence for $\lambda \leq d(3 \operatorname{diam}(G))^{-1}$, we know that $d(V_\lambda(x), G) \leq \|V(x) - x\| \leq d$, i.e., $V_\lambda(x)$ lies in G_d. With this restriction on λ, V is a nonexpansive mapping of G_d into G_d.

Since we choose $\lambda > 0$, U and V_λ have the same fixed points on G_0. If $U(x) = x$, then $V_\lambda(x) = \lambda x + (1 - \lambda)x = x$. On the other hand, if $V_\lambda(x) = x$, then $\lambda U(x) + (1 - \lambda)x = x$ which implies that $\lambda U(x) = \lambda x$. Since $\lambda > 0$, we obtain $U(x) = x$.

By Theorem (8.5), the nonexpansive mapping V_λ of G_d into itself has a nonempty fixed point set. Since fixed points of U and V_λ coincide, it follows that the set F_d of fixed points of U in G_d is nonempty for each d with $0 < d < \epsilon$. By Theorem (8.2), F_d is a closed convex subset of G_d and hence weakly compact. Hence $\bigcap_{0 < d < \epsilon} F_d \neq \emptyset$, but this intersection consists precisely of the fixed points of U in G. Hence U has a nonempty closed convex set of fixed points in G. q.e.d.

THEOREM (8.7). *Let X be a uniformly convex Banach space, G a closed convex subset of X. Let $\{U_\alpha : \alpha \in \Omega\}$ be a family of nonexpansive mappings of G into G, such that U_α commutes with U_β for each α and β in Ω.*

Then the mappings U_α have a common fixed point x_0, i.e., a point such that $U_\alpha(x_0) = x_0$ for each α in Ω.

PROOF OF THEOREM (8.7). For each α in Ω, $F(U_\alpha)$ is a closed convex subset of G by Theorem (8.2) which is bounded since G is bounded and nonempty by Theorem (8.5). In particular, since X is reflexive, $F(U_\alpha)$ is weakly compact. Hence, it follows that in order to show that $\bigcap_{\alpha \in \Omega} F(U_\alpha)$, the set of common fixed points of the U_α, is nonempty, it suffices to show that for any finite family of indices $\{\alpha_1, \ldots, \alpha_r\}$, $\bigcap_{j=1}^r F(U_{\alpha_j})$ is nonempty.

We shall prove this assertion by induction on r. For $r = 1$, we know already that $F(U_\alpha)$ is nonempty. Suppose that we have already proved that $\bigcap_{j=1}^r F(U_{\alpha_j})$ is nonempty and let α_{r+1} be another index in Ω. Since $U_{\alpha_{r+1}}$ and U_{α_j}, $j \leq r$, commute, it follows that for x in $F(U_{\alpha_j})$,

$$U_{\alpha_j}(U_{\alpha_{r+1}}(x)) = U_{\alpha_{r+1}}(U_{\alpha_j}(x)) = U_{\alpha_{r+1}}(x),$$

i.e., $U_{\alpha_{r+1}}(F(U_{\alpha_j})) \subset F(U_{\alpha_j})$. Thus if

$$G_0 = \bigcap_{j=1}^{r} F(U_{\alpha_j}),$$

it follows by the last remark that $U_{\alpha_{r+1}}$ maps the closed bounded convex subset G_0 of X into G_0. Hence by Theorem (8.5), $U_{\alpha_{r+1}}$ has a fixed point in G_0, i.e.,

$$\varnothing \neq G_0 \cap F(U_{\alpha_{r+1}}) = \bigcap_{j=1}^{r+1} F(U_{\alpha_j}),$$

and the induction step is complete. q.e.d.

THEOREM (8.8). *Let X be a uniformly convex Banach space, G a closed subset of X, $\{U_\alpha : \alpha \in \Omega\}$ a family of nonexpansive mappings of G into G such that for each α and β in Ω, U_α commutes with U_β. Suppose further that for each pair α and β in Ω, there exists ξ in Ω such that $U_\xi = U_\alpha U_\beta$.*

Then a necessary and sufficient condition that the mappings $\{U_\alpha : \alpha \in \Omega\}$ have a common fixed point in G is that there exists an element y_0 in G such that $\|U_\alpha y_0\|$ is uniformly bounded over all α in Ω.

PROOF OF THEOREM (8.8). Obviously, if the family $\{U_\alpha : \alpha \in \Omega\}$ has a common fixed point x_0, then $U_\alpha x_0 = x_0$ implies that we may choose $y_0 = x_0$.

Conversely, suppose that an element y_0 of G exists such that for a constant M, $\|U_\alpha y_0\| \leq M$ for all α in Ω. To show that the mappings $\{U_\alpha : \alpha \in \Omega\}$ have a common fixed point, it suffices by Theorem (8.7) to show that there exists a closed bounded convex subset G_0 of G which is mapped into itself by all the mappings U_α. We construct such a subset G_0 in the following way: Let $R = 2M$, and for any point x of X, let $B_R(x)$ denote the closed ball of radius R about x. We set

$$G_1 = \bigcup_{\alpha \in \Omega} \bigcap_{\beta \in \Omega} B_R(U_\alpha U_\beta y_0) \cap G.$$

We note first that if G_α is given by $\bigcap_{\beta \in \Omega} B_R(U_\alpha U_\beta y_0) \cap G$ then each G_α for α in Ω is a closed bounded convex subset of G which is nonempty because it contains the point y_0. Since $G_1 = \bigcup_{\alpha \in \Omega} G_\alpha$, G_1 is a nonempty subset of G which contains y_0 and is contained in $B_{2R}(y_0)$. We shall verify two further facts to complete the argument:

(1) G_1 is convex,
(2) U_ζ maps G_1 into G_1 for each ζ in Ω.

To prove that G_1 is convex, suppose x_0 and x_1 lie in G_1. Then x_0 lies in G_α and x_1 lies in G_β for some indices α and β in Ω. By hypothesis, there exists ξ in Ω such that $U_\xi = U_\alpha U_\beta = U_\beta U_\alpha$. Then

$$G_\alpha = \bigcap_{\lambda \in \Omega} B_R(U_\alpha U_\lambda y_0) \subset \bigcap_{\lambda \in \Omega} B_R(U_\alpha U_\beta U_\lambda y_0) = G_\xi$$

and similarly $G_\beta \subset G_\xi$. Since G_ξ is convex, the segment joining x_0 to x_1 lies within

G_ξ. On the other hand, $G_\xi \subset G_1$ so that the segment is contained in G_1 and G_1 is convex.

Let ζ be an index in Ω and α an index in Ω. By hypothesis there exists an index λ in Ω such that $U_\lambda = U_\zeta U_\alpha$. We shall show that U_ζ maps G_α into G_λ for this choice of λ, and it then follows that U_ζ maps $G_1 = \bigcup_{\alpha \in \Omega} G_\alpha$ into itself. Let y be a point of G_α. For each β in Ω, $\|y - U_\alpha U_\beta y_0\| \leq R$. Since U_ζ is a nonexpansive mapping of G into G, it follows that $\|U_\zeta y - U_\zeta U_\alpha U_\beta y_0\| \leq R$, i.e.,

$$\|U_\zeta y - U_\lambda U_\beta y\| = \|U_\zeta y - U_\zeta U_\alpha U_\beta y_0\| \leq R.$$

Since β was any index in Ω, it follows that

$$U_\zeta(y) \in \bigcap_{\beta \in \Omega} B_R(U_\lambda U_\beta y_0) = G_\lambda.$$

Finally, we let G_0 be the closure of G_1. Then G_0 is a closed bounded convex subset of G mapped into itself by all the nonexpansive mappings U_α and the family $\{U_\alpha : \alpha \in \Omega\}$ has a common fixed point in G_0. q.e.d.

An important special case of the general situation treated in Theorem (8.8) is that provided by the following:

DEFINITION (8.2). *Let X be a Banach space, G a closed convex subset of X. Then by a one-parameter semigroup of nonexpansive mappings of G, we mean a family $\{U_t, t \in R^+\}$ of nonexpansive mappings of G into G indexed by R^+, the nonnegative real numbers, such that the following conditions hold:*

(1) *For $s, t \geq 0$, $U_s U_t = U_{s+t}$.*

(2) *$U_0 = I$. For each x in G, the mapping of R^+ into G given by $t \to U_t x$ is continuous from R^+ to the strong topology of G.*

COROLLARY TO THEOREM (8.8). *Let X be a uniformly convex Banach space, G a closed convex subset of X, $\{U_t : t \in R^+\}$ a one-parameter semigroup of nonexpansive mappings of G. Then the mappings $\{U_t : t \in R^+\}$ have a common fixed point if there exists an element y_0 of G such that $\|U_t y_0\| \leq M$ for some constant $M > 0$ and all t in R^+.*

PROOF OF THE COROLLARY. The mappings U_t and U_s commute for s and t in R^+ since $U_s U_t = U_{s+t} = U_t U_s$, and the family $\{U_t : t \in R^+\}$ is closed under composition of mappings. q.e.d.

REMARK. We note that if the family $\{U_\alpha : \alpha \in \Omega\}$ of Theorem (8.8) has a common fixed point, then it follows that for every y_0 in G, $\|U_\alpha y_0\|$ is bounded over α in Ω. Indeed, if x_0 is the common fixed point, then

$$\|U_\alpha y_0 - x_0\| = \|U_\alpha y_0 - U_\alpha x_0\| - \|y_0 - x_0\|,$$

so that $\|U_\alpha y_0\| \leq \|x_0\| + \|y_0 - x_0\|$. This generalizes the fact that for a single nonexpansive mapping U of a closed bounded convex subset G of a uniformly convex space, U has a fixed point if and only if there exists for each point y_0 in G, a bounded invariant set containing y_0.

We have remarked at the opening of this section on the existence of important connections between the theory of nonexpansive mappings on the one hand and the theory of monotone and accretive mappings on the other. The simplest aspect of such a connection is given by the following:

THEOREM (8.9). *Let X be a Banach space, D a subset of X, U a nonexpansive mapping of D into X. Let $T = I - U : D \to X$. Then*
   (a) *If X is a Hilbert space H, T is a monotone mapping of D into H.*
   (b) *If X is a general Banach space, then T is an accretive mapping of X into X.*

PROOF OF THEOREM (8.9). Since a monotone mapping of a subset D of a Hilbert space H into H is accretive, it suffices to prove the assertion of part (b). Let J be a (single-valued) duality mapping of X into X^*, i.e., J is a mapping (not necessarily continuous) of X into X^* which satisfies the two conditions:
   (1) For each x in X, $(J(x), x) = \|x\|^2$.
   (2) For each x in X, $\|J(x)\| = \|x\|$.
We consider, for a pair of elements x and u in D,

$$(T(x) - T(u), J(x - u)) = (x - u, J(x - u)) - (U(x) - U(u), J(x - u))$$
$$\geq \|x - u\|^2 - \|U(x) - U(u)\| \cdot \|x - u\|$$
$$\geq \|x - u\| \, [\|x - u\| - \|U(x) - U(u)\|] \geq 0$$

since $\|U(x) - U(u)\| \leq \|x - u\|$ by the nonexpansiveness of U.

Hence T is an accretive mapping of D into X. q.e.d.

The converse of Theorem (8.9) is not valid, and it is therefore useful to introduce the following definition:

DEFINITION (8.3). *Let X be a Banach space, D a subset of X, U a mapping of D into X. Then U is said to be pseudo-contractive if $T = (I - U)$ is an accretive mapping of D into X.*

We consider fixed point theorems for pseudo-contractive mappings below, in the context of the discussion of accretive mappings, which extend the theory of nonexpansive mappings to this class.

One natural question which arises from the analogy between the fixed point theory of nonexpansive mappings in uniformly convex Banach spaces and the Contraction Mapping Principle for fixed points of strict contractions on complete metric spaces is the following: Can we obtain the fixed points of nonexpansive mappings which are asserted to exist in the fixed point theorems as the limits of determined sequences of approximants, and in particular of iterates of mappings? An affirmative answer to such questions is provided by the next several theorems. We begin with the following definitions:

DEFINITION (8.4). *Let X be a Banach space, G a closed convex subset of X, U a nonexpansive mapping of G into G, y_0 a given point of G. Then for $0 < \lambda < 1$, we let $U_\lambda(x) = (1 - \lambda)y_0 + \lambda U(x)$.*

DEFINITION (8.5). *Let X be a Banach space, G a closed convex subset of X, U a nonexpansive mapping of G into G. Then for $0 < \lambda < 1$, we set $U^{(\lambda)}(x) = (1 - \lambda)x + \lambda U(x)$.*

PROPOSITION (8.1). *Let X be a Banach space, G a closed convex subset of X, U a nonexpansive mapping of G into G, y_0 a point of G. Then*

(a) *For each λ with $0 < \lambda < 1$, U_λ is a strict contraction of G into G with $\|U_\lambda(x) - U_\lambda(u)\| \leq \lambda \|x - u\|$ and has a unique fixed point u_λ in G.*

(b) *For each λ with $0 < \lambda < 1$, $U^{(\lambda)}$ is a nonexpansive mapping of G into G which has the same fixed points as U.*

(c) *If X is uniformly convex, G is bounded, and if we define $u_n = (U^{(\lambda)})^n y_0$, then $(I - U)(u_n) \to 0$ as $n \to \infty$.*

PROOF OF PROPOSITION (8.1). PROOF OF (a). Since G is convex and $U_\lambda(x)$ is a convex linear combination of y_0 and $U(x)$, it follows that U_λ maps G into G. For each pair x and u in G, $U_\lambda(x) - U_\lambda(u) = \lambda(U(x) - U(u))$. Hence

$$\|U_\lambda(x) - U_\lambda(u)\| = \lambda \|U(x) - U(u)\| \leq \lambda \|x - u\|.$$

In particular, U_λ has exactly one fixed point u_λ in G by the Contraction Mapping theorem. q.e.d.

PROOF OF (b). $U^{(\lambda)}$ is a convex linear combination of the mappings I and U, both of which are nonexpansive. Since the family of nonexpansive mappings of G into X is convex, it follows that $U^{(\lambda)}$ is also nonexpansive.

If x is a fixed point of U, it follows that

$$U^{(\lambda)}(x) = (1 - \lambda)x + \lambda x = x,$$

and x is a fixed point of $U^{(\lambda)}$. If x is a fixed point of $U^{(\lambda)}$, $\lambda > 0$, then $(1 - \lambda)x + \lambda U(x) = x$, i.e., $\lambda U(x) = \lambda x$. Since $\lambda > 0$, $U(x) = x$. q.e.d.

PROOF OF (c). Since G is bounded, closed, and convex while X is uniformly convex, it follows from Theorem (8.5) that U has a fixed point x_0 in G. Then:

$$u_{n+1} - x_0 = (1 - \lambda)(u_n - x_0) + \lambda(U(u_n) - x_0).$$

Hence

$$\|u_{n+1} - x_0\| \leq (1 - \lambda) \|u_n - x_0\| + \lambda \|U(u_n) - x_0\| \leq \|u_n - x_0\|$$

since $\|U(u_n) - x_0\| = \|U(u_n) - U(x_0)\| \leq \|u_n - x_0\|$. Moreover

$$u_{n+1} - u_n = (1 - \lambda)u_n + \lambda U(u_n) - u_n = \lambda[U(u_n) - u_n].$$

Since $\|u_n - x_0\|$ is decreasing in n, it converges to a limit, $p_0 \geq 0$ as $n \to \infty$. If $p_0 = 0$, $u_n \to x_0$ in X and $\|u_{n+1} - u_n\| \to 0$. In this case,

$$\|(I - U)(u_n)\| = \lambda^{-1} \|u_{n+1} - u_n\| \to 0 \quad \text{as} \quad n \to \infty.$$

Suppose $p_0 > 0$. Then given $\epsilon > 0$, we may choose $N(\epsilon)$ such that for $n \geq N_\epsilon$, $p_0 \leq \|u_n - x_0\| \leq p_0 + \epsilon$. Thus $\|u_{n+1} - x_0\| \geq \epsilon$, while $u_{n+1} - x_0 = (1 - \lambda) \times [u_n - x_0] + \lambda[U(u_n) - x_0]$, where

$$\|u_n - x_0\| \leq p_0 + \epsilon, \quad \|U(u_n) - x_0\| \leq \|u_n - x_0\| \leq p_0 + \epsilon.$$

Since λ is a fixed constant with $0 < \lambda < 1$, it follows from the uniform convexity of X that $[U(u_n) - x_0] - [u_n - x_0] \to 0$ $(n \to \infty)$, i.e., $(I - U)(u_n) \to 0$. q.e.d.

THEOREM (8.10). *Let H be a Hilbert space, G a closed bounded convex subset of X, U a nonexpansive mapping of G into G. Let y_0 be a given point of G, and let U_λ for $0 < \lambda < 1$ be given by $U_\lambda(x) = (1 - \lambda)y_0 + \lambda U(x)$. Then U_λ has exactly one fixed point u_λ in G, and as $\lambda \to 1$, u_λ converges strongly in H to the fixed point x_0 of U in G closest to y_0.*

PROOF OF THEOREM (8.10). Since both the hypothesis and the conclusion of our theorem are invariant under translation, we may assume without loss of generality that 0 lies in G and that $y_0 = 0$. Let $0 < \lambda < \xi \le 1$. By Proposition (8.1)(a), U_λ has exactly one fixed point u_λ in G. Let u_ξ be a fixed point of U_ξ, the unique one if $\xi < 1$. By Theorem (8.9), $T = (I - U)$ is monotone, i.e.,

$$((I - U)(u_\lambda) - (I - U)(u_\xi), u_\lambda - u_\xi) \ge 0.$$

Since $y_0 = 0$, $u_\lambda = \lambda U(u_\lambda)$, $u_\xi = \xi U(u_\xi)$, so that

$$(I - U)(u_\lambda) = (1 - 1/\lambda)u_\lambda, \qquad (I - U)(u_\xi) = (1 - 1/\xi)u_\xi.$$

Hence $((1 - 1/\lambda)u_\lambda - (1 - 1/\xi)u_\xi, u_\lambda - u_\xi) \ge 0$ since

$$((1 - 1/\lambda)u_\lambda - (1 - 1/\xi)u_\xi, u_\lambda - u_\xi)$$
$$= (1 - 1/\xi)(u_\lambda - u_\xi, u_\lambda - u_\xi) + (1/\xi - 1/\lambda)(u_\lambda, u_\lambda - u_\xi)$$

with $(1 - 1/\xi) \le 0$, $(1/\xi - 1/\lambda) < 0$, it follows that

$$(u_\lambda, u_\lambda - u_\xi) \le (1/\lambda - 1/\xi)^{-1}(1 - 1/\xi) \|u_\lambda - u_\xi\|^2 \le 0.$$

Thus we have

$$\|u_\lambda\|^2 \le (u_\lambda, u_\xi) \le \|u_\lambda\| \cdot \|u_\xi\|; \|u_\lambda\| \le \|u_\xi\| \qquad (\lambda < \xi \le 1)$$

and the bounded increasing function $\|u_\lambda\|$ of λ converges to a limit c as $\lambda \to 1-$. Hence

$$\|u_\lambda - u_\xi\|^2 = (u_\xi, u_\xi - u_\lambda) + (u_\lambda, u_\lambda - u_\xi) \le (u_\xi, u_\xi - u_\lambda)$$
$$\le \|u_\xi\|^2 - (u_\lambda, u_\xi) \le \|u_\xi\|^2 - \|u_\lambda\|^2 \to 0 \qquad (\xi, \lambda \to 1-).$$

It follows that u_λ converges strongly to an element u of G as $\lambda \to 1$. Moreover, $U(u) = \lim_{\lambda \to 1} U(u_\lambda) = \lim_{\lambda \to 1} \lambda^{-1} u_\lambda = u$.

Let v be any fixed point of U. Then we may take $\xi = 1$, $u_\xi = v$ in the above calculations, and we obtain $\|u_\lambda\| \le \|v\|$. Taking the limit as $\lambda \to 1$, we find that $\|u\| \le \|v\|$. Hence u has the minimum norm of all the elements v of $F(U)$, i.e., it is the closest to $y_0 = 0$. The uniqueness of this nearest element follows from the convexity of $F(U)$ and the uniform convexity of the Hilbert space H. q.e.d.

THEOREM (8.11). *Let X be a uniformly convex Banach space, G a bounded closed convex subset of X, U a nonexpansive mapping of G into G. Let λ be a constant with $0 < \lambda < 1$, y_0 a given point of G. Suppose that U has at most one fixed point in G.*

Then as $n \to \infty$, $(U^{(\lambda)})^n y_0$ converges weakly to the unique fixed point x_0 of U in G, where $U^{(\lambda)} = (1 - \lambda)I + \lambda U$.

PROOF OF THEOREM (8.11). If we set $u_n = (U^{(\lambda)})^n y_0$, then it follows from Proposition (8.1)(c), that $(I - U)(u_n) \to 0$ as $n \to \infty$. Since G is a weakly compact subset of the reflexive space X, to show that u_n converges weakly, it suffices to assume that we have a weakly convergent infinite subsequence $\{u_{n_j}\}$ with weak limit w and to show that w is independent of the choice of the subsequence. However, since $(I - U)(u_{n_j})$ converges strongly to 0 in X, it follows from Theorem (8.4)(a) that $(I - U)(w) = 0$, i.e., w is a fixed point of U in G. Since such an element w is unique by hypothesis, u_n converges weakly to the unique fixed point x_0 of U in G. q.e.d.

THEOREM (8.12). *Let X be a uniformly convex Banach space, G a bounded closed convex subset of X, U a nonexpansive mapping of G into G. Let λ be a constant with $0 < \lambda < 1$, and set $U^{(\lambda)} = (1 - \lambda)I + \lambda U$. Suppose that $(I - U)$ is a closed mapping of G into X, i.e., $(I - U)$ maps closed sets of G on closed sets in X.*

Then for each y_0 in G, $(U^{(\lambda)})^n y_0$ converges strongly to a fixed point x_0 of U.

PROOF OF THEOREM (8.12). For any fixed point v of U in G, we know that v is a fixed point of $U^{(\lambda)}$ and hence that

$$\|u_{n+1} - v\| = \|U^{(\lambda)}(u_n) - U^{(\lambda)}(v)\| \leq \|u_n - v\|.$$

Hence if there exists a point v of $F(U)$ which is a strong limit point of an infinite subsequence of the sequence $\{u_n\}$, the whole sequence $\{u_n\}$ must converge strongly to v. To show that such an element v of $F(U)$ exists it suffices to show that there is a point v in $F(U) \cap \text{cl}\,(\bigcup_{n \geq 1} \{u_n\})$. However, if $C = \text{cl}\,(\bigcup_{n \geq 1} \{u_n\})$ (cl denotes closure), $(I - U)(C)$ is closed in X and contains a sequence converging to 0 strongly in X. Hence there exists v in C such that $(I - U)(v) = 0$, i.e., v lies in $F(U)$. q.e.d.

Under sharper restrictions on the character of the space X, we sharpen the result of Theorem (8.11) by eliminating the hypothesis that the mapping U has a unique fixed point. The additional hypothesis upon X which is involved is the following:

DEFINITION (8.3). *Let X be a Banach space. Then X is said to have a weakly continuous (generalized) duality mapping if there exists a continuous, strictly increasing function Φ from R^+ onto R^+ such that the corresponding duality mapping J_Φ is single-valued and continuous from the weak topology of X to the weak topology of X^*.*

We recall the definition of the generalized duality mapping J_Φ with gauge function Φ: J_Φ is a mapping of X into X^* such that for each u in X,

$$(J(u), u) = \|J(u)\| \cdot \|u\|, \qquad \|J(u)\| = \Phi(\|u\|).$$

PROPOSITION (8.2). *Let X be a Hilbert space or one of the sequence spaces p with $1 < p < \infty$. Then X is a uniformly convex space with a weakly continuous duality mapping.*

PROOF OF PROPOSITION (8.2). For the case when X is a Hilbert space H, the identity mapping I is a duality mapping which is continuous in the weak topology from H to $H^* = H$.

When X is the space l^p, we define the duality mapping J_Φ for $\Phi(r) = r^{p-1}$ by setting $J_\Phi(u)$ for $u = (u_1, u_2, \ldots, u_n, \ldots)$ to be equal to

$$J_\Phi(u) = (|u_1|^{p-2}u_1, \ldots, |u_n|^{p-2}u_n, \ldots).$$

It follows easily that

$$(J_\Phi(u), u) = \sum_{j=1}^\infty |u_j|^p = \|u\|_p^p,$$

and

$$\|J_\Omega(u)\|_{X^*} = \left(\sum_{j=1}^\infty |u_j|^{(p-1)p/(p-1)}\right)^{(p-1/p)} = \|u\|_X^{p-1}, \quad X^* = l^{p'}, \, p' = p(p-1)^{-1}.$$

Hence J_Φ is indeed the duality mapping of X into X^* with the given gauge function Φ (and is unique by the uniform convexity of the given space $X^* = l^{p'}$). To show that J_Φ is weakly continuous (i.e., continuous from the weak topology of X to the weak topology of X^*), we note that if a sequence of elements $u^{(k)}$ of X converges weakly to an element u of X, then for each integer j, the jth component of $u^{(k)}$ which we write as $u_j^{(k)}$ converges to the jth component of u. By the definition of J_Φ, the jth component of $J_\Phi(u)$ depends only upon the jth component of u, and $(J_\Phi(u^{(k)}))_j$ converges for each $j \geq 1$ to $(J_\Phi(u))_j$. On the other hand, weak convergence in either X or X^* is equivalent for these spaces to boundedness in the space and convergence of each component. Hence $J_\Phi(u^{(k)})$ converges weakly to $J_\Phi(u)$ in $l^{p'}$. q.e.d.

PROPOSITION (8.3). *Let X be a uniformly convex Banach space with a weakly continuous duality mapping J_Φ for some gauge function Φ. Then if $\{u_k\}$ is a sequence in X converging weakly to u_0 in X, it follows for each u in X with $u \neq u_0$,*

$$\limsup \|u_k - u_0\| < \limsup \|u_k - u\|.$$

PROOF OF PROPOSITION (8.3). By Proposition (7.6), if ψ is the primitive of Φ with $\psi(0) = 0$ (i.e., $\psi' = \Phi$), then J_Φ is the subgradient of the convex function $\psi(\|x\|)$ on X so that for each x and v in X, $\psi(\|v\|) \geq \psi(\|x\|) + (J_\Phi(x), v - x)$. If we set $v = u_k - u$, $x = u_k - u_0$ for a given integer k, we obtain

$$\psi(\|u_k - u\|) \geq \psi(\|u_k - u_0\|) + (J_\Phi(u_k - u_0), u_0 - u).$$

Since $u_k - u_0$ converges weakly to 0 in X and J_Φ is assumed to be continuous from the weak topology of X to the weak topology of X^*, $J_\Phi(u_k - u_0)$ converges weakly to 0 in X^* and $(J_\Phi(u_k - u_0), u_0 - u) \to 0$. Hence

$$\limsup \psi(\|u_k - u\|) \geq \limsup \psi(\|u_k - u_0\|).$$

Since ψ is strictly increasing and continuous, it has a continuous inverse, and it follows that $\limsup \|u_k - u\| \geq \limsup \|u_k - u_0\|$.

Let C be the set of all u in X for which the equality is assumed. Then C is convex for if v_0 and v_1 lie in C and $v_\lambda = (1 - \lambda)v_0 + \lambda v_1$ with $0 \leq \lambda \leq 1$, then $\|u_k - v_\lambda\| \leq (1 - \lambda)\|u_k - v_0\| + \lambda \|u_k - v_1\|$ so that

$$\limsup \|u_k - v_\lambda\| \leq (1 - \lambda) \limsup \|u_k - v_0\| + \lambda \limsup \|u_k - v_1\|$$
$$\leq \limsup \|u_k - u_0\|.$$

On the other hand, X is assumed to be uniformly convex so that for $v' = \frac{1}{2}(v_0 + v_1)$, and $\epsilon > 0$, there exists $\xi(\epsilon) > 0$ such that if

$$\|v' - x\| \geq \tfrac{1}{2}[\|v_0 - x\| + \|v_1 - x\|] - \xi(\epsilon)$$

with $\|v_0 - x\|$, $\|v_1 - x\| \leq N$ for a given constant M, then $\|v_0 - v_1\| \leq \epsilon$. We may choose $K(\epsilon)$ so large that for $k \geq K(\epsilon)$

$$\|v_0 - u_k\| \leq \limsup \|u_0 - u_k\| + \xi(\epsilon)/2,$$
$$\|v_1 - u_k\| \leq \limsup \|u_0 - u_k\| + \xi(\epsilon)/2.$$

Since $\limsup \|v' - u_k\| = \limsup \|u_0 - u_k\|$, we may choose $k \geq K(\epsilon)$ such that $\|v' - u_k\| \geq \limsup \|u_0 - u_k\| - \xi(\epsilon)/2$. For such an integer k, we obtain

$$\|v' - u_k\| \geq \tfrac{1}{2}[\|v_0 - u_k\| + \|v_1 - u_k\|] - \xi(\epsilon).$$

Hence $\|v_0 - v_1\| \leq \epsilon$. Since $\epsilon > 0$ is arbitrary, it follows that C is the one point set $\{u_0\}$. q.e.d.

PROPOSITION (8.4). *Let X be a uniformly convex Banach space with a weakly continuous duality mapping J_Φ for some gauge function Φ. Let F be a closed convex bounded subset of X, $\{u_k\}$ a bounded sequence in X such that for each x in F, $\|u_k - x\|$ does not increase with k. Suppose that any weak limit point of the sequence $\{u_k\}$ lies in F.*

Then there exists a point u_0 of F such that u_k converges weakly to u_0.

PROOF OF PROPOSITION (8.4). Since the sequence $\{u_k\}$ is bounded and X is reflexive, to show that $\{u_k\}$ converges weakly, it suffices to take any weakly convergent subsequence and show that the limit is independent of the subsequence. Since each weak limit point of the sequence lies in F, the weak limit u_0 of the sequence would then lie in F.

For each x in F, let $q(x) = \lim_{k \to \infty} \|u_k - x\|$. Since $\|u_k - x\|$ does not increase with k, $q(x)$ exists for every x in F. Since $q(x)$ can be considered as the

$$\limsup \|u_k - x\|,$$

$q(x)$ is a lower-semicontinuous convex function on F. Since F is a closed bounded convex subset of X, F is weakly compact. The function q is lower-semicontinuous and convex and hence weakly lower-semicontinuous on F. Hence, it assumes its minimum on F. Let u_0 be a point at which this minimum is assumed.

Consider an infinite weakly convergent subsequence $\{u_{k_j}\}$ of the sequence $\{u_k\}$ with weak limit u_1. Since u_1 is a weak limit point of the sequence $\{u_k\}$, it follows from the hypothesis that u_1 lies in F. By Proposition (8.3), if u_1 is different from the given element u_0 of F, then $\lim \|u_k - u_1\| < \lim \|u_k - u_0\|$, i.e., $q(u_1) < q(u_0)$. Since $q(x)$ assumes its minimum value on F at u_0, this is impossible. Hence $u_1 = u_0$, and therefore the whole sequence $\{u_k\}$ converges weakly to the element u_0 of F. q.e.d.

THEOREM (8.13). *Let X be a uniformly convex Banach space with a weakly continuous duality mapping J_Φ, G a closed bounded convex subset of X, U a nonexpansive mapping of G into G. For a given λ with $0 < \lambda < 1$, let $U^{(\lambda)} = (1 - \lambda)I + \lambda U$.*

Then for each y_0 in G, $(U^{(\lambda)})^n y_0$ converges weakly to a fixed point u_0 of U in G.

PROOF OF THEOREM (8.13). Let $u_k = (U^{(\lambda)})^k(y_0)$. By Proposition (8.1)(c), $(I - U)(u_k)$ converges strongly to 0 in X. By Theorem (8.4), each weak limit point of the sequence $\{u_k\}$ is contained in the fixed point set $F(U)$ of U in G. By Theorem (8.2), $F(U)$ is a closed bounded convex subset of G and is identical with the fixed point set of $U^{(\lambda)}$. For each x in $F(U)$,

$$\|u_{k+1} - x\| = \|U^{(\lambda)}(u_k) - U^{(\lambda)}(x)\| \leq \|u_k - x\|,$$

so that the sequence of norms $\|u_k - x\|$ is nonincreasing with k. Hence by Proposition (8.4), the sequence $\{u_k\}$ converges weakly to an element u_0 of $F(U)$. q.e.d.

The results of Theorems (8.11), 8.12), and (8.13) can be extended to more general results for compositions of a sequence of nonexpansive mappings, rather than iterations of the single nonexpansive mapping $U^{(\lambda)}$. Such an extension can be given in terms of the following definition:

DEFINITION (8.4). *Let X be a Banach space, G a closed convex subset of X, $\{V_j\}$ a sequence of mappings of G into G. Then we say that $\{V_j\}$ is an admissible sequence of mappings with respect to the point u_0 of G if u_0 is a fixed point of all the mappings V_j and if there exist two continuous strictly increasing functions ψ_0 and ψ_1 from R^+ to R^+ with $\psi_0(0) = \psi_1(0) = 0$ such that for all x in G and all $j \geq 1$,*

$$\psi_0(\|V_j(x) - u_0\|) + \psi_1(\|x - V_j(x)\|) \leq \psi_0(\|x - u_0\|).$$

DEFINITION (8.5). *Let $\{V_j\}$ and $\{U_k\}$ be two sequences of mappings of G into G. Then $\{V_j\}$ is said to be weakly recurrent with respect to the sequence $\{U_k\}$ if for each $\epsilon > 0$ and each $k \geq 1$, there exists a positive integer $m = m(\epsilon, k)$ such that from each block of m successive integers $[n+1, \ldots, n+m]$, we can find an integer j in the block such that for all x in G, $\|V_j(x) - U_k(x)\| < \epsilon$.*

THEOREM (8.14). *Let X be a uniformly convex Banach space, G a closed bounded convex subset of X, $\{V_j\}$ and $\{U_k\}$ two sequences of nonexpansive mappings of G into G having a common fixed point u_0. Suppose that the sequence $\{V_k\}$ is admissible with respect to u_0 in the sense of Definition (8.4), and that $\{V_j\}$ is weakly recurrent with respect to the sequence $\{U_k\}$ in the sense of Definition (8.5). Let $S_n = V_n V_{n-1} \cdots V_2 V_1$. Then*

(a) If u_0 is the unique common fixed point of the family of mappings $\{U_k\}$, then for each y_0 in G, $S_n(y_0)$ converges weakly to u_0.

(b) *If for one of the mappings U_k, $(I - U_k)$ is a closed mapping of G into X and $\bigcap_k F(U_k) \subset \bigcap_j F(V_j)$, then $S_n(y_0)$ converges strongly to a common fixed point u of the family $\{U_k\}$ as $n \to \infty$ for every y_0 in G.*

(c) *If X has a weakly continuous duality mapping J_Φ and $\bigcap_k F(U_k) \subset \bigcap_j F(V_j)$, then for each y_0 in G, $S_n(y_0)$ converges weakly in X to a common fixed point v_1 of the family $\{U_k\}$.*

PROOF OF THEOREM (8.14). We begin with some basic remarks which are used in the proofs for all three cases. Let $u_n = S_n(y_0)$. Then $u_{n+1} = V_{n+1}(u_n)$. By the hypothesis that the sequence $\{V_j\}$ is admissible with respect to u_0 in the sense of Definition (8.4), we have

$$\psi_1(\|u_n - u_{n+1}\|) \leq \psi_0(\|u_n - u_0\|) - \psi_0(\|u_{n+1} - u_0\|).$$

If we sum this inequality from $n = 1$ to m, we obtain

$$\sum_{n=1}^m \psi_1(\|u_n - u_{n+1}\|) \leq \psi_0(\|u_1 - u_0\|) - \psi_0(\|u_{m+1} - u_0\|)$$
$$\leq \psi_0(\|u_1 - u_0\|).$$

If we let $m \to \infty$, we find that

$$\sum_{j=1}^\infty \psi_1(\|u_n - u_{n+1}\|) \leq \psi_0(\|u_1 - u_0\|) < \infty.$$

In particular, it follows that $\psi_1(\|u_n - u_{n+1}\|) \to 0$ $(n \to \infty)$, and hence that $\|u_n - u_{n+1}\| \to 0$ $(n \to \infty)$.

In terms of the mappings V_j, this last fact can be written as $(I - V_{n+1})(u_n) \to 0$ strongly in X.

Let k be an integer, and let $\epsilon > 0$ be given. By the hypothesis that the sequence $\{V_j\}$ is weakly recurrent with respect to the sequence $\{U_k\}$, there exists an integer $m = m(\epsilon, k)$ such that in each block of successive integers of the form $[n, n+1, \ldots, n+m-1]$ there exists an integer j such that for all x in G, $\|U_k(x) - V_j(x)\| < \epsilon$. If n is sufficiently large and $j \geq n$, we know by the preceding discussion that $\|u_j - V_j(u_j)\| < \epsilon$. Hence

$$\|u_j - U_k(u_j)\| \leq \|u_j - V_j(u_j)\| + \|V_j(u_j) - U_k(u_j)\| < 2\epsilon.$$

On the other hand,

$$u_n - U_k(u_n) = (u_n - u_j) + (u_j - U_k(u_j)) + (U_k(u_j) - U_k(u_n))$$

while

$$\|u_n - u_j\| \leq \sum_{r=n}^{j-1} \|u_r - u_{r+1}\| \leq m \sup_{r \geq n} \|u_r - u_{r+1}\| \to 0 \quad (n \to \infty),$$

and $\|U_k(u_j) - U_k(u_n)\| \leq \|u_j - u_n\| \to 0$ $(n \to \infty)$.

Since $\epsilon > 0$ can be made arbitrarily small by taking m sufficiently large and then $\|u_j - u_n\|$ and $\|U_k(u_j) - U_k(u_n)\|$ can be made small for fixed m by taking n very large, it follows that for each fixed integer k, $\|(I - U_k)(u_n)\| \to 0$ $(n \to \infty)$. It follows from Theorem (8.4) since X is uniformly convex and U_k is nonexpansive, that any weak limit point u of the sequence $\{u_k\}$ is a fixed point of each of the mappings U_k, i.e., $u \in \bigcap_k F(U_k)$.

PROOF OF (a). Since G is a weakly sequentially compact set, to prove that $\{u_n\}$ converges weakly to u_0, it suffices to show that for any weakly convergent infinite subsequence $\{u_{n_j}\}$, its weak limit w must equal u_0. However, for each integer s, $(I - U_s)(u_{n_j}) \to 0$ as $j \to \infty$. Hence w lies in $\bigcap_s F(U_s) = \{u_0\}$ under the assumptions of part (a). Hence $w = u_0$. q.e.d.

PROOF OF (b). Let k_0 be an integer such that $(I - U_k)$ is a closed mapping of G into X. Since $(I - U_k)(u_n) \to 0$ strongly in X, it follows that there exists a strong limit point v_1 of the sequence $\{u_n\}$ which lies in $F(U_k)$. However, each strong limit point v_1 of the sequence $\{u_n\}$ must lie in $\bigcap_s F(U_s)$. Since $V_n(v_1) = v_1$ for each n, we note that $\|u_{n+1} - v_1\| = \|V_n(u_n) - V_n(v_1)\| \leq \|u_n - v_1\|$ since V_n is nonexpansive by hypothesis. If for an infinite subsequence $\|u_{n_j} - v_1\| \to 0$ $(j \to \infty)$, it follows that u_n converges strongly to v_1. q.e.d.

PROOF OF (c). For each x in $\bigcap_s F(U_s)$, $\|u_n - x\|$ is decreasing by the argument of the preceding proof since $V_n(x) = x$ for each n. Each weak limit point of the sequence $\{u_n\}$ must lie in $\bigcap_s F(U_s)$ by the argument of the beginning of the proof of the theorem. Since X is uniformly convex and has a weakly continuous duality mapping J_Φ and since $\bigcap_s F(U_s)$ is a closed bounded convex subset of G by Theorem (8.2), it follows from Proposition (8.4) that u_n converges weakly as $n \to \infty$ to some element v_1 of $\bigcap_s F(U_s)$. q.e.d.

We now note that some of the fixed point theorems obtained above for non-expansive mappings in uniformly convex Banach spaces can be obtained under a slightly weaker hypothesis by a different type of argument.

DEFINITION (8.6). *Let X be a Banach space, G a closed bounded convex subset of X. Then G is said to have normal structure if for each closed convex subset G_0 of G with more than one point, there exists a point y in G_0 such that $\sup_{x \in G_0} \|y - x\| <$ diam (G_0). Then U has a fixed point in G.*

We remark that each closed bounded convex subset G of a uniformly convex Banach space X has normal structure since if x_0 and x_1 are two distinct points of the closed convex subset G_0, then $y = \frac{1}{2}(x_0 + x_1)$ has the property that for each x in G_0

$$x - y = \tfrac{1}{2}[(x - x_0) + (x - x_1)]$$

and $\|x - y\| \leq \max(\|x - x_0\|, \|x - x_1\|) - d_0$ where $d_0 > 0$ depends upon $\|x_0 - x_1\|$ by the uniform convexity of X.

THEOREM (8.15). *Let X be a Banach space, G a weakly compact convex subset of X with normal structure, U a nonexpansive mapping of G into G.*

Then U has a fixed point in G.

PROOF OF THEOREM (8.15). Consider the family Ω of closed nonempty convex subsets G' of G which are invariant under the mapping U. Since each closed convex subset G' is weakly closed and G is weakly compact, each such G' is weakly compact. It follows that the intersection of a linearly ordered descending subfamily of elements of Ω also lies in Ω since it is closed, convex, invariant under U, and nonempty (the latter by the weak compactness of the elements of Ω).

If we apply Zorn's lemma, therefore, we obtain the existence of a minimal element G_0 of Ω. If G_0 consisted of a single point $\{u_0\}$, then $U(U_0) = u_0$ and the existence of a fixed point would be established. Hence it suffices to show that G_0 cannot have more than one point.

Suppose that G_0 does have more than one point, and let $d > 0$ be the diameter of G_0. Since G has normal structure, there exist a point y of G_0 and a constant $d_1 < d$ such that for all x in G_0, $\|y - x\| \le d_1$. We define a subset G_1 of G_0 by setting

$$G_1 = \{z \mid z \in G_0, \|z - x\| \le d_1 \text{ for all } x \text{ in } G_0\}.$$

Since y lies in G_1, G_1 is nonempty. Since we have $d_1 < d$, G_1 is not the whole of G_0. Since

$$G_1 = \bigcap_{x \in G_0} B_{d_1}(x),$$

where $B_{d_1}(x)$ is the closed ball of radius d_1 about x, it follows that G_1 is a closed convex subset of G_0.

Suppose z lies in G_1. Then for each x in G_0, $\|U(z) - U(x)\| \le \|x - z\| \le d_1$. Hence $U(z) \in \bigcap_{x \in G_0} B_{d_1}(U(x))$. Consider

$$G_2 = \{x \mid x \in G_0, \|x - U(z)\| \le d_1\}.$$

Then G_2 is a closed nonempty convex subset of G_0 which contains $U(G_0)$ and hence $U(G_2)$. Since G_2 is thereby invariant under U and since G_0 is a minimal nonempty closed convex subset of G invariant under U, $G_2 = G_0$. Hence $U(z) \in \bigcap_{x \in G_0} B_{d_1}(x) = G_2$, i.e., G_2 is invariant under U.

Thus G_2 is a proper subset of G_0 which is nonempty, closed, convex, and invariant under U. This contradicts the minimality of G_0 in Ω.   q.e.d.

We note without repeating the argument in detail that the following analogues of Theorems (8.6) and (8.7) are consequences of Theorem (8.15):

THEOREM (8.16). *Let X be a strictly convex, reflexive Banach space such that each bounded closed convex subset of X has normal structure. Let G be a closed bounded convex subset of X, G_0 an open subset of X which contains the ϵ-neighborhood of G for some $\epsilon > 0$. Let U be a mapping of G_0 into X which maps the boundary of G into G.*

Then U has a fixed point in G.

THEOREM (8.17). *Let X be a strictly convex, reflexive Banach space such that each bounded closed convex subset of X has normal structure. Let $\{U_\alpha : \alpha \in \Omega\}$ be a family of nonexpansive mappings of X into X such that $U_\alpha U_\beta = U_\beta U_\alpha$ for each α*

and β in Ω. Suppose that for each α and β in Ω, there exists ξ in Ω such that $U_\xi = U_\alpha U_\beta$.

Then a necessary and sufficient condition that the family $\{U_\alpha\}$ have a common fixed point in x is that there exist a point y_0 in X and a constant $M > 0$ such that $\|U_\alpha(y_0)\| \leq M$ $(\alpha \in \Omega)$.

It is possible to generalize the above results in part to multi-valued non-expansive mappings:

THEOREM (8.18). *Let X be a complete metric space, U a mapping of X into X such that for each x in X, $U(x)$ is a nonempty closed subset of X. If $d(u, v)$ is the metric on X and $d_H(A, B)$ is the Hausdorff metric on closed subsets of X, suppose that there exists a constant $k < 1$ such that for all x and u in X, $d_H(U(x), U(u)) \leq kd(x, u)$.*

Then there exists a fixed point x_0 of U in X, i.e., a point x_0 in X such that $x_0 \in U(x_0)$.

PROOF OF THEOREM (8.18). Following the pattern of the usual Picard contraction argument for single-valued mappings, we define a sequence $\{u_n\}$ of approximants for x_0 by a recursive definition. Let k_1 be a constant with $k < k_1 < 1$. We choose an arbitrary initial approximation u_0 in X and an arbitrary u_1 in $U(u_0)$. The remainder of the sequence of approximants $\{u_n\}$ is defined as follows: Suppose that for a given integer $n \geq 1$ that all the approximants $\{u_0, u_1, \ldots, u_{n-1}\}$ have been defined in such a way that for $1 \leq j \leq n$,

$$u_j \in U(u_{j-1}), \qquad d(u_j, u_{j-1}) \leq k_1^{j-1} d(u_1, u_0).$$

We now define u_n by noting that u_{n-1} lies in $U(u_{n-2})$ and that by the contractive hypothesis,

$$d_H(U(u_{n-1}), U(u_{n-2})) \leq kd(u_{n-1}, u_{n-2}).$$

If $d(u_{n-1}, u_{n-2}) = 0$, then u_{n-2} is the desired fixed point of U and the proof is trivialized. If not, there exists a point u_n in $U(u_{n-1})$ such that $d(u_n, u_{n-1}) \leq k_1 d(u_{n-1}, u_{n-2})$, where the right-hand side of this inequality is greater than $d_H(U(u_{n-1}), U(u_{n-2}))$. For such an element u_n, we have obviously

$$u_n \in U(u_{n-1}), \qquad d(u_n, u_{n-1}) \leq k_1 d(u_{n-1}, u_{n-2}) \leq k_1^{n-1} d(u_1, u_0).$$

Thus the sequence $\{u_0, u_1, \ldots, u_{n-1}, u_n\}$ satisfies the conditions of the recursive definition, and the construction of the infinite sequence $\{u_n\}$ can be carried through.

Since we have $k_1 < 1$, it follows that for $n > m$,

$$d(u_n, u_m) \leq \sum_{j=m+1}^{n} d(u_j, u_{j-1}) \leq \sum_{j=m+1}^{n} k_1^{j-1} d(u_1, u_0) \leq k_1^m (1 - k_1)^{-1} d(u_1, u_0),$$

so that $d(u_n, u_m) \to 0$ as $n, m \to \infty$. Since the metric space X is complete, there exists an element x_0 of X such that u_n converges to x_0 in X.

Let $\epsilon > 0$ be given. Then for n sufficiently large, $U(u_n)$ is contained in the ϵ-neighborhood of $U(x_0)$ so that u_{n+1} lies in the ϵ-neighborhood of $U(x_0)$. Similarly for n sufficiently large, x_0 is at distance at most ϵ from u_{n+1}. Hence for n sufficiently large, the ball of radius 2ϵ about x_0 intersects $U(x_0)$, a fact independent of n. Since this is true for each $\epsilon > 0$ and since $U(x_0)$ is closed in X, it follows that $x_0 \in U(x_0)$.

q.e.d.

We note explicitly that in the multi-valued case as opposed to the single-valued case, there is no guarantee that the fixed point x_0 is necessarily unique from the contractive hypothesis.

We shall extend the result of Theorem (8.18) to a class of multivalued non-expansive mappings rather than contractive mappings. We make use of the following definition:

DEFINITION (8.7). *Let X be a Banach space with X^* strictly convex, J the duality mapping of X into X^*. Then a mapping T of X into 2^X is said to be partially accretive if for each pair x and u in X with $T(x)$ and $T(u)$ nonempty and each y in $T(x)$, there exists w in $T(u)$ (depending upon x and y, in general) such that*

$$(y - w, J(x - u)) \geq 0.$$

PROPOSITION (8.5). *Let X be a reflexive Banach space with X^* strictly convex, J the duality mapping of X into X^*. Let T be a mapping of X into 2^X such that for each x in X, $T(x)$ is closed convex, and nonempty, while T is upper-semicontinuous from X to 2^X, with the range space given the weak topology.*

Let $[u_0, w_0]$ be a point of $X \times X$ such that for each x in X, there exists y in $T(x)$ such that $(w_0 - y, J(u_0 - x)) \geq 0$. Then $w_0 \in T(u_0)$.

PROOF OF PROPOSITION (8.5). Suppose that the conclusion of the proposition were false, then w_0 would lie outside of $T(u_0)$, and since $T(u_0)$ is a closed convex subset of X, there exists an element v of X^* such that $(v, w_0) > \sup_{x \in T(u_0)} (v, x)$. The duality mapping J of X into x^* is a monotone mapping of X into X^* and since X^* is strictly convex, J is continuous from the strong topology of X to the weak topology of X^*. (See Proposition (3.1) of §3.) Moreover, $(J(u), u) = \|u\|^2$ so that for each v in X^*,

$$(J(u) - v, u) = \|u\|^2 - (v, u) \geq \|u\|^2 - \|w\| \cdot \|u\| \to \infty,$$

so that J is coercive with respect to each v in X^*. Moreover J is monotone, since it is the subgradient of the convex function $\frac{1}{2}\|u\|^2$ on X (see Proposition (7.6)). By Theorem (7.5), since X is reflexive, J maps X onto X^*. In particular, there exists an element x_0 of X such that $J(x_0) = v$, where v is the element of X^* defined above.

For each real number $t > 0$, we let $x_t = u_0 + tx_0$. Then there exists an element y_t in $T(x_t)$ such that $(y_t - w_0, J(x_t - u_0)) \geq 0$, i.e.,

$$t(y_t - w_0, J(x_0)) = t(y_t - w_0, v) \geq 0.$$

Cancelling the positive factor t, we obtain $(y_t - w_0, v) \geq 0$.

Let V be the weak neighborhood of $T(u_0)$ in X given by $V = \{x \mid (v, x) < (v, w_0)\}$. Since T is upper-semicontinuous from X to 2^X with the range space given the weak topology, there exists a neighborhood N of u_0 in X, such that for u in N, $T(u) \subset V$. In particular, for $0 < t < t_0$, u_t is contained in N. Hence for y_t in

$T(u_t)$, $(y_t - w_0, v) < 0$. This contradicts the preceding inequality, thereby completing the proof of the proposition. q.e.d.

PROPOSITION (8.6). *Let X be a reflexive Banach space with strictly convex X^* and suppose that for some gauge function Φ, the corresponding duality mapping J_Φ is continuous from the weak topology of X to the weak topology of X^*. Let T be a partially accretive mapping of X into 2^X such that for each x in X, $T(x)$ is a nonempty compact convex subset of X while T is upper-semicontinuous from X to 2^X with the range space given the weak topology. Let G be a closed bounded convex subset of X.*

Then $T(G)$ is a closed subset of X.

PROOF OF PROPOSITION (8.6). Let $\{u_j\}$ be a sequence in G, $\{w_j\}$ a sequence in X such that $w_j \in T(u_j)$ for all j while w_j converges strongly to an element w_0 of X. By passing to an infinite subsequence using the sequential weak compactness of G, we may assume without loss of generality that u_j converges weakly to an element u_0 of G. Let v be any element of X. For each j, it follows from the partial accretiveness of T that there exists an element z_j of $T(v)$ such that $(w_j - z_j, J_\Phi(u_j - v)) \geq 0$. Since $T(v)$ is compact, if we pass to another infinite subsequence, we may assume that z_j converges strongly in X to an element z of $T(v)$. Since J_Φ is continuous from the weak topology of X to the weak topology of X^*, the fact that $(u_j - v)$ converges weakly in X to $(u_0 - v)$ implies that $J_\Phi(u_j - v)$ converges weakly in X^* to $J(u_0 - v)$. Hence

$$(w_j - z_j, J_\Phi(u_j - v)) \to (w_0 - z, J(u_0 - v)) \geq 0,$$

i.e., for each v in X, there exists z in $T(v)$ such that this last inequality is satisfied.

Applying Proposition (8.5), we have that $w_0 \in T(u_0) \subset T(G)$. Therefore $T(G)$ must be closed in X since it contains each of its limit points.

THEOREM (8.19). *Let X be a reflexive Banach space with strictly convex X^* such that for some gauge function Φ, J_Φ is continuous from the weak topology of X to the weak topology of X^*. Let U be a mapping of X into 2^X such that for each x in G, $U(x)$ is a nonempty compact convex subset of X, while for all x and u in X $d_H(U(x), U(u)) \leq d(x, u)$. Suppose that there exists a bounded closed convex subset G of X such that $U(G) \subset G$.*

Then U has a fixed point x_0 in G, i.e., a point x_0 such that $x_0 \in U(x_0)$.

PROOF OF THEOREM (8.19). We may assume without loss of generality that 0 lies in G. For each λ with $0 < \lambda < 1$, let U_λ be the mapping of G into 2^G given by $U_\lambda(x) = \lambda U(x)$. Each such U_λ is a strict contraction in the sense of Theorem (8.19) and hence has at least one fixed point u_λ in G. Since G is sequentially weakly sequentially compact, we may find a sequence $\lambda_j \to 0$ such that if $u_j = u_{\lambda_j}$, then u_j converges weakly in X to an element u_0 of G. Since $u_j \in U_{\lambda_j}(u_j)$, it follows that if we set $w_j = \lambda_j^{-1} u_j$, then $w_j \in U(u_j)$, $\|w_j - u_j\| \to 0$. Hence

$$u_j - w_j \in (I - U)(G), \qquad u_j - w_j \to 0 \quad \text{strongly in } X.$$

It follows that $(I - U)(G)$ contains 0 in its strong closure.

Let $T = (I - U)$, with U satisfying the nonexpansive condition of the hypothesis of Theorem (8.20). Let $[u, w]$ be a point of $G(U)$, and $[u, u - w]$ the corresponding point of $G(T)$. Let x be another point of X. Then there exists an element y of $T(x)$ such that $d(w, y) \leq d(u, x)$, the existence of such an element y following from the inequality on the Hausdorff metric together with the compactness of $T(x)$. Hence

$$((u - w) - (x - y), J_\Phi(u - x))$$
$$= (u - x, J_\Phi(u - x)) - (w - y, J_\Phi(u - x))$$
$$\geq \|u - x\| \Phi(\|u - x\|) - \|w - y\| \Phi(\|u - x\|)$$
$$\geq \|u - x\| \Phi(\|u - x\|) - \|u - x\| \Phi(\|u - x\|) \geq 0.$$

Thus T is partially accretive in the sense of Definition (8.7).

We may now apply Proposition (8.6) to conclude that $(I - U)(G)$ contains 0, i.e., there exists an element x_0 of G such that $0 \in x_0 - U(x_0)$, i.e., $x_0 \in U(x_0)$. q.e.d.

9. Accretive mappings and nonlinear equations of evolution. We have spoken in the preceding section of an essential connection between the theory of nonexpansive mappings on the one hand and the theories of monotone and accretive mappings on the other. It is our purpose in the present section to develop the most essential part of that connection by studying the existence and properties of solutions $u(t)$ ($t \in R^+$) of nonlinear differential equations of the form

$$du(t)/dt = -T(u(t)) \qquad (t > 0)$$

with the initial value condition $u(0) = v$ for a given element v of $D(T)$. Supposing that such solutions exist, we denote by the transition mapping $U(t)$, the mapping which carries the initial value v into $u(t) = U(t)v$, the value of the corresponding solution at time $t \geq 0$. As we see in a detailed form below, accretiveness for T corresponds to the nonexpansiveness of the transition mappings $U(t)$. Our principal objective for the discussion is to obtain the most general results (under varying smoothness or other hypotheses on the space X and the mapping T) for the existence of solutions for the initial value problem stated above.

DEFINITION (9.1). *Let X be a Banach space, T a mapping of X into 2^X. Then T is said to be g-accretive (generalized accretive) if there exists a mapping Φ (not necessarily continuous) of $G(T) \times G(T)$ into X^* with $\Phi(u, y, v, w) \in J(u - v)$ for each pair $[u, y]$ and $[v, w]$ in $G(T)$ and $(\Phi(u, y, v, w), y - w) \geq 0$. For single-valued mappings, this becomes $(\Phi(u, y, v, w), T(u) - T(v)) \geq 0$.*

In the terminology of §2, T is Φ-accretive with respect to the given mapping Φ, where we recall that the duality mapping J of X into 2^{X^*} is given by

$$J(x) = \{w \mid w \in X^*, \|w\| = \|x\|, (w, x) = \|x\|^2\}.$$

Obviously, if the space X^* is strictly convex so that $J(x)$ is a single point for each x in X, the above definition coincides with the definition of accretiveness given in §2, i.e., $(J(u - v), y - w) \geq 0$, or again for the single-valued case,

$(J(u - v), T(u) - T(w)) \geq 0$. We shall use the following definition for the general case:

DEFINITION (9.2). *Let X be a Banach space, J a single-valued duality mapping (i.e., a single-valued mapping from X to X^* which is a section, not necessarily continuous, of the multi-valued mapping J defined above). Then T is said to be accretive with respect to J (or more simply accretive, if J is fixed) if for each y in $T(u)$ and each w in $T(v)$, $(J(u - v), y - w) \geq 0$, or for the single-valued case,*

$$(J(u - v), T(u) - T(v)) \geq 0.$$

It is obvious from the structure of the definitions that each accretive mapping in the sense of Definition (9.2) is g-accretive in the sense of Definition (9.1).

THEOREM (9.1). *Let X be a Banach space, T a mapping of X into 2^X. Then T is g-accretive in the sense of Definition (9.1) if and only if for each $\lambda > 0$, $(I + \lambda T)^{-1}$ is a single-valued nonexpansive mapping of $R(I + \lambda T)$ into X.*

PROOF OF THEOREM (9.1). Suppose first that T is g-accretive. We wish to prove that for each u and v in X, each y in $T(u)$, each w in $T(v)$, and $\lambda > 0$,

$$\|(u + \lambda y) - (v + \lambda w)\| \geq \|u - v\|.$$

By Proposition (7.6) of §7, the mapping J of X into 2^{X^*} is the subgradient of the convex function $g(x) = \frac{1}{2}\|x\|^2$ on X. Hence for a given element $\Phi(u, y, v, w)$ in $J(u - v)$, we have

$$\|(u - v) + \lambda(y - w)\|^2 \geq \|u - v\|^2 + 2(\Phi(u, y, v, w), \lambda(y - w))$$

since by the definition of g-accretiveness of T, $(\Phi(u, y, v, w), y - w) \geq 0$. Hence

$$\|(u + \lambda y) - (v + \lambda w)\| \geq \|u - v\|.$$

Conversely, suppose that for each $\lambda > 0$, $(I + \lambda T)^{-1}$ is a single-valued nonexpansive mapping of $R(I + \lambda T)$ into X. For each element ψ_λ in $J(u - v + \lambda(y - w))$, we note that since J is the subgradient of $g(x) = \frac{1}{2}\|x\|^2$, we have

$$\|u - v\|^2 \geq \|u - v + \lambda(y - w)\|^2 - 2(\psi_\lambda, \lambda(y - w)),$$

while $\|u - v\|^2 \leq \|u - v + \lambda(y - w)\|^2$ by the assumption on $(I + \lambda T)^{-1}$. It follows that $(\psi_\lambda, y - w) \geq 0$ $(\lambda > 0)$. As $\lambda \to 0+$, the set $\{\psi_\lambda\}$ lies in a weak* compact subset of X^* and by the upper-semicontinuity of J from the strong topology of X to 2^{X^*}, with X^* given the weak* topology, it follows that there exists an element $\Phi(u, y, v, w)$ of $J(u - v)$ such that $(\Phi(u, y, v, w), y - w) \geq 0$. Hence T is g-accretive in the sense of Definition (9.1).   q.e.d.

An even more basic connection between g-accretiveness and nonexpansive mappings follows from the following theorem:

THEOREM (9.2). *Let X be a Banach space, T a g-accretive mapping from X to 2^X, d_0 a positive constant. Suppose that we are given two strongly continuous functions u and v from $[0, d_0]$ to X such that u and v are weakly once differentiable from*

$(0, d_0)$ to X with

$$-du(t)/dt \in T(u(t)), \qquad -dv(t)/dt \in T(v(t)) \qquad (0 < t < d_0).$$

Then $\|u(t) - v(t)\|$ is a nonincreasing function of t on $[0, d_0]$.

In the proof of Theorem (9.2), we shall employ the following simple result from real analysis:

PROPOSITION (9.1). *Let q be a continuous function from $[0, d_0]$ to the reals such that for each t in $(0, d_0)$,*

$$\varlimsup_{s \to t-} \left[\frac{q(t) - q(s)}{t - s} \right] \le 0.$$

Then $q(t)$ is a nonincreasing function of t on $[0, d_0]$.

PROOF OF PROPOSITION (9.1). Since q is continuous on $[0, d_0]$, it suffices to prove that it is nonincreasing on $(0, d_0)$. Let $\epsilon > 0$ be given. By the hypothesis, for each t in $(0, d_0)$, there exists $d_\epsilon(t) > 0$ such that for $t - d_\epsilon(t) \le s \le t$,

(9.1) $$q(s) \ge q(t) - \epsilon(t - s).$$

Let $K_\epsilon(t)$ be the largest connected subset of $(0, t]$ consisting of points s for which the inequality (9.1) is valid. For each s in $K_\epsilon(t)$, there exists $d_\epsilon(s) > 0$ such that for all r in $s - d_\epsilon(s) \le r \le s$, $q(r) \ge q(s) - \epsilon(s - r)$. Hence

$$q(r) \ge q(t) - \epsilon(t - s) - \epsilon(s - r) = q(t) - \epsilon(t - r).$$

Thus the interval $(s - d_\epsilon(s), s]$ is contained in $K_\epsilon(t)$ and $K_\epsilon(t)$ is open from below. We assert that $K_\epsilon(t)$ consists of $(0, t]$. If it did not, $K_\epsilon(t) = (d_1, t]$ with $d_1 > 0$. For each $s > d_1$, we have $q(s) \ge q(t) - \epsilon(t - s)$. Letting s tend to d_1 and using the continuity of q, we find that $q(d_1) \ge q(t) - \epsilon(t - d_1)$ so that d_1 lies in $K_\epsilon(t)$. Since this contradicts the assumption that $K_\epsilon(t) = (d, t]$, it follows that $K_\epsilon(t) = (0, t]$.

It follows that for each $\epsilon > 0$ and all $s < t$, we have $q(s) \ge q(t) - \epsilon(t - s)$. Letting $\epsilon \to 0$, we find that $q(s) \ge q(t)$ $(s < t)$. q.e.d.

PROOF OF THEOREM (9.2). Let $w(t) = u(t) - v(t)$, and set $q(t) = \|w(t)\|^2 = \|u(t) - v(t)\|^2$. We wish to show that $q(t)$ is nonincreasing in t.

For $0 < s < t < d_0$, we write $\Phi(u(t), du(t)/dt, v(t), dv(t)/dt)$ as $\Phi(u(t), v(t))$ and obtain

$$q(s) = \|w(s)\|^2 \ge \|w(t)\|^2 + 2(\Phi(u(t), v(t)), w(s) - w(t))$$

since $\Phi(u(t), v(t)) \in J(u(t) - v(t)) = J(w(t))$, and J is the subgradient of the convex function $\tfrac{1}{2} \|x\|^2$ on the Banach space X. Hence

$$\frac{q(t) - q(s)}{t - s} \le 2 \left(\Phi(u(t), v(t)), \frac{w(t) - w(s)}{t - s} \right),$$

where

$$\left(\Phi(u(t), v(t)), \frac{w(t) - w(s)}{t - s} \right) \to \left(\Phi(u(t), v(t)), \frac{du}{dt}(t) - \frac{dv}{dt}(t) \right) \le 0$$

as $s \to t-$, since
$$-du(t)/dt \in T(u(t)), \qquad -dv(t)/dt \in T(v(t)),$$
and T is g-accretive at the pair $[u(t), v(t)]$ with respect to the given element $\Phi(u(t), v(t))$ of $J(u(t) - v(t))$. Therefore
$$\varlimsup_{s \to t-} \left[\frac{q(t) - q(s)}{t - s}\right] \leq 0.$$

Applying Proposition (9.1), we see that $q(t)$ is a nonincreasing function of t on $[0, d_0]$. q.e.d.

A converse to Theorem (9.2) is the following:

THEOREM (9.3). *Let X be a Banach space, u and v two strongly continuous functions from the interval $[0, d_0]$ to X with u and v weakly once differentiable on the right from $[0, d_0)$ to X. Suppose that for $0 \leq s \leq t < d_0$, we know that*
$$\|u(t) - v(t)\| \leq \|u(s) - v(s)\|.$$

Then for each s with $0 < s < d_0$, and each element $\Phi(u(s), v(s))$ in $J(u(s) - v(s))$, we have
$$(\Phi(u(s), v(s)), du(s)/ds - dv(s)/ds) \leq 0.$$

PROOF OF THEOREM (9.3). Let $w(t) = u(t) - v(t)$. By hypothesis $\|w(s)\|^2 \geq \|w(t)\|^2$ ($s < t$). On the other hand,
$$\|w(t)\|^2 \geq \|w(s)\|^2 + 2(\Phi(u(s), v(s)), w(t) - w(s)).$$
Hence $(\Phi(u(s), v(s)), w(t) - w(s)) \leq 0$ ($s < t$). Dividing by $(t - s)$ and letting $t \to s+$, we obtain $(\Phi(u(s), v(s)), du(s)/ds - dv(s)/ds) \leq 0$. q.e.d.

As an important consequence of Theorem (9.3), we obtain the following result on the relation between accretiveness and g-accretiveness.

THEOREM (9.4). *Let X be a Banach space, T a mapping from a subset $D(T)$ of X to X. Suppose that for each v in $D(T)$, there exist an interval $[0, d_v]$ with $d_v > 0$ and a solution on that interval of the differential equation*
$$dv(t)/dt = -T(v(t)) \quad (0 \leq t < d_v), \quad v(0) = v,$$
where $v(t)$ is a strongly continuous, weakly once-differentiable function from $[0, d_v)$ to X with $v(t) \in D(T)$ for all t in $[0, d_v)$.

Then the following conditions are equivalent:

(a) *T is g-accretive.*
(b) *T is accretive with respect to any single-valued duality mapping J.*
(c) *For each u and v of $D(T)$ and each w in $J(u - v)$, $(w, T(u) - T(v)) \geq 0$.*

PROOF OF THEOREM (9.4). It follows from the definitions that (b) and (c) are equivalent and that they both imply (a). Suppose on the other hand that T satisfies (a) and that the hypothesized existence of solutions of initial value problems holds. Since T is g-accretive, it follows from Theorem (9.2) that for any

u and v in $D(T)$,

$$\|u(t) - v(t)\| \leq \|u(s) - v(s)\| \qquad (0 \leq s < t < \min(d_u, d_v))$$

where $u(t)$ and $v(t)$ are the corresponding solutions of the initial value problem with initial values u and v respectively. Hence by Theorem (9.3), for $s = 0$ and any w in $J(u(0) - v(0)) = J(u - v)$,

$$(w, du(0)/ds - dv(0)/ds) = -(w, T(u) - T(v)) \leq 0. \qquad \text{q.e.d.}$$

It follows from Theorem (9.4) that for those classes of single-valued g-accretive mappings T for which we obtain the existence of solutions of initial value problems postulated in Theorem (9.4), these mappings are in fact accretive in the sense of Definition (9.2). In particular, for a mapping T accretive with respect to a given duality mapping J, it will follow that whenever T falls into an existence class, it is accretive with respect to any duality mapping J.

We shall now begin the discussion of existence theorems. An essential rule in the proofs of these theorems is played by the following:

THEOREM (9.5). *Let X be a Banach space, T a g-accretive mapping from X to 2^X, d_0 a positive constant. Suppose that u is a strongly continuous function from $[0, d_0]$ to X which is weakly once-differentiable from $[0, d_0]$ to X with $-du(t)/dt \in T(u(t))$ for each t in $[0, d_0]$.*
Then

(a) *There exists a constant M such that*

$$\|du(t)/dt\| \leq M \quad (t \in [0, d_0]) \quad \left(M = \lim_{h \to 0+} \|h^{-1}[u(h) - u(0)]\| \right).$$

(b) *If $u(t)$ is strongly once-differentiable from the right at $t = 0$, then*

$$\|du(t)/dt\| \leq \|du(0)/dt\| \qquad (0 \leq t \leq d_0).$$

PROOF OF THEOREM (9.5). For each $h > 0$ sufficiently small, $v_h(t) = u(t + h)$ is also a solution of the differential equation $-dv(t)/dt \in T(v(t))$ on the interval $[0, d_0 - h]$ with $d_0 - h > 0$. We may therefore apply Theorem (9.2) to these two solutions, and we find that for $0 < t < d_0 - h$, we have $\|v_h(t) - u(t)\| \leq \|v_h(0) - u(0)\| = \|u(h) - u(0)\|$, i.e., $\|u(t + h) - u(t)\| \leq \|u(h) - u(0)\|$. We may divide by $h > 0$ and obtain

$$\|h^{-1}[u(t + h) - u(t)]\| \leq \|h^{-1}[u(h) - u(0)]\| \qquad (0 < h < d_0 - t).$$

As $h \to 0+$, $h^{-1}[u(h) - u(0)]$ converges weakly in X to $du(0)/dt$. Hence by the uniform boundedness theorem, there exists a constant $M > 0$ such that

$$\lim_{h \to 0+} \|h^{-1}[u(h) - u(0)]\| = M.$$

As $h \to 0+$, we know that $h^{-1}[u(t + h) - u(t)]$ converges weakly in X to $du(t)/dt$. Hence

$$\|du(t)/dt\| \leq \lim_{h \to 0+} \|h^{-1}[u(t + h) - u(t)]\| \leq \lim_{h \to 0+} \|h^{-1}[u(h) - u(0)]\| = M.$$

In particular, if $u(t)$ is strongly differentiable at $t = 0$, $M = \|du(0)/dt\|$. q.e.d.

THEOREM (9.6). *Let X be a Banach space, T a g-accretive mapping of X into X such that T is locally uniformly continuous (i.e., uniformly continuous on some neighborhood of each point of X). Then for each v in X, there exists one and only one C^1 solution u on R^+ of the differential equation*

$$du(t)/dt = -T(u(t)) \qquad (t \geq 0),$$

with $u(0) = v$.

For this solution, we have $\|T(u(t))\| \leq \|T(v)\|$ $(t \geq 0)$.

We shall derive Theorem (9.6) as a consequence of the corresponding local result, namely:

THEOREM (9.7). *Let X be a Banach space, v a point in X, N a neighborhood of v in X. Suppose that T is a uniformly continuous mapping of N into X with T g-accretive on N.*

Then there exist a constant $d > 0$ and a solution $u(t)$ for t in $[0, d]$ of the differential equation

$$du(t)/dt = -T(u(t)) \qquad (t \in [0, d]),$$

with the initial value $u(0) = v$.

The size of the interval $[0, d]$ (i.e., the choice of d) depends only upon the largest constant $r > 0$ such that the closed ball $B_r(v)$ of radius r about v in X is contained in N.

PROOF OF THEOREM (9.6) FROM THEOREM (9.7). We note that by Theorem (9.7), there exists a local solution $u(t)$ on some interval $[0, d]$ of the equation

$$du(t)/dt = -T(u(t)), \qquad u(0) = v,$$

and the constant $d > 0$ does not depend upon anything except the neighborhood N of v on which T is uniformly continuous.

Let $[0, d_0)$ be the maximal interval upon which such a solution exists. We note that the solution is unique as a consequence of Theorem (9.2) and that by Theorem (9.5), since $T(u(t))$ is strongly continuous in t, we know that

$$\|du(t)/dt\| = \|T(u(t))\| \leq \|du(0)/dt\| = \|T(v)\|$$

on the whole interval on which the solution exists.

Suppose $d_0 < \infty$. Then it follows that for each t with $0 < t < d_0$, we have $u(t) = v - \int_0^t T(u(s)) \, ds$, and for $0 < t < t_1 < d_0$,

$$\|u(t_1) - u(t)\| \leq \int_t^{t_1} \|T(u(s))\| \, ds \leq M(t_1 - t).$$

Therefore, by the completeness of X in its metric, $u(t)$ converges strongly to a limit as $t \to d_0-$, and if this limit element is denoted by v_0, we can construct a solution $u_1(t)$ of our differential equation with $u_1(d_0) = v_0$ on some interval of the form $[d_0, d_0 + d]$.

Let $v(t)$ be the function from $[0, d_0 + d]$ to X given by

$$v(t) = u(t), \qquad 0 \leq t < d_0,$$
$$= u_1(t), \qquad d_0 \leq y \leq d_0 + d.$$

By the construction of $u_1(t)$ and the fact that $u(t)$ converges strongly to v_0 as t converges to d_0 from below, it follows that v is a strongly continuous function from $[0, d_0 + d]$ to X. For $0 < t < d_0$, we know that $v(t) = v - \int_0^t T(v(s))\, ds$, while for $d_0 \leq t \leq d_0 + d$, $v(t) = v_0 - \int_d^t T(v(s))\, ds$. Since T is continuous in the strong topology of X, it follows from the first equation that $v_0 = v - \int_0^{d_0} T(v(s))\, ds$. Thus it follows that for all t in the interval $[0, d_0 + d]$,

$$v(t) = v - \int_0^t T(v(s))\, ds,$$

so that the extended function $v(t)$ is a solution of our initial value problem on the interval $[0, d_0 + d]$ contradicting the maximality of d_0. Hence $d_0 = +\infty$, and the conclusion of Theorem (9.6) has been verified. q.e.d.

PROOF OF THEOREM (9.7). Since T is uniformly continuous, we may assume T uniformly bounded on N, so that there exists a constant $M_0 > 0$ such that $\|T(u)\| \leq M_0$ for all u in N. We may assume without loss of generality that N contains the closed ball $B_r(v)$ of radius r about v in X. We shall construct a solution of the differential equation on the interval $[0, d]$ with $d > 0$, depending only upon r and the bound M_0 above.

For each $\epsilon > 0$, we consider the associated delay-differential equation

(9.2) $$du_\epsilon(t)/dt = -T(u_\epsilon(t - \epsilon)),$$

with initial data

(9.3) $$u_\epsilon(t) = v \qquad (t \leq 0).$$

Since the problem of integrating the equation (9.2) for a given $\epsilon > 0$ can be resolved trivially by successive integrations on intervals in t of length ϵ (since the right-hand side on each such interval depends upon the already constructed solution on the preceding interval), it follows that we can obtain an unique solution u_ϵ of the equation (9.3) with initial data given by the equation (9.3) on any interval $[0, d_\epsilon]$ for which the continuous process just described keeps $u_\epsilon(t)$ within $B_r(v)$. Let $[0, d_\epsilon]$ be the largest such interval. Then for some t in the interval $[d_\epsilon, d_\epsilon + \epsilon]$, $\|u_\epsilon(t - \epsilon) - v\| \geq r$. Hence

$$r \leq \int_0^{d_\epsilon - \epsilon} M_0\, ds = M_0(d_\epsilon - \epsilon), \qquad d_\epsilon \geq rM_0^{-1} + \epsilon.$$

Hence there exists a constant $d > 0$ independent of ϵ for $\epsilon > 0$ such that the solutions u_ϵ of the problems given by equations (9.2) and (9.3) exist on the intervals $[0, d + \epsilon]$, and on that interval satisfy the inequality

$$\|du_\epsilon(t)/dt\| \leq M_0 \qquad (t \in [0, d + \epsilon]).$$

We shall now show that $u_\epsilon(t)$ converges strongly to some continuous function

$u(t)$ from $[0, d]$ to X uniformly on $[0, d]$. Let $\epsilon \cdot \xi > 0$, and let
$$q_{\epsilon,\xi}(t) = \|u_\epsilon(t) - u_\xi(t)\|^2.$$
Then for $0 \leq s < t < d$,
$$q_{\epsilon,\xi}(s) \geq q_{\epsilon,\xi}(t) + 2(w, [u_\epsilon(s) - u_\epsilon(t)] - [u_\xi(s) - u_\xi(t)])$$
for any w in $J(u_\epsilon(t) - u_\xi(t))$. If we hold t fixed and let $s \to t-$ after dividing by $(t-s)$, we obtain
$$\varlimsup_{s \to t-}\left\{\frac{q_{\epsilon,\xi}(t) - q_{\epsilon,\xi}(s)}{t-s}\right\} \leq 2\left(w, \frac{du_\xi}{dt}(t) - \frac{du_\epsilon}{dt}(t)\right).$$

We consider the right-hand term of this last inequality with w given by $\Phi(u_\epsilon(t), u_\xi(t))$, the element of $J(u_\epsilon(t) - u_\xi(t))$ appearing in the g-accretiveness criterion for T at the pair of points $[u_\epsilon(t), u_\xi(t)]$. For this choice of w, we find that

$$(w, du_\xi(t)/dt - du_\epsilon(t)/dt) = (\Phi(u_\epsilon(t), u_\xi(t)), T(u_\xi(t)) - T(u_\epsilon(t)))$$
$$+ (\Phi(u_\epsilon(t), u_\xi(t)), T(u_\xi(t - \xi)) - T(u_\xi(t)))$$
$$+ (\Phi(u_\epsilon(t), u_\xi(t)), T(u_\epsilon(t)) - T(u_\epsilon(t - \epsilon)))$$

where the first term is nonpositive by the g-accretiveness of T while the last two terms are bounded by the following expression:
$$\|u_\epsilon(t) - u_\xi(t)\| \cdot \{\|T(u_\xi(t)) - T(u_\xi(t - \xi))\| + \|T(u_\epsilon(t)) - T(u_\epsilon(t - \epsilon))\|\}.$$

Since T is assumed to be uniformly continuous on $B_r(v)$, there exists a function $s(\lambda)$ of λ for $\lambda \geq 0$ such that $s(\lambda) \to 0$ as $\lambda \to 0$, while for any u and u_1 in $B_r(v)$ for which $\|u - u_1\| \leq \lambda$, we have $\|T(u) - T(u_1)\| \leq s(\lambda)$.

It follows from the above that
$$(w, du_\xi(t)/dt - du_\epsilon(t)/dt) \leq \|u_\epsilon(t) - u_\xi(t)\| \{s(M_0\xi) + s(M_0\epsilon)\}$$
since
$$\|u_\xi(t) - u_\xi(t - \xi)\| \leq \max_r \left\|\frac{du_\xi}{dt}(r)\right\| \xi \leq M_0\xi$$
and similarly $\|u_\epsilon(t) - u_\epsilon(t - \epsilon)\| \leq M_0\epsilon$.

Given a constant $\zeta > 0$, we can choose $\epsilon_0(\zeta) > 0$ so small that for $0 < \xi$, $\epsilon \leq \epsilon_0(\zeta)$, we have $2\{s(M_0\xi) + s(M_0\epsilon)\} \leq \zeta$. For such ϵ and ξ, we find by the above inequalities that
$$\varlimsup_{s \to t-}\left\{\frac{q_{\epsilon,\xi}(t) - q_{\epsilon,\xi}(s)}{t-s}\right\} \leq \zeta[q_{\epsilon,\xi}(t)]^{1/2}.$$

We now apply the following proposition:

PROPOSITION (9.2). *Let q be a nonnegative continuous function on the interval $[0, d_0]$ and suppose that for each t in $(0, d_0)$, we have*
$$\varlimsup \frac{q(t) - q(s)}{t - s} \leq c_0[q(t)]^{1/2}.$$

Then for all t in $[0, d_0]$, we have

$$q(t) \leq q(0) \exp(t) + c_0^2[\exp(t) - 1].$$

PROOF OF PROPOSITION (9.2). Since $c_0[q(t)]^{1/2} \leq c_0^2 + q(t)$, we have

$$\varlimsup_{s \to t-} \left\{ \frac{q(t) - q(s)}{t - s} \right\} \leq c_0^2 + q(t).$$

Let $p(t) = \exp(-t)q(t) + c_0^2 \exp(-t)$. Then for $0 < s < t < d_0$,

$$\frac{p(t) - p(s)}{t - s} = \frac{[\exp(-t) - \exp(-s)]}{t - s} q(t)$$

$$+ \exp(-s) \frac{[q(t) - q(s)]}{t - s} + c_0^2 \frac{[\exp(-t) - \exp(-s)]}{t - s}$$

so that

$$\varlimsup_{s \to t-} \left\{ \frac{p(t) - p(s)}{t - s} \right\} \leq -\exp(-t)q(t) + \exp(-t)[c_0^2 + q(t)] - c_0^2 \exp(-t) = 0.$$

By Proposition (9.1), it follows that $p(t)$ is nonincreasing in t on $[0, d_0]$ so that for all t in $[0, d_0]$, $p(t) \leq p(0)$, i.e.,

$$\exp(-t)q(t) + c_0^2 \exp(-t) \leq q(0) + c_0^2.$$

It follows that $q(t) \leq q(0) \exp(t) + c_0^2[\exp(t) - 1]$.   q.e.d.

PROOF OF THEOREM (9.7) COMPLETED. If we apply Proposition (9.2) to the inequality satisfied by $q_{\epsilon, \xi}$ above, we find that for $\epsilon, \xi < \epsilon_0(\zeta)$, we obtain

$$q_{\epsilon, \xi}(t) \leq q_{\epsilon, \xi}(0) \exp(t) + \zeta^2[\exp(t) - 1].$$

Since for all ϵ and ξ, $q_{\epsilon, \xi}(0) = 0$, it follows that

$$q_{\epsilon, \xi}(t) \leq \zeta^2 \exp(d) \qquad (0 \leq t \leq d).$$

Since ζ^2 can be made arbitrarily small by making ϵ and ξ small it follows that $q_{\epsilon, \xi}(t) = \|u_\epsilon(t) - u_\xi(t)\|^2 \to 0$ as $\epsilon, \xi \to 0$, uniformly for t in $[0, d]$. Since X is complete, there exists a continuous function u from $[0, d]$ to X such that $u_\epsilon(t)$ converges strongly to $u(t)$, uniformly for t in $[0, d]$.

For each $\epsilon > 0$, $u_\epsilon(t)$ satisfies the integral equation

$$u_\epsilon(t) = v - \int_0^t T(u_\epsilon(s - \epsilon))\, ds.$$

As $\epsilon \to 0$, $u_\epsilon(t)$ converges uniformly to $u(t)$ in X while $u_\epsilon(s - \epsilon)$ converges uniformly to $u_\epsilon(s)$ in X for s in $[0, d]$. Since T is uniformly continuous, $T(u_\epsilon(s - \epsilon))$ converges uniformly to $T(u(s))$ for s in $[0, d]$. Taking the limit of both sides of the last equation as $\epsilon \to 0$, we therefore obtain

$$u(t) = v - \int_0^t T(u(s))\, ds \qquad (t \in [0, d]).$$

It follows that $u(t)$ is once strongly differentiable in t on $[0, d]$, $du(t)/dt = -T(u(t))$ $(0 \leq t \leq d)$, and $u(0) = v$. q.e.d.

If we wish to weaken the continuity requirement upon the mapping T, we must replace it with a stronger assumption upon the space X, namely the existence of a single-valued duality mapping J from X to X^* which is uniformly continuous on bounded subsets of X. This assumption can be expressed in terms of the properties of the conjugate space X^* as follows:

PROPOSITION (9.3). *Let X be a Banach space. Then there exists a single-valued section J of the duality mapping of X into 2^{X^*} with J uniformly continuous on bounded subsets of X if and only if the conjugate space X^* of X is uniformly convex.*

PROOF OF PROPOSITION (9.3). Suppose first that X^* is uniformly convex. Then the duality mapping J of X into X^* is uniquely defined and is uniformly continuous in the strong topologies on bounded subsets of X by Proposition (3.1) of §3 above.

Suppose conversely that there exists a single-valued section J of the duality mapping which is uniformly continuous on bounded sets in X. By the theorem of Smulian [**427**], already quoted in §3, X^* is uniformly convex if and only if the norm of X is continuously once-differentiable on the complement of the origin with its derivative uniformly continuous on the unit sphere in X. This is obviously equivalent to the condition that the function $g(x) = \frac{1}{2}\|x\|^2$ is continuously Fréchet differentiable on X with its derivative uniformly continuous on bounded subsets of X. We shall show that this last condition is implied by the existence of a uniformly continuous J which is a single-valued section of the duality mapping from X to 2^{X^*}.

Let x and y be elements of X. For each λ in R^1, $g(x + \lambda y) \geq g(x) + \lambda(J(x), y)$, while $g(x) \geq g(x + \lambda y) - \lambda(J(x + \lambda y), y)$. Hence for $\lambda > 0$

$$\lambda^{-1}[g(x + \lambda y) - g(x)] \geq (J(x), y),$$

and

$$\lambda^{-1}[g(x + \lambda y) - g(x)] \leq (J(x + \lambda y), y).$$

As $\lambda \to 0+$, $J(x + \lambda y)$ converges strongly to $J(x)$. Hence

$$\lim_{\lambda \to 0+} \lambda^{-1}[g(x + \lambda y) - g(x)] = (J(x), y),$$

and

$$\lim_{\lambda \to 0-} \lambda^{-1}[g(x + \lambda y) - g(x)] = \lim_{\xi \to 0+} -\xi^{-1}[g(x - \xi y) - g(x)]$$

$$= -(J(x), -y) = (J(x), y).$$

It follows that for each x in X, $g'(x) = J(x)$, where by hypothesis J is uniformly continuous on bounded subsets of X from the strong topology of X to the strong topology of X^*. q.e.d.

THEOREM (9.8). *Let X be a Banach space with X^* uniformly convex, T a g-accretive mapping of X into X which is demicontinuous (i.e., continuous from the strong topology of X to the weak topology of X). Let v be an element of X.*

Then there exists one and only one strongly continuous, weakly C^1 solution, $u(t)$ on R^+ of the differential equation $du(t)/dt = -T(u(t))$ ($t \geq 0$) with $u(0) = v$.

For this solution, $\|T(u(t))\|$ is a bounded function of t on R^+.

We obtain Theorem (9.8) from the following local version of the result:

THEOREM (9.9). *Let X be a Banach space with X^* uniformly convex, v an element of X, T a demicontinuous accretive mapping of a neighborhood N of v into X.*

Then there exists an interval $[0, d)$ with $d > 0$ and a strongly continuous, weakly C^1 function u from $[0, d]$ to X such that

$$du(t)/dt = -T(u(t)) \qquad (t \in [0, d])$$

with $u(0) = v$. For each such solution, $\|T(u(t))\|$ is a bounded function of t on $[0, d)$.

PROOF OF THEOREM (9.8) FROM THEOREM (9.9). Under the hypotheses of Theorem (9.8), it follows from Theorem (9.9) that for each v in X, there exist a constant $d_v > 0$ and a solution of the differential equation $du(t)/dt = -T(u(t))$ ($0 \leq t < d_v$) with $u(0) = v$. The uniqueness of this solution follows from Theorem (9.2).

Let d_v be the largest constant such that the solution exists on the interval $[0, d_v)$. To prove Theorem (9.8), it suffices to show that $d_v = +\infty$. Suppose on the contrary that $d_v < +\infty$. Since, by Theorem (9.9), $\|T(u(t))\| \leq M$ for all t in $[0, d_v)$ for some $M < +\infty$, it follows that $\|du(t)/dt\| \leq M < +\infty$. For $0 < t < t_1 < d_v$, it follows that

$$\|u(t_1) - u(t)\| \leq \int_t^{t_1} \left\|\frac{du}{dt}(s)\right\| ds \leq M(t_1 - t).$$

Hence $u(t)$ converges strongly to a limit v_0 as $t \to d_v-$.

Applying Theorem (9.9) once more, we know that we may find a solution $u_1(t)$ of our given differential equation on the interval $[d_v, d_v + d_1]$ for some $d_1 > 0$ with $u_1(d_v) = v_0$. If we set

$$v(t) = u(t), \qquad 0 \leq t < d_v,$$
$$= u_1(t), \qquad d_v \leq t < d_v + d_1,$$

it follows as in the proof of Theorem (9.6) from Theorem (9.7) that v is a solution of our given differential equation on $[0, d_v + d_1)$ with $v(0) = v$. This contradicts the maximality of the interval $[0, d_v)$ and proves that $d_v = +\infty$. q.e.d.

PROOF OF THEOREM (9.9). Since N is a neighborhood of v in X, there exists $r > 0$ such that the closed ball $B_r(v)$ about v is contained in N. Since T is demicontinuous from N to X, we may choose $B_r(v)$ such that $T(B_r(v))$ is bounded in X.

(Indeed, otherwise, we could find a sequence x_j converging to 0 strongly in X while $\|T(x_j)\| \to +\infty$, contradicting the weak convergence of $T(x_j)$ to 0.)

For each $\epsilon > 0$, we consider the delay-differential equation $du_\epsilon(t)/dt = -T(u_\epsilon(t - \epsilon))$, with the initial data, $u_\epsilon(t) = v$ ($t \leq 0$). As in the proof of Theorem (9.7), the solution $u_\epsilon(t)$ of this problem can be obtained by successive integrations on intervals of length ϵ, and is uniquely determined by v. Moreover, there exists $d > 0$ independent of ϵ for $\epsilon > 0$ such that these solutions all exist on the interval $[0, d + \epsilon)$, with values in $B_r(v)$. Hence $\|T(u_\epsilon(t))\| \leq M_0$ for some $M_0 > 0$ which is independent of ϵ for $\epsilon > 0$. Therefore, $\|du_\epsilon(t)/dt\| \leq M_0$ ($t \in [0, d]$) and thus

$$\|u_\epsilon(t) - u_\epsilon(t - \epsilon)\| \leq \epsilon M \qquad (t \in [0, d]).$$

For $\epsilon, \xi > 0$, we let $q_{\epsilon,\xi}(t) = \|u_\epsilon(t) - u_\xi(t)\|^2$. Then for $0 \leq s < t \leq d$, we have

$$q_{\epsilon,\xi}(s) \geq q_{\epsilon,\xi}(t) + 2(J(u_\epsilon(t) - u_\xi(t)), [u_\epsilon(s) - u_\xi(s)] - [u_\epsilon(t) - u_\xi(t)]).$$

Hence

$$\frac{q_{\epsilon,\xi}(t) - q_{\epsilon,\xi}(s)}{t - s} \leq 2\left(J(u_\epsilon(t) - u_\xi(t)), \frac{[u_\epsilon(t) - u_\epsilon(s)]}{t - s} - \frac{[u_\xi(t) - u_\xi(s)]}{t - s}\right).$$

If we let $s \to t-$, we find that

$$\varlimsup_{s \to t-}\left\{\frac{q_{\epsilon,\xi}(t) - q_{\epsilon,\xi}(s)}{t - s}\right\} \leq 2\left(J(u_\epsilon(t) - u_\xi(t)), \frac{du_\epsilon}{dt}(t) - \frac{du_\xi}{dt}(t)\right).$$

For the right-hand term of the last inequality, we have

$(J(u_\epsilon(t) - u_\xi(t)), du_\epsilon(t)/dt - du_\xi(t)/dt)$

$= -(J(u_\epsilon(t) - u_\xi(t)), Tu_\epsilon(t - \epsilon) - Tu_\xi(t - \xi))$

$= -(J(u_\epsilon(t - \epsilon) - u_\xi(t - \xi)), Tu_\epsilon(t - \epsilon) - Tu_\xi(t - \xi)) + R_{\epsilon,\xi}(t) \leq R_{\epsilon,\xi}(t)$

by the accretiveness of the mapping T, where

$R_{\epsilon,\xi}(t) = (J(u_\epsilon(t - \epsilon) - u_\xi(t - \xi)) - J(u_\epsilon(t) - u_\xi(t)), Tu_\epsilon(t - \epsilon) - Tu_\xi(t - \xi))$

so that

$$|R_{\epsilon,\xi}(t)| \leq 2M_0 \|J(u_\epsilon(t - \epsilon) - u_\xi(t - \xi)) - J(u_\epsilon(t) - u_\xi(t))\|.$$

Since X^* is uniformly convex by assumption, J is uniformly continuous in the strong topologies on bounded subsets of X. Hence there exists a function $r(\lambda)$ for $\lambda \geq 0$ with $r(\lambda) \to 0$ as $\lambda \to 0$ such that for u and u_1 in the ball $B_{2r}(0)$ with $\|u - u_1\| \leq \lambda$, we have $\|J(u) - J(u_1)\| \leq r(\lambda)$. In particular, we obtain

$$\|J(u_\epsilon(t - \epsilon) - u_\xi(t - \xi)) - J(u_\epsilon(t) - u_\xi(t))\| \leq r(\epsilon M + \xi M)$$

since

$\|u_\epsilon(t - \epsilon) - u_\xi(t - \xi) - u_\epsilon(t) + u_\xi(t)\|$

$\leq \|u_\epsilon(t - \epsilon) - u_\epsilon(t)\| + \|u_\xi(t - \xi) - u_\xi(t)\| \leq \epsilon M + \xi M.$

It follows that $|R_{\epsilon,\xi}(t)| \le 2M_0 r(\epsilon M + \xi M) \to 0$ ($\epsilon, \xi \to 0$) uniformly on the interval $[0, d]$.

We now make use of the following proposition:

PROPOSITION (9.4). *Let X be a Banach space with X^* uniformly convex, J the uniquely determined normalized duality mapping of X into X^*. Then for any strongly continuous function w from an interval $[0, d]$ to X with w once weakly differentiable from $[0, d]$ to X, the real-valued function $\|w(t)\|^2$ is differentiable on $[0, d]$ and we have*

$$\frac{d}{dt}\{\|w(t)\|^2\} = 2\left(J(w(t)), \frac{dw}{dt}(t)\right),$$

where dw/dt is the weak derivative of w.

Furthermore, if, for a given constant $\beta \ge 0$, $2(w(t), dw(t)/dt) \le \beta$ ($t \in [0, d)$), then for all t in $[0, d]$, $\|w(t)\|^2 \le \|w(0)\|^2 + \beta t$.

PROOF OF PROPOSITION (9.4). The second assertion of the proposition follows in an obvious way from the first. To prove the first assertion we note that for $s < t$,

$$\|w(t)\|^2 \ge \|w(s)\|^2 + 2(J(w(s)), w(t) - w(s))$$

and

$$\|w(s)\|^2 \ge \|w(t)\|^2 + 2(J(w(t)), w(s) - w(t)).$$

Combining the two inequalities, we obtain

$$2\left(J(w(s)), \frac{w(t) - w(s)}{t - s}\right) \le \frac{\|w(t)\|^2 - \|w(s)\|^2}{t - s} \le 2\left(J(w(t)), \frac{w(t) - w(s)}{t - s}\right).$$

As $t \to s$ (or $s \to t$), the two extreme terms of the inequality converge to $2(J(w(s)), dw(s)/dt)$ (or $2(J(w(t)) dw(t)/dt)$, respectively) since the difference quotient $(w(t) - w(s))/(t - s)$ converges weakly and the other term $J(w(s))$ or $J(w(t))$ of the pairing is either fixed or converges strongly, since $w(t)$ converges strongly to $w(s)$ and J is continuous in the strong topologies. Hence

$$\frac{d}{dt}\{\|w(t)\|^2\} = 2\left(J(w(t)), \frac{dw}{dt}(t)\right). \quad \text{q.e.d.}$$

PROOF OF THEOREM (9.9) COMPLETED. If we apply Proposition (9.4) and our previous calculations, we find that

$$dq_{\epsilon,\xi}(t)/dt \le |R_{\epsilon,\xi}(t)| \le \beta \qquad (t \in [0, d]),$$

for a given $\beta > 0$ provided that $\epsilon, \xi \le \epsilon_0(\beta)$, for a suitable $\epsilon_0(\beta) > 0$. Hence, since $q_{\epsilon,\xi}(0) = 0$, it follows that $q_{\epsilon,\xi}(t) \le \beta t$ ($t \in [0, d]$) for such ϵ and ξ. Hence $\|u_\epsilon(t) - u_\xi(t)\| \to 0$ ($\epsilon, \xi \to 0$) uniformly on the interval $[0, d]$.

By the completeness of X in its metric, there exists a continuous function u from $[0, d]$ to X such that $u_\epsilon(t)$ converges strongly to $u(t)$ uniformly for t in $[0, d]$ as $\epsilon \to 0$. We shall show that u is the desired solution of the given initial value problem for the given differential equation. For each $\epsilon > 0$, it follows from the delay-differential equation for u_ϵ that for t in $[0, d]$,

$$u_\epsilon(t) = v - \int_0^t T(u_\epsilon(s - \epsilon)) \, ds \qquad (u_\epsilon(t) = v \text{ for } t \leq 0).$$

Since $\|u_\epsilon(s - \epsilon) - u_\epsilon(s)\| \leq \epsilon M$, it follows that $u_\epsilon(s - \epsilon)$ converges uniformly on $[0, d]$ to $u(s)$. Since T is demicontinuous, it follows that $T(u_\epsilon(s - \epsilon))$ converges weakly to $T(u(s))$ as $\epsilon \to 0$, uniformly for s in $[0, d]$. Hence

$$\int_0^t T(u_\epsilon(s - \epsilon)) \, ds \to \int_0^t T(u(s)) \, ds,$$

with the convergence being weak convergence for each t in $[0, d]$. It follows that

$$u(t) = v - \int_0^t T(u(s)) \, ds \qquad (t \in [0, d])$$

and therefore $u(0) = v$ and $du(t)/dt = -T(u(t))$ ($t \in [0, d]$).

The bound on $\|T(u(t))\|$ follows from Theorem (9.5) and does not depend upon any assumption that the values of T are uniformly bounded on the ball $B_r(v)$, but holds also if the solution $u(t)$ runs outside the ball as long as it continues to satisfy our differential equation. q.e.d.

We now derive a useful inequality which sharpens the conclusion of Theorem (9.5) under a slightly different hypothesis and which will be applied in several contexts in the later discussion.

THEOREM (9.10). *Let X be a Banach space, T a g-accretive mapping from X to 2^X, $h(t)$ a real-valued C^1 function on the interval $[0, d]$ ($d > 0$). Suppose that $w: [0, d] \to X$ is a strongly continuous, weakly once-differentiable function such that for each t in $[0, d]$,*

$$-dw(t)/dt \in T(w(t)) + h(t)w(t).$$

Then for each t in $[0, d]$,

$$\left\| \frac{dw}{dt}(t) \right\| \leq M_0 \exp(-H(t)) + M_1 \int_0^t |h'(r)| \exp(H(r) - H(t)) \, dr,$$

where

$$H(t) = \int_0^t h(s) \, ds, \qquad \|w(t)\| \leq M_1 \quad (t \in [0, d])$$

and

$$M_0 = \lim_{\xi \to 0+} \left\| \frac{w(\xi) - w(0)}{\xi} \right\| < \infty.$$

PROOF OF THEOREM (9.10). For any $\xi > 0$, we set $w_\xi(t) = w(t + \xi)$ for $0 \leq t \leq d - \xi$. Then w_ξ satisfies the condition that $-dw_\xi(t)/dt \in T(w_\xi(t)) + h(t + \xi)w_\xi(t)$. Let $q_\xi(t) = \|w_\xi(t) - w(t)\|^2$. Then for $0 \leq s < t \leq d_0$, we have

$$q_\xi(s) \geq q_\xi(t) + 2(\psi_t, [w_\xi(s) - w(s)] - [w_\xi(t) - w(t)])$$

for any ψ_t in $J(w_\xi(t) - w(t))$. Since $dw(t)/dt + h(t)w(t)$ lies in $-T(w(t))$, and $dw_\xi(t)/dt + h(t + \xi)w_\xi(t)$ lies in $-T(w_\xi(t))$, it follows from the g-accretiveness of T that there exists an element $\Phi(w_\xi(t), w(t))$ in $J(w_\xi(t) - w(t))$ such that

$$(\Phi(w_\xi(t), w(t)), dw_\xi(t)/dt - dw(t)/dt) \leq -(\Phi(w_\xi(t), w(t)), h(t + \xi)w_\xi(t) - h(t)w(t)).$$

We let $\psi_t = \Phi(w_\xi(t), w(t))$, and consider $s \to t-$. Then

$$\varlimsup_{s \to t-} \frac{q_\xi(t) - q_\xi(s)}{t - s} \leq \lim_{s \to t-} 2\left(\psi_t, \frac{w_\xi(t) - w_\xi(s)}{t - s} - \frac{w(t) - w(s)}{t - s}\right)$$

$$= 2(\Phi(w_\xi(t), w(t)), dw_\xi(t)/dt - dw(t)/dt)$$

$$\leq -2(\Phi(w_\xi(t), w(t)), h(t + \xi)w_\xi(t) - h(t)w(t))$$

$$\leq -2h(t)(\Phi(w_\xi(t), w(t)), w_\xi(t) - w(t))$$

$$+ 2[h(t) - h(t + \xi)](\Phi(w_\xi(t), w(t)), w_\xi(t)).$$

Since $(\Phi(w_\xi(t), w(t)), w_\xi(t) - w(t)) = q_\xi(t)$ while

$$(\Phi(w_\xi(t), w(t)), w_\xi(t)) \leq \|w_\xi(t) - w(t)\| M_1 \leq M_1[q_\xi(t)]^{1/2},$$

it follows that

$$\varlimsup_{s \to t-} \frac{q_\xi(t) - q_\xi(s)}{t - s} \leq -2h(t)q_\xi(t) + 2M_1 |h(t + \xi) - h(t)| [q_\xi(t)]^{1/2}.$$

We now prove the following proposition:

PROPOSITION (9.5). *Let q be a continuous nonnegative function on the interval $[0, d]$ for a given $d > 0$ such that for two continuous functions h and h_1 from $[0, d]$ to the reals $(h_1(t) \geq 0)$ and all t in $(0, d)$, we have*

$$\varlimsup_{s \to t-} \frac{q(t) - q(s)}{t - s} \leq -2h(t)q(t) + 2h_1(t)[q(t)]^{1/2}.$$

Then for all t in $[0, d]$,

$$[q(t)]^{1/2} \leq [q(0)]^{1/2} \exp(-H(t)) + \int_0^t h_1(r) \exp(H(r) - H(t)) \, dr,$$

where $H(t) = \int_0^t h(s) \, ds$.

PROOF OF PROPOSITION (9.5). We set

$$p(t) = \exp(H(t))[q(t)]^{1/2} - \int_0^t h_1(r) \exp(H(r)) \, dr.$$

It suffices to prove that for each t in $(0, d)$,

$$\varlimsup_{s \to t-} \left[\frac{p(t) - p(s)}{t - s}\right] \leq 0$$

for then by Proposition (9.1), $p(t)$ is nonincreasing in t on $[0, d]$ and for each t in $[0, d]$, $p(t) \leq p(0)$. If we translate this last statement into terms of $q(t)$, we obtain

$$[q(t)]^{1/2} \leq [q(0)]^{1/2} \exp(-H(t)) + \exp(-H(t)) \int_0^t h_1(r) \exp(H(r)) \, dr,$$

which is precisely our desired inequality.

Suppose first that for the given value of t, $q(t) = 0$. Then for $s < t$, $\exp(H(s))[q(s)]^{1/2} \geq 0$, so that

$$p(t) - p(s) \leq -\int_s^t h_1(r) \exp(H(r)) \, dr,$$

and

$$\varlimsup_{s \to t-} \frac{p(t) - p(s)}{t - s} \leq -\lim_{s \to t-} \frac{1}{t - s} \int_s^t h_1(r) \exp(H(r)) \, dr \leq 0.$$

In the other case, $q(t) > 0$ for the given point t. By the continuity of q, we need only consider $s < t$ such that $q(s) > 0$. Then

$$p(t) - p(s) = [\exp(H(t)) - \exp(H(s))][q(t)]^{1/2}$$

$$- \int_s^t h_1(r) \exp(H(r)) \, dr + \exp(H(s)) \frac{[q(t) - q(s)]}{[q(t)]^{1/2} + [q(s)]^{1/2}}.$$

Hence

$$\varlimsup_{s \to t-} \frac{p(t) - p(s)}{t - s} \leq \exp(H(t))h(t)q(t)^{1/2} - h_1(t) \exp(H(t))$$

$$+ \tfrac{1}{2}[q(t)]^{-1/2} \varlimsup_{s \to t-} \left[\frac{q(t) - q(s)}{t - s}\right] \exp(H(t)),$$

i.e.,

$$\varlimsup \frac{p(t) - p(s)}{t - s} \leq \exp(H(t))[h(t)[q(t)]^{1/2} - h_1(t) - h(t)[q(t)]^{1/2} + h_1(t)] \leq 0. \quad \text{q.e.d.}$$

PROOF OF THEOREM (9.10) COMPLETED. We apply Proposition (9.5) to the function $q_\xi(t)$ for each $\xi > 0$, and we obtain

$$\|w_\xi(t) - w(t)\| \leq \exp(-H(t)) \|w_\xi(0) - w(0)\|$$

$$+ M_1 \int_0^t |h(r + \xi) - h(r)| \exp(H(r) - H(t)) \, dr.$$

As $\xi \to 0+$, we know that

$$\lim_{\xi \to 0+} \left\| \frac{w(t + \xi) - w(t)}{\xi} \right\| \geq \left\| \frac{dw}{dt}(t) \right\|,$$

since $(w(t + \xi) - w(t))/\xi$ converges weakly to $dw(t)/dt$. Hence, we obtain

$$\left\| \frac{dw}{dt}(t) \right\| \leq M_0 \exp(-H(t)) + M_1 \int_0^t |h'(r)| \exp(H(r) - H(t)) \, dr$$

with
$$M_0 = \lim_{\xi \to 0+} \left\| \frac{w(\xi) - w(0)}{\xi} \right\|. \quad \text{q.e.d.}$$

An important special case of Theorem (9.10) is the following:

THEOREM (9.11). *Let X be a Banach space, T a g-accretive mapping of X into 2^X, $u: R^+ \to X$ a strongly continuous, weakly once-differentiable function such that for all t in R^+, $-du(t)/dt \in T(u(t)) + u(t)$.*
Then
(a) $\|du(t)/dt\|$ converges strongly to zero in X as $t \to \infty$, and indeed satisfies the inequality $\|du(t)/dt\| \le M_0 \exp(-t)$.
(b) $u(t)$ converges strongly to some element u_0 of X as $t \to \infty$.
(c) If the graph of T is closed in $X \times X$, then $0 \in T(u_0) + u_0$.

PROOF OF THEOREM (9.11). PROOF OF (a). It obviously suffices to prove the inequality stated in (a). This is a special case of Theorem (9.10), however, with $h(t) \equiv 1$. q.e.d.

PROOF OF (b). Since T is g-accretive, $(T + I)$ has an inverse which is a single-valued nonexpansive mapping. Since
$$-du(t)/dt \in (T + I)(u(t)),$$
and $du(t)/dt \to 0$ as $t \to \infty$, it follows that $u(t)$ converges strongly since
$$\|u(t_n) - u(t_m)\| \le \|du(t_m)/dt - du(t_n)/dt\| \to 0 \quad (t_n, t_m \to +\infty). \quad \text{q.e.d.}$$

PROOF OF (c). Since $u(t)$ converges strongly to u_0 and
$$-du(t)/dt \in (T + I)(u(t)), \quad du(t)/dt \to 0,$$
strongly in X, it follows from the hypothesis that the graph of T is strongly closed in $X \times X$ that $[u_0, 0]$ lies in $G(T)$, i.e., $0 \in (T + I)(u_0)$. q.e.d.

COROLLARY TO THEOREM (9.11). *Let X be a Banach space, T a g-accretive mapping of X into 2^X whose graph is closed in $X \times X$ in the strong topology. Suppose that for some element v in $D(T)$, there exists a strongly continuous, weakly once-differentiable solution u on R^+ of the equation $du/dt - (T + I)(u(t))$ with $u(0) = v$.*
Then 0 lies in the range of $(T + I)$.

THEOREM (9.12). *Let X be a reflexive Banach space, T a g-accretive mapping of X into X having the property that if u_j converges strongly to u and if $T(u_j)$ converges weakly to w in X, then $w = T(u)$. Suppose that for each v_0 in $D(T)$ and each w_0 in X, there exists a local solution of the differential equation*
$$du(t)/dt = -T(u(t)) - u(t) + w_0$$

with $u(0) = v_0$ on some interval $[0, d_{v_0, w_0}]$. Then:

(a) *For each v_0 and w_0, there exists a solution u on R^+ of the differential equation*

$$du(t)/dt = -T(u(t)) - u(t) + w_0 \quad (t \geq 0), \qquad u(0) = v_0.$$

(b) *The range of $(T + I)$ is all of X. In particular, T is hypermaximal accretive.*

PROOF OF THEOREM (9.12). PROOF OF (a). Consider for a given pair $[v_0, w_0]$ in $X \times X$ with v_0 lying in $D(T)$, the maximal interval $[0, d)$ of existence of the solution $u(t)$ of the differential equation

$$du(t)/dt = -T(u(t)) - u(t) + w_0, \qquad u(0) = v_0.$$

Such an interval will exist since any two solutions of this problem will be identical on the interval of their common existence by Theorem (9.2). We must show that $d = +\infty$.

Suppose that $d < +\infty$. By Theorem (9.5) since the mapping $T + I - w_0$ is g-accretive, we know that for the given solution $u(t)$ on $[0, d)$, $\|du(t)/dt\| \leq M$ for a suitable constant $M > 0$ and all t in $[0, d)$. It follows by the equation satisfied by $u(t)$ that

$$\|(T + I)(u(t))\| \leq M_1 \qquad (t \in [0, d)).$$

Since $(T + I)^{-1}$ is nonexpansive, it follows that $\|u(t)\| \leq M_2$ for a suitable constant M_2 and all t in $[0, d)$. Finally, we obtain $\|T(u(t))\| \leq M_3$ for a suitable constant M_3 and all t in $[0, d)$.

By the boundedness of $\|du(t)/dt\|$, we know that for $t < t_1 < d$, we have

$$\|u(t_1) - u(t)\| \leq \int_t^{t_1} \left\|\frac{du}{dt}(s)\right\| ds \leq M(t_1 - t) \to 0 \qquad (t, t_1 \to d-).$$

Hence $u(t)$ converges to an element u_0 strongly in X as $t \to \infty$. Since X is reflexive and $\|T(u(t))\|$ is bounded on the interval, to show that $T(u(t))$ converges weakly in X to $T(u_0)$, it suffices to assume that $T(u(t_j))$ converges weakly to w for a sequence $t_j \to d-$ and to prove that $w = T(u_0)$. This is true by hypothesis, however, and it follows that $u_0 \in D(T)$ and $T(u(t))$ converges weakly to $T(u_0)$ as $t \to d-$.

By hypothesis, there exists a solution $u_1(t)$ of the differential equation

$$du_1(t)/dt = -T(u_1(t)) - u_1(t) + w_0$$

on the interval $[d, d + \epsilon)$ for some $\epsilon > 0$ with $u_1(d) = u_0$. For this solution, we have

$$u_1(t) = u_0 + \int_d^t [-T(u_1(s)) - u_1(s) + w_0] ds \qquad (t \in [d, d + \epsilon)).$$

We now set

$$v(t) = \begin{cases} u(t), & 0 \leq t < d, \\ u_1(t), & d \leq t < d + \epsilon. \end{cases}$$

For $t < d$, we have
$$v(t) = v + \int_0^t [-T(v(s)) - v(s) + w_0]\, ds.$$
For $t = d$,
$$u_0 = v - \int_0^d [T(v(s)) + v(s) - w_0]\, ds,$$
since u_0 is the limit of $u(t)$ as $t \to d-$ and this limit is expressed by the integral on the right of a weakly continuous integrand on the interval $[0, d]$. Finally, for $d \leq t < d + \epsilon$, we have
$$v(t) = u_0 - \int_d^t [T(v(s)) + v(s) - w_0]\, ds$$
$$= v - \int_0^t [T(v(s)) + v(s) - w_0]\, ds.$$

Thus $v(t)$ is a solution of our initial value problem on a larger interval, thereby contradicting the maximality of the interval $[0, d)$ as the domain of existence of a solution. This contradiction proves that $d = +\infty$. q.e.d.

PROOF OF (b). We now apply Theorem (9.11), and note that because of the existence of solutions on R^+ proved in (a), the hypotheses of Theorem (9.11) are satisfied for $T + I - w_0$, i.e., there exists u_0 in $D(T)$ such that $T(u_0) + u_0 - w_0 = 0$. Since this is true for each w_0 in X, it follows that $R(T + I) = X$.

By Theorem (9.4), T is accretive. Hence T is hypermaximal accretive. q.e.d.

We now give some important special cases of the preceding results.

THEOREM (9.13). *Let X be a Banach space, T a g-accretive mapping of X into X which is locally uniformly continuous.*

Then $(T + I)$ maps X onto X, and T is hypermaximal accretive.

PROOF OF THEOREM (9.13). The conclusion of this theorem follows from Theorem (9.11) together with the existence theorem implied by Theorem (9.6) for solutions of the differential equations
$$du(t)/dt = -T(u(t)) - u(t) + w_0$$
with an initial value $u(0) = v_0$. Since the graph of any continuous mapping T from X to X is closed in $X \times X$, it follows that $w_0 = (T + I)(u_0)$ for some u_0 in X. Hence $R(T + I) = X$.

By Theorem (9.4), the existence theorem for solutions of the initial value problem for the equation
$$du(t)/dt = -T(u(t)), \quad u(0) = v_0,$$
implies that T is accretive, and hence hypermaximal accretive since $R(T + I) = X$. q.e.d.

THEOREM (9.14). *Let X be a Banach space with X^* uniformly convex, T a demicontinuous accretive mapping of X into X.*

Then $(T + I)$ has all of X as its range, and T is hypermaximal accretive.

PROOF OF THEOREM (9.14). This follows from Theorem (9.8) in the same fashion as the corresponding result in Theorem (9.13) follows from Theorem (9.6). The closedness of $G(T)$ in $X \times X$ follows immediately from the demicontinuity of T. Therefore, the conclusion of the present theorem follows from Theorem (9.11).
q.e.d.

THEOREM (9.15). *Let X be a Banach space with X^* uniformly convex, T an accretive mapping of X into X with $R(T + I) = X$ (i.e., T hypermaximal accretive).*

Then for each v_0 in $D(T)$, there exists one and only one solution u on R^+ of the differential equation

$$du(t)/dt = -T(u(t)) \qquad (t \geq 0),$$

with initial value $u(0) = v_0$, where the solution $u(t)$ is strongly continuous and weakly once-differentiable from R^+ to X.

For this solution, we have $\|du(t)/dt\|$ a nonincreasing function of t on R^+.

PROOF OF THEOREM (9.15). For each $\epsilon > 0$, we consider the mapping $(I + \epsilon T)$ with domain $D(T)$ and values in X. Since T is accretive and $\epsilon \geq 0$, it follows that $(I + \epsilon T)^{-1}$ is a nonexpansive mapping of $R(I + \epsilon T)$ into X. For $\epsilon = 1$, it follows from the hypothesis that $R(I + T) = X$. Hence it follows by Theorem (3.3) that $R(I + \epsilon T) = X$ for all $\epsilon \geq 0$. Hence $(I + \epsilon T)^{-1}$ is a well-defined nonexpansive mapping of X into X for all $\epsilon > 0$.

Since $(I + \epsilon T)^{-1}$ maps X into $D(T)$ for all $\epsilon > 0$, we may form the composite mapping $T_\epsilon = T(I + \epsilon T)^{-1}: X \to X$. For w in X, let $v_\epsilon = (I + \epsilon T)^{-1} w$. Then $v_\epsilon + \epsilon T(v_\epsilon) = w$, i.e.,

$$T_\epsilon(w) = T(v_\epsilon) = \epsilon^{-1}(w - v_\epsilon) = \epsilon^{-1}(I - (I + \epsilon T)^{-1})(w).$$

Since I and $(I + \epsilon T)^{-1}$ are nonexpansive mappings of X into X, it follows that for $\epsilon > 0$, T_ϵ is a Lipschitzian mapping of X into X with Lipschitz constant $\leq 2\epsilon^{-1}$.

We consider the solution u_ϵ on R^+ of the auxiliary problem defined for each $\epsilon > 0$ by $du_\epsilon(t)/dt = -T_\epsilon(u_\epsilon(t))$ $(t \geq 0)$, with initial value $u_\epsilon(0) = v_0$. The existence of these solutions is assured trivially by the Lipschitzian character of T_ϵ.

Since $(I + \epsilon T)^{-1}$ is a nonexpansive mapping of X into X, it follows from Theorem (8.9) that $I - (I + \epsilon T)^{-1}$ is an accretive mapping of X into X. Since any nonnegative multiple of an accretive mapping is also accretive, it follows that $T_\epsilon = \epsilon^{-1}(I - (I + \epsilon T)^{-1})$ is an accretive mapping of X into X for each $\epsilon > 0$. Hence by Theorem (9.5), we know that

$$\left\|\frac{du_\epsilon}{dt}(t)\right\| \leq \left\|\frac{du_\epsilon}{dt}(0)\right\| = \|T_\epsilon(v_0)\|.$$

Since v_0 lies in $D(T)$, however, it follows that $v_0 + \epsilon T(v_0) = v_0 + \epsilon T(v_0)$ implies that $v_0 = (I + \epsilon T)^{-1}(v_0 + \epsilon T v_0)$. If $v_\epsilon = (I + \epsilon T)^{-1}(v_0)$, we may apply the nonexpansiveness of $(I + \epsilon T)^{-1}$ to conclude that $\|v_0 - v_\epsilon\| \leq \|\epsilon T(v_0)\|$, and hence that

$$\|T_\epsilon v_0\| = \|\epsilon^{-1}(v_0 - v_\epsilon)\| \leq \|T(v_0)\| \qquad (\epsilon > 0).$$

In particular,
$$\|T_\epsilon(u_\epsilon(t))\| = \|du_\epsilon(t)/dt\| \leq \|T_\epsilon(v_0)\| \leq \|T(v_0)\|$$
for every $\epsilon > 0$ and all $t \geq 0$.

We consider $\epsilon, \xi > 0$, and let $q_{\epsilon,\xi}(t) = \|u_\epsilon(t) - u_\xi(t)\|^2$. By Proposition (9.4), $q_{\epsilon,\xi}$ is differentiable on R^+ with

$$\frac{d}{dt}(q_{\epsilon,\xi})(t) = 2\left(J(u_\epsilon(t) - u_\xi(t)), \frac{du_\epsilon}{dt}(t) - \frac{du_\xi}{dt}(t)\right).$$

Let
$$v_\epsilon(t) = (I + \epsilon T)^{-1}(u_\epsilon(t)); \qquad v_\xi(t) = (I + \xi T)^{-1}(u_\xi(t)).$$
Then
$$du_\epsilon(t)/dt = -T_\epsilon(u_\epsilon(t)) = -T(v_\epsilon(t)),$$
and
$$du_\xi(t)/dt = -T_\xi(u_\xi(t)) = -T(v_\xi(t)).$$

We know also that
$$\|u_\epsilon(t) - v_\epsilon(t)\| = \|(I - (I + \epsilon T)^{-1})(u_\epsilon(t))\| = \epsilon \|T_\epsilon(u_\epsilon(t))\| \leq \epsilon \|T(v_0)\|,$$
and similarly $\|u_\xi(t) - v_\xi(t)\| \leq \xi \|T(v_0)\|$.

Since X^* is uniformly convex by hypothesis, the duality mapping J of X into X^* is uniformly continuous on bounded subsets of X. Let $\rho > 0$ be given. Then there exists a function $r_\rho(\lambda)$ for $\lambda \geq 0$ such that if u and u_1 lie in the closed ball $B_\rho(0)$ with $\|u - u_1\| \leq \lambda$, then $\|J(u) - J(u_1)\| \leq r_\rho(\lambda)$, where $r_\rho(\lambda) \to 0$ as $\lambda \to 0$. Consider the interval $[0, N]$ of R^+ with $N > 0$. Since $\|du_\epsilon(t)/dt\| \leq \|T(v_0)\|$ for all $\epsilon > 0$ and $t \geq 0$, it follows that for t in $[0, N]$, $\|u_\epsilon(t)\| \leq \|v\| + \|T(v_0)\| N \leq M(N)$. Hence

$$\frac{d}{dt}\{q_{\epsilon,\xi}(t)\} = -2(J(v_\epsilon(t) - v_\xi(t)), T(v_\epsilon(t)) - T(v_\xi(t))) + R_{\epsilon,\xi}(t) \leq R_{\epsilon,\xi}(t),$$

where
$$R_{\epsilon,\xi}(t) = 2(J(v_\epsilon(t) - v_\xi(t)) - J(u_\epsilon(t) - u_\xi(t)), T(v_\epsilon(t)) - T(v_\xi(t))).$$

We apply the estimate
$$|R_{\epsilon,\xi}(t)| \leq 2 \|J(v_\epsilon(t) - v_\xi(t)) - J(u_\epsilon(t) - u_\xi(t))\| \cdot \|T_\epsilon(u_\epsilon(t))\| + \|T_\xi(u_\xi(t))\|.$$

For t in $[0, N]$, $\|u_\epsilon(t) - u_\xi(t)\| \leq 2M(N)$, and

$$\|v_\epsilon(t) - v_\xi(t)\| \leq \|u_\epsilon(t) - u_\xi(t)\| + \|v_\epsilon(t) - u_\epsilon(t)\| + \|v_\xi(t) - u_\xi(t)\|$$
$$\leq 2M(N) + \epsilon \|T(v_0)\| + \xi \|T(v_0)\| \leq M_1(N)$$

for $0 < \epsilon, \xi \leq 1$. Moreover,
$$\|[u_\epsilon(t) - u_\xi(t)] - [v_\epsilon(t) - v_\xi(t)]\| \leq (\epsilon + \xi) \|T(v_0)\|.$$
Therefore
$$\|J(v_\epsilon(t) - v_\xi(t)) - J(u_\epsilon(t) - u_\xi(t))\| \leq r_{M_1(N)}(\epsilon + \xi) \|T(v_0)\| \to 0$$

uniformly on $[0, N]$ as ϵ and $\xi \to 0$. On the other hand,
$$\|T_\epsilon(u_\epsilon(t))\| + \|T_\xi(u_\xi(t))\| \leq 2 \|T(v_0)\|.$$
It follows that $|R_{\epsilon,\xi}(t)| \to 0$ uniformly on each bounded interval $[0, N]$ as $\epsilon, \xi \to 0$.
Since for t in $[0, N]$,
$$q_{\epsilon,\xi}(t) \leq N \max_{t \in [0, N]} |R_{\epsilon,\xi}(t)|$$
it follows that $q_{\epsilon,\xi}(t) \to 0$ ($\epsilon, \xi \to 0$) uniformly on each bounded interval of R^+, so that there exists a continuous function $u: R^+ \to X$ such that $u_\epsilon(t)$ converges strongly to $u(t)$, uniformly for bounded t in R^+.

We now show that this limit function $u: R^+ \to X$ is indeed the desired solution of the given initial value problem for the differential equation $du/dt = -T(u)$. Since $u_\epsilon(0) = v_0$ for each $\epsilon > 0$, we know that $u(0) = v_0$. Since the solution is unique by Theorem (9.2), it suffices to show that $u(t)$ is a solution of the differential equation. This will follow if we demonstrate two facts:

(1) For each $t \geq 0$, $u(t)$ lies in $D(T)$ and the function $T(u(t))$ on R^+ is weakly continuous.

(2) For each $t > 0$, u satisfies the integral equation $u(t) = v_0 - \int_0^t T(u(s))\, ds$.

For each $\epsilon > 0$, we know that $\|v_\epsilon(t) - u_\epsilon(t)\| \leq \epsilon \|T(v_0)\|$. Hence $v_\epsilon(t)$ converges strongly to $u(t)$ in X, uniformly for t in bounded subsets of R^+. Since $\|T(v_\epsilon(t))\| = \|T_\epsilon(u_\epsilon(t))\| \leq \|T(v_0)\|$. We now assert that for each fixed t in R^+, $u(t)$ lies in $D(T)$ and $T(v_\epsilon(t))$ converges as $\epsilon \to 0$ to $T(u(t))$ in the weak topology on X. To show this, since X is reflexive, we may choose a sequence $\epsilon_k \to 0$ as $k \to 0$ for which $T(v_{\epsilon_k}(t))$ converges weakly in X to an element w_0 of X. For the $v_{\epsilon_k}(t)$ themselves, we know that the sequence $\{v_{\epsilon_k}(t)\}$ converges strongly in X as $k \to \infty$ to the element $u(t)$ of X. Since T is hypermaximal accretive, we now apply the conclusion of Theorem (3.16)(b) and conclude that w is an element of $T'(u(t))$, i.e., $u(t)$ lies in $D(T)$ and each weakly convergent subsequence of $\{T(v_{\epsilon_k}(t))\}$ converges to $T(u(t))$. Hence $T(v_\epsilon(t))$ converges weakly to $T(u(t))$ in X as $\epsilon \to 0$.

We assert furthermore that $T(u(t))$ is weakly continuous in t on R^+. Indeed, consider a sequence $\{t_j\}$ in R^+ converging to t. Since $u(t)$ is strongly continuous in t, it follows that $\{u(t_j)\}$ converges strongly to $u(t)$ in X. Since $T(u(t_j))$ is a bounded family in X, in order to prove that $T(u(t_j))$ converges weakly to $T(u(t))$ in X, it suffices to assume that $T(u(t_j))$ converges weakly to some element v_1 of X and to prove that this element v_1 must then be equal to $T(u(t))$. However, this follows once more from Theorem (3.16)(b), and it follows that $T(u(t))$ is weakly continuous in t on R^+.

Finally, for each $\epsilon > 0$, we have for all t in R^+:
$$u_\epsilon(t) = v_0 - \int_0^t T(v_\epsilon(s))\, ds.$$
For each y in X^* we have
$$(y, u_\epsilon(t)) = (y, v_0) - \int_0^t (y, T(v_\epsilon(s)))\, ds.$$

As $\epsilon \to 0$, $(y, u_\epsilon(t))$ converges to $(y, u(t))$, while for each s, $(y, T(v_\epsilon(s))) \to (y, T(u(s)))$ as $\epsilon \to 0$, with

$$|(y, T(v_\epsilon(s)))| \leq \|y\| \cdot \|T(v_\epsilon(s))\| \leq \|y\| \cdot \|T(v_0)\|.$$

Hence by the Lebesgue dominated convergence theorem,

$$\int_0^t (y, T(v_\epsilon(s))) \, ds \to \int_0^t (y, T(u(s))) \, ds.$$

Thus

$$(y, u(t)) = (y, v_0) - \int_0^t (y, T(u(s))) \, ds \qquad (t \geq 0),$$

i.e., $u(t) = v_0 - \int_0^t T(u(s)) \, ds$, and it follows that $u(t)$ does indeed satisfy the desired differential equation.

Since $T(v_\epsilon)(t) \to T(u(t))$ weakly in X as $\epsilon \to 0$, it follows that for all $t > 0$,

$$\|T(u(t))\| \leq \|T(v_0)\| = \|T(u(0))\|,$$

i.e., $\|du(t)/dt\| \leq \|du(0)/dt\|$, $t \geq 0$. We complete the proof of our theorem by showing that for any $0 < s < t$, we have the inequality $\|du(t)/dt\| \leq \|du(s)/dt\|$. To do this, we construct the solution $u^{(s)}(t)$ of the differential equation

$$du^{(s)}(t)/dt = -T(u^{(s)}(t)), \qquad t \geq s,$$

with the initial condition $u^{(s)}(s) = u(s)$. Since both u and $u^{(s)}$ are solutions of this differential equation on the interval $[s, \infty)$ with the same initial value at s, they must concide by Theorem (9.2). However, it follows from the argument of the preceding part of the proof that $\|T(u^{(s)}(t))\| \leq \|T(u^{(s)}(s))\|$ for $t \geq s$, i.e.,

$$\|T(u(t))\| \leq \|T(u(s))\| \qquad (s \leq t). \qquad\qquad \text{q.e.d.}$$

THEOREM (9.16). *Suppose that under the hypotheses of Theorem (9.15), we have the additional hypothesis that the space X itself is uniformly convex. Then $u(t)$ is strongly differentiable from the right for each $t \geq 0$, and $T(u(t))$ is strongly continuous from the right on R^+.*

PROOF OF THEOREM (9.16). It obviously suffices to prove that $T(u(t))$ is strongly continuous from the right. However, $T(u(t))$ is a weakly continuous function and $\|T(u(t))\|$ is nonincreasing in t on R^+. Hence, if $t_j \to t$ from above, we have $\|T(u(t_j))\| \leq \|T(u(t))\|$ and $T(u(t_j))$ converges weakly to $T(u(t))$. Since X is uniformly convex, $T(u(t_j))$ converges strongly to $T(u(t))$. q.e.d.

THEOREM (9.17). *Let X be a Banach space with X^* uniformly convex, T an accretive mapping of X into X with $R(T + I) = X$. For each $t > 0$ and each v_0 in $D(T)$, let $U(t)(v_0) = u(t)$, where u is the solution on R^+ of the differential equation $du(t)/dt = -T(u(t))$ $(t \geq 0)$ with $u(0) = v_0$. Then*

(a) The family of mappings $\{U(t) : t \geq 0\}$ forms a one-parameter semigroup of nonexpansive mappings of the subset $D(T)$ of X into itself, which can be extended by continuity to a one-parameter semigroup of nonexpansive self-mappings of X_0, the closure of $D(T)$ in X.

(b) *The mapping* $(-T)$ *is the infinitesimal generator of the semigroup* $\{U(t): t \geq 0\}$, *i.e.*, v *lies in* $D(T)$ *if and only if*

$$\frac{[U(t)-I]}{t}v$$

converges weakly in X *to an element* x *as* $t \to 0+$, *and then* $x = -T(v)$.

PROOF OF THEOREM (9.17). The nonexpansive character of the mapping $U(t)$ for $t > 0$ follows from the accretive character of the mapping T and the conclusion of Theorem (9.2). The semigroup-property follows from the fact that $U(t)(v)$ converges strongly to v as $t \to 0+$ and the uniqueness of the solution of the differential equation with a given initial value in $D(T)$. Thus the results of (a) follow without any complication and it suffices to verify the conclusion of part (b).

Let $T_1 = -A$, where A is the infinitesimal generator of the semigroup $\{U(t):t \geq 0\}$. Then T_1 is an extension of the mapping T, and we assert that T_1 itself is an accretive mapping of $D(T_1)$ in X into X. Indeed, suppose that v and v_1 lie in $D(T_1)$, and let $w = T_1(v)$, $w_1 = T_1(v_1)$. Then

$$\|v - v_1\|^2 \geq \|U(t)v - U(t)v_1\|^2$$
$$\geq \|v - v_1\|^2 + 2(J(v - v_1), [U(t)v - v] - [U(t)v_1 - v_1]).$$

Hence for all $t > 0$,

$$(J(v - v_1), t^{-1}[U(t)v - v] - t^{-1}[U(t)v_1 - v_1]) \leq 0.$$

Letting $t \to 0+$, we find that $(J(v - v_1), w - w_1) \geq 0$ so that T_1 is accretive. Since T is hypermaximal accretive by hypothesis and since by Theorem (3.16)(a), T is thereby maximal accretive, it follows that $T = T_1$. Hence $T = -A$.   q.e.d.

A simple converse to the preceding results culminating in Theorem (9.17) is provided by the following:

THEOREM (9.18). *Let* X *be a Banach space*, T *a mapping of a subset* $D(T)$ *of* X *into* X, X_0 *the closure of* $D(T)$ *in* X. *Suppose that for each* w_0 *in* X, *there exists a strongly continuous one-parameter semigroup* $\{V_{w_0}(t):t \geq 0\}$ *on* X *such that* $\|V_{w_0}(t)\|_{\text{Lip}} \leq \exp(-t)$, *with the infinitesimal generator of the semigroup given by* $(-T - I + w_0)$.

Then T *is hypermaximal accretive.*

PROOF OF THEOREM (9.18). For each given w_0 of X, each of the mappings $\{V_{w_0}(t)\}$ of X_0 into itself for $t > 0$ is a strict contraction and they commute with one another. It follows immediately that each such mapping has a unique fixed point x_0 in X_0 and that this fixed point is independent of t for $t > 0$. Hence

$$t^{-1}[V_{w_0}(t) - I](x_0) = 0 \to 0,$$

i.e., x_0 lies in the domain of the infinitesimal generator $(-T - I + w_0)$ of the semigroup and $(-T - I + w_0)(x_0) = 0$, i.e., $(T + I)(x_0) = w_0$. Since this is true for all w_0 in X, it follows that $R(T + I) = X$.

It follows from the preceding paragraph that in order to complete the proof of the theorem, it suffices to prove that the mapping T is accretive. Let $\{V(t):t \geq 0\}$

be one of the semigroups $\{V_{w_0}(t) : t \geq 0\}$ and let S be its infinitesimal generator (i.e., $S = -T - I + w_0$). For u and v in $D(S)$ we have $\|V(t)u - V(t)v\|^2 \leq \exp(-2t) \|u - v\|^2$, while for any ψ in $J(u - v)$,

$$\|V(t)u - V(t)v\|^2 \geq \|u - v\|^2 + 2(\psi, [V(t) - I](u) - [V(t) - I](v)).$$

Hence

$$(\psi, S(u) - S(v)) = \lim_{t \to 0+} t^{-1}(\psi, [V(t) - I](u) - [V(t) - I](v))$$

$$\leq \tfrac{1}{2} \lim_{t \to 0+} t^{-1}(\exp(-2t) - 1) \|u - v\|^2 = -\|u - v\|^2.$$

It follows that $(-S - I)$ is accretive, i.e., $T + w_0$ is accretive. Hence T is accretive. q.e.d.

A useful modification of Theorem (9.18) is obtained by modifying the hypotheses in the following fashion:

THEOREM (9.19). *Let X be a Banach space, T a g-accretive mapping of a subset $D(T)$ of X into X, X_0 the closure of $D(T)$ in X. Suppose that for each w_0 in X, $(-T - I + w_0)$ is the infinitesimal generator of a semigroup $\{V_{w_0}(t) : t \geq 0\}$ of continuous self-mappings of X_0 such that for each v in $D(T)$, $V_{w_0}(t)(v)$ is weakly differentiable in t on R^+.*

Then T is hypermaximal accretive.

PROOF OF THEOREM (9.19). By Theorem (9.10) with $h(t) \equiv -1$, we find that for u and v in $D(T)$, $\|V(t)u - V(t)v\| \leq \exp(-t) \|u - v\|$. This inequality extends by continuity to all u and v in X_0. Hence Theorem (9.18) may be applied and yields the conclusion of the present theorem. q.e.d.

By combining Theorems (9.15) and (9.18), we obtain the following nonlinear extension of the well-known generation theorem of Hille-Yosida for one-parameter semigroups of linear mappings in a Banach space:

THEOREM (9.20). *Let X be a Banach space with X^* uniformly convex and let T be a mapping with $D(T)$ in X and values in X. Let X_0 be the closure of $D(T)$ in X. Suppose that for each $\lambda > 0$, the mapping $(T + \lambda I)$ is a one-to-one mapping of $D(T)$ onto X with a Lipschitzian single-valued inverse satisfying the inequality*

$$\|(T + \lambda I)^{-1}\|_{\mathrm{Lip}} \leq 1/\lambda \qquad (\lambda > 0).$$

Then $(-T)$ is the infinitesimal generator on a one-parameter strongly continuous semigroup of nonexpansive mappings $\{V(t) : t \geq 0\}$ of X_0 into X_0.

PROOF OF THEOREM (9.20). Since $(T + I)$ has all of X as its range, it suffices to prove that the inequality of the hypothesis of Theorem (9.20) implies that T is accretive. This follows, however, from Theorem (9.1). q.e.d.

We now proceed to derive more detailed existence theorems for solutions of nonlinear equations of evolution. The first of our more detailed results is the following combination and sharpening of Theorems (9.8) and (9.15):

THEOREM (9.21). *Let X be a Banach space with uniformly convex conjugate space X^*, T a hypermaximal accretive mapping with domain $D(T)$ in X and values in X, T_0 a demicontinuous accretive mapping of X into X. Then*

(a) *$T + T_0$ is a hypermaximal accretive mapping.*

(b) *For each v in $D(T)$, there exists one and only one solution $u: R^+ \to X$ of the differential equation*

$$du(t)/dt = -T(u(t)) - T_0(u(t)) \quad (t \in R^+),$$

with $u(0) = v$.

(c) *For this solution given in part* (b),

$$\|du(t)/dt\| \le \|du(s)/dt\| \quad (0 \le s \le t)$$

(*i.e.*, $\|(T + T_0)(u(t))\| \le \|(T + T_0)(u(s))\|$).

PROOF OF THEOREM (9.21). Following the general lines of the proof of Theorem (9.15), we consider for each $\epsilon > 0$, the approximating differential equation

$$du_\epsilon(t)/dt = -T(I + \epsilon T)^{-1}(u_\epsilon(t)) - T_0(u_\epsilon(t)) \quad (t \ge 0)$$

with the initial value condition $u_\epsilon(0) = v$. For each $\epsilon > 0$, $T_\epsilon = T(I + \epsilon T)^{-1}$ is a Lipschitzian mapping of X into X with Lipschitz norm at most $2\epsilon^{-1}$, and hence the mapping $-T_\epsilon - T_0$ is a demicontinuous mapping of X into X. Since $(T_\epsilon + T_0)$ is accretive, it follows from Theorem (9.14) that $(T_\epsilon + T_0)$ is hypermaximal accretive. Hence by Theorem (9.15), there exists one and only one solution $u_\epsilon: R^+ \to X$ of the approximating problem and for this solution, $\|du_\epsilon(t)/dt\|$ is a nonincreasing function of t on R^+. In particular, for $t > 0$, we have

$$\|du_\epsilon(t)/dt\| \le \|du_\epsilon(0)/dt\| = \|T(I + \epsilon T)^{-1}(u_\epsilon(0)) + T_0(u_\epsilon(0))\|$$
$$= \|T(I + \epsilon T)^{-1}(v) + T_0(v)\| \le \|T(v)\| + \|T_0(v)\|.$$

If we set $v_\epsilon(t) = (I + \epsilon T)^{-1}(u_\epsilon(t))$, we find as in the proof of Theorem (9.15) that

$$\|v_\epsilon(t) - u_\epsilon(t)\| \le \epsilon \|T_\epsilon(u_\epsilon(t))\| \le \epsilon \left\{ \left\| \frac{du}{dt}\epsilon(t) \right\| + \|T_0(v)\| \right\} \le \epsilon \{\|T(v)\| + 2\|T_0(v)\|\}$$

so that $u_\epsilon(t) - v_\epsilon(t) \to 0$ strongly in X, uniformly for t in R^+.

For $\epsilon, \xi > 0$, consider the function $q_{\epsilon,\xi}$ from R^+ to R^+ given by $q_{\epsilon,\xi}(t) = \|u_\epsilon(t) - u_\xi(t)\|^2$ $(t \ge 0)$. By Proposition (9.4), $q_{\epsilon,\xi}$ is once-differentiable at each point t of R^+, and its derivative is given by

$$dq_{\epsilon,\xi}(t)/dt = 2(J(u_\epsilon(t) - u_\xi(t)), du_\epsilon(t)/dt - du_\xi(t)/dt)$$
$$= -2(J(u_\epsilon(t) - u_\xi(t)), T(v_\epsilon(t)) + T_0(v_\epsilon(t)) - T(v_\xi(t)) - T_0(u_\xi(t)).$$

The right-hand term of the last equation may be rewritten in the form

$$-2(J(u_\epsilon(t) - u_\xi(t)), T_0(u_\epsilon(t)) - T_0(u_\xi(t)))$$
$$+ \{-2(J(v_\epsilon(t) - v_\xi(t)), T(v_\epsilon(t)) - T(v_\xi(t))) + R_{\epsilon,\xi}(t)\} \le R_{\epsilon,\xi}(t)$$

by the accretiveness of T and T_0, where

$$R_{\epsilon,\xi}(t) = 2(J(v_\epsilon(t) - v_\xi(t)) - J(u_\epsilon(t) - u_\xi(t)), T(v_\epsilon(t)) - T(v_\xi(t))).$$

Hence

$$|R_{\epsilon,\xi}(t)| \leq 2 \, \|J(v_\epsilon(t) - v_\xi(t)) - J(u_\epsilon(t) - u_\xi(t))\| \cdot \|T(v_\epsilon(t)) - T(v_\xi(t))\|.$$

Since X^* is uniformly convex and J is thereby uniformly continuous on bounded subsets of X, and since

$$\|[v_\epsilon(t) - v_\xi(t)] - [u_\epsilon(t) - u_\xi(t)]\| \leq \|v_\epsilon(t) - u_\epsilon(t)\| + \|v_\xi(t) - u_\xi(t)\| \to 0$$

uniformly on R^+, while $\|u_\epsilon(t)\|$, $\|u_\xi(t)\|$, $\|v_\epsilon(t)\|$, and $\|v_\xi(t)\|$ are uniformly bounded on bounded subsets of R^+, it follows that on each bounded interval $[0, N]$ of R^+,

$$\|J(v_\epsilon(t) - v_\xi(t)) - J(u_\epsilon(t) - u_\xi(t))\| \to 0$$

uniformly for t in $[0, N]$. On the other hand, since T_0 is assumed to be demicontinuous, there exists a ball $B_r(v)$ about v in K with $r > 0$ such that for x in $B_r(v)$, $\|T_0(x)\| \leq M$ for some $M > 0$. Since

$$\|u_\epsilon(t) - v\| \leq \{\|T(v)\| + \|T_0(v)\|\} t \leq r,$$

for $0 \leq t \leq r, (\|T(v)\| + \|T_0(v)\|)^{-1} = d_0$, it follows that for each $\epsilon > 0$ and the constant $d > 0$ independent of ϵ, we know that for $0 \leq t \leq d$, $u_\epsilon(t)$ lies in $B_r(v)$ and $\|T_0(u_\epsilon(t))\| \leq M$. For such t, it follows that

$$\|T(v_\epsilon(t)) - T(v_\xi(t))\| \leq \|du_\epsilon(t)/dt - du_\xi(t)/dt\| + \|T_0(u_\epsilon(t)) - T_0(u_\xi(t))\|$$
$$\leq 2[\|T(v)\| + \|T_0(v)\| + M] = M_1.$$

Thus for t in the interval $[0, d]$ with $d > 0$, we find that as $\epsilon, \xi \to 0$

$$|R_{\epsilon,\xi}(t)| \leq M_1 \, \|J(v_\epsilon(t) - v_\xi(t)) - J(u_\epsilon(t) - u_\xi(t))\| \to 0$$

uniformly for t in $[0, d]$. Since $[dq_{\epsilon,\xi}(t)/dt] \leq R_{\epsilon,\xi}(t)$ $(t \geq 0)$, with $q_{\epsilon,\xi}(0) = 0$, it follows that $q_{\epsilon,\xi}(t) \to 0$ as $\epsilon, \xi \to 0$, uniformly on the interval $[0, d]$. It follows that $\|u_\epsilon(t) - u_\xi(t)\| \to 0$ as $\epsilon, \xi \to 0$ uniformly on this interval. Hence, by the completeness of the Banach space X in its metric, there exists a strongly continuous function $u : [0, d] \to X$ such that $u_\epsilon(t)$ converges strongly to $u(t)$ as $\epsilon \to 0$, uniformly on the interval $[0, d]$. Moreover, it follows that $v_\epsilon(t)$ converges strongly to $u(t)$ uniformly on the same interval as $\epsilon \to 0$.

Since

$$\|T(v_\epsilon(t))\| \leq \|du_\epsilon(t)/dt\| + \|T_0(u_\epsilon(t))\| \leq \|T(v)\| + \|T_0(v)\| + M$$

on this interval, it follows from Theorem (3.16)(b) that as $\epsilon \to 0$, $T(v_\epsilon(t))$ converges weakly to $T(u(t))$ for each t in $[0, d]$, where for each t in this interval, $u(t)$ must lie in $D(T)$. Furthermore, since T_0 is demicontinuous, it follows from the strong convergence of $u_\epsilon(t)$ to $u(t)$ that $T_0(u_\epsilon(t))$ converges weakly to $T_0(u(t))$ as $\epsilon \to 0$

for each t in $[0, d]$. For each t in R^+, we have the integral equation

$$u_\epsilon(t) = v - \int_0^t [T(v_\epsilon(s)) + T_0(u_\epsilon(s))] \, ds.$$

For each s in $[0, d]$, $T(v_\epsilon(s)) + T_0(u_\epsilon(s))$ converges weakly to $T(u(s)) + T_0(u(s))$ and the norm of $\|T(v_\epsilon(s)) + T_0(u_\epsilon(s))\epsilon\|$ is uniformly bounded on R^+. Hence we may apply the Lebesgue dominated convergence theorem to the weak integral on the right-hand side of the integral equation and the integral converges weakly as $\epsilon \to 0$ to $\int_0^t [T(u(s)) + T_0(u(s))] \, ds$, for each t in $[0, d]$. Hence

$$u(t) = v - \int_0^t [T(u(s)) + T_0(u(s))] \, ds \qquad (0 \le t \le d),$$

and it follows immediately that $u : [0, d] \to X$ is a solution on this interval of the differential equation

$$du(t)/dt = -T(u(t)) - T_0(u(t)) \qquad (0 \le t < d)$$

with $u(0) = v$.

Thus, we have shown the existence of a local solution of the given initial value problem for each v in $D(T)$, and by Theorem (9.2), this local solution is uniquely determined by its initial value v. Hence, we may assume the existence of this solution on a maximal interval $[0, d)$ with $d > 0$, and it suffices to show that $d = +\infty$.

Suppose that d is finite. We note that since for each $\epsilon > 0$,

$$\|du\epsilon(t)/dt\| \le \|T(v)\| + \|T_0(v)\| = M_2,$$

it follows that for any t and t_1 with $0 \le t < t_1 < d$, we have $\|u_\epsilon(t_1) - u_\epsilon(t)\| \le M_2(t_1 - t)$. If we take the limit as $\epsilon \to 0$, we obtain the inequality $\|u(t_1) - u(t)\| \le M_2(t_1 - t)$. It follows that as $t \to d-$, $u(t)$ converges strongly to an element u_0 of X, and u can be extended to a strongly continuous mapping of the closed interval $[0, d]$ into X. Since T_0 is demicontinuous, $T_0(u(t))$ is weakly continuous in t on $[0, d]$ and hence there exists a constant $M_3 > 0$ such that $\|T_0(u(t))\| \le M_3$ for t in $[0, d]$. Hence

$$\|T(u(t))\| \le \|du(t)/dt\| + \|T_0(u(t))\| \le \|T_0(v)\| + \|T(v)\| + M_3$$

on $[0, d]$ since

$$du(t)/dt = -[T(u(t)) + T_0(u(t))] = \underset{\epsilon \to 0}{\text{weak lim}} \, [-T(v_\epsilon(t)) - T_0(u_\epsilon(t))]$$

with

$$\| T(v_\epsilon(t)) + T_0(u_\epsilon(t))\| = \|du\epsilon(t)/dt\| \le \|T(v)\| + \|T_0(v)\|.$$

If we apply Theorem (3.16)(b) once more, it follows that $u_0 = u(d)$ lies in $D(T)$ and that $T(u(t))$ converges weakly to $T(u(d))$ as $t \to d-$. Since $u_0 = u(d)$ lies in $D(T)$, we can find a local solution of our differential equation on an interval $[d, d + \beta)$ ($\beta > 0$) with the initial value $u(d) = u_0$. Combining this solution with the solution u on the interval $[0, d)$, we obtain an extension of the solution u to the interval $[0, d + \beta)$ with $\beta > 0$, contradicting the maximality of d. Hence $d = +\infty$, and the solution u exists on R^+.

We thus have completed the proof of part (b) of Theorem (9.21). We shall derive the proofs of parts (a) and (c) of the theorem from the result of part (b).

PROOF OF (a). We apply the result of part (b) to the accretive operator $(T + T_1)$, with $T_1 = T_0 + I - w_0$ for a given element w_0 of X. By the result of part (b) derived above, there exists a solution $u: R^+ \to X$ of the differential equation

$$du(t)/dt = -T(u(t)) - T_1(u(t)) \qquad (t \geq 0)$$

with $u(0) = v$ for each element v of $D(T)$. If we fix such an element v, then it follows from Theorem (9.11) that as $t \to \infty$, $(T + T_1)(u(t)) \to 0$ strongly in X, i.e., $(T + T_0 + I)(u(t)) \to w_0$ strongly in X as $t \to \infty$. Since $(T + T_0 + I)^{-1}$ is a single-valued nonexpansive mapping of $R(T + T_0 + I)$ into X by the accretiveness of $(T + T_0)$, it follows that $u(t)$ converges strongly in X as $t \to \infty$ to some element u_1 of X. Since T_0 is demicontinuous, $T_0(u(t))$ converges weakly to $T_0(u_1)$ as $t \to \infty$. Hence $(T + I)(u(t)) \to w_0 - T_0(u_1)$ as $t \to \infty$, while $u(t)$ converges strongly to u_1. Since T is hypermaximal accretive by hypothesis (i.e., $R(T + I) = X$ and T is accretive), it follows from Theorem (3.16)(b) that $(T + I)(u_1) = w_0 - T_0(u_1)$, i.e., $w_0 = (T + T_0 + I)(u_1)$. Since w_0 was an arbitrary element of X, it follows that $R(T + T_0 + I) = X$ and $(T + T_0)$ is hypermaximal accretive. Hence the proof of part (a) is complete. q.e.d.

PROOF OF (c). Since $(T + T_0)$ has just been shown to be hypermaximal accretive, it follows by Theorem (9.15) that for the solutions constructed in part (b) (and indeed for all solutions of the given differential equation), $\|du(t)/dt\|$ is a nonincreasing function of t on R^+. q.e.d.

THEOREM (9.22). *Let X be a Banach space with X^* uniformly convex, T and T_0 two hypermaximal accretive mappings with domain and range in X. Suppose that $D(T) \subset D(T_0)$ and that the following condition is satisfied:*

(α) *For each v in X, there exist a neighborhood N of v in X and two constants k and k_1 with $0 \leq k < 1$ and $k_1 \geq 0$, such that for all u in $D(T) \cap N$, $\|T_0(u)\| \leq k \|T(u)\| + k_1$.*

Then:

(a) *For each v in $D(T)$, there exists exactly one solution $u: R^+ \to X$ of the differential equation*

$$du(t)/dt = -T(u(t)) - T_0(u(t)) \qquad (t \geq 0)$$

with $u(0) = v$.

(b) *$(T + T_0)$ is hypermaximal accretive.*

(c) *For each solution $u: R^+ \to X$ of the differential equation of part (a), $\|du(t)/dt\|$ is a nonincreasing function of t on R^+.*

PROOF OF THEOREM (9.22). PROOF OF (a). For each $\epsilon > 0$, we consider the approximating equation

$$du_\epsilon(t)/dt = -T(u_\epsilon(t)) - T_0(I + \epsilon T_0)^{-1}(u_\epsilon(t)) \qquad (t \geq 0)$$

with the initial value $u_\epsilon(0) = v$. Since T_0 is hypermaximal accretive, $(I + \epsilon T_0)^{-1}$

is a well-defined nonexpansive self-mapping of X and $T_{0,\epsilon} = T_0(I + \epsilon T_0)^{-1}$ is a Lipschitzian mapping of X into X. Since T is assumed to be hypermaximal accretive, it follows from Theorem (9.21) that for each $\epsilon > 0$ and each v in $D(T)$, there exists exactly one solution $u_\epsilon : R^+ \to X$ of the approximating equation.

For this solution u_ϵ, we know moreover that

$$\|du\epsilon(t)/dt\| \leq \|du\epsilon(0)/dt\| = \|T(v) + T_0(I + \epsilon T_0)^{-1}(v)\| \leq \|T(v)\| + \|T_0(v)\|.$$

Let $v_\epsilon(t) = (I + \epsilon T_0)^{-1}(u_\epsilon(t))$. Then $\|u_\epsilon(t) - v_\epsilon(t)\| \leq \epsilon \|T_0(v_\epsilon(t))\|$.

We proceed first to the proof of the existence of a local solution. By hypothesis, there exists a ball $B_r(v)$ about v in X with $r > 0$ contained in the neighborhood N of condition (α) of the hypothesis, and thereby constants k and k_1 with $0 \leq k < 1$ and $k_1 \geq 0$ such that for all u in $D(T)$ with $\|u - v\| \leq r$, we have $\|T_0(u)\| \leq k \|T(u)\| + k_1$. If we set $M_0 = \|T(v)\| + \|T_0(v)\|$, then it follows from the inequalities above that for all t in R^+, and all $\epsilon > 0$, $\|u_\epsilon(t) - v\| \leq M_0 t \leq r$ if $0 \leq t \leq r(M_0)^{-1} = d$ $(d > 0)$. For t in the interval $[0, d]$ and all $\epsilon > 0$, we have

$$\|T_0(u_\epsilon(t))\| \leq k \|T(u_\epsilon(t))\| + k_1.$$

On the other hand,

$$\|T_0(v_\epsilon(t))\| = \|T_0(I + \epsilon T_0)^{-1}(u_\epsilon(t))\| \leq T_0(u_\epsilon(t))\| \leq k \|T(u_\epsilon(t))\| + k_1.$$

Since

$$\|T(u_\epsilon(t))\| \leq \|du\epsilon(t)/dt\| + \|T_0(v_\epsilon(t))\| \leq M_0 + k \|T(u_\epsilon(t))\| + k_1$$

for all t in $[0, d]$, it follows that for t in this interval

$$\|T(u_\epsilon(t))\| \leq (1 - k)^{-1}(k_1 + M_0) = M_1 \qquad (0 \leq t \leq d).$$

Furthermore, we then obtain

$$\|T_0(v_\epsilon(t))\| \leq k \|T(u_\epsilon(t))\| + k_1 \leq kM_1 + k_1 = M_2 \qquad (0 \leq t \leq d).$$

Finally,

$$\|u_\epsilon(t) - v_\epsilon(t)\| \leq \epsilon \|T_0(v_\epsilon(t))\| \leq \epsilon M_2 \qquad (0 \leq t \leq d).$$

Let ϵ, ξ be two positive constants, and for t in $[0, d]$, set $q_{\epsilon,\xi}(t) = \|u_\epsilon(t) - u_\xi(t)\|^2$. Then for each t in $[0, d)$, we have

$$dq_{\epsilon,\xi}(t)/dt = 2J((u_\epsilon(t) - u_\xi(t)), du\epsilon(t)/dt - du\xi(t)/dt),$$

where the right-hand side of this equation can be rewritten in the form

$$-2(J(u_\epsilon(t) - u_\xi(t)), T(u_\epsilon(t)) - T(u_\xi(t)))$$
$$-2(J(u_\epsilon(t) - u_\xi(t)), T_0(v_\epsilon(t)) - T_0(v_\xi(t)))$$
$$\leq -2(J(u_\epsilon(t) - u_\xi(t)), T_0(v_\epsilon(t)) - T_0(v_\xi(t)))$$

by the accretiveness of T. For the last term, we find that

$$-2(J(u_\epsilon(t) - u_\xi(t)), T_0(v_\epsilon(t)) - T_0(v_\xi(t)))$$
$$= -2(J(v_\epsilon(t) - v_\xi(t)), T_0(v_\epsilon(t)) - T_0(v_\xi(t))) + R_{\epsilon,\xi}(t) \leq R_{\epsilon,\xi}(t)$$

by the accretiveness of T_0, where

$$R_{\epsilon,\xi}(t) = 2(J(v_\epsilon(t) - v_\xi(t)) - J(u_\epsilon(t) - u_\xi(t)), T_0(v_\epsilon(t)) - T_0(v_\xi(t)))$$

$$\leq 2 \|J(v_\epsilon(t) - v_\xi(t)) - J(u_\epsilon(t) - u_\xi(t))\| \cdot \{\|T_0(v_\epsilon(t))\| + \|T_0(v_\xi(t))\|\}$$

$$\leq 4M_2 \|J(v_\epsilon(t) - v_\xi(t)) - J(u_\epsilon(t) - u_\xi(t))\| \to 0$$

uniformly for all t in $[0, d]$ as $\epsilon, \xi \to 0$. Since $dq_{\epsilon,\xi}(t)/dt \leq |R_{\epsilon,\xi}(t)|$, $q_{\epsilon,\xi}(0) = 0$, it follows that $q_{\epsilon,\xi}(t) \to 0$ uniformly on $[0, d]$, i.e., $\|u_\epsilon(t) - u_\xi(t)\| \to 0$ uniformly on $[0, d]$ as $\epsilon, \xi \to 0$.

By the completeness of the Banach space X in its metric, there exists a strongly continuous function $u:[0, d] \to X$ such that $u_\epsilon(t)$ converges strongly to $u(t)$ as $\epsilon \to 0$, uniformly for t in $[0, d]$. It follows that $v_\epsilon(t)$ converges strongly to $u(t)$ as $\epsilon \to 0$, uniformly for t in $[0, d]$. Since $\|T(u_\epsilon(t))\| \leq M_1$ on $[0, d]$, if follows from the hypermaximal accretiveness of T that by Theorem (3.16)(b), $T(u_\epsilon(t))$ converges weakly as $\epsilon \to 0$ to $T(u(t))$, where $u(t)$ lies in $D(T)$ for each t in $[0, d]$ and $T(u(t))$ is weakly continuous in t on this interval. Similarly, since $\|T_0(v_\epsilon(t))\| \leq M_2$ on $[0, d]$ and T_0 is assumed to be hypermaximal accretive, it follows by Theorem (3.16)(b) that $T_0(v_\epsilon(t))$ converges weakly as $\epsilon \to 0$ to $T_0(u(t))$ and $T_0(u(t))$ is weakly continuous in t on $[0, d]$.

For each $\epsilon > 0$ and all t in R^+, we have the integral equation:

$$u_\epsilon(t) = v - \int_0^t [T(u_\epsilon(s)) + T_0(v_\epsilon(s))] \, ds.$$

Since for each s in $[0, d]$, the integrand converges weakly as $\epsilon \to 0$ to $(T(u(s)) + T_0(u(s)))$ and with a fixed bound for the norms for all s in $[0, d]$, it follows that as $\epsilon \to 0$,

$$u(t) = v - \int_0^t [T(u(s)) + T_0(u(s))] \, ds.$$

Hence $u:[0, d) \to X$ is a strongly continuous, weakly once continuously differentiable solution of the differential equation

$$du(t)/dt = -T(u(t)) - T_0(u(t)) \qquad (t \in [0, d)),$$

with $u(0) = v$. Hence, we have proved the existence of local solutions for each v in $D(T)$, while the uniqueness of each such solution for a given v in $D(T)$ follows from Theorem (9.2).

To show the existence of a solution over R^+, it suffices to assume the existence of a solution over a maximal interval of the form $[0, d]$ and prove that $d = +\infty$. Suppose that d is finite. Since for all $\epsilon > 0$ we have

$$\|u_\epsilon(t) - u_\epsilon(t_1)\| \leq M_1(t_1 - t) \qquad (0 \leq t < t_1 < d),$$

it follows that in the limit $\|u(t) - u(t_1)\| \leq M_1(t_1 - t)$ $(0 \leq t < t_1 < d)$. It follows that as $t \to d-$, $u(t)$ converges strongly in X to an element u_0 of X. By the condition (a) of the hypothesis, there exist constants k and k_1 with $0 \leq k < 1$

and $k_1 \geq 0$ such that for a suitable neighborhood N of u_0 and all u in $N \cap D(T)$, we have $\|T_0(u)\| \leq k \|T(u)\| + k_1$. In particular, this is true for all points $u(t)$ with $|t - d| \leq r_1$ where $r_1 > 0$ is chosen from the conditions that $B_r(u_0) \subset N$ by the inequality $M_1 r_1 \leq r$. Then $\|T_0(u(t))\| \leq k \|T(u(t))\| + k_1$, for such t, so that $\|T(u(t))\| \leq \|du(t)/dt\| + \|T_0(u(t))\|$ implies that

$$\|T(u(t))\| \leq (1 - k)^{-1}(k_1 + M_0) = M_1,$$
$$\|T_0(u(t))\| \leq kM_1 + k_1 = M_2$$

for all t in $[d - r_1, d)$. It follows from Theorem (3.16)(b) that $T(u(t))$ converges weakly to $T(u_0)$ as $t \to d-$, that u_0 indeed does lie in $D(T)$, and that $T_0(u(t))$ converges weakly to $T_0(u_0)$ as $t \to d-$. If we now apply the local existence result obtained above to solve the differential equation on the interval $[d, d + \beta)$ for some $\beta > 0$ with the initial value $u(d) = u_0$, we can combine this local solution with the given solution u over the interval $[0, d)$ to obtain a solution over the larger interval $[0, d + \beta)$. This contradicts the maximality of d, thereby proving that $d = +\infty$. Hence the proof of part (a) of the theorem is complete. q.e.d.

PROOF OF (b). Let w_0 be an arbitrary point of X, and let $T_1 = T_0 + I - w_0$. By the result of part (a), we can find a solution $u: R^+ \to X$ of the differential equation $du(t)/dt = -T(u(t)) - T_1(u(t))$ for some given initial value $u(0) = v$, $v \in D(T)$. By Theorem (9.11), for this solution, $(T + T_1)(u(t)) \to 0$ strongly in X as $t \to \infty$, i.e., $(T + T_0 + I)(u(t))$ converges strongly to w_0 as $t \to +\infty$. Since $T + T_0$ is accretive, it follows that $(T + T_0 + I)^{-1}$ is uniformly continuous (and in fact, nonexpansive) as a mapping of $R(T + T_0 + I)$ into X. Hence $u(t)$ converges strongly to some element u_0 of X as $t \to \infty$. By the hypothesis (α), we may assume that for $t \geq T$, $u(t)$ lies in the neighborhood N of condition (α) so that $\|T_0(u(t))\| \leq k \|T(u(t))\| + k_1$ for $t \geq T$ with $k < 1$ and some $k_1 \geq 0$. Hence for such t,

$$\|T(u(t))\| \leq \|(T + T_0 + I)(u(t))\| + \|T_0(u(t))\| + \|u(t)\|$$
$$\leq M + k \|T(u(t))\| + k_1$$

which implies that

$$\|T(u(t))\| \leq (1 - k)^{-1}(M + k_1), \quad \|T_0(u(t))\| \leq k(1 - k)^{-1}(M + k_1) + k_1.$$

It follows from Theorem (3.16)(b) that u_0 lies in $D(T)$, $T(u(t))$ converges weakly as $t \to \infty$ to $T(u_0)$ and $T_0(u(t))$ converges weakly as $t \to \infty$ to $T_0(u_0)$. Hence $w_0 = T(u_0) + T_0(u_0) + u_0 = (T + T_0 + I)(u_0)$. Since w_0 is an arbitrary element of X, it follows that $R(T + T_0 + I) = X$ and $(T + T_0)$ is hypermaximal accretive.
 q.e.d.

PROOF OF (c). This follows from part (b) and Theorem (9.15). q.e.d.

We now turn to results involving multi-valued accretive mappings and the corresponding equations of evolution.

THEOREM (9.23). *Let X be a uniformly convex Banach space with X^* uniformly convex, T a mapping of X into 2^X which is hypermaximal accretive, v an element of $D(T)$, the effective domain of T.*

Then

(a) *There exists one and only one solution* $u: R^+ \to X$ *of the equation*

$$-du(t)/dt \in T(u(t)) \quad (t \geq 0),$$

with $u(0) = v$.

(b) *For this solution, we have* $u(t)$ *is strongly continuous and strongly once-differentiable from the right at each* t *in* R^+ *with* $du(t)/dt$ *strongly continuous from the right in* t *on* R^+ *and* $\|du(t)/dt\|$ *nonincreasing in* t. *For each* u *in* $D(T)$, *if we set*

$$\tilde{T}(u) = \{w \mid w \in T(u), \|w\| \leq \|y\| \text{ for all } y \text{ in } T(u)\}$$

then $\tilde{T}$ *is a single-valued accretive mapping of* $D(T)$ *into* X *for each* t *in* R^+, $du(t)/dt = -\tilde{T}(u(t))$ $(t \geq 0)$.

PROOF OF THEOREM (9.23). Since T is hypermaximal accretive, $(T + I)^{-1}$ is a well-defined nonexpansive mapping of X into X. We know by Theorem (3.5) that for each $\epsilon > 0$, $(I + \epsilon T)^{-1}$ is a well-defined single-valued nonexpansive mapping of X into X.

For each $\epsilon > 0$, we set $T_\epsilon = \epsilon^{-1}(I - (I + \epsilon T)^{-1})$. It follows from Theorem (8.9)(b) that T_ϵ is an accretive mapping of X into X, while it follows from the fact that I and $(I + \epsilon T)^{-1}$ are both nonexpansive, that for each $\epsilon > 0$, T_ϵ is Lipschitzian.

We consider the approximate problem $du_\epsilon(t)/dt = -T_\epsilon(u_\epsilon(t))$ $(t \geq 0)$, with $u_\epsilon(0) = v$. Since T_ϵ is accretive and continuous, it follows that $\|du_\epsilon(t)/dt\|$ is nonincreasing in t on R^+, i.e.,

$$\left\| \frac{du_\epsilon}{dt}(t) \right\| \leq \left\| \frac{du_\epsilon}{dt}(0) \right\| = \|T_\epsilon(v)\| \quad (t \geq 0).$$

For any u in $D(T)$, let $v_\epsilon = (I + \epsilon T)^{-1}(u)$. Then there exists w_ϵ in $T(v_\epsilon)$ such that $v_\epsilon + \epsilon w_\epsilon = u$. For any w_0 in $T(u)$, we have $u + \epsilon w_0 = u + \epsilon w_0$ i.e., $u = (I + \epsilon T)^{-1}(u + \epsilon w_0)$. Using the fact that $(I + \epsilon T)^{-1}$ is nonexpansive, it follows that

$$\|u - v_\epsilon\| = \|(I + \epsilon T)^{-1}(u + \epsilon w_0) - (I + \epsilon T)^{-1}(u)\|$$
$$\leq \|(u + \epsilon w_0) - u\| = \epsilon \|w_0\|.$$

Hence

$$\|T_\epsilon(u)\| = \|\epsilon^{-1}(u - v_\epsilon)\| \leq \|w_0\| \quad (w_0 \in T(u)).$$

For each u in $D(T)$, the set $T(u)$ is a closed convex subset of the uniformly convex space X since T, being hypermaximal accretive, is maximal accretive by Theorem (3.16) while by Theorem (3.11), $T(u)$ is closed and convex. Since X is uniformly convex, the minimum of the norm of elements w of $T(u)$ is assumed at exactly one point of $T(u)$, which we denote by $\tilde{T}(u)$. Then our argument of the preceding paragraph implies that for each $\epsilon > 0$,

$$\|T_\epsilon(u)\| \leq \|\tilde{T}(u)\| \quad (u \in D(T), \epsilon > 0).$$

In particular, $\|du\epsilon(t)/dt\| \leq \|\tilde{T}(v)\|$ $(t \geq 0)$. Let $v_\epsilon(t) = (I + \epsilon T)^{-1}(u_\epsilon(t))$. Then

$$\|v_\epsilon(t) - u_\epsilon(t)\| = \epsilon\|T_\epsilon(u_\epsilon(t))\| = \epsilon\|du\epsilon(t)/dt\| \leq \epsilon\|\tilde{T}(v)\| \to 0 \qquad (\epsilon \to 0).$$

For each $\epsilon > 0$ and each t in R^+, there exists $w_\epsilon(t) \in T(v_\epsilon(t))$ such that $v_\epsilon(t) + \epsilon w_\epsilon(t) = u_\epsilon(t)$, i.e.,

$$T_\epsilon(u_\epsilon(t)) = w_\epsilon(t) \in T(v_\epsilon(t)) \qquad (\epsilon > 0, t \in R^+).$$

Let $\epsilon, \xi > 0$. We set $q_{\epsilon,\xi}(t) = \|u_\epsilon(t) - u_\xi(t)\|^2$. Then for each t in R^+,

$$dq_{\epsilon,\xi}(t)/dt = 2(J(u_\epsilon(t) - u_\xi(t)), du_\epsilon(t)/dt - du_\xi(t)/dt)$$
$$= -2(J(v_\epsilon(t) - v_\xi(t)), w_\epsilon(t) - w_\xi(t)) + R_{\epsilon,\xi}(t) \leq R_{\epsilon,\xi}(t)$$

by the accretiveness of T, where

$$R_{\epsilon,\xi}(t) = 2(J(u_\epsilon(t) - u_\xi(t)) - J(v_\epsilon(t) - v_\xi(t)), w_\epsilon(t) - w_\xi(t))$$
$$\leq 2 \|J(u_\epsilon(t) - u_\xi(t)) - J(v_\epsilon(t) - v_\xi(t))\| \{2\|\tilde{T}(v)\|\} \to 0$$

as $\epsilon, \xi \to 0$ uniformly over each bounded subinterval of R^+. Since $q_{\epsilon,\xi}(0) = 0$, it follows as in the preceding proofs that $q_{\epsilon,\xi}(t) \to 0$ as $\epsilon, \xi \to 0$, uniformly for t in each bounded subinterval of R^+. By the completeness of X in its metric, there exists a strongly continuous function $u: R^+ \to X$ such that $u_\epsilon(t)$ converges strongly to $u(t)$ as $\epsilon \to 0$, uniformly for t in each bounded subinterval of R^+.

Since for each fixed t in R^+, $u_\epsilon(t) - v_\epsilon(t)$ converges strongly to 0 as $\epsilon \to 0$, it follows that $v_\epsilon(t)$ converges to $u(t)$ strongly in X as $\epsilon \to 0$. Moreover, $\|w_\epsilon(t)\|$ is uniformly bounded for all $\epsilon > 0$ by $\|\tilde{T}(v)\|$. Hence, it follows from the hyper-maximal accretiveness of T and Theorem (3.16)(b) that for each t in R^+, $u(t)$ lies in $D(T)$. For each s in R^+ and each $\epsilon > 0$, we define another family of approximants $u_\epsilon^{(s)}(t)$ for t in $[s, \infty)$, by letting $u_\epsilon^{(s)}$ be the solution of the differential equation

$$du_\epsilon^{(s)}(t)/dt = -T_\epsilon(u_\epsilon^{(s)}(t)) \qquad (t \geq s),$$

with the initial condition $u_\epsilon^{(s)}(s) = u(s)$. Since T_ϵ is accretive, it follows from Theorem (9.2) that

$$\|u_\epsilon^{(s)}(t) - u_\epsilon(t)\| \leq \|u(s) - u_\epsilon(s)\| \qquad (t \geq s),$$

and since $u_\epsilon(s)$ converges strongly to $u(s)$ as $\epsilon \to 0$, it follows that $u_\epsilon^{(s)}(t)$ converges to $u(t)$ uniformly on each bounded interval of $[s, \infty)$ as $\epsilon \to 0$. Moreover, we know that for $t \geq s$, $\|T_\epsilon(u_\epsilon^{(s)}(t))\| \leq \|\tilde{T}(u(s))\|$. If we choose a subsequence of values of ϵ tending to 0 so that $T_\epsilon(u_\epsilon^{(s)}(s))$ converges weakly in X to an element $w(t)$, then by Theorem (3.16) we know that $w(t)$ is an element of $T(u(t))$ while $\|w(t)\| \leq \|\tilde{T}(u(s))\|$. Hence

$$\|\tilde{T}(u(t))\| \leq \|w(t)\| \leq \|\tilde{T}(u(s))\| \qquad (t \geq s),$$

so that the function $\|\tilde{T}(u(t))\|$ is a nonincreasing function of t on R^+.

It follows from this last fact that $T(u(t))$ is strongly continuous from the right in t on R^+. Indeed, consider a sequence $t_j \to t+$. Then $u(t_j)$ converges strongly to

$u(t)$, while for any subsequence such that $T(u(t_j))$ converges weakly to w in X, w lies in $T(u(t))$ by Theorem (3.16). On the other hand

$$\|w\| \leq \lim \|\tilde{T}(u(t_j))\| \leq \|\tilde{T}(u(t))\|.$$

Thus $w = \tilde{T}(u(t))$ and $\tilde{T}(u(t_j))$ must converge strongly to $\tilde{T}(u(t))$ as $j \to \infty$, both by the uniform convexity of the space X.

To complete the proof of Theorem (9.23), we shall make use of the following auxiliary result:

PROPOSITION (9.6). *Let X be a uniformly convex Banach space, T a mapping of a subset $D(T)$ of X into 2^X such that for each x in $D(T)$, $T(x)$ is a nonempty closed convex subset of X. Suppose that the following condition holds for T:*

(β) For any strongly convergent sequence $\{u_j\}$ in X with u_j in $D(T)$ for each j, and for any sequence $\{w_j\}$ in X with $w_j \in T(u_j)$ for all j, if u_j converges strongly to u_0 and if w_j converges weakly to w_0 in X, then u_0 lies in $D(T)$ and $w_0 \in T(u_0)$.

Let $\tilde{T}$ be the mapping of $D(T)$ into X given by

$$\tilde{T}(u) = \{w \mid w \in T(u), \|w\| \leq \|y\|, y \in T(u)\}.$$

Then given $\epsilon > 0$ and u in $D(T)$, there exists $\lambda(\epsilon) > 0$ such that if v is any element in $D(T)$ with $\|v - u\| < \lambda(\epsilon)$ and w is an element of $T(v)$ with $\|w\| \leq \|\tilde{T}(u)\| + \lambda(\epsilon)$, then $\|w - \tilde{T}(u)\| < \epsilon$.

PROOF OF PROPOSITION (9.6). Suppose the conclusion of the proposition were false. Then we could find an infinite sequence $\{v_j\}$ in $D(T)$ converging strongly to u in X and a corresponding sequence $\{w_j\}$ in X with $w_j \in D(T)$ for each j such that $\|w_j\| \to \|\tilde{T}(u)\|$ while for a given $\epsilon > 0$, $\|w_j - T(u)\| \geq \epsilon$. Passing to an infinite subsequence, we may assume that w_j converges weakly to an element w of the reflexive Banach space X. By the property (β) of the hypothesis w must be an element of $T(u)$. Since $\|w\| \leq \lim \|w_j\| = \|T(u)\|$, it follows that $w = \tilde{T}(u)$ and that w_j converges strongly to $w = \tilde{T}(u)$. This contradicts the assumption that $\|w_j - \tilde{T}(u)\| \geq \epsilon > 0$. q.e.d.

PROOF OF THEOREM (9.23) COMPLETED. To complete the proof, it suffices to show that for the function $u: R^+ \to X$ which we have constructed, $du(t)/dt = -\tilde{T}(u(t))$. Consider a given s in R^+ and the corresponding approximants $u_\epsilon^{(s)}(t)$ ($t \geq s$). We know that

$$\frac{u_\epsilon^{(s)}(t) - u_\epsilon^{(s)}(s)}{t - s} = \frac{-1}{t - s} \int_s^t T_\epsilon(u_\epsilon^{(s)}(r)) \, dr,$$

where $\|T_\epsilon(u_\epsilon^{(s)}(r))\| \leq \|\tilde{T}(u(s))\|$. Moreover

$$T_\epsilon(u_\epsilon^{(s)}(r)) \in T(I + \epsilon T)^{-1}(u_\epsilon^{(s)}(r))$$

where

$$\|(I + \epsilon T)^{-1}(u_\epsilon^{(s)}(r)) - u_\epsilon^{(s)}(r)\| \leq \epsilon \|\tilde{T}(u(s))\|,$$

while

$$\|u_\epsilon^{(s)}(r) - u(s)\| \leq \|\tilde{T}(u(s))\| (r - s) \leq \|\tilde{T}(u(s))\| \cdot (t - s).$$

We consider $\epsilon > 0$ and $\alpha > 0$ so small that $\|\tilde{T}(u(s))\| (\alpha + \epsilon) < \lambda(\xi)$ for a given $\xi > 0$, where $\lambda(\xi)$ is the function defined in Proposition (9.6) for the element $u(s)$ if $D(T)$. (We note that the condition (β) holds for the hypermaximal accretive mapping T in the space X with X^* uniformly convex by the application of Theorem (3.16)(b).) It follows that $\|T_\epsilon(u_\epsilon^{(s)}(r)) - \tilde{T}(u(s))\| < \xi$ for ϵ and α sufficiently small, i.e., $\|T(u(s))\| (\alpha + \epsilon) < \lambda(\xi)$, with $|t - s| < \alpha$. In particular

$$\|(t - s)^{-1}(u_\epsilon^{(s)}(t) - u_\epsilon^{(s)}(s)) + \tilde{T}(u(s))\| < \xi.$$

Letting $\epsilon \to 0$, we find that

$$\|(t - s)^{-1}[u(t) - u(s)] + \tilde{T}(u(s))\| < \xi$$

for $|t - s| < \lambda(\xi) \|\tilde{T}(u(s))\|^{-1}$.

Hence u is strongly differentiable from the right at each point s in R^+, and we have $du(s)/dt = -\tilde{T}(u(s))$ ($s \in R^+$). q.e.d.

We now turn to the extension of the preceding results to the case where the differential equation involved has its right-hand term dependent upon the variable t.

THEOREM (9.24). *Let X be a Banach space with X^* uniformly convex, and let $\{T_t : t \in R^+\}$ be a family of hypermaximal accretive mappings with a fixed domain $D(T)$ and ranges in X. Let $c : R^+ \to R^1$ be a Lipschitzian function on R^+ and suppose that for some function $\psi : R^+ \to R^+$ and a continuous function $c_0 : R^+ \to R^+$, we have the inequality*

$$\|T_{t+h}(u) - T_t(u)\| \leq c_0(t)h[\|T_t(u)\| + \psi(\|u\|)]$$

for all u in $D(T)$.

Then for each v in $D(T)$, there exists one and only one solution $u : R^+ \to X$ of the differential equation

$$du(t)/dt = -T_t(u(t)) + c(t)u(t) \qquad (t \geq 0)$$

with $u(0) = v$.

For this solution, u is strongly continuous and weakly once continuously differentiable from R^+ to X, with $T_t(u(t))$ weakly continuous in t on R^+.

PROOF OF THEOREM (9.24). For each $\epsilon > 0$, we let $T_{\epsilon,t}(u) = T_t(I + \epsilon T_t)^{-1}$, and we consider the approximation equations

$$du_\epsilon(t)/dt = -T_{\epsilon,t}(u_\epsilon(t)) + c(t)u_\epsilon(t)$$

with the initial value $u_\epsilon(0) = v$. As in the preceding arguments, we shall obtain the desired solution u as the limit of the u_ϵ as $\epsilon \to 0+$.

For each fixed t in R^+, $T_{\epsilon,t}$ is an accretive mapping of X into X which is Lipschitzian with a Lipschitz constant $\leq 2\epsilon^{-1}$. Since $T_{\epsilon,t} = \epsilon^{-1}(I - (I + \epsilon T_t)^{-1})$, $T_{\epsilon,t}$ is continuous in t on R^+ in the strong topology if $(I + \epsilon T_t)^{-1}$ is strongly continuous in t on R^+. We note that for $h > 0$, u in $D(T)$,

$(I + \epsilon T_{t+h})^{-1}u - (I + \epsilon T_t)^{-1}u$
$= (I + \epsilon T_{t+h})^{-1}u - (I + \epsilon T_{t+h})^{-1}[u + \epsilon(T_{t+h} - T_t)(I + \epsilon T)^{-1}u].$

Since for each t in R^+, $(I + \epsilon T_t)^{-1}$ is a nonexpansive mapping, it follows that

$$\|(I + \epsilon T_{t+h})^{-1}u - (I + \epsilon T_t)^{-1}u\| \leq \epsilon \|(T_{t+h} - T_t)(I + \epsilon T_t)^{-1}u\|.$$

It follows from the assumed Lipschitzian character of T_t in t with respect to the graph norm of T_t that $(I + \epsilon T_t)^{-1}u$ satisfies a Lipschitzian condition of the following form:

$$\|(I + \epsilon T_{t+h})^{-1}u - (I + \epsilon T_t)^{-1}u\| \leq c_0(t, \epsilon)h\psi_\epsilon(\|u\|).$$

(We shall treat this inequality more precisely below.)

It follows that the approximate solutions u_ϵ exist on R^+ and are uniquely determined by v. We shall now proceed to obtain some essential a priori bounds for these approximate solutions.

(1) *There exists a continuous function k_0 from R^+ to R^+ (with k_0 independent of ϵ for $\epsilon > 0$) such that for all $\epsilon > 0$, $\|u_\epsilon(t)\| \leq k_0(t)$.*

PROOF OF (1). Let $q_\epsilon(t) = \|u_\epsilon(t)\|^2$. Then

$$dq_\epsilon(t)/dt = 2(J(u_\epsilon(t)), du_\epsilon(t)/dt) = -2(J(u_\epsilon(t)), T_{\epsilon,t}(u_\epsilon(t))) + c(t)\|u_\epsilon(t)\|^2,$$

where

$$-2(J(u_\epsilon(t)), T_{\epsilon,t}(u_\epsilon(t))) = -2(J(u_\epsilon(t) - 0), T_{\epsilon,t}(u_{\epsilon,t}) - T_{\epsilon,t}(0)) - 2(J(u_\epsilon(t)), T_{\epsilon,t}(0))$$
$$\leq -2(J(u_{\epsilon,t}), T_{\epsilon,t}(0))$$

while

$$-2(J(u_{\epsilon,t}), T_{\epsilon,t}(0)) \leq 2\|u_{\epsilon,t}\| \cdot \|T_{\epsilon,t}(0)\| \leq 2\|u_{\epsilon,t}\| \cdot \|T_t(0)\| \leq c_2(t)q_{\epsilon,t}^{1/2}$$

for a continuous function $c_2(t)$. (We have assumed here without loss of generality that 0 lies in $D(T)$, since otherwise we can shift the origin of coordinates without altering the terms of the theorem.)

Hence

$$dq_\epsilon(t)/dt \leq c(t)q_\epsilon(t) + c_2(t)^2 + q_\epsilon(t), \qquad q_\epsilon(0) = \|v\|^2,$$

and it follows by the standard argument on differential inequalities that there exists a continuous function k_0 on R^+, depending only upon $\|v\|$, c, and c_2, such that $q_\epsilon(t) \leq [k_0(t)]^2$ ($t \in R^+$). q.e.d.

(2) *There exists a continuous function $k_1: R^+ \to R^+$, with k_1 independent of ϵ for $\epsilon > 0$, such that for all t in R^+ and all $\epsilon > 0$,*

$$\|du_\epsilon(t)/dt\| \leq k_1(t).$$

PROOF OF (2). For $\epsilon > 0$ and t in R^+, with $h > 0$, we set

$$q_{\epsilon,h}(t) = \|u_\epsilon(t + h) - u_\epsilon(t)\|^2.$$

Then

$$\frac{d}{dt}[q_{\epsilon,h}](t) = 2(J(u_\epsilon(t+h) - u_\epsilon(t)), du_\epsilon(t+h)/dt - du_\epsilon(t)/dt)$$
$$= -2(J(u_\epsilon(t+h) - u_\epsilon(t)), T_{\epsilon,t+h}(u_\epsilon(t+h)) - T_{\epsilon,t}(u_\epsilon(t))) + 2c(t)q_{\epsilon,h}(t).$$

We note that

$$-2(J(u_\epsilon(t+h) - u_\epsilon(t)), T_{\epsilon,t+h}(u_\epsilon(t+h)) - T_{\epsilon,t}(u_\epsilon(t)))$$
$$= -2(J(u_\epsilon(t+h) - u_\epsilon(t)), T_{\epsilon,t+h}(u_\epsilon(t+h)) - T_{\epsilon,t+h}(u_\epsilon(t)))$$
$$\quad -2(J(u_\epsilon(t+h) - u_\epsilon(t)), [T_{\epsilon,t+h} - T_{\epsilon,t}](u_\epsilon(t)))$$
$$\leq -2(J(u_\epsilon(t+h) - u_\epsilon(t)), [T_{\epsilon,t+h} - T_{\epsilon,t}](u_\epsilon(t)))$$
$$\leq 2 \|u_\epsilon(t+h) - u_\epsilon(t)\| \cdot \|T_{\epsilon,t+h}(u_\epsilon(t)) - T_{\epsilon,t}(u_\epsilon(t))\|$$
$$\leq 2[q_{\epsilon,h}(t)]^{1/2} \|T_{\epsilon,t+h}(u_\epsilon(t)) - T_{\epsilon,t}(u_\epsilon(t))\|.$$

For the norm in the last term of the inequality, we have

$$\|[T_{\epsilon,t+h} - T_{\epsilon,t}](u_\epsilon(t))\|$$
$$= \epsilon^{-1} \|(I - (I + \epsilon T_{t+h})^{-1})(u_\epsilon(t)) - (I - (I + \epsilon T_t)^{-1})(u_\epsilon(t))\|$$
$$= \epsilon^{-1} \|(I + \epsilon T_t)^{-1} u_\epsilon(t) - (I + \epsilon T_{t+h})^{-1} u_\epsilon(t)\|$$
$$\leq \epsilon^{-1} \epsilon \|(T_{t+h} - T_t)(I + \epsilon T_t)^{-1} u_\epsilon(t)\|$$
$$\leq c_0(t) h [\|T_t(I + \epsilon T_t)^{-1}(u_\epsilon(t))\| + \psi(\|(I + \epsilon T_t)^{-1}(u_\epsilon(t))\|)].$$

We let

$$p_\epsilon(t) = \|T_t(I + \epsilon T_t)^{-1}(u_\epsilon(t))\| = \|T_{\epsilon,t}(u_\epsilon(t))\| \quad \text{and} \quad s_\epsilon(t) = \psi(\|(I + \epsilon T_t)^{-1}(u_\epsilon(t))\|).$$

Then we obtain from the inequalities above

$$dq_{\epsilon,h}(t)/dt \leq c(t) q_{\epsilon,h}(t) + c_0(t)^2 q_{\epsilon,h}(t) + h^2 [p_\epsilon(t) + s_\epsilon(t)]^2.$$

If we let $C(t) = \int_0^t [2c(r) + c_0(r)^2] \, dr$, then we obtain

$$q_{\epsilon,h}(t) \leq q_{\epsilon,h}(0) \exp(C(t)) + \int_0^t h^2 \exp(C(t) - C(r))[p_\epsilon(r) + s_\epsilon(r)] \, dr.$$

Dividing both sides by h^2 and letting $h \to 0+$, we obtain

$$\|du_\epsilon(t)/dt\|^2 \leq \|du_\epsilon(0)/dt\|^2 \exp(C(t)) + \int_0^t 2 \exp(C(t) - C(r))[p_\epsilon(r)^2 + s_\epsilon(r)^2] \, dr,$$

where $\|du_\epsilon(0)/dt\| = \|T_{\epsilon,0}(v)\| \leq \|T_0(v)\|$, while

$$p_\epsilon(t) = \|T_{\epsilon,t}(u_\epsilon(t))\| \leq \left\|\frac{du}{dt}\epsilon(t)\right\| + |c(t)| \cdot \|u_\epsilon(t)\| \leq \left\|\frac{du}{dt}\epsilon(t)\right\| + c_3(t)$$

for all t in R^+ and a suitable continuous function c_3 from R^+ to R^+ independent of ϵ for $\epsilon > 0$. Moreover, if we set $w_\epsilon(t) = \epsilon T_t(0)$, then $(I + \epsilon T_t)^{-1}(w_\epsilon(t)) = 0$, so

that
$$\|(I + \epsilon T_t)^{-1}(u_\epsilon(t)) - 0\| \leq \|u_\epsilon(t) - w_\epsilon(t)\| \leq \|u_\epsilon(t)\| + \|T_t(0)\| \leq c_4(t)$$

for all t in R^+ and $0 < \epsilon \leq 1$. Hence $s_\epsilon(t) \leq \psi(c_4(t)) = c_5(t)$.

Combining the preceding inequalities with the inequality derived above involving $du_\epsilon(t)/dt$, we obtain for $m_\epsilon(t) = \|du_\epsilon(t)/dt\|^2$, the integral inequality

$$m_\epsilon(t) \leq c_6(t) + \int_0^t \exp\left(C(t) - C(r)\right)[m_\epsilon(r) + c_6(r)]\, dr.$$

If we introduce the new dependent variable

$$a_\epsilon(t) = \int_0^t \exp\left(C(t) - C(r)\right)[m_\epsilon(r) + c_6(r)]\, dr,$$

then we obtain the relation

$$da_\epsilon(t)/dt = m_\epsilon(t) + c_6(t) + \int_0^t C'(t) \exp\left(C(t) - C(r)\right)[m_\epsilon(r) + c_6(r)]\, dr$$

and hence the differential inequality

$$da_\epsilon(t)/dt = m_\epsilon(t) + c_6(t) + C'(t)a_\epsilon(t) \leq 2c_6(t) + [C'(t) + 1]a_\epsilon(t).$$

Hence,
$$\frac{d}{dt}[a_\epsilon(t) \exp(-C(t) - t)] \leq 2c_6(t) \exp(-C(t) - t)$$

and since $a_\epsilon(0) = 0$, we obtain

$$a_\epsilon(t) \leq \exp(C(t) + t) \int_0^t 2c_6(r) \exp(-C(r) - r)\, dr \leq c_7(t).$$

Finally
$$\|du\epsilon(t)/dt\|^2 = m_\epsilon(t) \leq c_6(t) + a_\epsilon(t) \leq c_6(t) + c_7(t). \quad \text{q.e.d.}$$

(3) *There exists a strongly continuous function* $u: R^+ \to X$ *such that* $u_\epsilon(t)$ *converges strongly in X to $u(t)$ as $\epsilon \to 0$, uniformly for t in bounded subsets of R^+.*

PROOF OF (3). Let $\epsilon, \xi > 0$, and set $q_{\epsilon,\xi}(t) = \|u_\epsilon(t) - u_\xi(t)\|^2$ ($t \geq 0$). Then for each t in R^+,

$$\frac{d}{dt}[q_{\epsilon,\xi}](t) = 2\left(J(u_\epsilon(t) - u_\xi(t)), \frac{du}{dt}\epsilon(t) - \frac{du}{dt}\xi(t)\right).$$

We set
$$v_\epsilon(t) = (I + \epsilon T_t)^{-1}(u_\epsilon(t)), \qquad v_\xi(t) = (I + \xi T_t)^{-1}(u_\xi(t)).$$

Then
$$\|v_\epsilon(t) - u_\epsilon(t)\| \leq \epsilon \|T_{\epsilon,t}(u_\epsilon(t))\|$$
$$\leq \epsilon \left\|\frac{du}{dt}\epsilon(t)\right\| + \epsilon |c(t)| \|u_\epsilon(t)\| \leq \epsilon c_9(t) \to 0 \qquad (\epsilon \to 0+)$$

uniformly on each bounded interval of R^+, and similarly $\|v_\xi(t) - u_\xi(t)\| \leq \xi c_9(t) \to 0$ ($\xi \to 0+$) uniformly on each bounded interval of R^+. It follows that $\|(v_\epsilon(t) - v_\xi(t)) - (u_\epsilon(t) - u_\xi(t))\| \to 0$ ($\epsilon, \xi \to 0+$), uniformly on bounded subsets of R^+. On the other hand

$$2(J(u_\epsilon(t) - u_\xi(t)), du\epsilon(t)/dt - du\xi(t)/dt)$$
$$= 2(J(u_\epsilon(t) - u_\xi(t)), [-T_t(v_\epsilon(t)) + c(t)u_\epsilon(t)] - [-T_t(v_\xi(t)) + c(t)u_\xi(t)])$$
$$= 2c(t)q_{\epsilon,\xi}(t) - 2(J(v_\epsilon(t) - v_\xi(t)), T_t(v_\epsilon(t)) - T_t(v_\xi(t))) + R_{\epsilon,\xi}(t)$$
$$\leq 2c(t)q_{\epsilon,\xi}(t) + R_{\epsilon,\xi}(t),$$

where

$$R_{\epsilon,\xi}(t) = 2(J(v_\epsilon(t) - v_\xi(t)) - J(u_\epsilon(t) - u_\xi(t)), T_t(v_\epsilon(t)) - T_t(v_\xi(t)))$$
$$\leq 2 \|J(v_\epsilon(t) - v_\xi(t)) - J(u_\epsilon(t) - u_\xi(t))\| \cdot [\|T_t(v_\epsilon(t))\| + \|T_t(v_\xi(t))\|] \to 0$$

uniformly on bounded subsets of R^+ as $\epsilon, \xi \to 0$ since

$$\|J(v_\epsilon(t) - v_\xi(t)) - J(u_\epsilon(t) - u_\xi(t))\| \to 0$$

on bounded subsets of R^+ by the uniform continuity of J on bounded subsets of X, while $\|T_t(v_\epsilon(t))\| + \|T_t(v_\xi(t))\|$ remains uniformly bounded on bounded subsets of R^+ as $\epsilon, \xi \to 0$.

Since $dq_{\epsilon,\xi}(t)/dt - 2c(t)q_{\epsilon,\xi}(t) \leq R_{\epsilon,\xi}(t)$, it follows from the fact that $q_{\epsilon,\xi}(0) = 0$ that

$$q_{\epsilon,\xi}(t) \leq \int_0^t \exp(2C(t) - 2C(r)) R_{\epsilon,\xi}(r) \, dr \to 0$$

uniformly on bounded subsets of R^+. Since $\|u_\epsilon(t) - u_\xi(t)\| \to 0$ uniformly on bounded subsets of R^+, it follows from the completeness of the Banach space X that $u_\epsilon(t)$ converges to $u(t)$ strongly in X as $\epsilon \to 0$, where u is a strongly continuous function from R^+ to X and the convergence is uniform on each bounded subset of R^+. q.e.d.

(4) *The function* $u: R^+ \to X$ *is the desired solution of our differential equation with* $u(0) = v$.

PROOF OF (4). For each t in R^+, we have $u_\epsilon(t)$ converging strongly to $u(t)$ with $\|T_t(v_\epsilon(t))\|$ uniformly bounded as $\epsilon \to 0$. Since $u_\epsilon(t) - v_\epsilon(t)$ converges strongly to 0 in X, it follows that $v_\epsilon(t) \to u(t)$ strongly in X as $\epsilon \to 0$. Hence by Theorem (3.16)(b) since T_t is a hypermaximal accretive mapping and X^* is uniformly convex, $u(t)$ lies in $D(T_t)$ and $T_{\epsilon,t}(u_\epsilon(t)) = T_t(v_\epsilon(t))$ converges weakly to $T_t(u(t))$ as $\epsilon \to 0$. Since this convergence is bounded over finite subintervals of R^+, we can take the limit as $\epsilon \to 0$ in the integral relation

$$u_\epsilon(t) = v + \int_0^t [-T_r(v_\epsilon(r)) + c(r)u_\epsilon(r)] \, dr \qquad (t \geq 0)$$

to obtain the new relation

$$u(t) = v + \int_0^t [-T_r(u(r)) + c(r)u(r)] \, dr \qquad (t \geq 0).$$

It follows since $u(t)$ is strongly continuous in t on R^+ with $\|T_t(u(t))\|$ uniformly bounded on bounded subintervals of R^+, that

$$\|T_t(u(t+h)) - T_{t+h}(u(t+h))\|$$
$$= \|[(T_t - T_{t+h})(I + T_{t+h})^{-1}]u(t+h) + T_{t+h}(u(t+h))\|$$
$$\leq c_0(t+h)h\psi_2(t) \to 0 \quad (h \to 0).$$

Hence by Theorem (3.16)(b), $T_t(u(t))$ is weakly continuous in t on R^+, and u is a solution of the differential equation:

$$du(t)/dt = -T_t(u(t)) + c(t)u(t) \quad (t \geq 0).$$

Thus the proof of Theorem (9.24) is complete. q.e.d.

We conclude the discussion of the present section with a similar result for the case of T_t multi-valued:

THEOREM (9.25). *Let X be a uniformly convex Banach space with X^* uniformly convex, and let $\{T_t : t \in R^+\}$ be a family of hypermaximal accretive mappings of X into 2^X with $D(T_t)$ independent of t. Suppose that the family $\{T_t\}$ satisfies the following Lipschitzian condition in t on R^+: There exists continuous function c_0 from R^+ to R^+ such that for each t in R^+, $h > 0$, and $\epsilon > 0$,*

$$\|(I + \epsilon T_{t+h})^{-1}(u) - (I + \epsilon T_t)^{-1}(u)\| \leq \epsilon c_0(t)h[\|(I + \epsilon T_t)^{-1}(u)\| + \psi(\|u\|)]$$

for a given function ψ from R^+ to R^+ and all u in X.

Then for each v in $D(T)$, there exists one and only one strongly continuous, strongly once right differentiable function $u: R^+ \to X$ which satisfies the equation $-du(t)/dt \in T_t(u(t))$ ($t \geq 0$) with the initial condition $u(0) = v$.

This solution satisfies the single-valued differential equation

$$du(t)/dt = -\tilde{T}_t(u(t)) \quad (t \geq 0)$$

where for each t in R^+, u in $D(T) = D(T_t)$,

$$\tilde{T}_t(u) = \{w \mid w \in T_t(u), \|w\| \leq \|y\| \text{ for all } y \text{ in } T_t(u)\}.$$

PROOF OF THEOREM (9.25). We set $T_{\epsilon,t} = \epsilon^{-1}(I - (I + \epsilon T_t)^{-1})$, and form the approximate equations $du_\epsilon(t)/dt = -T_{\epsilon,t}(u_\epsilon(t))$ ($t > 0$) with the initial condition $u_\epsilon(0) = v$. For each t and ϵ, $T_{\epsilon,t}$ is accretive. Hence we may repeat the proof of assertion (1) in the proof of Theorem (9.24) and obtain the conclusion that there exists a continuous function k_0 from R^+ to R^+ such that for all $\epsilon > 0$ and all t in R^+, $\|u_\epsilon(t)\| \leq k_0(t)$.

The assumption made in the hypothesis on the Lipschitzian character of the dependence of T_t on t makes it possible to carry through the proof of the assertion (2) along the lines given in the proof of Theorem (9.24) and to obtain the conclusion that there exists a continuous function k_1 from R^+ to R^+ such that for all $\epsilon > 0$ and all t in R^+, $\|du_\epsilon(t)/dt\| \leq k_1(t)$.

Similarly (if we replace each occurrence of the terms $T_t(v_\epsilon(t))$ by an element of the latter set as well as doing the same for $T_t(v_\xi(t))$), the proof of assertion (3) in

Theorem (9.24) remains valid in the present case, and assures us that the sequence of approximants $u_\epsilon(t)$ converges strongly to a strongly continuous function $u(t)$ as $\epsilon \to 0$, uniformly on finite subintervals of R^+.

The principal alteration in the present proof from that of Theorem (9.24) lies in the verification of the final assertion (4) which becomes in this case:

(4)' *The function* $u: R^+ \to X$ *satisfies the differential equation* $du(t)/dt = -\tilde{T}_t(u(t))$ $(t \geq 0)$.

PROOF OF (4)'. Let s be a given point of R^+ at which we propose to verify the validity of the differential equation. We consider an auxiliary family of approximants $u_\epsilon^{(s)}$ which satisfy the differential equation

$$du_\epsilon^{(s)}(t)/dt = -T_{\epsilon,t}(u_\epsilon^{(s)}(t)) \qquad (t \geq s)$$

with the initial condition $u_\epsilon^{(s)}(s) = u(s)$. Since $u_\epsilon^{(s)}$ and u_ϵ satisfy the same differential equation on $[s, \infty)$ with $T_{\epsilon,t}$ accretive for each t, it follows that

$$\|u_\epsilon^{(s)}(t) - u_\epsilon(t)\| \leq \|u(s) - u_\epsilon(s)\| \to 0 \qquad (\epsilon \to 0+)$$

for all $t \geq s$. Hence $u_\epsilon^{(s)}(t)$ converges strongly in X to $u(t)$ as $\epsilon \to 0$, uniformly on bounded subsets of $[s, \infty)$.

By the proof of assertion (2) in the proof of Theorem (9.24), there exist a continuous function $K(t, s)$ for $t \geq s$, which is independent of ϵ for $\epsilon > 0$, and a continuous function C_1 from R^+ to R^1 such that for $t \geq s$,

$$\|du_\epsilon^{(s)}(t)/dt\| \leq du^{(s)}(s)/dt\| \exp(C_1(t) - C_1(s)) + K(t, s),$$

while $K(t, s) \to 0$ $(t \to s+)$. On the other hand,

$$\|du_\epsilon^{(s)}(s)/dt\| = \|T_{\epsilon,s}(u(s))\| \leq \|\tilde{T}_\epsilon(u(s))\|.$$

Hence it follows that

$$\|T_{\epsilon,t}(u_\epsilon^{(s)}(t))\| = \|[I - (I + \epsilon T_t)^{-1}](u_\epsilon^{(s)}(t))\| \leq \|\tilde{T}_s(u(s))\| + \lambda(\xi)$$

for $s \leq t < s + d(\xi)$ for a given $\xi > 0$.

Consider a sequence $\{t_j\}$ converging to s from above as $j \to \infty$, and let $u_j = u(t_j)$ for each j, where $\{u_j\}$ converges strongly to $u(s)$ in X as $j \to +\infty$. For each j let

$$w_j = T_{\epsilon_j, t_j}(u_j) = \epsilon_j^{-1}[(I + \epsilon_j T_{t_j})^{-1}](u_j)$$

for a sequence $\{\epsilon_j\}$ of constants tending to zero as $j \to \infty$. Suppose that w_j converges weakly in X to an element w of X. By the hypothesis of the Lipschitzian character of T_t in t, we see immediately that

$$\|T_{\epsilon_j,t_j}(u_j) - T_{\epsilon_j,s}(u_j)\| \leq c_0(t_j)(t_j - s)\{\|(I + \epsilon_j T_{t_j})^{-1}(u_j)\| + \psi(\|u_j\|)\}$$

so that $\|T_{\epsilon_j,t_j}(u_j) - T_{\epsilon_j,s}(u_j)\| \to 0$ $(j \to \infty)$. It follows that if $y_j = T_{\epsilon_j,s}(u_j)$, then $y_j \to w$ $(j \to \infty)$ weakly in X. On the other hand,

$$u_j - \epsilon_j y_j = v_j = (I + \epsilon_j T_s)^{-1}(u_j),$$

and there exists z_j in $T_s(v_j)$ such that $v_j + \epsilon_j z_j = u_j$. Hence $y_j = z_j \in T_s(v_j)$, while $\|v_j - u_j\| \le \epsilon_j$, $\|T_{\epsilon_j,s}(u_j)\| \to 0$ $(j \to \infty)$.

If we apply Theorem (3.16)(b), we find that w lies in $T_s(u(s))$. Thus the hypotheses of Proposition (9.6) are fulfilled, and it follows that $T_{\epsilon,t}(u_\epsilon^{(s)}(t))$ will differ from $T_s(u(s))$ by 2ξ at most for $\epsilon > 0$ sufficiently small, uniformly for all t with $|t - s|$ sufficiently small. It follows as in the proof of Theorem (9.23) that $du(s)/ds = -\tilde{T}(u(s))$. q.e.d.

An important special case of Theorem (9.25) is the following:

THEOREM (9.26). *The smoothness conditions of Theorem (9.25) are satisfied for a family $\{T_t\}$ of hypermaximal accretive mappings with $T_t = T + T_{0,t}$, where T is a fixed mapping of X into 2^X and $\{T_{0,t}\}$ is a family of single-valued mappings of $D(T)$ into X such that for each t in R^+ and $0 < h \le 1$*

$$\|(T_{0,t+h} - T_{0,t})(I + \epsilon T_t)^{-1} u\| \le c_0(t) h \, \|(I + \epsilon T_t)^{-1} u\| + \psi(\|u\|)).$$

PROOF OF THEOREM (9.26). We note that for each u in X,

$$(I + \epsilon T_{t+h})^{-1} u - (I + \epsilon T_t)^{-1} u$$
$$= (I + \epsilon T_{t+h})^{-1} u - (I + \epsilon T_{t+h})^{-1}(u + \epsilon(T_{0,t+h} - T_{0,t})(I + \epsilon T_t)^{-1} u),$$

so that

$$\|(I + \epsilon T_{t+h})^{-1} u - (I + \epsilon T_t)^{-1} u\| \le \epsilon \, \|(T_{0,t+h} - T_{0,t})(I + \epsilon T_t)^{-1} u\|$$
$$\le \epsilon c_0(t) h [\|(I + \epsilon T_t)^{-1} u\| + \psi(\|u\|)]. \quad \text{q.e.d.}$$

10. Existence theorems involving accretive mappings. If we combine the results of §9 on the existence of solutions of nonlinear equations of evolution involving accretive mappings from a Banach space X to itself with the fixed point theorems for nonexpansive mappings established in §8, we obtain general existence theorems for solutions u_0 of the equation $T(u_0) = f$ for an accretive mapping T defined on a subset of the Banach space X.

The principle underlying these existence theorems, which has already been applied in the discussion of §9, can be stated precisely in the following form:

THEOREM (10.1). *Let X be a Banach space, T a mapping of a subset $D(T)$ of X into X, X_0 the closure of $D(T)$ in X. Suppose that for each v in $D(T)$, there exists one and only one solution of the equation*

$$du(t)/dt = -T(u(t)), \quad u(0) = v,$$

on the interval $[0, d_0]$ with $d_0 > 0$ independent of the choice of v in $D(T)$. Let $U(t)$ be the mapping of $D(T)$ into itself given by

$$U(t)v = u(t), \quad 0 \le t \le d_0,$$

where u is the solution described above, and suppose that $U(t)$ can be extended by continuity to a continuous mapping of X_0 into itself. Suppose finally that $(-T)$ is the *infinitesimal generator of the local semigroup* $\{U(t): 0 \leq t \leq d_0\}$ *on* X_0, *i.e., if*

$$\text{weak limit}_{t \to 0+} t^{-1}(U(t)v - v) = A(v)$$

exists for an element v *in* X_0, *then* v *lies in* $D(T)$ *and* $A(v) = -T(v)$.

Then: An element of v *of* X_0 *is a common fixed point of all the mappings* $\{U(t): 0 \leq t \leq d_0\}$ *if and only if* v *lies in* $D(T)$ *and* $T(v) = 0$.

PROOF OF THEOREM (10.1). Obviously if v lies in $D(T)$ and $T(v) = 0$, then the constant function $u(t) = v$ is a solution of the equation $du(t)/dt = -T(u(t))$ with $u(0) = v$, and by assumption, must be the unique solution of this differential equation with the given initial value. Therefore, by the definition of the mappings $U(t)$, $U(t)v = v$ for all t in $[0, d_0]$.

Suppose on the other hand that v is a common fixed point of all the mappings $\{U(t): 0 \leq t \leq d_0\}$. Then

$$\text{weak limit}_{t \to 0+} t^{-1}(U(t)v - v)$$

exists and is equal to 0. Hence v lies in $D(T)$, and $T(v) = 0$, since $A(v) = 0$.
q.e.d.

An extremely simple case in which Theorem (10.1) can be applied occurs when $D(T)$, the domain of T, is closed in X and coincides with X_0. Then the mapping $(-T)$ is automatically the infinitesimal generator of the local semigroup $\{U(t)\}$ since then for each v in X_0,

$$\text{weak limit}_{t \to 0+} t^{-1}(U(t)v - v) = \frac{d(U(t)v)}{dt}\bigg|_{t=0} = \frac{du}{dt}(0) = -T(v),$$

since no process of continuous extension of the domain of the mappings $U(t)$ needs to be invoked.

We give a number of interesting special cases in which we may apply the general principle presented in Theorem (10.1):

THEOREM (10.2). *Let* X *be a uniformly convex Banach space,* T *an accretive mapping of* X *into* X *with* T *locally uniformly continuous in the strong topology of* X. *Suppose that* T^{-1} *is locally bounded, i.e., each point* x_0 *of* X *has a neighborhood* N *such that* $T^{-1}(N)$ *is bounded in* X.

Then T *maps* X *onto* X, *i.e.,* $R(T)$, *the range of* T, *is all of* X.

PROOF OF THEOREM (10.2). To prove that $R(T) = X$, it suffices since X is connected to show that each point of cl $(R(T))$ (the closure of $R(T)$ in X) lies in the interior of $R(T)$. (Indeed, it then follows that $R(T)$ is a nonempty open and closed subset of X and hence must be equal to X.)

Suppose w_0 lies in cl $(R(T))$. Then by hypothesis, there exists a neighborhood N of w_0 such that $T^{-1}(N)$ is bounded in X. The remainder of the proof of Theorem (10.2) then follows from the following localized version of the result:

THEOREM (10.3). *Let* X *be a uniformly convex Banach space,* T *an accretive mapping of* X *into* X *with* T *locally uniformly continuous. Suppose that for a point*

w_0 in cl $(R(T))$, there exists a neighborhood N of w_0 such that $T^{-1}(N)$ is bounded in X. Then there exists a neighborhood N_1 of w_0 contained in $R(T)$.

PROOF OF THEOREM (10.3). If we replace the mapping T by the new accretive mapping T_{w_0} given by $T_{w_0}(u) = T(u) - w_0$, we immediately reduce the theorem to the case in which $w_0 = 0$. We can assume that the neighborhood N of the hypothesis is the ball $B_d(0)$ of radius $d > 0$ about 0. We shall show that for each u in X with $\|u\| < d/2$, u lies in $R(T)$.

We consider the semigroup $\{U(t): t \geq 0\}$ of nonexpansive self-mappings of X generated according to Theorem (9.6) by the solutions $u: R^+ \to X$ of the differential equation $dw(t)/dt = -T(w(t)) + u$ with initial value $w(0) = v$ for an arbitrary element v of X, by setting

$$U(t)(v) = w(t) \qquad (t \geq 0)$$

where w is the solution for the given initial value v. Since $T_u(w) = T(w) - u$ for each fixed element u of X yields an accretive, locally uniformly continuous self-mapping T_u of X, this semigroup $\{U(t): t \geq 0\}$ is indeed a well-defined strongly continuous one-parameter semigroup of nonexpansive mappings of X into X. We may find elements v_k of X such that $\|T(v_k)\| \to 0$ as $k \to 0$ so that for k sufficiently large

$$\|T_u(v_k)\| = \|T(v_k) - u\| \leq \|T(v_k)\| + \|u\| < d/2.$$

By Theorem (9.6), we know that for all $t \geq 0$,

$$\|T_u(U(t)v_k)\| \leq \|T_u(v_k)\| < d/2$$

for k sufficiently large, i.e., $\|T(U(t)v_k) - u\| < d/2$. Hence for all t in R^+ and $k \geq k_0$, we have

$$\|T(U(t)v_k)\| \leq \|T(U(t)v_k) - u\| + \|u\| < d.$$

By hypothesis, there exists $R < 0$ such that if $\|T(z)\| \leq d$, then $\|z\| \leq R$. Applying this fact to the conclusion of the preceding paragraph, we see that for all $t \geq 0$ and any $k \geq 0$, $\|U(t)(v_k)\| \leq R$. By the Corollary to Theorem (8.8), it follows that the family of mappings $\{U(t): t \geq 0\}$ has a common fixed point x_0.

For this fixed point x_0, however, it follows since $(-T_u)$ is the infinitesimal generator of the semigroup $\{U(t)\}$, that by Theorem (10.1), $T_u(x_0) = 0$, i.e., $T(x_0) = u$.   q.e.d.

The proof of Theorem (10.3) completes the proof of Theorem (10.2).

THEOREM (10.4). *Let X be a uniformly convex Banach space with X^* uniformly convex and consider three single-valued accretive mappings T, T_0 and T_1 with dense domains in X and with ranges in X. Suppose that T and T_1 are hypermaximal accretive in X with $D(T) \subset D(T_1)$ while T_0 is defined on all of X and is demicontinuous. Suppose further that for each point x_0 in X, there is a neighborhood N_0 of x_0 and constants $k_{x_0} < 1$ and $k'_{x_0} \geq 0$ such that for all u in $D(T) \cap N$, $\|T_1(u)\| \leq k_{x_0}\|T(u)\| + k'_{x_0}$.*

Then:

(a) *The mapping $S = (T + T_1 + T_0)$ is hypermaximal accretive in X.*

(b) *If for each point u_0 in X, there is a neighborhood N of u_0 such that $S^{-1}(N)$ is bounded in X, then S has all of X as its range.*

PROOF OF THEOREM (10.4). PROOF of (a). By Theorem (9.12), T_0 is hypermaximal accretive. By Theorem (9.22), $T + T_1$ is hypermaximal accretive. By Theorem (9.21), finally, $S = (T + T_1) + T_0$ is hypermaximal accretive.

PROOF OF (b). It suffices as in the proof of Theorem (10.2) to show that each point w_0 of cl $(R(S))$ lies in the interior of $R(S)$. Since we can replace the mapping S by the mapping S_{w_0} where

$$S_{w_0}(u) = S(u) - w_0 = T(u) + T_1(u) + (T_0(u) - w_0),$$

we may assume without loss of generality that $w_0 = 0$. By hypothesis, there exists a ball $B_d(0)$ with $d > 0$ such that $S^{-1}(B_d(0))$ is contained in the ball $B_R(0)$ for some $R > 0$. For any u in X with $\|u\| < d/2$, it suffices to show that u lies in $R(S)$.

Let S_u be the mapping given by

$$S_u(v) = S(v) - u \qquad (v \in D(S) = D(T)).$$

Since S is hypermaximal accretive, it follows trivially that S_u is hypermaximal accretive. Hence by Theorem (9.15) for each v in $D(S_u) = D(T)$, a dense subset of X, there exists exactly one strongly continuous, strongly once right differentiable solution $w: R^+ \to X$ of the differential equation $dw(t)/dt = -S_u(w(t))$ $(t \geq 0)$ with initial value $w(0) = v$. We set $U(t)v = w(t)$ $(t \geq 0)$, where $w(t)$ is the value of this solution with the given initial value v at the time $t \geq 0$. Then by Theorem (9.17), we know that the family of mappings $\{U(t)\}$ can be extended to a strongly continuous, one-parameter semigroup $\{U(t): t \geq 0\}$ of nonexpansive mappings of X into X, and that, moreover, $(-S_u)$ is the infinitesimal generator of this semigroup.

Since 0 lies in cl $(R(S))$, it follows that we may find an element v of $D(S)$ such that $\|S(v)\| < d/2 - \|u\|$. Then

$$\|S_u(v)\| = \|S(v) - u\| \leq \|S(v)\| + \|u\| < d/2.$$

By Theorem (9.15)(b), we know that for all $t \geq 0$, $\|S_u(U(t)v)\| \leq \|S_u(v)\| < d/2$. Hence

$$\|S(U(t)(v))\| \leq \|S_u(U(t)(v))\| + \|u\| < d,$$

and therefore $U(t)(v)$ lies in $S^{-1}(B_d(0)) \subset B_R(0)$, i.e., $\|U(t)(v)\| \leq R$ $(t \geq 0)$.

Since X is uniformly convex, we may apply the Corollary to Theorem (8.8) and conclude that the family $\{U(t)\}$ of nonexpansive self-mappings of X has a common fixed point x_0 in X. By Theorem (10.1), this common fixed point x_0 must lie in the domain of the infinitesimal generator A of the semigroup $\{U(t): t \geq 0\}$ and $A(x_0) = 0$. Since $A = -S_u$, it follows that x_0 lies in $D(S)$ and $S(x_0) = u$.   q.e.d.

REMARK. The result of Theorem (10.4) obviously included the special cases in which one or several of the mappings T, T_0, or T_1 may not be present.

THEOREM (10.5). *Let X be a Banach space, T a locally uniformly continuous accretive mapping of X into X. Suppose that T^{-1} is locally bounded as a mapping of X into 2^X.*

Then the range of T is dense in X.

PROOF OF THEOREM (10.5). For each $\epsilon > 0$, we consider the mapping $T_\epsilon = T + \epsilon I$. If we can show that for each such T_ϵ, $R(T_\epsilon) = X$ then it follows from the result of Theorem (5.1) that since each T_ϵ is continuously invertible from $R(T_\epsilon)$ to X, then $R(T)$ must be dense in X.

Since we can add an arbitrary constant element to T without altering the hypotheses, it suffices to show that for each fixed $\epsilon > 0$, 0 lies in $R(T_\epsilon)$. To do this, we construct the semigroup $\{U(t): t \geq 0\}$ of nonexpansive self-mappings of X, where for each v in X, $U(t)v = u(t)$ and $u: R^+ \to X$ is the solution of the differential equation

$$du(t)/dt = -T(u(t)) - \epsilon u(t) \qquad (t \geq 0)$$

with initial value $u(0) = v$. It follows from Theorem (9.10) that for each pair of elements v and v_1 of X we have

$$\|U(t)(v) - U(t)v_1\| \leq \exp(-\epsilon t) \|v - v_1\|.$$

Hence $u(t)$ for each fixed $t > 0$ is a strict contraction mapping the Banach space X into itself and hence has exactly one fixed point x_t in X. Since the mappings $U(t)$ and $U(s)$ commute for $0 < s < t$, it follows that x_t is independent of t for $t > 0$ and hence is a common fixed point of the whole family $\{U(t): t > 0\}$. Hence, if we set x_0 equal to this x_t,

$$-(T(x_0) + \epsilon x_0) = \lim_{t \to 0+} t^{-1}(U(t)(x_0) - x_0) = 0. \quad \text{q.e.d.}$$

THEOREM (10.6). *Let X be a uniformly convex Banach space, G a bounded closed convex subset of X. Suppose that G_0 is an open subset of X containing G such that G has positive distance from $X - G_0$ and let T be a locally uniformly continuous mapping of G_0 into X which is accretive on G_0 and bounded on G.*

Then $T(G)$ is a closed subset of X.

PROOF OF THEOREM (10.6). Since T is locally uniformly continuous from G_0 to X, then by Theorem (9.7) for each v in G_0, there exist an interval $[0, d_v)$ with $d_v > 0$ for each v, and a solution $u: [0, d_v) \to X$ of the differential equation

$$du(t)/dt = -T(u(t)) \qquad (t \in [0, d_v))$$

with the initial value $u(0) = v$. This solution is uniquely determined by v, it can be continued till $u(t)$ converges to some point y_0 of $X - G_0$ as $t \to d_v-$, and for all t in $[0, d_v)$, $\|du(t)/dt\| \leq \|T(v)\|$.

If we set $U(t)(v) = u(t)$ for each such v, we obtain a family of mappings $U(t)$ of subsets of G_0 into G_0, each of which is nonexpansive for any pair of points for which it is defined. For each v in G, $\|T(v)\| \leq M$, where M is a positive constant independent of v in G. Hence for all solutions $u_v(t)$ starting at points v of G,

$\|duv(t)/dt\|$ is uniformly bounded. Since G has positive distance from $X - G_0$, it follows that there exists a constant $d > 0$ such that all these solutions are defined on the fixed interval $[0, d]$, i.e., $U(t)$ is a well-defined nonexpansive mapping of G into G_0 for all t in $[0, d]$.

We wish to show that $T(G)$ is closed in X. Since we can add an arbitrary fixed element w_0 of X to T without changing the hypotheses of the theorem, it suffices to assume that we are given a sequence v_k in G with $T(v_k)$ converging strongly to 0 in X and to show that there exists x_0 in G such that $T(x_0) = 0$.

Let $u_k(t) = U(t)(v_k)$ for each k. Then for each fixed $t > 0$,

$$U(t)(v_k) - v_k = -\int_0^t (TU(r))(v_k)\,dr,$$

where $\|T(U(r))(v_k)\| \leq \|T(v_k)\|$. Hence

$$\|U(t)(v_k) - v_k\| \leq t\,\|T(v_k)\| \to 0 \qquad (k \to +\infty).$$

Thus for each fixed $t > 0$, $0 \in$ closure $((I - U(t))(G))$.

Since X is uniformly convex and $U(t)$ is a nonexpansive mapping of G into X, it follows from Theorem (8.4) that $(I - U(t))(G)$ is closed for the closed bounded convex subset G of X. Hence $U(t)$ has a fixed point in G for each $t > 0$.

We may apply the above argument to any closed convex subset G_1 of G which contains an infinite subsequence of the sequence $\{v_k\}$ and obtain a fixed point of $U(t)$ for each t with $0 < t < d$ in that set G_1. Since the space X is reflexive, we may consider weakly convergent subsequences (again denoted by $\{v_k\}$) such that v_k converges weakly to v_0 in G. If we set G_m to be the convex closure of $\{v_m, v_{m+1}, \cdots\}$ in G, the fixed point sets F_m of $U(t)$ in G_m are a decreasing sequence of nonempty closed convex subsets of G and are weakly compact. Hence their intersection is nonempty. However, $\bigcap_m G_m = \{v_0\}$. Hence the set of weak limit points of the sequence $\{v_k\}$ is nonempty and each of these limit points is a fixed point of each mapping $U(t)$ for $t > 0$.

Finally, for each weak limit point v_0 of the sequence $\{v_k\}$,

$$-T(v_0) = \lim_{t \to 0+} t^{-1}(U(t) - I)(v_0) = 0,$$

i.e., v_0 lies in $T^{-1}(0)$. q.e.d.

The last part of the proof of Theorem (10.6) actually gave the following sharper result:

THEOREM (10.7). *Let X be a uniformly convex Banach space, G a closed bounded convex subset of X. Suppose that for an open subset G_0 of X containing G, T is a locally uniformly continuous mapping of G_0 into X which is accretive. Suppose that T is bounded on G.*

Then for any weakly convergent sequence $\{v_k\}$ in G with weak limit v_0 such that $T(v_k)$ converges strongly to an element w_0 of X, we have $T(v_0) = w_0$.

Applying the same line of proof combined with the other existence theorems (Theorem (9.9) and Theorem (9.15)) of §9), we have the following results:

THEOREM (10.8). *Let X be a uniformly convex Banach space with X^* uniformly convex, G a closed bounded convex subset of X. Let G_0 be an open subset of X which contains G and let T be a demicontinuous accretive mapping of G_0 into X which is bounded on G. Suppose that G has bounded distance from $X - G_0$.*
Then:
(a) For any weakly convergent sequence $\{v_k\}$ in G with weak limit v_0 and with $T(v_k)$ converging strongly to w_0 in X, we have $T(v_0) = w_0$.
(b) $T(G)$ is a closed subset of X.

THEOREM (10.9). *Let X be a uniformly convex Banach space with X^* uniformly convex, G a closed bounded convex subset of X. Let T, T_1, and T_0 be three accretive mappings with domains in X such that the following conditions all hold:*
(i) T and T_1 are hypermaximal accretive, $D(T) \subset D(T_1)$, and $D(T)$ is dense in G.
(ii) T_0 is defined on all of X and is a demicontinuous self-mapping of X.
(iii) For each x_0 in X, there are a neighborhood N and constants $k_{x_0} < 1$ and $k'_{x_0} \geq 0$ such that for all u in $D(T) \cap N$, $\|T_1(u)\| \leq k_{x_0} \|T(u)\| + k'(x_0)$.
Then:
(a) If $S = T + T_1 + T_0$ with $D(S) = D(T)$, then for any sequence $\{v_k\}$ in $D(S) \cap G$ with v_k converging weakly to v_0 and $S(v_k)$ converging strongly to w_0, v_0 lies in $D(S)$ and $S(v_0) = w_0$.
(b) The image under S of $D(S) \cap G$ is a closed subset of X.

PROOF OF THEOREM (10.8). This is a repetition of the proof of Theorems (10.6) and (10.7) in which we apply the local existence theorem, Theorem (9.9) instead of Theorem (9.7). q.e.d.

PROOF OF THEOREM (10.9). By Theorem (9.22), $T + T_1$ is hypermaximal accretive in X. By Theorem (9.21), $S = (T + T_1) + T_0$ is hypermaximal accretive in X. Hence by Theorem (9.15), for each v in $D(S) = D(T)$, we can find one and only one solution $u_v : R^+ \to X$ of the differential equation

$$du_v(t)/dt = -S(u_v(t)) + w_0 \quad (t \geq 0)$$

satisfying the initial condition $u_v(0) = v$. If we let $U(t)(v) = u_v(t)$ for $t \geq 0$, we obtain a semigroup of nonexpansive mappings of $D(S) = D(T)$ into X which we can extend by continuity to the strong closure of $D(T)$ in X. In particular, we note that $D(T)$ is dense in G by hypothesis. Hence, we obtain a family $\{U(t) : t \geq 0\}$ of nonexpansive mappings of G into X. In addition, we note that for v in $D(S)$,

$$\|S(U(t)(v)) - w_0\| \leq \|S(v) - w_0\| \quad (t \geq 0).$$

To prove the assertion of (a), we note that

$$\|S(U(t)(v_k)) - w_0\| \leq \|S(v_k)\| - w_0 \to 0$$

uniformly for t in R^+. Hence, for each $t > 0$,

$$\|(I - U(t))(v_k)\| \leq \int_0^t \|S(U(r)(v_k)) - w_0\| \, dr \to 0$$

as $k \to +\infty$. Hence by part (a) of Theorem (8.4),

$$(I - U(t))(v_0) = 0 \qquad (t > 0).$$

Thus, since the mapping $-S_{w_0}$ is the infinitesimal generator of the semigroup $\{U(t): t \geq 0\}$ by Theorem (9.17), where $S_{w_0}(v) = S(v) - w_0$, it follows that

$$\lim_{t \to 0+} t^{-1}(U(t) - I)(v_0) = 0$$

implies that v_0 lies in $D(S)$ and $S(v_0) = w_0$. Hence the proof of (a) is complete.

To prove the assertion of part (b), we note that if w_0 lies in the strong closure of $S(G \cap D(S))$, then there exists a sequence $\{v_k\}$ in $D(S) \cap G$ such that $S(v_k)$ converges strongly to w_0. Since G is weakly sequentially compact, we may assume that v_k converges weakly to v_0 in G as $k \to \infty$. Hence by part (a), v_0 lies in $D(S) \cap G$ and $S(v_0) = w_0$. Thus w_0 lies in the image of $G \cap D(S)$ under S. q.e.d.

Another existence theorem which follows from the techniques of argument we have applied above is the following:

THEOREM (10.10). *Let X be a uniformly convex Banach space, and let B be a closed ball about the origin in X. Let G_0 be an open subset of X which contains B and consider an accretive mapping T with domain in X and values in X of one of the following three types:*

(1) *T is a locally uniformly continuous mapping of G_0 into X.*
(2) *X^* is uniformly convex and T is a demicontinuous mapping of G_0 into X.*
(3) *X^* is uniformly convex and $T = (T_1 + T_2 + T_0)$ where T_1 and T_2 are hypermaximal accretive in X, $D(T_1)$ is dense in X and contained in $D(T_2)$, $D(T_0) = X$ and T_0 is a demicontinuous self-mapping of X, and for each point x_0 of X, there exist a neighborhood N of x_0 in X and two constants $k_{x_0} < 1$ and $k'_{x_0} \geq 0$ such that for all u in $D(T_1) \cap N$, $\|T_2(u)\| \leq k_{x_0} \|T_1(u)\| + k'_{x_0}$.*

Suppose further that for each point u in $D(T) \cap S$ for the bounding sphere S of B, $(J(u), T(u)) \geq 0$ where J is a positively homogeneous duality mapping of X into X^*.

Then there exists a point x_0 in $D(T) \cap B$ such that $T(x_0) = 0$.

PROOF OF THEOREM (10.10). Under any of the three assumptions (1), (2), or (3), it follows from the results of §9 that for each v in $D(T) \cap G_0$, there exist an interval $[0, d_v)$ with $d_v > 0$ and a solution $u_v: [0, d_v) \to X$ of the differential equation

$$du_v(t)/dt = -T(u_v(t)) \qquad (t \in [0, d_v))$$

such that $u_v(0) = v$. For this solution, we know that if $d_v < +\infty$ then the solution u_v can be continued in t until $u_v(t)$ converges to a point in $X - G_0$ as $t \to d_v-$.

We shall show first that under the hypotheses of our theorem, it follows that for each v in B, u_v exists on the whole of R^+. To prove this, it suffices to show that for each v in B and each t in $[0, d_v)$, $u_v(t)$ lies in B, since B is closed and it is therefore impossible for $u_v(t)$ to converge to a point of $X - G_0$ as $t \to d_v-$. For this purpose, it suffices to show that for any v in B, there exists $d_v' > 0$ such that the solution u_v with initial value v remains in B for $0 \leq t < d_v'$.

We prove the latter assertion separately for the cases (1) and (2) on the one hand, and (3) on the other. In cases (1) and (2), T is defined at each point of S and for each u in S, there exists a neighborhood N of u in X such that T is defined as a mapping of N into X. We may assume without loss of generality that v lies in S and that the solution $u_v(t)$ remains in N through the interval $[0, d_v']$. If $B = B_R(0)$, then for each point u in N, the point $r(u) = Ru/\|u\|$ lies in S. We may assume N chosen without loss of generality so that $r(u)$ lies in N for each u in N. Since T is accretive and solutions of initial value problems exist for each initial value in G_0, it follows from Theorem (9.4) that T is accretive on G_0 with respect to any single-valued duality mapping. In particular, T is accretive on G_0 with respect to the duality mapping J used in the hypothesis on the boundary behavior of T on B. Thus for any u in $N - B$, $(J(u - r(u)), T(u) - T(r(u))) \geq 0$. Since

$$u - r(u) = \lambda u, \quad \lambda = (1 - R\|u\|^{-1}) > 0$$

for each such u, it follows that

$$(J(u), T(u)) \geq (J(u), T(r(u))) = \|u\| R^{-1}(J(r(u)), T(r(u))) \geq 0.$$

Hence for each u in $(N - B) \cup S$, we have

$$\varlimsup_{t \to s-} (t-s)^{-1} \|u(t) - u(s)\|^2 \leq 0$$

on any interval $[0, d]$ on which a solution of the differential equation $du(t)/dt = -T(u(t))$ lies on or outside S in N. The norm inequality above, however, implies by Proposition (9.1) that $\|u(t)\|^2$ is nonincreasing on that interval. If $u(0)$ lies in S, then it follows that for the whole interval, $u(t)$ lies in B. Thus we have shown in cases (1) and (2) that any solution emanating from a point in B must always remain in B.

We now carry through a somewhat different proof of the same fact in the case (3). For this purpose, we let T_ϵ be a new mapping defined for each $\epsilon > 0$ by setting $T_\epsilon(u) = T(u) + \epsilon u$, with $D(T_\epsilon) = D(T)$. We note that for each $\epsilon > 0$, T_ϵ satisfies the same conditions as T does, and we can therefore establish the existence of solutions $u_{v,\epsilon}$ of the differential equation

$$du_{v,\epsilon}(t)/dt = -T_\epsilon(u_{v,\epsilon}(t)) \quad (t \geq 0)$$

with the initial value $u_{v,\epsilon}(0) = v$, with these solutions defined on all of R^+ and uniquely determined by the initial value v. For any initial value v in B, we assert that the solution $u_{v,\epsilon}(t)$ remains in B for all t in R^+. Indeed, if it ever intersects the sphere S bounding B for a given time value t_0, then at that time value, we have for $u(t) = u_{v,\epsilon}(t)$,

$$d\|u(t_0)\|^2/dt = -2(J(u(t_0)), T(u(t_0)) + \epsilon u(t_0)) \leq -2\|u(t_0)\|^2 < 0.$$

Since
$$d\,\|u(t)\|^2/dt = -2(J(u(t)),\,T_\epsilon(u(t)))$$
is continuous in t, it follows that such a solution must reenter B (and indeed can only intersect S if $t_0 = 0$). Furthermore, for each $\epsilon > 0$, we have
$$\frac{d}{dt}\|u_v(t) - u_{v,\epsilon}(t)\|^2 = -2(J(u_v(t) - u_{v,\epsilon}(t)), T(u_v(t)) - T(u_{v,\epsilon}(t)))$$
$$+ 2(J(u_v(t) - u_{v,\epsilon}(t)), \epsilon u_{v,\epsilon}(t)).$$
Hence
$$\frac{d}{dt}\|u_{v,\epsilon}(t) - u_v(t)\|^2 \leq 2\epsilon R\,\|u_{v,\epsilon}(t) - u_v(t)\|$$
$$\leq \epsilon^2 R^2 + \|u_{v,\epsilon}(t) - u_v(t)\|^2.$$
It follows that $\|u_{v,\epsilon}(t) - u_v(t)\|^2 \leq \epsilon^2 R^2 \exp(t)$ since $u_{v,\epsilon}(0) = v = u_v(0)$. It follows that for each fixed t in $[0, d_v)$, $u_{v,\epsilon}(t)$ converges strongly to $u_v(t)$ and hence the latter element lies in B for each t in $[0, d_v)$. Thus the proof for case (3) of the fact that solutions emanating from B remain in B is complete.

Since it follows that all solutions $u_v(t)$ are defined for all t in R^+ for v in B and remain in B for all $t \geq 0$, we may define a semigroup of nonexpansive self-mappings of B by setting $U(t)(v) = u_v(t)$ ($t \geq 0$). It follows from Theorem (8.8) that the family of mappings $\{U(t): t > 0\}$ of B into B which commute with one another must have a common fixed point x_0 in B. If A is the infinitesimal generator of the semigroup $U(t)$ it follows that $A(x_0) = 0$. However, as in the preceding proofs, $A = -T$ and we obtain $T(x_0) = 0$. q.e.d.

In the remainder of this section, we shall consider another question related to those treated above, namely the problem of obtaining solutions of $T(x_0) = 0$ as the limits as $t \to \infty$ of solutions of approximating equations of evolutions of the form
$$du(t)/dt = -T(u(t)) - h(t)u(t) \qquad (t \geq 0)$$
with an arbitrary initial value v in $D(T)$.

THEOREM (10.11). *Let X be a uniformly convex Banach space with X^* uniformly convex, T a hypermaximal accretive mapping of $D(T)$ in X into X such that for all u in X with $\|u\| > R$ for some given $R > 0$ and all w in $T(u)$, $(w, u) > 0$.*

Let $h: R^+ \to R^+$ be a monotone nonincreasing function of class C^1 with $h(t) \to 0$ as $t \to \infty$, and $\int_0^\infty h(s)\,ds = +\infty$.

Then:

(a) For each v in $D(T)$, there exists exactly one solution $u: R^+ \to X$ of the differential equation
$$du(t)/dt = -T(u(t)) - h(t)u(t) \qquad (t \geq 0)$$
with initial value $u(0) = v$. For this solution, we have $\|T(u(t))\| \to 0$ ($t \to \infty$).

(b) For any weakly convergent subsequence $\{u(t_k)\}$ with weak limit u_0 as $t_k \to \infty$, $T(u_0) = 0$.

(c) If T^{-1} is single-valued and continuous from $R(T)$ to X, then $u(t)$ converges strongly as $t \to \infty$. If T^{-1} is demicontinuous from $R(T)$ to X, then $u(t)$ converges weakly as $t \to \infty$. In both cases, the limit lies in $T^{-1}(0)$.

PROOF OF THEOREM (10.11). PROOF OF (a). Let v be a given element of $D(T)$. The family of mappings $T + h(t)I$ satisfies the conditions of Theorem (9.26) and hence there follows the existence of one and only one solution u on R^+ of the differential equation $du(t)/dt = -T(u(t)) - h(t)u(t)$ with the initial value $u(0) = v$. We apply to this solution the estimate derived in Theorem (9.10) in §9, which in this case becomes

$$\begin{aligned}\|T(u(t))\| &\leq \|du(t)/dt\| + \|u(t)\| \cdot h(t) \\ &\leq \|T(v) + h(0)v\| \exp(-H(t)) \\ &\quad + M_1 \int_0^t |h'(r)| \exp(H(r) - H(t))\, dr + M_1 h(t)\end{aligned}$$

where $M_1 = \sup \|u(t)\|$, $H(t) = \int_0^t h(s)\, ds$. We note first that if $\|u(t)\| > R$, then

$$\frac{d}{dt}\{\|u(t)\|^2\} = -2(J(u(t)), T(u(t)) + h(t)u(t)) < 0.$$

Hence for all t in R^+,

$$\|u(t)\| \leq \max(R, \|v\|) = M_1.$$

We note next that by hypothesis, $h(t) \to 0$ as $t \to \infty$ and $\int_0^{+\infty} h(s)\, ds = +\infty$. In particular, $M_1 h(t) \to 0$ $(t \to \infty)$ and $H(t) \to +\infty$ $(t \to +\infty)$. Thus

$$\|T(v) + h(0)v\| \exp(-H(t)) \to 0 \quad (t \to +\infty).$$

Finally, $h'(r) \leq 0$ for all r since $h(t)$ is nonincreasing in t. Hence

$$\int_0^t |h'(r)| \exp(H(r) - H(t))\, dr = -\int_0^t h'(r) \exp(H(r))[\exp(-H(t))]\, dr$$

where

$$-\int_0^t h'(r) \exp(H(r))\, dr = -[h(s)]^2 \exp(H(s))\Big|_0^t + \int_0^t [h(r)]^2 \exp(H(r))\, dr$$

$$\leq \int_0^t [h(r)]^2 \exp(H(r))\, dr.$$

If $N > 0$,

$$\int_0^t [h(r)]^2 \exp(H(r))\, dr = \left[\int_0^N + \int_N^t\right](h(r))^2 \exp(H(r))\, dr$$

$$\leq [h(0)]^2 \exp(H(N)) + [h(N)]^2 \exp(H(t)).$$

Hence

$$\int_0^t |h'(r)| \exp(H(r) - H(t))\, dr \leq [h(0)]^2 \exp(H(N) - H(t)) + [h(N)]^2.$$

If for a given $\epsilon > 0$, we choose N so large that $[h(N)]^2 < \epsilon/2$, and then choose

$M > 0$ so large that for the given N,

$$[h(0)]^2 \exp(H(N) - H(t)) < \epsilon/2 \qquad (t > M),$$

then for $t > M$, $\int_0^t |h'(r)| \exp(H(r) - H(t))\, dr < \epsilon$. Thus

$$\int_0^t |h'(r)| \exp(H(r) - H(t))\, dr \to 0 \qquad (t \to +\infty).$$

Thus $\|T(u(t))\| \to 0$ as $t \to \infty$. q.e.d.

PROOF OF (b). Suppose that for a given sequence $\{t_k\}$ with $t_k \to +\infty$ as $k \to \infty$, we have $u(t_k)$ converging weakly to u_0. Since $T(u(t_k))$ converges strongly to 0 in X, it follows from Theorem (10.9) that $T(u_0) = 0$. q.e.d.

PROOF OF (c). Suppose that T^{-1} is either continuous or demicontinuous from $R(T)$ to X. Since X is reflexive, there exists an element u_0 in the weak limit set of the elements $u(t)$ as $t \to \infty$. Thus $T(u(t))$ converges strongly to $T(u_0)$. Hence $u(t)$ converges to u_0 in X, weakly or strongly according as to whether T^{-1} is demicontinuous or continuous. q.e.d.

THEOREM (10.12). *Let H be a Hilbert space, T a maximal monotone mapping from H to 2^X such that for some $R > 0$, for each u in X with $\|u\| > R$ and all w in $T(u)$, $(w, u) > 0$. Let $D(T)$ be the effective domain of T. Consider a function $h : R^+ \to R^+$ which is nonincreasing and of class C^1 with $h(t) \to 0$ as $t \to \infty$ and $\int_0^\infty h(s)\, ds = +\infty$. Then:*

(a) *For each v in $D(T)$, there exists exactly one solution $u : R^+ \to R^+$ of the differential equation*

$$[du(t)/dt] \in \{-T(u(t)) - h(t)u(t)\},$$

with initial value $u(0) = v$.

For this solution, we have $\|du(t)/dt + h(t)\| \to 0$ as $t \to +\infty$.

(b) *As $t \to \infty$, $u(t)$ converges strongly in H to the solution u_0 of the equation $0 \in T(u_0)$ of minimum norm in H.*

PROOF OF THEOREM (10.12). PROOF OF (a). We note that since T is maximal monotone (i.e., maximal accretive with respect to the duality mapping I of H into H), T is accretive and by Theorem (7.9), T is hypermaximal accretive from H to 2^H.

Hence, we apply Theorem (9.26) to obtain a solution of the desired equation with the desired initial value. We apply the estimates of Theorem (9.10) once more, as in the proof of Theorem (10.11) to obtain the conclusion that

$$\|du(t)/dt + h(t)u(t)\| \to 0.$$

In particular, it follows that for any weakly convergent subsequence $\{u(t_k)\}$ with weak limit u_0 in H as $t_k \to \infty$, we have $0 \in T(u_0)$ by Theorem (3.18)(b). q.e.d.

PROOF OF (b). It follows from our final remark in the proof of part (a) that since H is reflexive and the norm of $u(t)$ is uniformly bounded for all t in R^+, we have at least one solution u_0 of the equation $0 \in T(u_0)$.

Let $q(t) = \|u(t) - u_0\|^2$ for one such solution u_0. Then
$$\begin{aligned} dq(t)/dt &= +2(u(t) - u_0, du(t)/dt) \\ &= +2(u(t) - u_0, (-w(t) - h(t)u(t)) \\ &\quad - (-w_0 - h(t)u_0)) - 2(u(t) - u_0, h(t)u_0) \\ &\leq -2h(t)(u(t) - u_0, u_0) - 2h(t)q(t) \end{aligned}$$

where $w_0 = 0 \in T(u_0)$, $w(t)$ is an element of $T(u(t))$. Let u_0 be the solution of least norm of $T(u) = 0$. Such an element u_0 exists since $T^{-1}(0)$ is nonempty under our hypotheses and by Theorem (3.18) is a closed bounded convex subset of the reflexive space H. Since the Hilbert space H is uniformly convex, u_0 is unique.

For any u_1 in $T^{-1}(0)$, we have $(u_0, u_0 - u_1) \leq 0$. Hence
$$(u(t) - u_0, u_0) = (u(t) - u_1, u_0) + (u_1 - u_0, u_0) \geq (u(t) - u_1, u_0)$$
and therefore
$$-2h(t)(u(t) - u_0, u_0) \leq -2h(t)(u(t) - u_1, u_0).$$
It follows that
$$\frac{dq}{dt}(t) \leq \inf_{u_1 \in T^{-1}(0)} \{[-2h(t)(u(t) - u_1, u_0)]\} - 2h(t)q(t),$$
for all $t \geq 0$.

We now apply the following results:

LEMMA (10.1). *Under the hypotheses of Theorem (10.12), we have for the function s on R^+ given by*
$$s(t) = \inf_{u_1 \in T^{-1}(0)} [-(u(t) - u_1, u_0)],$$
the following relation:
$$\varlimsup_{t \to +\infty} s(t) \leq 0.$$

PROOF OF LEMMA (10.1). Suppose the assertion of the lemma were false. Then we could find a sequence $\{t_k\}$ with $t_k \to +\infty$ as $k \to +\infty$ such that there exists $d > 0$ and for all u_1 in $T^{-1}(0)$ and for all k, $(u(t_k) - u_1, u_0) \geq d > 0$. Since $u(t_k)$ is a bounded set in H, we can extract a weakly convergent subsequence, which we identify with the original sequence. Let v_0 be the weak limit of this subsequence. Then $(u(t_k) - v_0, u_0) \to 0$. On the other hand, it follows from our previous remarks that $T(v_0) = 0$. Hence $(u(t_k) - v_0, u_0) \geq d > 0$, which is a contradiction proving the lemma. q.e.d.

PROOF OF THEOREM (10.12) COMPLETED. For the function
$$q(t) = \|u(t) - u_0\|^2 \quad (t \geq 0)$$
we have the differential inequality $dq(t)/dt \leq -2h(t)q(t) + s(t)h(t)$ with $h(t)$, $s(t) \to 0$ as $t \to \infty$, $\int_0^\infty h(s)\,ds = +\infty$. It follows that if we set $H(t) = \int_0^t h(s)\,ds$, then $H(t) \to +\infty$ as $t \to +\infty$, and the differential inequality may be written in

the form

$$\frac{d}{dt}[q(t)\exp(H(t))] \leq s(t)h(t)\exp(H(t)).$$

Integrating the inequality from 0 to t, we obtain the following:

$$q(t)\exp(H(t)) - q(0) \leq \int_0^t s(r)h(r)\exp(H(r))\,dr,$$

i.e.,

$$q(t) \leq q(0)\exp(-H(t)) + \int_0^t s(r)h(r)\exp(H(r)-H(t))\,dr.$$

For the first term on the right-hand side of the inequality, we know that

$$q(0)\exp(-H(t)) \to 0 \quad (t \to +\infty).$$

For the second term, we have for each $N > 0$ and for $t > N$,

$$\int_0^t s(r)h(r)\exp(H(r))\,dr \leq \left(\int_0^N + \int_N^t\right)[s(r)h(r)\exp(H(r))\,dr]$$

$$\leq M_0 \int_0^N h(r)\exp(H(r))\,dr + M(N)\int_N^t h(r)\exp(H(r))\,dr$$

with $M_0 = \sup_{r \in R}[h(r)]^+$ and $M(r) = \sup_{\lambda \geq r}[h(\lambda)]^+$ (where $[k]^*$ denotes the nonnegative part of k). We note that $M(r) \to 0$ as $r \to +\infty$.

We make the following estimates:

$$\int_0^N h(r)\exp(H(r))\,dr = \exp(H(N)) - 1 \leq \exp(H(N)),$$

$$\int_N^t h(r)\exp(H(r))\,dr = \exp(H(t)) - \exp(H(N)) \leq \exp(H(t)).$$

Hence

$$\int_0^t s(r)h(r)\exp(H(r)-H(t))\,dr \leq M_0\exp(H(N)-H(t)) + M(N).$$

For a fixed $\epsilon > 0$, we first choose N so large that $M(N) < \epsilon/2$. Having fixed N, we then choose T so large that for $t \geq T$, $M_0\exp(H(N)-H(T)) < \epsilon/2$. Hence for $t \geq N$,

$$\int_0^t s(r)h(r)\exp(H(r)-H(t))\,dr \leq \epsilon.$$

Thus

$$q(t) \leq q(0)\exp(-H(t)) + \int_0^t s(r)h(r)\exp(H(r)-H(t))\,dr \to 0$$

as $t \to +\infty$, i.e., $u(t) - u_0 \to 0$ strongly in H as $t \to +\infty$. q.e.d.

11. Nonlinear interpolation. It is the purpose of the present section to present some general results concerning the interpolation of nonexpansive nonlinear mappings in different Banach spaces, and to derive as consequences results on interpolation of nonlinear semigroups of nonexpansive mappings and of accretive mappings.

DEFINITION (11.1). *Let (X, X_1, X_2) be a triple of Banach spaces, each contained, with a continuous injection mapping, in some given separated locally convex topological vector space E and such that $X_0 = X \cap X_1 \cap X_2$ is dense in each of the three spaces. Then we say that this triple is a linear interpolation system for self-mappings if for each linear mapping L of X_0 into X_0 such that for a given constant $M > 0$,*

$$\|Lu\|_{X_1} \leq M \|u\|_{X_1}, \quad \|Lu\|_{X_2} \leq M \|u\|_{X_2} \quad (u \in X_0),$$

it follows that $\|Lu\|_X \leq M \|u\|_X$ $(u \in X_0)$.

THEOREM (11.1). *Let (X, X_1, X_2) be a linear interpolation system in the sense of Definition (11.1) and suppose that the following two conditions are both satisfied:*

(1) *For any sequence $\{u_k\}$ in X_0 converging in both the X_1- and X_2-norms to an element u of X_0 and for which $\|u_k\|_X \leq R$ for a constant $R > 0$ and all k, we have $\|u\|_X \leq R$.*

(2) *There exist a family $\{F_\alpha : \alpha \in \Omega\}$ of finite-dimensional subspaces of X_0 which is a directed set under inclusion and a corresponding family $\{P_\alpha\}$ of linear projection mappings of X_0 on F_α such that for all u in X_0,*

$$\|P_\alpha u\|_{X_1} \leq \|u\|_{X_1}, \quad \|P_\alpha u\|_{X_2} \leq \|u\|_{X_2},$$

while, in addition, the union of the spaces F_α of this family is dense in X_0 in the norm obtained by adding the three norms of X, X_1, and X_2.

Then each (possibly nonlinear) mapping T of X_0 into X_0 which is nonexpansive with respect to both the X_1-norm and the X_2-norm must be nonexpansive with respect to the X-norm.

PROOF OF THEOREM (11.1). We begin with the remark that it suffices to prove the desired inequality $\|T(u) - T(v)\|_X \leq \|u - v\|_X$ not for all u and v in X_0 but merely for all u and v in the subset X' of X_0, where $X' = \bigcup_\alpha F_\alpha$. Indeed, suppose that the above inequality did hold for all u and v in X'. For a pair of elements u and v of X_0, it follows from the last part of condition (2) of the hypothesis of our theorem that there exist sequences $\{u_k\}$ and $\{v_k\}$ in X' such that

$$\|u - u_k\|_{X_1} + \|u - u_k\|_{X_2} + \|u - u_k\|_X \to 0 \quad (k \to +\infty)$$

and similarly

$$\|v - v_k\|_{X_1} + \|v - v_k\|_{X_2} + \|v - v_k\|_X \to 0 \quad (k \to +\infty).$$

Since we know by hypothesis that for all u and v in X_0,

$$\|T(u) - T(v)\|_{X_1} \leq \|u - v\|_{X_1}, \quad \|T(u) - T(v)\|_{X_2} \leq \|u - v\|_{X_2},$$

it follows that $T(u_k)$ converges to $T(u)$ in both the X_1-norm and the X_2-norm and similarly $T(v_k)$ converges to $T(v)$ in both the X_1-norm and the X_2-norm. Hence $T(u_k) - T(v_k)$ converges to $T(u) - T(v)$ in both the X_1- and X_2-norms while for each k, $\|T(u_k) - T(v_k)\|_X \leq \|u_k - v_k\|_X$. Since $\|u_k - v_k\|_X \to \|u - v\|_X$ as $k \to +\infty$, it follows from condition (1) of the hypothesis of the theorem that $\|T(u) - T(v)\|_X \leq \|u - v\|_X$. Thus nonexpansiveness of T in X-norm on X_0 follows from the same fact on X'.

Let u and v be two elements of X'. Each of these elements lies in one of the spaces F_α of our given family. Since the family $\{F_\alpha\}$ is a directed set under inclusion, it follows that the two elements u and v lie in a common subspace F_α of the family. We fix such an α, and we assume throughout the rest of the proof of the present theorem that we are concerned only with elements of the given F_α. Let n be the dimension of the given F_α.

Let ψ be a nonnegative C^∞ real-valued function on F_α with compact support such that $\int_{F_\alpha} \psi(x) \, dx = 1$ where the integration is taken with respect to Lebesgue n-measure on F_α. For each $\epsilon > 0$, we consider the function g_ϵ from F_α to X_0 given by

$$g_\epsilon(x) = \int_{F_\alpha} \psi(\epsilon^{-1}[x - y])\epsilon^{-n} T(y) \, dy \qquad (x \in F_\alpha)$$

as well as the corresponding mapping f_ϵ of X_0 into X_0 given by $f_\epsilon(u) = g_\epsilon(P_\alpha u)$ ($u \in X_0$). Since T is continuous in the X_1- and X_2-norms on X_0, for each fixed $\epsilon > 0$, g_ϵ is strongly continuously once-differentiable from F_α to X_0 with respect to both the X_1- and X_2-norms on X_0. Since P_α is continuous in both norms and linear, it follows that for each fixed $\epsilon > 0$, f_ϵ is a strongly continuously once-differentiable mapping from X_0 to X_0 with respect to both the X_1-norm and the X_2-norm.

If we set

$$f'_\epsilon(x)[y] = \lim_{t \to 0+} (t^{-1}[f_\epsilon(x + ty) - f_\epsilon(x)]) \qquad (x, y \in X_0),$$

then for each u and v in X_0, we have

$$f_\epsilon(u) - f_\epsilon(v) = \int_0^1 f'_\epsilon(v + t(u - v))[u - v] \, dt.$$

We note that

$$f_\epsilon(x + ty) - f_\epsilon(x) = g_\epsilon(P_\alpha x + tP_\alpha y) - g_\epsilon(P_\alpha(x))$$
$$= \int_{F_\alpha} [T(P_\alpha x + tP_\alpha y - z) - T(P_\alpha x - z)]\epsilon^{-n}\psi(\epsilon^{-1}z) \, dz.$$

Hence, for $j = 1$ and 2,

$$\|f_\epsilon(x + ty) - f_\epsilon(x)\|_{X_j} \leq \int_{F_\alpha} \|T(P_\alpha x + tP_\alpha y - z) - T(P_\alpha x - z)\|_{X_j} \epsilon^{-n} \psi(\epsilon^{-1}z) \, dz$$
$$\leq t \|P_\alpha y\|_{X_j} \int_{F_\alpha} \epsilon^{-n} \psi(\epsilon^{-1}z) \, dz \leq t \|y\|_{X_j}.$$

In particular, it follows that

$$\|f'_\epsilon(x)[y]\|_{X_j} = \lim_{t \to 0+} t^{-1} \|f_\epsilon(x + ty) - f_\epsilon(x)\|_{X_j} \leq \|y\|_{X_j},$$

i.e., the linear mapping L given by $L(y) = f'_\epsilon(x)[y]$ of X_j into X_j for $j = 1$ and 2, for each $\epsilon > 0$ and each x in X_0, is a mapping of norm ≤ 1 with respect to both the X_1-norm and the X_2-norm. Since the triple (X, X_1, X_2) is a linear interpolation system in the sense of Definition (11.1), it follows that for each x and y in X_0, $\|f'_\epsilon(x)[y]\|_X \leq \|y\|_X$ for each $\epsilon > 0$. Thus, by the equation above, for all u and v in X_0,

$$\|f_\epsilon(u) - f_\epsilon(v)\|_X \leq \int_0^1 \|f'_\epsilon(v + t(u-v))[u-v]\|_X \, dt \leq \|u - v\|_X,$$

i.e., f_ϵ is nonexpansive in the X-norm on X_0. For u and v in F_α, however, $f_\epsilon(u) = g_\epsilon(u)$ and $f_\epsilon(v) = g_\epsilon(v)$. Since T is continuous from F_α to X_0 in both the X_1-norm and the X_2-norm, it follows by a standard argument on convergence of regularizations that as $\epsilon \to 0$, $g_\epsilon(u)$ converges strongly in both the X_1-norm and X_2-norm to $T(u)$. Similarly $g_\epsilon(v)$ for each v in F_α converges strongly as $\epsilon \to 0$ in both the X_1-norm and the X_2-norm to $T(v)$. Hence $g_\epsilon(u) - g_\epsilon(v)$ converges to $T(u) - T(v)$ in both the X_1- and X_2-norms while $\|g_\epsilon(u) - g_\epsilon(v)\|_X \leq \|u - v\|_X$ for all $\epsilon > 0$. Hence by condition (1) in the hypothesis of our theorem, we have

$$\|T(u) - T(v)\|_X \leq \|u - v\|_X. \quad \text{q.e.d.}$$

THEOREM (11.2). *Suppose that the Banach space X is reflexive. Then condition (1) of the hypothesis of Theorem (11.1) follows from the definition of a linear interpolation system.*

PROOF OF THEOREM (11.2). Suppose that $\{u_k\}$ is an infinite sequence in X_0 converging to an element u of X_0 in both the X_1- and X_2-norms with $\|u_k\|_X \leq R$. If we pass to an infinite subsequence (which we identify with the original sequence), we may assume that u_k converges weakly in X to some element v of X. However, X, X_1, and X_2 are all assumed to be imbedded continuously in some separated locally convex topological vector space E. Hence if u_k converges strongly to u in X_1 and weakly to v in X, it follows that $v = u$. Therefore,

$$\|u\|_X \leq \liminf \|u_k\|_X \leq R. \quad \text{q.e.d.}$$

THEOREM (11.3). *Let (X, H_1, H_2) be a linear interpolation system consisting of a reflexive Banach space X and two Hilbert spaces H_1 and H_2 such that $H_1 \subset X \subset H_2$ in the sense that the injection mappings are continuous and have dense images while the injection mapping of H_1 into H_2 is compact.*

Let U be a mapping of H_2 into H_2 such that $U(H_1) \subset H_1$, while U is nonexpansive in both the H_1- and H_2-norms.

Then U maps X into X and is nonexpansive in the X-norm.

PROOF OF THEOREM (11.3). Since X is reflexive, it suffices by Theorem (11.2) to verify condition (2) of the hypothesis of Theorem (11.1) on the existence of the directed set $\{F_\alpha\}$ of finite-dimensional subspaces of H_1 and of the corresponding projections $\{P_\alpha\}$.

Since the injection mapping of H_1 into H_2 is compact and has dense image in H_2, we may define a compact selfadjoint linear mapping L of H_1 into H_1 such that for each pair of elements u and v of H_1, we have $(u, L(v))_{H_1} = (u, v)_{H_2}$. Since $(L(u), u)_{H_1} = \|u\|_{H_2}^2 > 0$ $(u \neq 0)$ and L is compact and selfadjoint, there exists a complete orthonormal family of eigenfunctions $\{h_j\}$ of L in H_1 such that $L(h_j) = \lambda_j h_j$. This family of eigenfunctions is automatically countable, so we can define an increasing sequence of finite-dimensional subspaces $\{F_n\}$ of H_1 whose union is dense in both H_1 and H_2 by letting F_n be the span of the first n-eigenfunctions ordered by decreasing eigenvalues λ_j.

We define the corresponding projection mapping P_n by letting P_n be the orthogonal projection mapping upon F_n in H_1, i.e.,

$$P_n(u) = \sum_{j=1}^{n} (u, h_j)_{H_1} h_j = \sum_{j=1}^{n} \lambda_j^{-1}(u, h_j)_{H_2} h_j.$$

The second definition can be applied to every u in H_2 and yields a projection mapping (again denoted by P_n) of the larger space H_2 upon F_n. It is obvious that the norm of P_n as a mapping in H_1 is equal to 1. We must verify that the same is true for P_n as a linear self-mapping of H_2.

However,

$$(h_j, h_k)_{H_2} = (h_j, L(h_k))_{H_1} = \lambda_k (h_j, h_k)_{H_1} = \lambda_k \delta_k^j \qquad (\delta_k^j = \text{the Kronecker delta}).$$

In particular $\{\lambda_j^{-1/2} h_j = f_j\}$ is a complete orthonormal set in H_2 and the orthogonal projection on F_n in H_2 is given by

$$Q_n(u) = \sum_{j=1}^{n} (u, f_j)_{H_2} f_j = \sum_{j=1}^{n} \lambda_j^{-1}(u, h_j)_{H_2} h_j = \sum_{j=1}^{n} \lambda_j^{-1}(u, L(h_j))_{H_1} h_j = P_n(u).$$

Hence P_n has norm 1 as a bounded linear mapping of H_2. q.e.d.

THEOREM (11.4). *Let (X, X_1, X_2) be a linear interpolation system of Banach spaces with X uniformly convex and such that the triple satisfies condition (2) of the hypothesis of Theorem (11.1). Let C be a nonempty closed bounded convex subset of X, and such that U is a continuous mapping of X into X which maps $X \cap X_1 \cap X_2$ into itself and maps the boundary of C in X into C. Suppose further that U is nonexpansive on $X \cap X_1 \cap X_2$ with respect to both the X_1-norm and the X_2-norm.*

Then:

(a) *U is a nonexpansive self-mapping of X with respect to the X-norm.*
(b) *U has a fixed point in C.*

PROOF OF THEOREM (11.4). PROOF OF (a). By hypothesis, U is a nonexpansive mapping on $X_0 = X \cap X_1 \cap X_2$ with respect to both the X_1- and X_2-norms. Since X is uniformly convex and hence reflexive, it follows from Theorem (11.2) that condition (1) of the hypothesis of Theorem (11.1) holds for the triple (X, X_1, X_2). By the hypothesis of the present theorem, condition (2) of the hypothesis of Theorem (11.1) holds for the triple (X, X_1, X_2).

Hence by Theorem (11.1) U is a nonexpansive self-mapping of X_0 with respect to the X-norm. Since U is a continuous self-mapping of X and X_0 is assumed to be dense in X, it follows immediately that U is a nonexpansive self-mapping of the Banach space X. q.e.d.

PROOF of (b). By our hypothesis, U maps the boundary of the closed bounded convex subset C of X into C. X is assumed to be uniformly convex, and by the result of part (a), U is a nonexpansive self-mapping of X. Hence if we apply Theorem (8.6), we see that U has a fixed point in C. q.e.d.

THEOREM (11.5). *Let (X, X_1, X_2) be a linear interpolation system of Banach spaces satisfying condition (2) of the hypothesis of Theorem (11.1) (i.e., having a system of finite-dimensional subspaces $\{F_\alpha\}$ and projections $\{P_\alpha\}$ satisfying that condition). Suppose that X^*, X_1^*, and X_2^* are all uniformly convex, and that $X_1 \subset X \subset X_2$, where each injection mapping is continuous and has dense image.*

Let T_2 be a hypermaximal accretive mapping in X_2 whose domain $D(T_2)$ is dense in X_2, and let T and T_1 be the restrictions of T_2 to X and X_1, respectively, i.e.,

$$D(T) = \{u \mid u \in X \cap D(T_2), T_2(u) \in X\},$$
$$D(T_1) = \{u \mid u \in X_1 \cap D(T_2), T_2(u) \in X_1\},$$

with

$$T(u) = T_2(u) \text{ for } u \in D(T), \quad T_1(u) = T_2(u) \text{ for } u \in D(T_1).$$

Suppose that T_1 is hypermaximal accretive in X_1 with $D(T_1)$ dense in X_1.
Then:

(a) *T is hypermaximal accretive in X.*

(b) *$(-T)$ is the infinitesimal generator of a strongly continuous one-parameter semigroup $\{U(t): t \geq 0\}$ of nonexpansive self-mappings of X. $(-T_2)$ is the infinitesimal generator of a strongly continuous one-parameter semigroup $\{V(t): t \geq 0\}$ on X_2. For each $t \geq 0$, $V(t)$ maps X into X and its restriction to X coincides with $U(t)$.*

PROOF OF THEOREM (11.5). PROOF of (a). If X is a Banach space with X^* uniformly convex, then it follows from Theorem (9.1) that a mapping T with domain $D(T)$ in X and mapping $D(T)$ into X is hypermaximal accretive in X if and only if for each $\lambda > 0$, $(\lambda T + I)^{-1}$ is a well-defined nonexpansive mapping from X to X (i.e., $(\lambda T + I)$ is an injective mapping of X onto X with a nonexpansive inverse).

By hypothesis, T_2 and its restriction T_1 to X_1 are hypermaximal in X_2 and X_1, respectively. Hence for $\lambda > 0$, $(\lambda T_2 + I)^{-1}$ is a well-defined nonexpansive mapping of X_2 into X_2, and $(\lambda T_1 + I)^{-1}$ is a well-defined nonexpansive mapping of X_1 into X_1. Since $(\lambda T_1 + I)$ is a restriction of $(\lambda T_2 + I)$ and since $(\lambda T_2 + I)^{-1}$ is well defined, it follows that $(\lambda T_1 + I)^{-1}$ is the restriction to X_1 of $(\lambda T_2 + I)^{-1}$, and the latter mapping carries X_1 into X_1 and is a nonexpansive mapping in the X_1-norm. Hence, it follows from Theorems (11.1) and (11.2) that $(\lambda T_2 + I)^{-1}$ maps X into X and is nonexpansive in the X-norm.

However, by the definition of the restriction T of T_2 to X, it follows that $(\lambda T + I)^{-1}$ maps X into $D(T)$, while on $D(T)$, $(\lambda T + I)$ coincides with $(\lambda T_2 + I)$. Hence the restriction of $(\lambda T_2 + I)^{-1}$ to X coincides with $(\lambda T + I)^{-1}$, where the latter mapping is well defined and is a nonexpansive mapping of X into X. Since this is true for each $\lambda > 0$, it follows from Theorem (9.1) that T is hypermaximal accretive in X. q.e.d.

PROOF of (b). Since T and T_2 are hypermaximal accretive in X and X_2, respectively, it follows from Theorem (9.15) and Theorem (9.18) that $(-T)$ and $(-T_2)$ are the infinitesimal generators of strongly continuous semigroups $\{U(t) : t \geq 0\}$ and $\{V(t) : t \geq 0\}$ of nonexpansive self-mappings of X and X_2, respectively. (We note that the domain $D(T_2)$ is assumed to be dense in X_2, while since $D(T_1) \subset D(T)$ and $D(T_1)$ is assumed dense in X_1 where X_1 has a continuous injection on a dense subset of X, $D(T)$ is thereby dense in X.) It remains to show that for each $t > 0$, $V(t)$ maps X into X and that its restriction to X coincides with $U(t)$.

For each v in $D(T)$, $u(t) = U(t)(v)$ satisfies the differential equation $du(t)/dt = -T(u(t))$, $t \geq 0$, which coincides with the equation $du(t)/dt = -T_2(u(t))$, $t \geq 0$, since T is the restriction of T_2 to X. Hence $U(t)(v) = V(t)v$ for all v in the dense subset $D(T)$ of X. For a sequence $\{v_k\}$ in $D(T)$ converging to v in X, we have $U(t)(v_k)$ converging to $U(t)(v)$ in X and $V(t)(v_k)$ converging to $V(t)(v)$ in X_2. Since the topologies are compatible, it follows that $V(t)(v) = U(t)(v)$ for all v in X. q.e.d.

THEOREM (11.6). *Let (X, H_1, H_2) be a linear interpolation system where H_1 and H_2 are Hilbert spaces, X is a Banach space with X^* uniformly convex, and $H_1 \subset X \subset H_2$ with the injection mappings continuous and having dense image while the injection of H_1 in H_2 is compact. Let T_2 be a maximal monotone single-valued mapping in H_2 and let T and T_1 be the restrictions of T_2 to X and H_1 respectively. Suppose that $D(T_2)$ is dense in H_2 and that T_1 is maximal monotone and has dense domain in H_1.*

Then:

(a) T is hypermaximal accretive in X.

(b) $(-T)$ is the infinitesimal generator of a strongly continuous one parameter semigroup $\{U(t) : t \geq 0\}$ of nonexpansive self-mappings of X. $(-T_2)$ is the infinitesimal generator of a strongly continuous one-parameter semigroup $\{V(t) : t \geq 0\}$ of nonexpansive self-mappings of H_2. For each $t > 0$, $V(t)$ maps X into X and the restriction of $V(t)$ to X coincides with $U(t)$.

PROOF OF THEOREM (11.6). By hypothesis, T_2 is a maximal monotone mapping in H_2. Hence by Theorem (7.2) $(T_2 + I)$ maps onto H_2, and therefore T_2 is a hypermaximal accretive mapping in H_2. Similarly, T_1 is a hypermaximal accretive mapping in H_1. X is reflexive since X^* is uniformly convex and reflexive. Hence by Theorem (11.3), the triple (X, H_1, H_2) satisfies condition (2) of Theorem (11.1). Since $D(T_2)$ is dense in H_2 and $D(T_1)$ is dense in H_1, we may apply Theorem (11.5) and obtain the conclusion of the present theorem. q.e.d.

12. Generalizations of the topological degree of a mapping.

Let X and Y be topological spaces, G an open subset of X whose closure in X we denote by cl (G) and whose boundary in X, we denote by bdry (G). If f is a continuous mapping of cl (G) into Y and y is a point of Y which does not lie in the image of bdry (G) under f, then the degree of the mapping f on G over y is supposed to be (whenever it is defined) an algebraic count of the number of times that the mapping f assumes the value y at points x of G.

For the classical case in which X and Y are oriented Euclidean spaces R^n of the same finite dimension, the concept of the topological degree was first introduced explicitly by Brouwer in 1912. Its classical extension to an infinite-dimensional context was carried through by Leray and Schauder in 1934, who extended the Brouwer degree to the case in which $X = Y$ is a Banach space and the mapping f is of the form $f = I - C$, with I the identity mapping and C a compact mapping (i.e., with $C(\text{cl}(G))$ relatively compact in X).

It is our purpose in the present section to develop several extensions of the topological degree in much more general contexts for mappings f in Banach spaces which need not be of the form $f = I - C$. In the following three sections, we shall show how this extended degree can be applied to obtain a variety of interesting fixed point and mapping theorems which lie outside of the framework of the classical theory of compact nonlinear mappings in a Banach space X.

In our discussion, we shall apply the basic properties of the Leray-Schauder degree which we state immediately below. We do not give an independent derivation of these properties here, primarily because of the unreasonable addition to the length of the present discussion which such a treatment would impose. In addition, we may refer to the very precise and complete exposition of the technical details of the classical theory given in the paper by Nagumo [354]. Let us merely note that the crucial feature of the technical discussion of the theory of the Leray-Schauder degree for mappings of the form $f = I - C$ with C compact is that each compact mapping C can be approximated uniformly on cl (G) by a sequence of mappings $\{C_n\}$ with finite-dimensional range (and indeed, that this characterizes the class of compact mappings C). If we denote by $\deg_B$ the Brouwer degree, then the Leray-Schauder degree of $f = I - C$ which we write as $\deg_{LS}(f, G, y)$ is defined to be the common value for all sufficiently large n of $\deg_B(f_n, G \cap X_n, y)$ where X_n is a finite-dimensional subspace of X for each n which contains the point y as well as the image of cl (G) under the mapping C_n, and f_n is the mapping $I - C_n$ restricted to cl $(G) \cap X_n$.

To focus on the properties of the generalized degree functions which we shall define below, we consider the following basic properties of the Leray-Schauder degree which will be carried over under the extensions:

(1) *Let X be a Banach space, G an open subset of X, f a continuous mapping of* cl (G) *into X of the form $f = I - C$, with I the identity mapping and C compact. Let y be a point of $X - f(\text{bdry}(G))$. Then an integer* $\deg_{LS}(f, G, y)$ *is defined (where this integer may be positive, negative, or zero). If* $\deg_{LS}(f, G, y) \neq 0$, *then there exists at least one point x in G such that $f(x) = y$.*

(2) *The degree is additive in the open set G, i.e., if $G = G_1 \cup G_2$ with G_1 and G_2 open subsets of X and if y is a point of $X - f(G')$, where*

$$G' = (G_1 \cap G_2) \cup \mathrm{bdry}\,(G_1) \cup \mathrm{bdry}\,(G_2),$$

then

$$\deg_{\mathrm{LS}} (f, G, y) = \deg_{\mathrm{LS}} (f|_{G_1}, G_1, y) + \deg_{\mathrm{LS}} (f|_{G_2}, G_2, y).$$

(3) *The degree is invariant under homotopies in which it remains defined, i.e., if W is an open subset of $X \times [0, 1]$, C a continuous mapping of $\mathrm{cl}\,(W)$ into a relatively compact subset of X, and if for each t in $[0, 1]$, we let*

$$G_t = \{u \mid u \in X, [u, t] \in W\},$$

and

$$C_t : \mathrm{cl}\,(G_t) \to X, \qquad C_t(u) = C(u, t) \qquad \textit{for } u \in \mathrm{cl}\,(G_t),$$

$$f_t = I - C_t,$$

then for any continuous curve

$$P = \{y(t) : 0 \leq t \leq 1\}$$

in X such that for each t in $[0, 1]$, $y(t)$ does not lie in $f_t(\mathrm{bdry}\,(G_t))$, we have $\deg_{\mathrm{LS}} (f_t, G_t, y(t))$ well defined for t in $[0, 1]$ and independent of t in $[0, 1]$.

(4) *If f is a homeomorphism of G into X of the form $f = I - C$, with C a compact mapping of $\mathrm{cl}\,(G)$ into X, then for any point y in $f(G) - f(\mathrm{bdry}\,(G))$,*

$$\deg_{\mathrm{LS}} (f, G, y) = \pm 1.$$

(5) *The degree function, $\deg_{\mathrm{LS}} (f, G, y)$, depends only upon the behavior of f on the boundary of G.*

A sixth property of the Leray-Schauder degree, the multiplicativity property, which does not carry over automatically in our extensions, is given in detail below during our later discussion.

Our principal objective in the construction of an extended degree theory is to extend the concept of the degree to mappings f of a more general type than $f = I - C$. The simplest example of the more general type of mapping f which we shall consider is that of the form $f = h - C$, with h a homeomorphism and C compact (or more generally, with the homeomorphism h and the mapping C intertwined in a fashion which we describe precisely below). More generally still, we shall consider mappings which are locally of this form, or finally, we shall consider locally uniform limits of sequences of these latter mappings (and thereby pass to the case of degenerate homeomorphisms h).

Since the concept of homeomorphism is obviously a central point in this discussion, we begin with a precise formulation of what we shall mean by a homeomorphism in the following argument:

DEFINITION (12.1). *Let X and Y be topological spaces, G an open subset of X, h a continuous mapping of $\mathrm{cl}\,(G)$ into Y. Then h is said to be a permissible homeomorphism of $\mathrm{cl}\,(G)$ into Y if h is a homeomorphism of G on an open subset $h(G)$ of Y*

such that h maps cl (G) homeomorphically onto cl $(h(G))$ (where the latter denotes the closure of $h(G)$ in Y).

In the treatment given below, we give several definitions of degree for mappings f having representations of various forms with respect to a given family M of homeomorphisms from open sets in X to Y. One crucial point in every case is that we shall not (at least initially) define these degrees as integer-valued functionals of mappings but rather of *representations of mappings*. To put this point more sharply, we shall take up the viewpoint that the generalized degree should be considered in the most general context as a quantity defined for a representation of the mapping f within a given class of representations, and not initially as being defined in terms of the mapping f itself. It then becomes the objective of the degree theory to prove *theorems* concerning each of these generalized degree functions that under various general hypotheses the degree and its properties depend only on the mapping f and its properties, and not upon the representations involved.

A basic piece of data in each of the subsequent definitions of the generalized degree function is the following:

DEFINITION (12.2). *Let X and Y be topological spaces. Then by a class M of homeomorphisms from X to Y, we mean a family of mappings h where each h is a permissible homeomorphism of cl (G) into Y for an open subset G of X. We assume that for each open subset G_1 of G, the restriction of h to cl (G_1) is also an element of M.*

DEFINITION (12.3). *Let M be a class of homeomorphisms from X to Y in the sense of Definition (12.2). Then for each open subset G of X, we let M_G be the subfamily of all h in M defined on cl (G). If Y is a metric space, each M_G can be considered as a pseudo-metric space with pseudo-metric given by*

$$d(h, h_1) = \sup_{x \in \text{cl}(G)} d(h(x), h_1(x)),$$

where the function d on the right is the given metric on Y. If Y is assumed to be a bounded metric space, the above pseudo-metric becomes an ordinary metric. If Y is a (possibly) unbounded metric space, the topology of uniform convergence in M_G is that given by a bounded metric d_0 on Y which induces an equivalent metric to d on bounded subsets of Y.

DEFINITION (12.4). *Let X and Y be topological spaces, with Y a topological vector space, and let M be a class of homeomorphisms from X to Y in the sense of Definition (12.2). Then M is said to be convex if for each open subset G of X and any pair of elements h_0 and h_1 in M_G, the mapping h_λ of cl (G) into Y given for λ in $[0, 1]$ by*

$$h_\lambda(x) = (1 - \lambda)h_0(x) + \lambda h_1(x) \qquad (x \in \text{cl}(G))$$

also lies in M_G.

For the narrowest class of extended representations of mappings f, we give the following first definition of the generalized degree:

DEFINITION (12.5). FIRST DEFINITION OF THE GENERALIZED DEGREE. *Let X and Y be Banach spaces, M a class of permissible homeomorphisms from X to Y in the sense of Definition (12.2). Suppose that G is an open subset of X, f is a continuous* mapping of cl (G) into Y such that $f = h - C$, where h is a homeomorphism of cl (G) into Y lying in the class M and C maps cl (G) into a relatively compact subset of Y. Let y be a point of $Y - f(\text{bdry}(G))$. Then, we set

$$\deg_1([f, h], G, y) = \deg_{LS}(fh^{-1}, h(G), y)$$

where the degree on the right is taken for the mapping fh^{-1} of the open subset $h(G)$ of Y into Y.

THEOREM (12.1). *Let X and Y be Banach spaces, M a class of homeomorphisms from X to Y in the sense of Definition (12.2). Let f be a continuous mapping of cl (G) into Y for an open subset G of X with f having a representation of the form $f = h - C$, with h a homeomorphism of cl (G) into Y lying in the class M and C compact. Then*

(a) *The degree $\deg_1([f, h], G, y)$ is well defined for each y in $Y - f(\text{bdry}(G))$.*

(b) *The degree function $\deg_1([f, h], G, y)$ is independent of representation of f in the form $f = h - C$ up to a change of sign for G connected.*

(c) *In the general case, $\deg_1$ has the following properties:*

(1) *If $\deg_1([f, h], G, y)$ is well defined and nonzero, then there exists x in G such that $f(x) = y$.*

(2) *The degree is additive in G in the sense that if $G = G_1 \cup G_2$ with G_1 and G_2 open sets in X and if $\deg_1([f, h], G, y)$ is well defined while y lies in $Y - f(G')$ for $G' = (G_1 \cap G_2) \cup \text{bdry}(G_1) \cup \text{bdry}(G_2)$, then*

$$\deg_1([f, h], G, y) = \deg_1([f|_{G_1}, h|_{G_1}], G_1, y) + \deg_2([f|_{G_2}, h|_{G_2}], G_2, y).$$

(3) *The degree function $\deg_1$ is invariant under permissible homotopies of representations in the following sense: Let G be an open subset of X, f a continuous mapping of cl $(G) \times [0, 1]$ into Y, and for each t in $[0, 1]$, let f_t be the mapping of cl (G) into Y given by $f_t(x) = f(x, t)$ ($x \in \text{cl}(G), t \in [0, 1]$). Suppose that there exist a continuous mapping C of cl $(G) \times [0, 1]$ into a relatively compact subset of Y and a continuous function h from $[0, 1]$ to M_G (with the topology of uniform convergence on cl (G)) such that for each t in $[0, 1]$, $f_t = h_t - C_t$ with $C_t(x) = C(x, t)$ ($x \in \text{cl}(G), t \in [0, 1]$). Let $P = \{y(t): t \in [0, 1]\}$ be a continuous curve in Y such that for each t in $[0, 1]$, $y(t) \in Y - f_t(\text{bdry}(G))$. Then $\deg_1([f_t, h_t], G, y(t))$ is well defined and independent of t in $[0, 1]$.*

(4) *If $\deg_1([f, h], G, y)$ is well defined and f is a homeomorphism of G on an open subset of Y such that y lies in $f(G)$, then $\deg_1([f, h], G, y) = \pm 1$.*

(5) *The degree function $\deg_1([f, h], G, y)$ depends only upon the behavior of f and h on the boundary of G.*

REMARK. We note that by property (b) the degree function $\deg_1$ in particular would define an unambiguous degree function if taken mod 2.

The proof of Theorem (12.1) depends upon the basic properties of the Leray-Schauder degree as well as upon the multiplicativity property of the Leray-Schauder degree to which we have already referred above. This is the following:

(6) *Let X be a Banach space, G an open subset of X, f a continuous mapping of cl (G) into X of the form $f = I - C$, where $C(\text{cl}(G))$ is relatively compact in X.*

Let H be an open subset of X which contains $f(\mathrm{cl}\,(G))$ and consider the components $\{H_j\}$ of the open set $H - f(\mathrm{bdry}\,(G))$. Let g be a continuous mapping of $\mathrm{cl}\,(H)$ into X of the form $g = I - C_1$, with $C_1(\mathrm{cl}\,(H))$ relatively compact in X. Suppose finally that y is a point of X outside of $g(\mathrm{bdry}\,(H)) \cup gf(\mathrm{bdry}\,(G))$. Then

$$\deg_{\mathrm{LS}} (gf, G, y) = \sum_j \deg_{\mathrm{LS}} (g, H_j, y) \deg_{\mathrm{LS}} (f, G, z_j),$$

where for each j, z_j is an arbitrary point of H_j.

We use the product formula for the Leray-Schauder degree given in this property (6) to obtain the following auxiliary result to be applied in the proof of Theorem (12.1):

PROPOSITION (12.1). *Let X be a Banach space, G a connected open subset of X, f a continuous mapping of $\mathrm{cl}\,(G)$ into a second Banach space Y. Suppose that f can be represented in two ways in the form $f = h - C = h_1 - C_1$ where h and h_1 are permissible homeomorphisms of $\mathrm{cl}\,(G)$ into Y and C and C_1 are compact mappings of $\mathrm{cl}\,(G)$ into Y. Then:*

(a) $|\deg_{\mathrm{LS}} (I - Ch^{-1}, h(G), y)| = |\deg_{\mathrm{LS}} (I - C_1 h_1^{-1}, h_1(G), y)|$ *for any y in* $Y - f(\mathrm{bdry}\,(G))$.

(b) *For any y in $Y - f(\mathrm{bdry}\,(G))$, $\deg_1 ([f, h], G, y)$ is well defined and $|\deg_1 ([f, h], G, y)|$ is independent of the representation of f in the form $f = h - C$.*

PROOF OF PROPOSITION (12.1). By hypothesis $h - C = h_1 - C_1$. Hence on $h(\mathrm{cl}\,(G)) = \mathrm{cl}\,(h(G))$, $I - Ch^{-1} = h_1 h^{-1} - C_1 h^{-1}$, i.e., $h_1 h^{-1} = I - Ch^{-1} + C_1 h^{-1} = I - C_2$ where $-C_2 = C_1 h^{-1} - Ch^{-1}$ is a compact mapping of $\mathrm{cl}\,(h(G))$ into Y. Since $h_1 h^{-1}$ is a homeomorphism of $h(G)$ onto $h_1(G)$, it follows from property (4) of the Leray-Schauder degree that $\deg_{\mathrm{LS}} (h_1 h^{-1}, h(G), z)$ is well defined and equal to ± 1 for any z in $h_1(G)$. On the other hand, we have

$$(I - Ch^{-1}) = fh^{-1} = (I - C_1 h_1^{-1}) h_1 h^{-1} = (I - C_1 h_1^{-1})(h_1 h^{-1}).$$

By hypothesis, G is connected, and therefore the homeomorphs $h(G)$ and $h_1(G)$ of G are connected as well. Both h and h_1 are permissible homeomorphisms in the sense of Definition (12.1) and map G homeomorphically onto $h(G)$ and $h_1(G)$, respectively, and $\mathrm{cl}\,(G)$ homeomorphically onto $\mathrm{cl}\,(h(G))$ and $\mathrm{cl}\,(h_1(G))$, respectively. Hence h maps $\mathrm{bdry}\,(G)$ homeomorphically onto $\mathrm{bdry}\,(h(G))$ and h_1 maps $\mathrm{bdry}\,(G)$ homeomorphically onto $\mathrm{bdry}\,(h_1(G))$. It follows that $h_1 h^{-1}$ maps $\mathrm{bdry}\,(h(G))$ onto $\mathrm{bdry}\,(h_1(G))$, while $h_1(G)$ is connected and therefore is contained in a single component of $Y - (h_1 h^{-1})(\mathrm{bdry}\,(h(G))) = Y - \mathrm{bdry}\,(h_1(G))$. Indeed, $h_1(G)$ coincides with a single component of $Y - \mathrm{bdry}\,(h_1(G))$ since otherwise the component of $Y - \mathrm{bdry}\,(h_1(G))$ which contains $h_1(G)$ would contain a point of $\mathrm{bdry}\,(h_1(G))$.

We now apply the product formula of property (6) of the Leray-Schauder degree with $f = h_1 h^{-1}$ and $g = I - C_1 h_1^{-1}$. We then obtain

$$\deg_{\mathrm{LS}} (I - Ch^{-1}, h(G), y) = \deg_{\mathrm{LS}} (I - C_1 h_1^{-1}, h_1(G), y) \cdot \deg_{\mathrm{LS}} (h_1 h^{-1}, h(G), z)$$

for any point z in the single component of

$$Y - (h_1 h^{-1})(\text{bdry }(h(G))) = Y - \text{bdry }(h_1(G))$$

which contains points of $(h_1 h^{-1})(h(G)) = h_1(G)$. Since $\deg_{LS}(h_1 h^{-1}, h(G), z) = \pm 1$ for any point z in $h_1(G)$, it follows that

$$\deg_{LS}(I - Ch^{-1}, h(G), y) = \pm \deg_{LS}(I - C_1 h_1^{-1}, h_1(G), y),$$

and the proof of part (a) is complete.

To prove part (b), we note first of all that $fh^{-1} = (h - C)h^{-1} = I - Ch^{-1}$. If y lies in $Y - f(\text{bdry }(G))$, it follows that y lies in $Y - (fh^{-1})(\text{bdry }(h(G)))$ since h^{-1} maps bdry $(h(G))$ onto bdry (G). Hence

$$\deg_1([f, h], G, y) = \deg_{LS}(I - Ch^{-1}, h(G), y)$$

is well defined. By the result of part (a), it is independent of the representation of f in the form $f = h - C$ up to a change of sign. q.e.d.

PROOF OF THEOREM (12.1). PROOF OF (a). This follows from the conclusion of part (b) of Proposition (12.1).

PROOF OF (b). This also follows from the conclusion of part (b) of Proposition (12.1).

PROOF OF (c). We divide this proof into the separate proofs of each of the properties stated in part (c).

PROOF OF PROPERTY (1). Suppose that $f = h - C$ and that y lies in

$$Y - f(\text{bdry }(G)).$$

Then

$$\deg_1([f, h], G, y) = \deg_{LS}(fh^{-1}, h(G), y),$$

where $fh^{-1} = I - Ch^{-1}$. Hence if $\deg_1([f, h], G, y) \neq 0$, we have

$$\deg_{LS}(I - Ch^{-1}, h(G), y) \neq 0,$$

and it follows from property (1) of the Leray-Schauder degree that there exists a point y_0 of $h(G)$ such that $(fh^{-1})(y_0) = y$. Let $x = h^{-1}(y_0)$. Then x lies in G and $f(x) = fh^{-1}(y_0) = y$. q.e.d.

PROOF OF PROPERTY (2). Suppose that $G = G_1 \cup G_2$, where G_1 and G_2 are open subsets of G and suppose that y lies in $Y - f(G')$, with

$$G' = (G_1 \cap G_2) \cup \text{bdry }(G_1) \cup \text{bdry }(G_2).$$

If we set $g = fh^{-1}$, then it follows that y lies in $Y - g(G'')$, where

$$G'' = (h(G_1) \cap h(G_2)) \cup \text{bdry }(h(G_1)) \cup \text{bdry }(h(G_2)).$$

Since $g = I - Ch^{-1}$, we may apply the additivity property of the Leray-Schauder degree and obtain the equation

$$\deg_{LS}(g, h(G), y) = \deg_{LS}(g, h(G_1), y) + \deg_{LS}(g, h(G_2), y).$$

Translating this equation into the terms of the degree function $\deg_1$, we find that

$$\deg_1([f, h], G, y) = \deg_1([f, h], G_1, y) + \deg_1([f, h], G_2, y). \quad \text{q.e.d.}$$

PROOF OF PROPERTY (3). Let W be the subset of $Y \times [0, 1]$ given by

$$W = \{[u, t] \mid u \in Y, t \in [0, 1], u \in h_t(G)\}.$$

For each t in $[0, 1]$, we have

$$\deg_1([f_t, h_t], G, y(t)) = \deg_{LS}(f_t h_t^{-1}, h_t(G), y(t)) = \deg_{LS}(I - C_t h_t^{-1}, W_t, y(t))$$

with W_t the slice of W at the level t, i.e., $W_t = \{u \mid u \in Y, [u, t] \in W\}$. To show that $\deg_1([f_t, h_t], G, y(t))$ is well defined for each t in $[0, 1]$ and is independent of the choice of t in $[0, 1]$, it suffices to show by property (3) of the Leray-Schauder degree that the following facts all hold:

(i) W is an open subset of $Y \times [0, 1]$.
(ii) The boundary of W in $Y \times [0, 1]$ consists of the set

$$W' = \{[u, t] \mid u \in Y, t \in [0, 1], u \in h_t(\text{bdry}(G))\}.$$

(iii) The mapping S of cl (W) into Y given by $S(u, t) = C_t h_t^{-1}(u)$ is continuous and maps cl (W) into a relatively compact subset of Y.

We derive properties (i) and (ii) from the following proposition:

PROPOSITION (12.2). *Let X and Y be Banach spaces, G an open subset of X, Z a topological space, and consider a function $z \to h_z$ from Z to the family of permissible homeomorphisms of* cl (G) *into Y (in the sense of Definition (12.1)). Suppose that the mapping $z \to h_z$ is continuous from Z to the space of homeomorphisms with the latter given the topology of uniform convergence on* cl (G). *Consider the subsets W_0 of $Y \times Z$ given by*

$$W_0 = \{[u, z] \mid u \in Y, z \in Z, u \in h_z(G)\}.$$

Then:
(a) W_0 *is an open subset of $Y \times Z$.*
(b) *The boundary of W_0 in $Y \times Z$ is given by*

$$W_0' = \{[u, z] \mid u \in Y, z \in Z, u \in h_z(\text{bdry}(G))\}.$$

PROOF OF PROPOSITION (12.2). PROOF OF (a). Suppose that $[u_0, z_0]$ is a point of W_0, i.e., $u_0 \in h_{z_0}(G)$. Since h_{z_0} is a homeomorphism, there exists a unique point x_0 in G such that $u_0 = h_{z_0}(x_0)$. Since G is open there exists a closed ball $B(x_0)$ with center at x_0 in X which is contained in G. Let B_0 be the interior of this ball, so that $B(x_0) = $ cl (B_0). Since h_{z_0} is a permissible homeomorphism of cl (G) into Y in the sense of Definition (12.1), h_{z_0} maps G homeomorphically onto the open subset $h_{z_0}(G)$ of Y and cl (G) homeomorphically onto cl $(h_{z_0}(G))$. Hence if F is any subset of G whose closure in X is contained in G, $h_{z_0}(F)$ is a closed subset of Y. In particular, $h_{z_0}(B(x_0))$ is a closed subset of Y which contains the open set $h_{z_0}(B_0)$ of Y, and h_{z_0} maps $B(x_0)$ homeomorphically onto $h_{z_0}(B(x_0))$. Since cl $(B_0) = B(x_0)$, it follows that $h_{z_0}(B(x_0))$ is contained in the closure of $h_{z_0}(B_0)$ and hence coincides with this closure. Thus

$$h_{z_0}(\text{bdry }(B(x_0))) = \text{bdry }(h_{z_0}(B(x_0))).$$

By the last remark, u_0 lies in the open set $h_{z_0}(B_0)$ which is disjoint from $h_{z_0}(\text{bdry }(B(x_0)))$. Therefore, there exists a positive number $d > 0$ such that the ball $B_d(u_0)$ is disjoint from $h_{z_0}(\text{bdry }(B(x_0)))$. Since the mapping $z \to h_z$ is continuous from Z to the space of homeomorphisms from cl (G) to Y with the topology of uniform convergence imposed upon the latter space, there exists a neighborhood N_0 of z_0 in Z such that for z in N_0, $\|h_z(x) - h_{z_0}(x)\| < d/2$. It follows that for any z in N_0,

$$h_z(\text{bdry }(B(x_0))) \cap B_{d/2}(u_0) = \varnothing.$$

Moreover, $h_z(x_0)$ lies in $B_{d/2}(u_0)$ for all z in N_0.

Since the open ball B_0 is connected and h_z is continuous, $h_z(B_0)$ is a connected set for each z in N_0 and hence contained in a single component C_0 of the open set $Y - h_z(\text{bdry }(B(x_0))) = Y - \text{bdry }(h_z(B(x_0)))$. Moreover, $h_z(B_0)$ must coincide with the whole component C_0 since otherwise there would be a point y in C_0 which also lies in bdry $(h_z(B_0)) = \text{bdry }(h_z(B(x_0))) = h_z(\text{bdry }(B(x_0)))$, which is excluded by the fact that h_z is a homeomorphism on $B(x_0)$ and $C_0 \subset Y - \text{bdry }(h_z(B(x_0)))$. Since $B_{d/2}(u_0)$ is a connected set in $Y - h_z(\text{bdry }(B(x_0)))$ and contains at least one point $h_z(x_0)$ of $h_z(B_0)$ for each z in N_0, it follows that for z in N_0, $h_z(B_0)$ contains $B_{d/2}(u_0)$. In other words, $B_{d/2}(u_0) \times N_0 \subset W_0$, i.e., W_0 is an open subset of $Y \times Z$.

PROOF OF (b). We note first that each point $[u, z]$ of the set W_0' as defined in the hypothesis of the proposition does indeed lie in the boundary of W_0. Indeed, for such a point, there exists a sequence $\{x_n\}$ in G converging to a point x in bdry (G) such that $y_n = h_z(x_n)$ converges to u in Y. Then for each n, $[z, y_n]$ lie in W_0 so that $[z, u]$ lies in the closure of W_0. On the other hand, since h_z is a homeomorphism of cl (G) into Y, u does not lie in $h_z(G)$ and $[u, z]$ is not a point of W_0 itself. Therefore, $[u, z]$ is a point of bdry (W_0) and we have shown that $W_0' \subset$ bdry (W_0).

Suppose conversely that $[u, z]$ is a point of $Y \times Z$ which lies in the boundary of the set W_0. Then for each $\epsilon > 0$ and each neighborhood N of z, there exists a point $[u_\epsilon, z_\epsilon]$ of W_0 such that $\|u - u_\epsilon\| < \epsilon$, and z_ϵ lies in N. In particular, there exists a point x_ϵ of G such that $u_\epsilon = h_{z_\epsilon}(x_\epsilon)$. For each given $\epsilon > 0$, we choose N as a neighborhood of z so small that for all z' in N and all x of G, $\|h_z(x) - h_{z'}(x)\| < \epsilon$. Then we find that

$$\|h_z(x_\epsilon) - u\| \leq \|h_z(x_\epsilon) - h_{z_\epsilon}(x_\epsilon)\| + \|u_\epsilon - u\| < 2\epsilon.$$

Since $\epsilon > 0$ is arbitrary, it follows that u lies in bdry $(h_z(G))$. Since h_z is a permissible homeomorphism of cl (G) into Y, it follows that $h_z(\text{bdry }(G)) = \text{bdry }(h_z(G))$. Hence u lies in $h_z(\text{bdry }(G))$, and we know that $[u, z] \in W_0'$. Hence $W_0' = \text{bdry }(W_0)$.

q.e.d.

To prove property (iii) stated above in the proof of property (3), we apply the following further proposition:

PROPOSITION (12.3). *Let X and Y be Banach spaces, G an open subset of X, Z a topological space, and consider a function $z \to h_z$ from Z to the family of permissible homeomorphisms of $\mathrm{cl}\,(G)$ into Y. Suppose that the mapping $z \to h_z$ is continuous from Z to the space of homeomorphisms from $\mathrm{cl}\,(G)$ to Y. Let C be a continuous mapping of $\mathrm{cl}\,(G) \times Z$ into Y with relatively compact image in Y, and for each z in Z, let C_z be the mapping of $\mathrm{cl}\,(G)$ into Y given by $C_z(x) = C(x, z)$.*

Then if C' is the mapping of $\mathrm{cl}\,(W_0)$ into Y given by

$$C'([u, z]) = C(h_z^{-1}(u), z) \qquad ([u, z] \in \mathrm{cl}(W_0)),$$

where W_0 is the subset of $Y \times Z$ defined in Proposition (12.2), then C' is a continuous mapping of $\mathrm{cl}\,(W_0)$ into Y with relatively compact image in Y.

PROOF OF PROPOSITION (12.3). Suppose first that $\{[u_k, z_k]\}$ is a sequence in $\mathrm{cl}\,(W_0)$ converging to an element $[u, z]$ of $\mathrm{cl}\,(W_0)$. We wish to show that $C'([u_k, z_k])$ converges strongly in Y to $C'([u, z])$, i.e., that

$$C(h_z^{-1}(u), z) = \lim_k C(h_{z_k}^{-1}(u_k), z_k).$$

Since C is continuous from $\mathrm{cl}\,(G) \times Z$ into Y, it suffices to show that $h_{z_k}^{-1}(u_k) \to h_z^{-1}(u)$. Since the mapping $z \to h_z$ is continuous from Z to the space of homeomorphisms from $\mathrm{cl}\,(G)$ to Y with the topology of uniform convergence on $\mathrm{cl}\,(G)$, given $\epsilon > 0$, there exists an index $k(\epsilon)$ such that for $k > k(\epsilon)$ and all x in $\mathrm{cl}\,(G)$, $\|h_{z_k}(x) - h_z(x)\| < \epsilon$. If we set $x_k = h_{z_k}^{-1}(u_k)$, $x = h_z^{-1}(u)$, we have $h_{z_k}(x_k) = u_k$, $h_z(x) = u$, and for $k > k(\epsilon)$, $\|h_z(x_k) - u_k\| < \epsilon$. Since for $k > k_1(\epsilon)$ for a suitable $k_1(\epsilon)$, we have $\|u_k - u\| < \epsilon$, it follows that for $k > \max(k(\epsilon), k_1(\epsilon))$, $\|h_z(x_k) - u\| < 2\epsilon$. Since $\epsilon > 0$ is arbitrary, it follows that $h_z(x_k) \to u$ in Y as $k \to \infty$. Since h_z^{-1} is continuous on $\mathrm{cl}\,(h_z(G)) = h_z(\mathrm{cl}\,(G))$, it follows that $x_k \to h_z^{-1}(u) = x$. Hence the continuity of C' is established.

To complete the proof of Proposition (12.3), it suffices to prove that the image of the mapping C' is relatively compact in Y. However, the image of $\mathrm{cl}\,(W_0)$ under C' is contained by the definition of C' in the image under the mapping C of the set $\mathrm{cl}\,(G) \times Z$. By hypothesis, this latter image is relatively compact in Y.

q.e.d.

PROOF OF THEOREM (12.1) CONTINUED. We now complete the proof of property (3) for $\deg_1$. This follows from Propositions (12.2) and (12.3), since as we observed above, it suffices to establish the properties (i), (ii), and (iii), while (i) and (ii) follow from Proposition (12.2) and property (iii) from Proposition (12.3). Hence property (3) has been established.

PROOF OF PROPERTY (4) OF THE DEGREE FUNCTION $\deg_1$. Let G be an open subset of X, f a continuous mapping of $\mathrm{cl}\,(G)$ into Y where $f = h - C$ with h a permissible homeomorphism of $\mathrm{cl}\,(G)$ into Y and C a compact mapping of $\mathrm{cl}\,(G)$ into Y. Suppose that y is a point of $Y - f(\mathrm{bdry}\,(G))$ and that f is a homeomorphism of G on an open subset of Y with y lying in $f(G)$. Then $fh^{-1} = I - Ch^{-1}$ is a homeomorphism of $h(G)$ onto $f(G)$ with y lying in $f(G)$, and by property (4) of the Leray-Schauder degree $\deg_{\mathrm{LS}}(fh^{-1}, h(G), y) = \pm 1$. Therefore, $\deg_1([f, h], G, y) = \deg_{\mathrm{LS}}(fh^{-1}, h(G), y) = \pm 1$.

q.e.d.

PROOF OF PROPERTY (5) FOR THE DEGREE FUNCTION $\deg_1$. Since

$$\deg_1([f, h], G, y) = \deg_{LS}(fh^{-1}, h(G), y)$$

by its definition, it follows from property (5) of the Leray-Schauder degree that $\deg_1([f, h], G, y)$ depends only upon the behaviour of the mapping fh^{-1} on the boundary of $h(G)$. Since h maps bdry $h(G)$ into bdry (G), it follows that the behaviour of the mapping fh^{-1} on bdry $(h(G))$ is determined by the action of the mapping h^{-1} on bdry $(h(G))$ (i.e., the action of the mapping h on bdry (G)), and upon the action of the mapping f on bdry (G).   q.e.d.

With this last verification, the proof of Theorem (12.1) is now complete.
    q.e.d.

For this relatively simple extension of the degree function (as for the wider extensions we shall consider in the later discussion), we now observe in detail the great simplification of the basic results that follows if we assume that our permissible homeomorphism h is chosen from a convex family M of permissible homeomorphisms in the sense of Definition (12.4).

THEOREM (12.2). *Let X and Y be Banach spaces, M a convex family of permissible homeomorphisms from X to Y. Suppose that G is an open subset of X, f a continuous mapping of* cl (G) *into Y which is representable in the form $f = h - C$ with h an element of M and C a compact mapping of* cl (G) *into Y. Let y be a point of $Y - f$(bdry (G)). Then:*

(a) *The degree $\deg_1([f, h], G, y)$ is independent of the choice of the representation of f in the form $f = h - C$ with h in M, and we may write its value as $\deg_{1, M}(f, G, y)$.*

(b) *In this case, the degree function $\deg_{1, M}$ has the following property in addition to those given in Theorem (12.1). Let G be an open subset of X and for each t in $[0, 1]$, let f_t be a mapping of* cl (G) *into Y with f_t representable in the form $f_t = h_t - C_t$ for h_t a homeomorphism of* cl (G) *into Y which is an element of M and C_t a compact mapping of* cl (G) *into Y. Suppose that the mapping of $[0, 1]$ into the space of mappings of* cl (G) *into Y given by $t \to f_t$ is continuous (with the space of mappings given the topology of uniform convergence on* cl (G)*). Suppose that we have a continuous curve $\{y(t), 0 \le t \le 1\}$ in Y with $y(t)$ lying in $Y - f_t$(bdry (G)) for each t in $[0, 1]$.*

Then $\deg_{1, M}(f_t, G, y(t))$ is independent of t in $[0, 1]$.

(c) *The degree function $\deg_{1, M}(f, G, y)$ is dependent only upon the behaviour of f on the boundary of G.*

PROOF OF THEOREM (12.2). PROOF OF (a). Suppose that a given mapping f of cl (G) into Y has two representations, $f = h - C = h_1 - C_1$ with h and h_1 in M and C, C_1 compact mappings of cl (G) into Y. We set

$$h_t = (1 - t)h + th_1, \qquad C_t = (1 - t)C + tC_1$$

for each t in $[0, 1]$. Since h and h_1 are assumed to lie in the convex family M of homeomorphisms, $\{h_t\}$ is a continuous family of permissible homeomorphisms of cl (G) into Y for t in $[0, 1]$, and the family $\{C_t\}$ yields a continuous mapping of cl $(G) \times [0, 1]$ into Y with relatively compact image in Y. Hence we may apply property (3) of Theorem (12.1) to the family $f = h_t - C_t$ of representations varying continuously with t in $[0, 1]$. We obtain the conclusion that $\deg_1 ([f, h_t], G, y)$ is independent of t in $[0, 1]$, and in particular that

$$\deg_1 ([f, h], G, y) = \deg_1 ([f, h_1], G, y). \quad \text{q.e.d.}$$

PROOF OF (b). To prove that $\deg_{1,M} (f_t, G, y(t))$ is independent of t in $[0, 1]$, it suffices to show that for each t_0 in $[0, 1]$, this function is constant for t near t_0. Hence we may assume without loss of generality that for a prescribed constant $\epsilon > 0$, we have $\|f_t(x) - f_0(x)\| < \epsilon$ for all t in $[0, 1]$ and all x in cl (G). We may also assume that $\|y(t) - y(0)\| < \epsilon$ for all t in $[0, 1]$. We choose the constant $\epsilon > 0$ so small that the ball $B_{3\epsilon}(y(0))$ does not intersect $f_0(\text{bdry } (G))$, which it is possible to do since $f_0(\text{bdry } (G)) = (I - C_0 h_0^{-1})(\text{bdry } (h_0(G)))$ is a closed subset of Y which does not contain $y(0)$.

Under the above conditions, it suffices to show that $\deg_{1,M} (f_0, G, y) = \deg_{1,M} (f_1, G, y)$. We choose representations for f_0 and f_1 in the form:

$$f_0 = h_0 - C_0, \quad f_1 = h_1 - C_1,$$

with h_0 and h_1 permissible homeomorphisms of cl (G) into Y which are elements of the convex family M and with C_0 and C_1 compact mappings of cl (G) into Y. For each t in $[0, 1]$, we form

$$h_t = (1 - t)h_0 + th_1, \quad C_t = (1 - t)C_0 + tC_1.$$

Then if we set $g_t = h_t - C_t$ $(t \in [0, 1])$, we have a continuous homotopy of representations in the sense of property (3) of Theorem (12.1). Furthermore, if we set $z(t) = (1 - t)y_0 + ty_1$, it will follow that $\deg_1 ([g_t, h_t], G, z(t))$ is independent of t in $[0, 1]$, provided that for all t in $[0, 1]$, $z(t)$ lies outside of $g_t(\text{bdry } (G))$.

However, $\|z(t) - y_0\| \leq \|y_1 - y_0\| < \epsilon$, while for any x in bdry (G),

$$\|g_t(x) - y_0\| \geq 3\epsilon - \|f_1(x) - f_0(x)\| \geq 2\epsilon.$$

Hence $\|g_t(x) - z(t)\| > \epsilon > 0$ for every t in $[0, 1]$ and each x in bdry (G).

In particular, it follows from the above that

$$\deg_{1,M} (f_0, G, y(0)) = \deg_1 ([g_0, h_0], G, Z(0))$$
$$= \deg_1 ([g_1, h_1], G, Z(1)) = \deg_{1,M} (f_1, G, y(1)). \quad \text{q.e.d.}$$

PROOF OF (c). Since $\deg_{1,M} (f, G, y) = \deg_1 ([f, h], G, y)$ depends only upon the behaviour of f and h upon the boundary of G and does not depend upon h under our given hypotheses, the conclusion of (c) follows. q.e.d.

THEOREM (12.3). *Let X and Y be Banach spaces, M a convex class of permissible homeomorphisms from X to Y. Let G be an open subset of X, f a continuous mapping of cl (G) into Y, y a point of Y such that y has positive distance from $f(\text{bdry } (G))$.*

Then:

(a) *Suppose there exists a sequence $\{f_k\}$ of mappings from cl (G) into Y converging uniformly to f on cl (G) with each f_k representable in the form $f_k = h_k - C_k$ with h_k a permissible homeomorphism of cl (G) into Y which belongs to the class M and C_k a compact mapping of cl (G) into Y. Then for all sufficiently large k, $\deg_{1,M}(f_k, G, y)$ is well defined and independent of k. The common value for k sufficiently large is independent of the choice of the sequence $\{f_k\}$. We denote this value by $\deg_{1,M}(f, G, y)$.*

(b) *The degree function thus defined for the class of mappings f of cl (G) into Y which are uniformly approximable by mappings $\{f_k\}$ of the form $f_k = h_k - C_k$ with $h_k \in M$ and C_k compact has the following properties:*

(1) *If $\deg_{1,M}(f, G, y)$ is well defined as above and $\deg_{1,M}(f, G, y) \neq 0$ then y lies in the closure of $f(G)$ in Y. In particular, if $f(\mathrm{cl}\,(G))$ is closed in Y, there exists x in G such that $f(x) = y$.*

(2) *The degree function is additive in the domain G, i.e., if $G = G_1 \cup G_2$ with G_1 and G_2 open subsets of G and if $y \notin \mathrm{cl}\,(f(G'))$, where $G' = (G_1 \cap G_2) \cup \mathrm{bdry}\,(G_1) \cup \mathrm{bdry}\,(G_2)$, then $\deg_{1,M}(f, G, y) = \deg_{1,M}(f, G_1, y) + \deg_{1,M}(f, G_2, y)$.*

(3) *The degree function is invariant under permissible homotopies of the mapping, i.e., if G is an open subset of X, f a continuous mapping of cl $(G) \times [0, 1]$ into Y such that for each t in $[0, 1]$, the mapping f_t of cl (G) into Y given by $f_t(x) = f(x, t)$ is uniformly approximable on cl (G) by mappings $\{f_k\}$ of the form $f_k = h_k - C_k$ with h_k a permissible homeomorphism of cl (G) into Y which are elements of M, C_k compact mappings of cl (G) into Y, and if $\{y(t): t \in [0, 1]\}$ is a continuous curve in Y such that for each t in $[0, 1]$, $y(t)$ does not lie in cl $(f_t(\mathrm{bdry}\,(G)))$, then $\deg_{1,M}(f_t, G, y(t))$ is well defined for all t in $[0, 1]$ and is independent of t.*

(4) *If $\deg_{1,M}(f, G, y)$ is well defined and f is a homeomorphism of G on an open subset of Y such that y lies in $f(G)$, then $\deg_{1,M}(f, G, y) = \pm 1$.*

(5) *The degree function $\deg_{1,M}(f, G, y)$ depends only upon $f\big|_{\mathrm{bdry}\,(G)}$.*

PROOF OF THEOREM (12.3). PROOF OF (a). Let f be a mapping of cl (G) into Y such that there exists a sequence $\{f_k\}$ converging uniformly to f on cl (G) where each f_k is of the form $f_k = h_k - C_k$ with h_k in the convex class of homeomorphisms M and C_k compact. Let y be a point of Y outside of cl $(f(\mathrm{bdry}\,(G)))$. Then there exists $d > 0$ such that the closed ball $B_{2d}(y)$ does not intersect $f(\mathrm{bdry}\,(G))$. For each $\epsilon > 0$, there exists an index $k(\epsilon)$ such that for $k > k(\epsilon)$, $\|f(x) - f_k(x)\| < \epsilon$ for all x in cl (G). If we choose $\epsilon > 0$, so that $\epsilon < d$, we find that for x in bdry (G) and all $k > k(\epsilon)$, we have $\|f_k(x) - y\| > d$, while for k_1 also greater than $k(\epsilon)$, $\|f_k(x) - f_{k_1}(x)\| < \epsilon < d$. If we construct the linear deformation $\{g_t: 0 \leq t \leq 1\}$ of f_k into f_{k_1} by setting $g_t(x) = (1 - t)f_k(x) + tf_{k_1}(x)$, then it follows from the above inequalities that for all t in $[0, 1]$ and all x in bdry (G),

$$\|g_t(x) - y\| \geq \|f_k(x) - y\| - \|g_t(x) - f_k(x)\| > d - d = 0.$$

Since, in addition, $g_t = h_t - C_t$ where $h_t = (1 - t)h_k + th_{k_1} \in M$ and $C_t = (1 - t)C_k + tC_{k_1}$ is continuous and has its images for all t in $[0, 1]$ contained in a fixed relatively compact subset of Y, it follows from property (3) of the degree function $\deg_{1,M}$ as defined in Theorem (12.2) that $\deg_{1,M}(g_t, G, y)$ is independent of t in $[0, 1]$. Hence $\deg_{1,M}(f_k, G, y) = \deg_{1,M}(f_{k_1}, G, y)$ for all $k, k_1 > k(\epsilon)$, and we have verified the eventual stability of the sequence $\deg_{1,M}(f_k, G, y)$.

For two sequences $\{f_k\}$ and $\{g_k\}$ of mappings converging uniformly to f on cl (G), we can intersperse the two sequences to obtain a single sequence converging uniformly to f on cl (G). Applying the argument of the preceding part of the proof to this interspersed sequence, we see that for k sufficiently large $\deg_{1,M}(f_k, G, y) = \deg_{1,M}(g_k, G, y)$ so that the limit value of $\deg_{1,M}(f_k, g, y)$ is independent of the choice of the approximating sequence. Thus, we may define $\deg_{1,M}(f, G, y)$ for any mapping f of cl (G) uniformly approximable by a sequence $\{f_k\}$ of maps of the form $f_k = h_k - C_k$ with $h_k \in M$ and C_k compact, and the resulting definition is independent of the choice of the approximating sequence. q.e.d.

PROOF OF (b). PROOF OF PROPERTY (1). Suppose that f is approximable in the appropriate sense described above by the sequence $\{f_k\}$, uniformly on cl (G). Then for $k > k_0$, $\deg_{1,M}(f, G, y) = \deg_{1,M}(f_k, G, y)$ by the definition of the left-hand side. If we assume that $\deg_{1,M}(f, G, y) = 0$, then for each $k > k_0$, it follows by property (1) of the degree function considered in Theorem (12.1) that there exists a point x_k in G such that $f_k(x_k) = y$. Hence

$$\|f(x_k) - y\| \le \|f_k(x_k) - y\| + \|f_k(x_k) - f(x_k)\| = \|f_k(x_k) - f(x_k)\| \to 0$$

as $k \to \infty$ since f_k converges uniformly to f on cl (G). Thus, y lies in the closure of $f(G)$ in Y.

In particular, if $f(\text{cl}(G))$ is closed in Y, $f(\text{cl}(G))$ is a closed set which contains $f(G)$ and hence contains cl $(f(G))$. It then follows that y lies in $f(\text{cl}(G))$. However, we know by assumption that y does not lie in $f(\text{bdry}(G))$. Hence y lies in $f(G)$. q.e.d.

PROOF OF PROPERTY (2). Suppose that $G = G_1 \cup G_2$, where G_1 and G_2 are open subsets of G, and that there exists $d > 0$ such that $B_d(y)$ does not intersect $f(G')$, with

$$G' = (G_1 \cap G_2) \cup \text{bdry}(G_1) \cup \text{bdry}(G_2).$$

(This is obviously equivalent to saying that y does not lie in cl $(f(G'))$.) Let $\{f_k\}$ be a sequence of mappings converging uniformly to f on cl (G) with each f_k of the form $f_k = h_k - C_k$, $h_k \in M$, C_k compact. Then for $k > k_0$ for a suitable integer k_0, we know that $\|f_k(x) - f(x)\| < d$ for all x in cl (G). It follows that for such k, y does not lie in $f_k(G')$. Hence applying property (2) of the degree function $\deg_1$ in Theorem (12.1), we see that for $k > k_0$, we have

$$\deg_{1,M}(f_k, G, y) = \deg_{1,M}(f_k, G_1, y) + \deg_{1,M}(f_k, G_2, y).$$

Taking the limit on this equation as $k \to \infty$, we find that

$$\deg_{1,M}(f, G, y) = \deg_{1,M}(f, G_1, y) + \deg_{1,M}(f, G_2, y). \quad \text{q.e.d.}$$

PROOF OF PROPERTY (3). By subdividing the interval $[0, 1]$ into a finite number of small subintervals $[t_k, t_{k+1}]$ and considering each subinterval separately, we may assume without loss of generality that for a given $\epsilon > 0$, all x in cl (G) and all t in $[0, 1]$, $\|f_t(x) - f_0(x)\| < \epsilon$, $\|y(t) - y(0)\| < \epsilon$. It suffices to prove that $\deg_{1,M}(f_1, G, y(1)) = \deg_{1,M}(f_0, G, y(0))$.

For this latter purpose, we note that since $y(0)$ does not lie in cl $(f_0 (\text{bdry } (G)))$, there exists $d > 0$ such that the closed ball $B_{5d}(y_0)$ does not intersect $f_0 (\text{bdry } (G))$. We let ϵ satisfy: $0 < \epsilon < d$. Then, we may choose a mapping g_0 approximating f_0 of the form $g_0 = h_0 - C_0$ with $h_0 \in M$ and C_0 mapping cl (G) into a relatively compact subset of Y so that for all x in cl (G), $\|g_0(x) - f_0(x)\| < d$ while

$$\deg_{1,M} (f_0, G, y(0)) = \deg_{1,M} (g_0, G, y(0)).$$

Similarly, we may choose g_1 approximating f_1 on cl (G) of the form $g_1 = h_1 - C_1$ with $h_1 \in M$ and C_1 compact so that for all x in cl (G) $\|g_1(x) - f_1(x)\| < d$ while

$$\deg_{1,M} (f_1, G, y(1)) = \deg_{1,M} (g_1, G, y(1)).$$

It suffices therefore to prove that

$$\deg_{1,M} (g_1, G, y(1)) = \deg_{1,M} (g_0, G, y(0)).$$

To prove the equality of the last two degrees, we construct the homotopy between g_0 and g_1 given by $\{g_t : 0 \leq t \leq 1\}$, where

$$g_t(x) = (1 - t)g_0(x) + tg_1(x) \qquad (x \in \text{cl }(G), t \in [0, 1]),$$

and a corresponding curve in Y given by

$$z(t) = (1 - t)y(0) + ty(1) \qquad (t \in [0, 1]).$$

It obviously suffices to prove that $\deg_{1,M} (g_t, G, z(t))$ is independent of t in $[0, 1]$. By property (3) of Theorem (12.1), it suffices since

$$g_t = h_t - C_t, \quad h_t = (1 - t)h_0 + th_1 \in M, \quad C_t = (1 - t)C_0 + tC_1$$

to prove that $\deg_{1,M} (g_t, G, z(t))$ is well defined for each t in $[0, 1]$. To prove the latter fact, it suffices to establish that for each t in $[0, 1]$, $z(t)$ does not lie in $g_t (\text{bdry } (G))$. However, for x in bdry (G), we have

$$\|g_t(x) - z(t)\| \geq \|g_0(x) - y(0)\| - \|g_0(x) - g_t(x)\| - \|z(t) - y(0)\|$$
$$\geq \|f_0(x) - y(0)\| - \|f_0(x) - g_0(x)\| - \|g_0(x) - g_1(x)\| - d.$$

For each x in bdry (G), we have $\|f_0(x) - y(0)\| > 5d$, while for each x in cl (G), $\|f_0(x) - g_0(x)\| < d$, and

$$\|g_0(x) - g_1(x)\| \leq \|g_0(x) - f_0(x)\| + \|f_0(x) - f_1(x)\| + \|f_1(x) - g_1(x)\| < 3d.$$

Hence for all x in bdry (G), and all t in $[0, 1]$, $\|g_t(x) - z(t)\| > 5d - 5d = 0$. q.e.d.

With the proof of property (3), the proof of part (b) of Theorem (12.3) is complete, and with it, the proof of Theorem (12.3). q.e.d.

We now turn to a second and different extension of the concept of the topological degree, which in this case is defined for representations of mappings f in the following more general form:

DEFINITION (12.6). *Let X and Y be Banach spaces, G an open subset of X, f a continuous mapping of G into Y. Then by an intertwined representation for f with respect to a given class M of permissible homeomorphisms from X to Y, we mean a* pair (W, X) where W is an open subset of $X \times X$ and S is a continuous mapping of W into Y for which all of the following conditions are satisfied:

(a) For each x in G, the element $[x, x]$ of $X \times X$ lies in W and $S(x, x) = f(x)$. If $[x, x] \in W$, then x lies in G.

(b) For each v in G, let $W_v = \{u \mid u \in X, [u, v] \in W\}$, and for u in W_v, let $S_v(u) = S(u, v)$. Then S_v is assumed to be a homeomorphism of W_v on an open subset of Y.

(c) For each point $[u_0, v_0]$ of W, there exists a product neighborhood $N \times N_1$ with closure contained in W, such that for each v in cl (N_1) the restriction of S_v to cl (N) lies in the class of permissible homeomorphisms M, while the mapping $v \to S_v$ of cl (N_1) into the space of permissible homeomorphisms of cl (N) into Y is continuous and has relatively compact image.

(We consider the topology of uniform convergence on cl (N) for the space of homeomorphisms.)

THEOREM (12.4). *Let X and Y be Banach spaces, G an open subset of X, f a continuous mapping of G into Y having an intertwined representation in the sense of Definition (12.6) with respect to a given class M of permissible homeomorphisms from X to Y. Then*

(a) *For each y in Y, the subset G_y of X given by*

$$G_y = \{v \mid v \in X, S_v(W_v) \text{ contains } y\}$$

is an open subset of X, while the mapping C_y of G_y into X given by $C_y(v) = S_v^{-1}(y)$ $(v \in G_y)$ is a continuous, locally compact mapping of G_y into X.

(b) *The fixed points of C_y in G_y coincide with the solution x of the equation $S(x, x) = y$, i.e., with points x in G such that $f(x) = y$.*

(c) *Suppose that for a given point y of Y, $f^{-1}(y)$ is compact. Then we may define a new degree function for the representation (W, S) of the mapping f by setting $\deg_2([f, S], G, y) = \deg_{LS}(I - C_y, G'_y, 0)$, where G'_y is an open subset of G_y which contains the compact set $f^{-1}(y)$ such that $C_y(G'_y)$ is relatively compact in X. Such subsets G'_y exist, and the degree thus defined is independent of the choice of G'_y.*

PROOF OF THEOREM (12.4). PROOF OF PART (a). Let v_0 be a point of G_y, i.e., $S_{v_0}(W_{v_0})$ contains y. Let $u_0 = C_y(v_0)$, so that $S(u_0, v_0) = y$. We note that by the definition of the intertwined representation (W, S) for f given in Definition (12.6), there exists a product neighborhood $N \times N_1$ of the point $[u_0, v_0]$ in $X \times X$ such that for each v in cl (N_1), the mapping S_v is defined on cl (N) and is a permissible homeomorphism of cl (N) into Y. In particular, $S_v(\text{bdry}(N))$ is a closed subset of Y for each v in N_1 and $S_{v_0}(\text{bdry}(N))$ does not contain the point y since y lies in $S_{v_0}(N)$ itself. Thus there exists a constant $d > 0$ such that the ball $B_{2d}(y)$ lies in $S_{v_0}(N)$, which is an open subset of Y containing y. Since the mapping $v \to S_v$ is assumed to be continuous from N_1 to the space of homeomorphisms of cl (N) into Y, with the latter space given the topology of uniform convergence on cl (N), there exists a neighborhood N_2 of v_0 contained in N_1 such that for v in N_2, $\|S_v(u) - S_{v_0}(u)\| < d$ $(u \in \text{cl}(N))$. It follows that for v in N_2, $S_v(u_0)$ lies in the ball $B_d(y)$ while $B_d(y)$ does not intersect $S_v(\text{bdry}(N))$.

We may assume without loss of generality that N is a connected neighborhood of u_0 in X. Then for v in N_2, $S_v(N)$ is a connected subset of Y which is therefore contained in one component of $Y - S_v(\text{bdry }(N))$. Since S_v is a permissible homeomorphism of cl (N) into Y, $S_v(\text{bdry }(N)) = \text{bdry }(S_v(N))$. Hence $S_v(N)$ must coincide with the component of $Y - S_v(\text{bdry }(N))$ which contains it, since otherwise that component would contain a point of bdry $(S_v(N)) = S_v(\text{bdry }(N))$, and this is excluded by the definition of the component. In particular, the connected set $B_d(y)$ contains at least one point $S_v(u_0)$ of $S_v(N)$ and is contained in $Y - S_v(\text{bdry }(N))$. Hence $B_d(y)$ is contained in $S_v(N)$. A fortiori, y lies in $S_v(N)$ for v in N_2, i.e., $N_2 \subset G_y$. Hence G_y is an open subset of X.

To complete the proof of part (a), we must show that the mapping C_y of G_y into X given by $C_y(v) = S_v^{-1}(y)$ is both continuous and locally compact. To establish the continuity of C_y, we must show that for a given point v_0 in G_y, for $u_0 = C_y(v_0) = S_{v_0}^{-1}(y)$, and for a given neighborhood N of u_0, there exists a neighborhood N_3 of v_0 in G_y such that C_y maps N_3 into N. We may assume without loss of generality that N is the first component of a product neighborhood $N \times N_1$ in $X \times X$ whose closure is contained in W and on which the mapping S satisfies condition (c) of Definition (12.6). Arguing as in the preceding paragraph, there then exists a neighborhood N_2 of v_0 in X which is contained in G_y and such that for each v in N_2, $S_v(N)$ contains y. It follows since S_v is a homeomorphism of W_v into Y that $S_v^{-1}(N_2) \subset N$. Hence, if we set $N_3 = N_2$, it follows that C_y maps N_3 into N. Therefore, C_y is a continuous mapping from G_y into X.

To show that C is locally compact, we again choose a product neighborhood $N \times N_1$ in $X \times X$ of the point $[u_0, v_0]$, such that $C_y(v_0) = u_0$ satisfying condition (c) of Definition (12.6). In particular, the mapping $v \to S_v$ of cl (N_1) into the space of homeomorphisms of cl (N) into Y is relatively compact. Let $\{v_k\}$ be an infinite sequence in cl (N_1). Passing to an infinite subsequence, we may assume that S_{v_k} converges uniformly on cl (N) as $k \to +\infty$ to a permissible homeomorphism h of cl (N) into Y. Let $u_k = C_y(v_k)$ for each k. Then $S(u_k, v_k) = y$, i.e., $S_{v_k}(u_k) = y$. Since

$$\|h(u_k) - S_{v_k}(u_k)\| \leq \sup_{u \in \text{cl}(N)} \|h(u) - S_{v_k}(u)\| = \epsilon_k \to 0 \quad (k \to \infty),$$

it follows that $y = \lim_{k \to \infty} h(u_k)$, so that y lies in cl $(h(\text{cl }(N)))$. Since h is permissible, $h(\text{cl }(N))$ is closed in Y. Hence y lies in $h(\text{cl }(N))$ and $u_k = h^{-1}(h(u_k)) \to h^{-1}(y)$ by the continuity of h^{-1} on $h(\text{cl }(N))$. It follows since $C_y(v_k)$ has been shown to have a strongly convergent subsequence, that the mapping C_y is locally compact. q.e.d.

PROOF OF PART (b). If x is a point of G such that $f(x) = y$, it follows from the condition (a) of Definition (12.6) that $[x, x] \in W$ and $S(x, x) = f(x) = y$. Hence x lies in G_y and $C_y(x) = S_x^{-1}(y) = x$. On the other hand, suppose that x lies in G_y and is a fixed point of C_y. Then $S_x^{-1}(x) = C_y(x) = x$, i.e., $S(x, x) = y$. Hence by condition (a) of Definition (12.6), x lies in G and $f(x) = S(x, x) = y$. q.e.d.

PROOF OF PART (c). If the set $f^{-1}(y)$ is a compact subset of G, then by the result of part (b), the fixed point set $F(C_y)$ of the mapping C_y of G_y into X is a compact subset of G_y which coincides with $f^{-1}(y)$. By the result of part (a), the mapping C_y is both continuous and locally compact from G_y to X. Hence each point v of $F(C_y)$ has an open neighborhood N_v in G_y such that C_y is a continuous compact mapping of N_v into X. Since $F(C_y)$ is compact, it can be covered by a finite number of such neighborhoods. If G_1 is the union of this family of neighborhoods, then G_1 is an open subset of G_y and it follows immediately that C_y is a continuous compact mapping of G_1 into X. Shrinking G_1 slightly (using the normality of the metric space X), we can assume that C_y is a continuous compact mapping of cl (G_1) into X (and that cl (G_1) is contained in G_y).

By the preceding discussion, it follows that we can find open subsets G'_y of G_y with $F(C_y) \subset G'_y \subset$ cl $(G'_y) \subset G_y$ such that C_y is a compact mapping of cl (G'_y) into X. Hence the Leray-Schauder degree $\deg_{LS} (I - C_y, G'_y, 0)$ is well defined and is independent of the choice of the particular G'_y since all fixed points of C_y are contained in the intersection of all such G'_y. q.e.d.

With the proof of the assertion of part (c), the proof of Theorem (12.4) is complete. q.e.d.

THEOREM (12.5). *Let X and Y be Banach spaces, G an open subset of X, f a continuous mapping of G into Y having an intertwined representation in the sense of Definition (12.6) with respect to a given class M of permissible homeomorphisms from X to Y. Let $\deg_2 ([f, S], G, y)$ be the degree function defined in part (c) of the assertion of Theorem (12.4). Then this degree function has the following properties:*

(a) *If $f^{-1}(y)$ is a compact subset of G, then $\deg_2 ([f, S], G, y)$ is well defined. If $\deg_2 ([f, S], G, y) \neq 0$, then $f^{-1}(y)$ is nonempty, i.e., there exists a point x in G such that $f(x) = y$.*

(b) *The degree function $\deg_2$ is additive in G, i.e., if G contains two open subsets G_1 and G_2 such that $f^{-1}(y)$ is a compact subset of $(G_1 \cup G_2) -$ cl $(G_1 \cap G_2)$, then*

$$\deg_2 ([f, S], G, y) = \deg_2 ([f, S], G_1, y) + \deg_2 ([f, S], G_2, y).$$

(c) *The degree function $\deg_2$ is invariant under permissible homotopies of representations in the following sense: Let $\{f_t : 0 \leq t \leq 1\}$ be a one-parameter family of continuous mappings of the one-parameter family $\{G_t : 0 \leq t \leq 1\}$ of open subsets of X into Y, and let $\{y(t) : 0 \leq t \leq 1\}$ be a continuous curve in Y. Suppose that there exist an open subset G' in $X \times [0, 1]$ such that for each t in $[0, 1]$, $G_t = \{x \mid x \in X, [x, t] \in G'\}$ and a compact subset K of G' such that for each t in $[0, 1]$, $f_t^{-1}(y(t)) \subset K$. Suppose that there exists an open subset W' of $X \times X \times [0, 1]$ such that $G' = \{[x, t] \mid [x, x, t] \in W'\}$ while there exists a continuous mapping S' of W' into Y such that $f_t(x) = S(x, x, t)$ for any $[x, t]$ in G', while the following conditions hold:*

(1) *If for a given $[v, t]$ in $X \times [0, 1]$, we set*

$$W_{v,t} = \{u \mid u \in X, [u, v, t] \in W\}$$

and

$$S_{v,t}(u) = S(u, v, t), u \in W_{v,t}$$

then $S_{v,t}$ is a homeomorphism of $W_{v,t}$ into Y.

(2) *For each point $[u_0, v_0, t_0]$ in W', there exists a product neighborhood $N \times N_1 \times N_0$ whose closure is contained in W', such that for each $[v, t]$ in cl $(N_1) \times N_0$,*

$S_{v,t}$ is a permissible homeomorphism of cl (N) into Y lying in the class M, and the mapping $[v, t] \to S_{v,t}$ of cl $(N_1) \times N_0$ into the space of permissible homeomorphisms of cl (N) into Y is continuous and has relatively compact image.

Then $\deg_2 ([f_t, S_t], G_t, y(t))$ is well defined where S_t is the intertwined representation of f_t on G_t given by

$$W_t = \{[u, v] \mid [u, v] \in X \times X, [u, v, t] \in W'\},$$
$$S_t(u, v) = S(u, v, t) \qquad ([u, v] \in W_t).$$

This degree $\deg_2 ([f_t, S_t], G_t, y(t))$ is independent of t in $[0, 1]$.

(d) If f has a representation S such that $S(u, v)$ is independent of v, then

$$\deg_2 ([f, S], G, y) = +1$$

if $f^{-1}(y)$ is a nonempty compact subset of G.

PROOF OF THEOREM (12.5). PROOF OF (a). We have already shown in part (c) of Theorem (12.4) that if $f^{-1}(y)$ is compact for a mapping f having an intertwined representation in the sense of Definition (12.6), then $\deg_2 ([f, S], G, y)$ is well defined by setting

$$\deg_2 ([f, S], G, y) = \underset{\text{LS}}{\deg} (I - C_y, G'_y, 0)$$

where G'_y and C_y are defined in Theorem (12.4). If we know that $\deg_2 ([f, S], G, y) \neq 0$, then it follows from property (1) of the Leray-Schauder degree that C_y has a fixed point in G'_y. By part (b) of Theorem (12.4), for such a fixed point x, x lies in G and $f(x) = y$. Hence $f^{-1}(y)$ is nonempty. q.e.d.

PROOF OF (b). Let $G_3 = (G_1 \cup G_2) - $ cl $(G_1 \cap G_2)$. Then G_3 is the union of the two open subsets G'_1 and G'_2, where

$$G'_1 = G_1 - \text{cl } (G_1 \cap G_2), \qquad G'_2 = G_2 - \text{cl } (G_1 \cap G_2),$$

and cl $(G'_1) \cap$ cl $(G'_2) = 0$. We note first that

$$\deg_2 ([f, S], G, y) = \deg_2 ([f, S], G_3, y)$$

and similarly,

$$\deg_2 ([f, S], G_1, y) = \deg_2 ([f, S], G'_1, y),$$
$$\deg_2 ([f, S], G_2, y) = \deg_2 ([f, S], G'_2, y).$$

Indeed, let G_y and $G_{3,y}$ be defined as in Theorem (12.4) with respect to the open sets G and G_3, respectively. Then $G_{3,y}$ is an open subset of G_y and contains all the points of $f^{-1}(y)$, i.e., all the fixed points of the mapping C_y. If $C_{3,y}$ is the mapping of $G_{3,y}$ into X given by $C_{3,y}(u) = S_v^{-1}(y)$ for v in $G_{3,y}$, it is obvious that $C_{3,y}$ is simply the restriction of C_y to the subset $G_{3,y}$. If $G'_{3,y}$ is an open subset of $G_{3,y}$ which contains $f^{-1}(y)$ and on which $C_{3,y}$ is compact, it follows by their definitions that

$$\deg_2 ([f, S], G_3, y) = \underset{\text{LS}}{\deg} (I - C_y, G'_{3,y}, 0) = \deg_2 ([f, S], G, y).$$

A similar argument holds for the other two equalities.

Hence it suffices to prove the assertion of (b) under the hypothesis that $\text{cl}(G_1) \cap \text{cl}(G_2) = \emptyset$. In that case, for the given y,

$$G_y \supset G_{1,y} \cup G_{2,y}$$

with $G_{1,y}$ and $G_{2,y}$ disjoint and C_y having all its fixed points in $G_{1,y} \cup G_{2,y}$. Hence

$$\deg_{LS}(I - C_y, G_y, 0) = \deg_{LS}(I - C_y, G_{1,y}, 0) + \deg_{LS}(I - C_y, G_{2,y}, 0),$$

by property (2) of the Leray-Schauder degree. Thus

$$\deg_2([f, S], G, y) = \deg_2([f, S], G_1, y) + \deg_2([f, S], G_2, y). \quad \text{q.e.d.}$$

PROOF OF (c). For each t in $[0, 1]$, we form the intertwined representations S_t of f_t on G_t defined in part (c) above. Let $G_{y(t),t}$ be the corresponding open subset of X given by

$$G_{y(t),t} = \{v \mid v \in X, S_{v,t}(W_{v,t}) \text{ contains } y(t)\}$$

and C_t, the locally compact mapping of $G_{y(t),t}$ into X given by $C_t(v) = S_{v,t}^{-1}(y(t))$. If we apply property (3) of the Leray-Schauder degree to the present situation, it suffices to prove that the following three properties all hold:

The subset G'' of $X \times [0, 1]$ given by

(i) $G'' = \{[v, t] \mid v \in G_{y(t),t}\}$ is open in $X \times [0, 1]$, and the subset $K'' = \{[v, t] \mid C_t(v) = v\}$ is a compact subset of G''.

(ii) The mapping C'' of G'' into X given by $C''(v, t) = C_t(v)$ is continuous and locally compact.

Indeed, suppose that properties (i) and (ii) have been established. For each t in $[0, 1]$, we may therefore choose an open subset $G'_{y(t),t}$ with closure contained in $G_{y(t),t}$ such that C_t is a compact mapping of $G'_{y(t),t}$ into X while the fixed points of C_t are a compact subset of $G'_{y(t),t}$. For each t in $[0, 1]$, therefore, the degree function $\deg_2([f_t, S_t], G_t, y(t))$ is given by

$$\deg_2([f_t, S_t], G_t, y(t)) = \deg_{LS}(I - C_t, G'_{y(t),t}, 0).$$

We must show that the degree function on the right side of this last equation is independent of t in $[0, 1]$. For this purpose, it suffices to show that for each t_0 in $[0, 1]$, there exists $\epsilon(t_0) > 0$ such that this degree function is constant on the interval $[t_0 - \epsilon(t_0), t_0 + \epsilon(t_0)]$. It follows immediately that we lose no generality if we show that

$$\deg_{LS}(I - C_t, G'_{y(t),t}, 0) = \deg_{LS}(I - C_{t_0}, G'_{y(t_0),t_0}, 0)$$

for $t_0 \leq t \leq t_0 + \epsilon(t_0)$.

Let K_t be the fixed point set of C_t in $G'_{y(t),t}$. By property (i), K_t is compact for each t in $[0, 1]$, and the set K'' generated in the cylinder by the different subsets K_t has positive distance $d > 0$ from $X \times [0, 1] - G''$. Let $G'_{y(t_0),t_0}$ be an open subset of $G_{y(t_0),t_0}$ on which C_{t_0} is compact. We assert that for t near t_0, K_t must be contained in $G'_{y(t_0),t_0}$. Otherwise, we could find a sequence of points $[v_k, t_k]$ in K'' with $C_{t_k}(v_k) = v_k$ and $t_k \to t_0$ as $k \to \infty$ such that v_k all lie outside of $G'_{y(t_0),t_0}$. Since K''

is compact, we may assume that v_k converges to a point v_0 as $k \to \infty$. Obviously v_0 must lie in K_{t_0}, which contradicts the fact that all the elements v_k lie in the closed set $X - G'_{y(t_0),t_0}$ which contains no points of K_{t_0}.

Since K_t is contained in $G'_{y(t_0),t_0}$ for t in some interval $[t_0, t_0 + \epsilon]$ while $G'_{y(t_0),t_0}$ may be chosen as such a small neighborhood of K_{t_0} that $G'_{y(t_0),t_0} \times [t_0, t_0 + \epsilon]$ is contained in G'', it follows that by a suitable choice,

$$\deg_{LS} (I - C_{t_0}, G'_{y(t_0),t_0}, 0) = \deg_{LS} (I - C_t, G'_{y(t_0),t_0}, 0)$$

for t in $[t_0, t_0 + \epsilon]$ since the mapping C'' is locally compact from G'' to X. On the other hand,

$$\deg_{LS} (I - C_t, G'_{y(t_0),t_0}, 0) = \deg_2 ([f_t, S_t], G_t, y(t))$$

whenever C_t is a compact mapping of $G'_{y(t_0),t_0}$ into X, a fact which we have applied (using property (ii) for C'') to guarantee the meaningfulness of

$$\deg_{LS} (I - C_t, G'_{y(t_0),t_0}, 0)$$

by making $G'_{y(t_0),t_0}$ a suitably small neighborhood of the given compact set K_{t_0}.

The above remarks establish the fact that the verification of properties (i) and (ii) above suffices to complete the proof of the assertion of part (c) of Theorem (12.5). We now proceed to the proof of these properties:

PROOF OF PROPERTY (i). Let $[v_0, t_0]$ be a point of G''. Then v_0 lies in $G_{y(t_0),t_0}$, i.e., $S(u_0, v_0, t_0) = y(t_0)$ for some point u_0 in W_{v_0,t_0}. Using the definition of W_{v_0,t_0}, this last fact is equivalent to the condition that for some u_0 in X, $[u_0, v_0, t_0]$ lies in W' and $S(u_0, v_0, t_0) = y(t_0)$. By condition (2) of the hypothesis of Theorem (12.5) on the representation S for the permissible homotopy of the representations, there exists a product neighborhood $N \times N_1 \times N_0$ of $[u_0, v_0, t_0]$ with closure contained in W' such that for each $[v, t]$ in cl $(N_1) \times N_0$, $S_{v,t}$ is a permissible homeomorphism of cl (N) into Y, depending continuously upon $[v, t]$ in cl $(N_1) \times N_0$. Since S_{v_0,t_0} maps bdry (N) on a closed subset of Y not containing the point $y(t_0)$, there exists $d > 0$ such that the closed ball $B_{3d}(y(t_0))$ is disjoint from $S_{v_0,t_0}(\text{bdry } (N))$ If we take N_1 and N_0 sufficiently small, we may ensure by the continuity properties of $S_{v,t}$ in $[v, t]$ that for $[v, t]$ in $N_1 \times N_0$, we have $\|S_{v,t}(u) - S_{v_0,t_0}(u)\| < d$. Similarly, we may ensure that for t in N_0, $\|y(t) - y(t_0)\| < d$. In particular, we have

$$\|S_{v,t}(u_0) - y(t)\| \leq \|S_{v,t}(u_0) - S_{v_0,t_0}(u_0)\| + \|y(t_0) - y(t)\| \leq 2d.$$

Thus, $S_{v,t}(u_0)$ lies in $S_{v,t}(N) \cap B_{2d}(y(t_0))$, for $[v, t]$ in $N_1 \times N_0$.

Finally, $S_{v,t}(N)$ is a connected subset of $Y - S_{v,t}(\text{bdry } (N))$ if we choose N connected, as we may by the local connectedness of the Banach space X. Hence it is contained in a single component C of $Y - S_{v,t}(\text{bdry } (N))$, and indeed coincides with the component C since otherwise C would have to contain a point of bdry $(S_{v,t}(N))$, and since $S_{v,t}$ is assumed to be a permissible homeomorphism of cl (N) into Y, $S_{v,t}(\text{bdry } (N)) = \text{bdry } (S_{v,t}(N))$ for $[v, t]$ in $N_1 \times N_0$. In particular, C contains a point of $B_{2d}(y(t_0))$, and this latter ball is a connected set which is entirely disjoint from $S_{v,t}(\text{bdry } (N))$ for $[v, t]$ in $N_1 \times N_0$, since $S_{v,t}(\text{bdry } (N))$

consists for such $[v, t]$ entirely of points at distance less than d from points of $S_{v_0, t_0}(\text{bdry}(N))$, and the whole of the latter set is contained in the complement of $B_{3d}(y(t_0))$. Hence $B_{2d}(y(t_0))$ is entirely contained in $S_{v,t}(N)$ for $[v, t]$ in $N_1 \times N_0$. Since $y(t)$ for t in N_0 is contained in $B_d(y(t_0))$, it follows that for $[v, t]$ in $N_1 \times N_0$, $y(t)$ lies in $S_{v,t}(W_{v,t})$, i.e., $N_1 \times N_0$ is contained in G''. Hence G'' is an open subset of $X \times [0, 1]$.

To complete the proof of property (i), we must show that K'' is a compact subset of G''. However, $[v, t]$ lie in K'' if and only if $f_t(v) = y(t)$. Hence K'' is a compact subset of $X \times [0, 1]$ by the hypothesis of Theorem (12.5). We need only prove that K'' is a subset of G''. This follows immediately, however, since $[v, t] \in K''$ implies that $S(v, v, t) = y(t)$ so that $[v, t]$ lies in $G_{y(t),t}$. q.e.d.

PROOF OF PROPERTY (ii). We show first that C'' is continuous. To do this, suppose that $[v_0, t_0]$ lies in G'' and that $C''(v_0, t_0) = u_0$. Then $[u_0, v_0, t_0]$ is a point of W' and $S(u_0, v_0, t_0) = y(t_0)$. Let N be a neighborhood of u_0 in X. We must show that there exist neighborhoods N_1 and N_0 of v_0 and t_0, respectively, in X and in $[0, 1]$ such that $C''(N_1 \times N_0) \subset N$. We may assume that N is chosen so small as to be the first component of a product neighborhood $N \times N_1 \times N_0$ of $[u_0, v_0, t_0]$ in W' on which the homotopy S satisfies the condition (2) of the hypothesis of Theorem (12.5). In particular, for $[v, t]$ in $N_1 \times N_0$, $S_{v,t}$ is a permissible homeomorphism of cl (N) into Y. In particular, S_{v_0, t_0} maps bdry (N) on a closed subset of Y not containing $y(t_0)$. Hence (as in the argument for property (i)), there exists $d > 0$ such that $B_{3d}(y(t_0))$ does not contain any point of $S_{v_0, t_0}(\text{bdry}(N))$. If we choose $N_1 \times N_0$ still smaller so that for $[v, t]$ in $N_1 \times N_0$, $\|S_{v,t}(u) - S_{v_0, t_0}(u)\| < d$ for all u in cl (N), it follows as in the proof of property (i) that for $[v, t]$ in $N_1 \times N_0$, $y(t)$ lies in $S_{v,t}(N)$. However, $S_{v,t}$ is a homeomorphism of $W_{v,t}$ into Y. Hence $S_{v,t}^{-1}(y(t))$ lies in N for $[v, t]$ in $N_1 \times N_0$, i.e., $C''(N_1 \times N_0) \subset N$. Hence C'' is continuous.

To show that C'' is locally compact, we again choose a neighborhood $N \times N_1 \times N_0$ of $[u_0, v_0, t_0]$ in W' such that $S(u_0, v_0, t_0) = y(t_0)$, while S has the properties described in condition (2) of the hypothesis on the closure of this product neighborhood. In particular, the mapping $[v, t] \to S_{v,t}$ is assumed to map $N_1 \times N_0$ into a relatively compact subset of the space of permissible homeomorphisms of cl (N) into Y. Suppose that $\{[v_k, t_k]\}$ is an infinite sequence in $N_1 \times N_0$. By passing to an infinite subsequence, we may assume that S_{v_k, t_k} converges uniformly on cl (N) to a permissible homeomorphism h of cl (N) into Y. Let $u_k = C''(v_k, t_k)$. Then $S_{v_k, t_k}(u_k) = y(t_k)$, where we may assume that $t_k \to t$, and that $y(t_k)$ converges strongly in Y to $y(t)$ as $k \to \infty$. Hence $h(u_k) - y(t_k) \to 0$, so that $h(u_k)$ converges to a point y' of cl $(h(\text{cl}(N))) = h(\text{cl}(N))$. Finally, $u_k = h^{-1}(h(u_k))$ converges in X to $h^{-1}(y')$ as $k \to \infty$. Thus C'' is locally compact and the proof of property (ii) is complete.

With the proof of property (ii), the proof of part (c) is complete. q.e.d.

PROOF OF PART (d). If $S(u, v)$ is independent of v, then C_y is a constant map and $I - C_y$ has degree $+1$ with respect to 0 if 0 lies in $(I - C_y)(G_y)$. q.e.d.

With the four parts of Theorem (12.5) having been established, the proof of Theorem (12.5) is now complete. q.e.d.

In order to apply the degree function $\deg_2$ defined in Theorem (12.4) in various specialized contexts, it is both useful and necessary to restrict the generality of the intertwined representations defined in Definition (12.6) and to apply the more special representations described in the following:

DEFINITION (12.7). *Let X and Y be Banach spaces, G an open subset of X, f a continuous mapping of cl (G) into Y. Then f is said to have a restricted intertwined representation S with respect to a given class M of permissible homeomorphisms from X to Y if S is a continuous mapping of cl $(G) \times$ cl (G) into Y having the following properties:*

(a) *For each x in cl (G), $f(x) = S(x, x)$.*

(b) *For each v in cl (G), the mapping S_v of cl (G) into Y given by $S_v(u) = S(u, v)$ ($u \in$ cl (G)) is a permissible homeomorphism of cl (G) into Y lying in the class M.*

(c) *The mapping $v \to S_v$ of cl (G) into the space of permissible homeomorphisms of cl (G) into Y (with the latter space being given the topology of uniform convergence on cl (G)) is continuous and has relatively compact image.*

THEOREM (12.6). *Let X and Y be Banach spaces, G an open subset of X, f a continuous mapping of cl (G) into Y having a restricted intertwined representation S in the sense of Definition (12.7) above with respect to a given class of permissible homeomorphisms M from X to Y. Then:*

(a) *The representation S in the sense of Definition (12.7) is also a representation in the sense of Definition (12.6) if W is taken to be $G \times G$.*

(b) *The condition that $f^{-1}(y)$ should be a compact subset of G for a mapping f having an intertwined representation in the restricted sense of Definition (12.7) is equivalent to the weaker condition that $f(\text{bdry } (G))$ does not contain y.*

(c) *If $f^{-1}(y)$ includes no points of bdry (G), it follows from (b) that $\deg_2([f, S], G, y)$ is well defined by Theorem (12.4) and has the properties given in Theorem (12.4). In this case, it has the following simpler homotopy property for homotopies involving restricted representations:*

(h) *Let $\{f_t : 0 \leq t \leq 1\}$ be a family of continuous mappings of cl (G) into Y, $\{y(t) : 0 \leq t \leq 1\}$ a continuous curve in Y such that $f_t(h) \neq y(t)$ for any x in bdry (G) and any t in $[0, 1]$. Suppose that we are given a continuous deformation of restricted representations for the family $\{f_t\}$ in the form of a mapping S of cl $(G) \times$ cl $(G) \times [0, 1]$ into Y having the property that for each $[v, t]$ in cl $(G) \times [0, 1]$, the mapping $S_{v,t}$ of cl (G) into Y given by $S_{v,t}(u) = S(u, v, t)$ ($u \in$ cl (G)) is a restricted representation of the mapping f_t in the sense of Definition (12.7), while the mapping $[v, t] \to S_{v,t}$ of cl $(G) \times [0, 1]$ into the space of permissible homeomorphisms of cl (G) into Y is continuous and has relatively compact image.*

Then $\deg_2([f_t, S_t], G, y(t))$ is well defined for each t in $[0, 1]$ and is independent of t in $[0, 1]$.

PROOF OF THEOREM (12.6). PROOF OF (a). The conditions imposed upon S in Definition (12.6) are automatically satisfied if the more stringent conditions of Definition (12.7) are imposed. In particular, we may take a single neighborhood $N \times N_1$ for condition (2) in Definition (12.6) with $N = N_1 = G$. q.e.d.

PROOF of (b). Obviously if $f^{-1}(y)$ is a compact subset of G, then it does not intersect bdry (G). Suppose on the other hand that $f^{-1}(y)$ does not intersect bdry (G) and hence is a subset of G. We shall show then that $f^{-1}(y)$ is compact. We note that for each y, the mapping C_y of $G_y = \{v \mid v \in X, S_v(G) \text{ contains } y\}$ is a continuous mapping of G_y into X. We assert that C_y extends by continuity to a continuous mapping of cl (G_y) into X which maps bdry (G_y) into bdry (G) and has relatively compact image in X. We may show this by setting

$$G'_y = \{v \mid v \in \text{cl}(G), S_v(\text{cl } G)) \text{ contains } y\}$$

and for v in G'_y, we set $C'_y(v) = S_v^{-1}(y)$, obtaining thereby a mapping of G'_y into cl (G). Since G'_y contains G_y and C'_y is an extension of C_y to the larger domain G'_y, it suffices to prove that cl (G_y) is contained in G'_y and that C'_y is a continuous mapping of cl (G_y) into cl (G) with relatively compact image. For the first of these points, suppose that v lies in cl (G_y). Then there exists a sequence $\{v_k\}$ in G_y with v_k converging strongly to v in X as $k \to \infty$. For each k, there exists an element u_k of G such that $S_{v_k}(u_k) = y$, i.e., $S(u_k, v_k) = y$. Since the mapping $v \to S_v$ is continuous from cl (G) to the space of permissible homeomorphisms of cl (G) into Y, it follows that $S(u_k, v) \to y$ as $k \to \infty$, i.e., y lies in cl $(S_v(G)) = S_v(\text{cl}(G))$. Hence $y = S_v(u)$ for an uniquely defined u in cl (G) and $u_k = S_v^{-1}(S_v(u_k)) \to S_v^{-1}(y) = u$. In particular, cl (G_y) is contained in G'_y and $C'_y(v) = S_v^{-1}(y)$ defined a continuous mapping C'_y of cl (G_y) into cl (G).

We show next that C'_y maps cl (G_y) into a relatively compact subset of cl (G). Indeed, suppose that $\{v_k\}$ is an infinite sequence in cl (G_y) and let $u_k = C'_y(v_k)$ for each k. Since the mapping $v \to S_v$ maps cl (G) into a relatively compact subset of the space of permissible homeomorphisms of cl (G) into Y, we may choose an infinite subsequence of the sequence $\{v_k\}$ (which we identify with the original sequence) such that S_{v_k} converges uniformly on cl (G) as $k \to \infty$ to the permissible homeomorphism h of cl (G) into Y. Since for each k, $S(u_k, v_k) = y$, it follows that $h(u_k) \to y$ as $k \to \infty$, so that in particular, y lies in cl $(h(\text{cl}(G))) = h(\text{cl}(G))$, and $u_k = h^{-1}(h(u_k))$ converges strongly as $k \to \infty$ to $h^{-1}(y)$. Hence $C'_y(\text{cl}(G_y))$ is relatively compact.

We verify next that C'_y maps bdry (G_y) into bdry (G). Indeed, it maps cl (G_y) into cl (G) and G_y into G. If a point v in cl (G_y) is mapped by C'_y into G, it follows from the definition of C_y that v lies in G_y. Hence $C'_y(\text{bdry}(G_y)) \subset \text{bdry}(G)$.

To complete the proof of the assertion of part (b), we note that $f(v) = y$ for an element v of cl (G) if and only if $S(v, v) = y$, i.e., v lies in cl (G_y) and $C'_y(v) = v$. All such points v lie in G, however, and coincide with the fixed points of the compact mapping C'_y. Hence $f^{-1}(y)$ is compact.   q.e.d.

PROOF OF (c). The fact that $\deg_2([f, S], G, y)$ is well defined if $f^{-1}(y) \subset G$ follows from the results of part (b). We need only verify the homotopy invariance property (h) for the homotopies of restricted representations.

Let S be a mapping of cl $(G) \times \text{cl}(G) \times [0, 1]$ giving a homotopy of representations of the family of mappings $\{f_t : 0 \leq t \leq 1\}$ in the sense of the condition (h).

For each t, we define a mapping C_t of the subset G_t of G given by

$$G_t = \{v \mid v \in G, S_{v,t}(G) \text{ contains } y(t)\}$$

with

$$C_t(v) = S_{v,t}^{-1}(y(t)) \qquad (v \in G_t).$$

Then by the argument of part (b), it follows that for each t in $[0, 1]$, cl $(G_t) \subset G'_t$, where

$$G'_t = \{v \mid v \in \text{cl } (G), S_{v,t}(\text{cl } (G)) \text{ contains } y(t)\}$$

and C_t can be extended to the continuous mapping C'_t of cl (G_t) into cl (G) given by $C'_t(v) = S_{v,t}^{-1}(y(t))$. C'_t maps each cl (G_t) continuously into a relatively compact subset of cl (G), and by hypothesis $(I - C'_t)$ has no fixed points on the boundary of G_t since each such fixed point v satisfies the condition that $C'_t(v) = v$ must lie in bdry (G) and satisfy the equation $f_t(x) = y(t)$.

Hence

$$\deg_2 ([f_t, S_t], G, y(t)) = \deg_{\text{LS}} (I - C_t, G_t, 0)$$

is well defined. We need to show that it is independent of t in $[0, 1]$. For this purpose, we need to verify the following two properties which enable us to apply the basic homotopy property (property (3)) for the Leray-Schauder degree:

(i) The subset G' of $X \times [0, 1]$ given by

$$G' = \{[v, t] \mid [v, t] \in X \times [0, 1], v \in G_t\}$$

is open in $X \times [0, 1]$.

(ii) The mapping C' of cl (G') into X given by

$$C'(v, t) = C'_t(v) \qquad (t \in [0, 1], v \in \text{cl } (G_t))$$

is continuous and has relatively compact image in X.

PROOF OF PROPERTY (i). Suppose that $[v_0, t_0]$ lies in G'. Then there exists u_0 in G such that $S(u_0, v_0, t_0) = y(t_0)$. We choose a product neighborhood $N \times N_1 \times N_0$ of $[u_0, v_0, t_0]$ in $G \times G \times [0, 1]$ with N connected. Then the mapping $[v, t] \to S_{v,t}$ gives a continuous and compact mapping of $N_1 \times N_0$ into the space of permissible homeomorphisms of cl (N) into Y. Using the same arguments as applied in the proof of Theorem (12.5), it follows that for N_1 and N_0 sufficiently small, $[v, t]$ in $N_1 \times N_0$ implies that $S_{v,t}(N)$ contains $y(t)$, i.e., $N_1 \times N_0$ is contained in G_t and $C_t(N_1 \times N_0) \subset N$. Hence G' is open in $X \times [0, 1]$, and C' is continuous from G' to X.

PROOF OF (b). The argument applied in the proof of part (a) yields the continuity of C' on cl (G') to X, as well as its continuity on G'. We need therefore only to verify that C' has a relatively compact image in cl (G). Suppose that $\{[v_k, t_k]\}$ is an infinite sequence in cl (G') and let $u_k = C'(v_k, t_k)$. Passing to an infinite subsequence, we may assume without loss of generality that S_{v_k, t_k} converges uniformly on cl (G) as $k \to \infty$ to a permissible homeomorphism h of cl (G) into Y. We may assume that $t_k \to t$ and that $y(t_k)$ converges strongly to $y(t)$ as $k \to \infty$. Then

$h(u_k) - y(t_k) \to 0$ $(k \to \infty)$ so that $y(t)$ lies in cl $(h(\mathrm{cl}\,(G))) = h(\mathrm{cl}\,(G))$. By the continuity of h^{-1} on $h(\mathrm{cl}\,(G)) = \mathrm{cl}\,(h(G))$, we have $u_k = h^{-1}(h(u_k)) \to h^{-1}(y(t))$. Hence $u_k \to h^{-1}(y(t))$ and C' is a compact mapping, proving property (ii).

Hence the proof of part (c) is complete, and with it, the proof of Theorem (12.6).
q.e.d.

THEOREM (12.7). *Let X and Y be Banach spaces, G an open subset of X, f a continuous mapping of cl (G) into Y having a restricted intertwined representation S in the sense of Definition (12.7) with respect to a given class M of permissible homeomorphisms from X to Y. Suppose that M is convex. Then*

(a) $\deg_2([f, S], G, y)$ *is defined whenever $f^{-1}(y)$ is contained in G and is independent of the choice of representation S for f with respect to the class M in the sense of Definition (12.7). We denote this common value by* $\deg_{2,M}(f, G, y)$.

(b) *This degree function* $\deg_{2,M}(f, G, y)$ *has the basic properties:*

(1) $\deg_{2,M}(f, G, y) \neq 0$ *implies that there exists x in G such that $f(x) = y$.*

(2) $\deg_{2,M}(f, G, y)$ *is additive in G in the usual sense.*

(3) $\deg_{2,M}(f, G, y)$ *is invariant under permissible homotopies of the mapping f, i.e., suppose that $\{f_t : 0 \leq t \leq 1\}$ is a continuous path in the space of mappings of cl (G) into Y such that each f_t for t in $[0, 1]$ has a restricted representation in the sense of Definition (12.7) with respect to a given convex class M of permissible homeomorphisms from X to Y. Let $\{y(t) : 0 \leq t \leq 1\}$ be a continuous curve in Y such that $y(t)$ does not lie in $f_t(\mathrm{bdry}\,(G))$ for any t in $[0, 1]$. Then* $\deg_{2,M}(f_t, G, y(t))$ *is well defined for each t in $[0, 1]$ and is independent of t in $[0, 1]$.*

PROOF OF THEOREM (12.7). PROOF OF (a). For each representation S of f with respect to the class M, $\deg_2([f, S], G, y)$ is well defined. We must show that this degree function is independent of the choice of the representation S. Let S_1 be another representation. Then for t in $[0, 1]$, we define $S_t(u, v) = (1 - t)S(u, v) + tS_1(u, v)$. Then each $S_{v,t}$ is a homeomorphism of cl (G) into Y lying in the class M since M is convex by assumption, and it follows immediately that the family of representations $\{S_t : 0 \leq t \leq 1\}$ is a continuous homotopy of representations for the single mapping f in the sense of Theorem (12.6)(c). Hence $\deg_2([f, S_t], G, y)$ is independent of t in $[0, 1]$ by Theorem (12.6), so that $\deg_2([f, S], G, y) = \deg_2([f, S_1], G, y)$. Thus the values of the degree function $\deg_2([f, S], G, y)$ are independent of the choice of the representation S, and we denote the common value by $\deg_{2,M}(f, G, y)$.
q.e.d.

PROOF OF (b). Properties (1) and (2) of $\deg_{2,M}(f, G, y)$ follow without further argument from the corresponding properties of $\deg_2([f, S], G, y)$ obtained in Theorem (12.6). It therefore suffices to establish the homotopy invariance property for the degree, property (3). Since $\deg_{2,M}(f_t, G, y(t))$ is well defined under the assumptions of property (3) for each t in $[0, 1]$, it is only necessary to show that this degree is independent of t in $[0, 1]$. Since $[0, 1]$ is a connected topological space, it suffices to show that for each t_0 in $[0, 1]$, $\deg_{2,M}(f_t, G, y(t))$ is independent of t in some interval about t_0. Hence, we can assume without loss of generality that if we are given $\epsilon > 0$, then $\|f_t(u) - f_{t_0}(u)\| < \epsilon$ $(u \in \mathrm{cl}\,(G))$, and

$\|y(t) - y(t_0)\| < \epsilon$ ($t \in [0, 1)$. We may assume also that $t_0 = 0$, and we need to prove only that

$$\deg_{2,M}(f_0, G, y(0)) = \deg_{2,M}(f_1, G, y(1)).$$

To prove this last equality, we note that $f_0(\text{bdry}(G))$ is a closed subset of Y which does not contain the point $y(0)$. (Indeed, it follows from our hypothesis that $y(0)$ is not a point of $f_0(\text{bdry}(G))$.) To show that $f_0(\text{bdry}(G))$ is closed in Y, it suffices to show that f_0 is a closed mapping of $\text{cl}(G)$ into Y, and this assertion will follow if we can show that f_0 is a proper mapping of $\text{cl}(G)$ into Y, i.e., $f_0^{-1}(K)$ is compact for each compact subset K of Y. Suppose that $\{u_k\}$ is a sequence in $\text{cl}(G)$ with $f(u_k)$ in K for each k. We may assume by passing to an infinite subsequence that $f(u_k) \to y$ as $k \to \infty$. Since $f(x) = S(x, x)$ ($x \in \text{cl}(G)$), for a given representation S of f in restricted intertwined form according to Definition (12.7), it follows that $S_{u_k}(u_k) = f(u_k) \to y$. By passing to another infinite subsequence, we may assume that S_{u_k} converges uniformly on $\text{cl}(G)$ to a permissible homeomorphism h of $\text{cl}(G)$ into Y. Hence $h(u_k) - f(u_k) \to 0$, i.e., $h(u_k) \to y$ as $k \to \infty$. Hence y lies in $\text{cl}(h(\text{cl}(G))) = h(\text{cl}(G))$, and $u_k = h^{-1}(h(u_k)) \to h^{-1}(y)$. Hence the original sequence $\{u_k\}$ was relatively compact in $\text{cl}(G)$.

Hence there exists $d > 0$ such that $B_{2d}(y)$ does not intersect $f_0(\text{bdry}(G))$. We take $\epsilon > 0$ so that $\epsilon < d$. Then we define a homotopy of representations between any given representation S_0 for f_0 and another given representation S_1 for f_1 by setting

$$g_t = (1-t)f_0 + tf_1,$$

$$S_t(u, v) = (1-t)S_0(u, v) + tS_1(u, v).$$

We note that for each $[v, t]$ in $\text{cl}(G) \times [0, 1]$, $S_{v,t} = (1-t)S_{0,v} + tS_{1,v}$ in a permissible homeomorphism of $\text{cl}(G)$ into Y lying in the class M, by the assumed convexity of the class M. We verify immediately that all the conditions for a permissible homotopy of representations in the sense of Theorem (12.6) will be satisfied if we show that $g_t(\text{bdry}(G))$ does not contain $z(t) = (1-t)y(0) + ty(1)$, and we will then obtain $\deg_2([g_t, S_t], G, z(t))$ independent of t in $[0, 1]$. This proves the desired assertion if we take the special cases $t = 0$ and 1.

On the other hand, suppose that x lies in $\text{bdry}(G)$. Then

$$\|g_t(x) - z(t)\| \geq \|f_0(x) - y(0)\| - \|f_0(x) - g_t(x)\| - \|z(t) - y(0)\|$$
$$> 2d - \|f_0(x) - f_1(x)\| - \|y(1) - y(0)\|$$
$$> 2d - 2\epsilon > 0. \quad \text{q.e.d.}$$

THEOREM (12.8). *Let X and Y be Banach spaces, G an open subset of X, f a continuous mapping of $\text{cl}(G)$ into Y, y a point of Y such that y does not lie in $\text{cl}(f(\text{bdry}(G)))$. Suppose that there exists an infinite sequence $\{f_k\}$ of continuous mappings of $\text{cl}(G)$ into Y converging uniformly to f on $\text{cl}(G)$, where each f_k has a restricted representation of the form described in Definition (12.7) with respect to a fixed convex class M of permissible homeomorphisms from X to Y.*

Then:

(a) *The degrees* $\deg_{2,M}(f_k, G, y)$ *are well defined and independent of k for $k > K$ for some K. This limit value is independent of the choice of the sequence $\{f_k\}$ approximating the mapping f, and we denote it by* $\deg_{2,M}(f, G, y)$

(b) *The degree function defined in part* (a) *for mappings f which are approximable by mappings $\{f_k\}$ which have restricted representations with respect to the convex class M of permissible homeomorphisms has the following properties:*

(1) *Suppose that* $\deg_{2,M}(f, G, y)$ *is well defined and* $\neq 0$. *Then y lies in* cl $(f(G))$. *If $f(\text{cl}(G))$ is closed in Y, it follows that $y \in f(G)$.*

(2) $\deg_{2,M}(f, G, y)$ *is additive in the domain G in the usual sense.*

(3) $\deg_{2,M}(f, G, y)$ *is invariant under permissible homotopies in the following sense: Let $\{f_t: 0 \leq t \leq 1\}$ be a continuous curve in the space of continuous mappings of* cl (G) *into Y, $\{y(t): 0 \leq t \leq 1\}$ a continuous curve in Y such that for each t, in $[0, 1]$, $y(t)$ does not lie in* cl $(f_t(\text{bdry }(G)))$. *Suppose that each f_t, for t in $[0, 1]$, is uniformly approximable on* cl (G) *by mappings $\{f_k\}$ each of which has a restricted intertwined representation on* cl (G) *in the sense of Definition* (12.7) *with respect to the convex class M of permissible homeomorphisms from X to Y. Then* $\deg_{2,M}(f_t, G, y(t))$ *is well defined and independent of t in $[0, 1]$.*

PROOF OF THEOREM (12.8). PROOF OF (a). By hypothesis, there exists a constant $d > 0$ such that $B_d(y)$ does not intersect $f(\text{bdry }(G))$. Since the mappings f_k approximate f uniformly on cl (G), there exists an index K such that for $k > K$, $\|f_k(x) - f(x)\| < d$ ($x \in$ cl (G)). Suppose that $j, k > K$. We consider representations S and S_1 for f_j and f_k, respectively, in the sense of Definition (12.7), and for t in $[0, 1]$, we set

$$S_t(u, v) = (1 - t)S(u, v) + tS_1(u, v),$$

so that for each t in $[0, 1]$ and v in cl (G),

$$S_{v,t}(u) = (1 - t)S_v + tS_{1,v}.$$

Since M is a convex class, it follows that $\{S_t: 0 \leq t \leq 1\}$ is a homotopy of representations and the mappings $g_t(u) = S_t(u, u)$ have the property that $\deg_{2,M}(g_t, G, y)$ is independent of t in $[0, 1]$ provided that $g_t(x)$ is never equal to y for any point x in bdry (G). However, for x in bdry (G), we know that $g_t(x)$ is a convex linear combination of $f_k(x)$ and $f_j(x)$, both of which lie in the convex set $B_d(f(x))$. Hence $g_t(x)$ lies in $B_d(f(x))$ for t in $[0, 1]$, while y lies outside of $B_d(f(x))$ by our assumption above. Hence $g_t(x) \neq y$ for any x in bdry (G) and any t in $[0, 1]$. It follows that $\deg_{2,M}(g_0, G, y) = \deg_{2,M}(g_1, G, y)$, i.e., $\deg_{2,M}(f_j, G, y) = \deg_{2,M}(f_k, G, y)$ for any $j, k > K$. Thus we have shown that $\deg_{2,M}(f_k, G, y)$ is independent of k for k sufficiently large for any sequence $\{f_k\}$ converging uniformly to f, where each f_k has a representation with respect to the convex class M in the sense of Definition (12.7).

Finally, if we have two sequences $\{f_k\}$ and $\{g_k\}$ converging to f uniformly on G, with each sequence satisfying the above conditions, we may intersperse the sequence to obtain a single sequence converging to f. Applying the above argument to this sequence, we see that

$$\lim_{k \to \infty} \deg_{2,M}(f_k, G, y) = \lim_{k \to \infty} \deg_{2,M}(g_k, G, y),$$

and we denote this common value (which is thereby obviously independent of the choice of the approximating sequence) by $\deg_{2,M}(f, G, y)$. q.e.d.

PROOF OF (b). PROOF OF PROPERTY (1). Suppose that $\deg_{2,M}(f, G, y)$ is defined as above, and that $\deg_{2,M}(f, G, y) \neq 0$ If we choose a sequence of mappings $\{f_k\}$ with representations in the sense of Definition (12.7) with respect to the class M, such that f_k converges uniformly to f on cl (G), then for a sufficiently large K and all $k > K$, $\deg_{2,M}(f_k, G, y) \neq 0$. Hence by Theorem (12.7), for each such k, there exists a point u_k in G such that $y = f_k(u_k)$. Since

$$\|f(u_k) - f_k(u_k)\| \leq \sup_{x \in \text{cl}(G)} \|f(x) - f_k(x)\| = \epsilon_k \to 0 \quad (k \to \infty),$$

it follows that $f(h_k) \to y$ as $k \to \infty$. Thus y lies in cl $(f(G))$.

If $f(\text{cl}(G))$ is a closed subset of Y, it contains $f(G)$ and hence cl $(f(G))$. Therefore, y lies in $f(\text{cl}(G))$. However, by hypothesis, y does not lie in $f(\text{bdry}(G))$. Hence y lies in $f(G)$. q.e.d.

PROOF OF PROPERTY (2). Suppose that G contains two open subsets G_1 and G_2 such that y does not lie in cl $(f(G'))$, where

$$G' = [\text{cl}(G) - (G_1 \cup G_2)] \cup \text{cl}(G_1 \cap G_2).$$

Then there exists a constant $d > 0$ such that $B_d(y)$ does not intersect $f(G')$. Consider a sequence $\{f_k\}$ converging to f uniformly on cl (G), where each f_k has a restricted intertwined representation with respect to the class M in the sense of Definition (12.7). For k sufficiently large, we know that

$$\deg_{2,M}(f, G, y) = \deg_{2,M}(f_k, G, y),$$
$$\deg_{2,M}(f, G_1, y) = \deg_{2,M}(f_k, G_1, y),$$
$$\deg_{2,M}(f, G_2, y) = \deg_{2,M}(f_k, G_2, y),$$

while $\|f(x) - f_k(x)\| < d$ $(x \in \text{cl}(G))$. In particular, it follows that $f_k(G')$ does not contain the point y. Hence

$$\deg_{2,M}(f_k, G, y) = \deg_{2,M}(f_k, G_1, y) + \deg_{2,M}(f_k, G_2, y).$$

Applying the equalities above, we see that

$$\deg_{2,M}(f, G, y) = \deg_{2,M}(f, G_1, y) + \deg_{2,M}(f, G_2, y). \quad \text{q.e.d.}$$

PROOF OF PROPERTY (3). To show that $\deg_{2,M}(f_t, G, y(t))$ is independent of t in $[0, 1]$ under the hypotheses of property (3) for a permissible homotopy of mappings, it suffices to show that this degree is locally constant on $[0, 1]$ (since $[0, 1]$ is indeed a connected topological space).

Let $t_0 \in [0, 1]$. For t in $[t_0, t_0 + \xi]$, we may assume that

$$\|f_t(x) - f_{t_0}(x)\| < \epsilon, \qquad \|y(t) - y(t_0)\| < \epsilon$$

for all x in cl (G) for a prescribed $\epsilon > 0$ by choosing $\xi > 0$ sufficiently small. Hence, it suffices without loss of generality to assume that for a given $\epsilon > 0$, we have for all t in $[0, 1]$ and all x in cl (G)

$$\|f_t(x) - f_0(x)\| < \epsilon, \qquad \|y(t) - y(0)\| < \epsilon,$$

and to prove under such a hypothesis that $\deg_{2,M}(f_1, G, y(1)) = \deg_{2,M}(f_0, G, y(0))$.

The choice of ϵ is made as follows: By our initial hypothesis, we know that $f_{t_0}(\text{bdry}(G))$ does not have $y(t_0)$ in its closure. Hence there exists a constant $d > 0$ such that $B_{3d}(y)$ does not intersect $f_{t_0}(\text{bdry}(G))$. We then choose $\epsilon = d$, and assume that $\|f_0(x) - y(0)\| > 4d$ for all x in bdry (G).

We choose sequences of uniform approximants $\{f_k\}$ for f_0 on cl (G), $\{g_k\}$ for f_1 on cl (G), with each f_k and g_k having a representation in the sense of Definition (12.7) with respect to the convex class M. We may assume that for all k, and all x in cl (G),

$$\|f_k(x) - f_0(x)\| < \epsilon, \qquad \|g_k(x) - f_1(x)\| < \epsilon,$$

so that in particular

$$\deg_{2,M}(f_0, G, y(0)) = \deg_{2,M}(f_k, G, y(0)),$$
$$\deg_{2,M}(f_1, G, y(1)) = \deg_{2,M}(g_k, G, y(1)).$$

Hence it suffices to prove that

$$\deg_{2,M}(f_k, G, y(0)) = \deg_{2,M}(g_k, G, y(1)).$$

To prove this last equality, we construct a linear homotopy between f_k and g_k for a fixed k, by setting $w_t = (1-t)f_k + tg_k$ ($t \in [0, 1]$). The mappings w_t each have a representation in the sense of Definition (12.7) with respect to the convex class M of permissible homeomorphisms from X to Y. For each x in bdry (G) and each t in $[0, 1]$, if we set $z(t) = (1-t)y(0) + ty(1)$, then

$$\|w_t(x) - z(t)\| \geq \|f_0(x) - y(0)\| - \|w_t(x) - f_0(x)\| - \|z(t) - y(0)\|,$$

while $\|f_0(x) - y(0)\| > 3d$, and

$$\|w_t(x) - f_0(x)\| \leq \max(\|f_k(x) - f_0(x)\|, \|g_k(x) - f_0(x)\|)$$
$$\leq \max(\epsilon, \|g_k(x) - f_1(x)\| + \|f_1(x) - f_0(x)\|)$$
$$\leq 2\epsilon.$$

Moreover, $\|z(t) - y(0)\| \leq \|y(1) - y(0)\| < \epsilon$. Hence

$$\|w_t(x) - z(t)\| > 3d - 3\epsilon > 0.$$

Since $w_t(x) \neq z(t)$ for any x in bdry (G), it follows from Theorem (12.7) that $\deg_{2,M}(w_t, G, z(t))$ is well defined for each t in $[0, 1]$ and independent of t in

[0, 1]. Hence $\deg_{2,M}(w_0, G, z(0)) = \deg_{2,M}(w_1, G, z(1))$, i.e., $\deg_{2,M}(f_k, G, y(0)) = \deg_{2,M}(g_k, G, y(1))$. q.e.d.

A concept related to that of the topological degree of a compact mapping in a Banach space is that of the local fixed point index for compact mappings of metric absolute neighborhood retracts. We present a brief summary of the theory of the fixed point index in this general context below, primarily in order to apply one of the results of this theory to the study of the generalized degree theories presented in the previous portion of this section.

The basic result on the local fixed point index which we shall use is that contained in the following proposition, for whose proof we refer to the discussion in Browder [**60**]:

PROPOSITION (12.4). *Let R be a compact absolute neighborhood retract (ANR), V an open subset of R, g a continuous mapping of V into R whose fixed point set $F_g = \{x \mid x \in V, f(x) = x\}$ is a compact subset of V. Then there exists an integer-valued function* ind (g, V, R) *which is called the index of g on V with respect to R, which has the following five properties and is uniquely determined by these properties:*

(1) *If* ind (g, V, R) *is well defined and $\neq 0$, then g must have a fixed point in V.*

(2) *Suppose that V contains two open subsets V_1 and V_2 while for a given mapping g of V into R, F is a compact subset of $(V_1 \cup V_2) - \text{cl}\,(V_1 \cap V_2)$. Then*

$$\text{ind}\,(g, V, R) = \text{ind}\,(g|_{V_1}, V_1, R) + \text{ind}\,(g|_{V_2}, V_2, R).$$

(3) *The fixed point index is invariant under homotopies in the following sense: Let $\{g_t : 0 \leq t \leq 1\}$ be a continuous path in the space of mappings of V into R, and suppose that there exists a fixed compact subset K_0 of V such that $F_{g_t} \subset K_0$ for all t in $[0, 1]$. Then* ind (g_t, V, R) *is independent of t in $[0, 1]$.*

(4) *For a mapping g of R into R_1,* ind (g, R, R) *is the Lefschetz number $\Lambda(g)$ (i.e., the alternating sum of the traces of the homology endomorphisms induced by g for the homology groups of R with rational coefficients).*

(5) *If R and R_1 are two compact absolute neighborhood retracts, V an open subset of R, V_1 an open subset of R_1, and if g is a continuous mapping of V into R_1, g_1 a continuous mapping of V_1 into R, such that the fixed point set of the composite mapping $g_1 g$ on the set $g^{-1}(V_1)$ is compact, then the fixed point set of the composite mapping $g g_1$ in the other order on $g_1^{-1}(V)$ is also compact, and we have*

$$\text{ind}\,(g_1 g, g^{-1}(V_1), R) = \text{ind}\,(g g_1, g_1^{-1}(V), R_1).$$

For the special case of Banach spaces, we have the following further property: If X is a Banach space, U an open subset of X, and if C is a compact mapping of cl (U) *into X whose fixed point set is a compact subset of U, then for any compact absolute neighborhood retract R which contains the image of C (and in particular for any R which is the finite union of compact convex subsets of X), we have*

$$\deg_{LS}(I - C, U, O) = \text{ind}\,(C|_{U \cap R}, U \cap R, R).$$

We note the following simple consequences of the properties given in Proposition (12.4):

PROPOSITION (12.5). *The local fixed point index described in Proposition (12.4) has the following further properties:*

(6) *If g is a constant mapping of V into R, then $\mathrm{ind}\,(g, V, R) = +1$ if $g(V)$ is a point of V.*

(7) *If R is contained in a second compact absolute neighborhood retract R_1, V_1 is an open subset of R_1, and if g_1 is a continuous mapping of V_1 into R such that F_{g1} is compact, then setting $V = V_1 \cap R$ and $g = g_1|_V$, we have $\mathrm{ind}\,(g, V, R) = \mathrm{ind}\,(g_1, V_1, R_1)$.*

PROOF OF PROPOSITION (12.5). PROOF OF PROPERTY (6). Suppose that g maps V into the point x_0 of V, and let R_0 be a one-point space. We define a mapping f of R_0 into R by setting $f(R_0) = x_0$. Then by property (5), if we let g_0 be the unique mapping of V into R_0, we have $fg_0 = g$, and

$$\mathrm{ind}\,(g, V, R) = \mathrm{ind}\,(I, R_0, R_0) = +1. \quad \text{q.e.d.}$$

PROOF OF PROPERTY (7). Let f be the injection mapping of R into R_1. Then by property (5)

$$\mathrm{ind}\,(g_1, V_1, R_1) = \mathrm{ind}\,(fg_1, V_1, R_1)$$
$$= \mathrm{ind}\,(g_1 f, V_1 \cap R, R) = \mathrm{ind}\,(g, V, R). \quad \text{q.e.d.}$$

In the discussion of the main part of the present section, we have introduced two apparently distinct functions $\deg_1$ and $\deg_2$, both of which are extensions of the usual concept of the topological degree for mappings more general than the class $f = I - C$ with C compact. It is very natural to ask how these two types of degree function are related for the class of mappings for which both can be defined, namely mappings of the form $f = h - C$, with h a permissible homeomorphism and C compact. The surprisingly general answer to this question is provided by the next theorem with the help of the theory of the local fixed point index which we have just introduced.

THEOREM (12.9). *Let X and Y be Banach spaces, M a class of permissible homeomorphisms from X to Y which has the property that for any h in M and any fixed element y_0 of Y, the translation of h by y_0 also lies in M. Let G be an open subset of X, f a continuous mapping of $\mathrm{cl}\,(G)$ into Y such that $f = h - C$, where h is a permissible homeomorphism of $\mathrm{cl}\,(G)$ into Y with $h \in M$, C a continuous mapping of $\mathrm{cl}\,(G)$ into a relatively compact subset of Y. Then*

(a) *The mapping S of $\mathrm{cl}\,(G) \times \mathrm{cl}\,(G)$ into Y given by*

$$S(u, v) = h(u) - C(v) \qquad ([u, v] \in \mathrm{cl}\,(G) \times \mathrm{cl}\,(G))$$

is a representation of f in the form described in Definition (12.7) with respect to the class M. Hence, we may define $\deg_2([f, S], G, y)$ for any point y in $Y - f(\mathrm{bdry}\,(G))$.

(b) *For each y in $Y - f(\mathrm{bdry}\,(G))$, we have $\deg_2([f, S], G, y) = \deg_1([f, h], G, y)$. In particular, if we know in addition that M is a convex class, then $\deg_{1, M}(f, G, y) = \deg_{2, M}(f, G, y)$.*

PROOF OF THEOREM (12.9). PROOF OF (a). For each v in cl (G), the mapping S_v of cl (G) into Y is given by $S_v(u) = h(u) - C(v)$ and is therefore a permissible homeomorphism of cl (G) into Y, with $S_v \in M$ since h lies in M and M is closed under translation by fixed elements of Y. Moreover, since C is compact, it follows that the mapping $v \to S_v$ is a compact mapping of cl (G) into the space of homeomorphisms M.
q.e.d.

PROOF OF (b). By the result of part (a), we know that for any y in $Y - f(\text{bdry } (G))$, $\deg_2 ([f, S], G, y)$ is well defined and is given by

$$\deg_2 ([f, S], G, y) = \deg_{LS} (I - C_y, G_y, 0),$$

where

$$G_y = \{v \mid v \in G, h(u) = y + C(v) \text{ for some } u \text{ in } G\},$$

and

$$C_y(v) = h^{-1}(C(v) + y) \qquad (v \in G_y = (C + y)^{-1}h(G)).$$

Hence

$$\deg_2 ([f, S], G, y) = \deg_{LS} (I - h^{-1}(C + y), (C + y)^{-1}h(G), 0).$$

On the other hand,

$$\deg_1 ([f, H], G, y) = \deg_{LS} (I - Ch^{-1}, h(G), y) = \deg_{LS} (I - Ch^{-1} - y, h(G), 0)$$
$$= \deg_{LS} (I - (C + y)h^{-1}, h(G), 0).$$

We assume without loss of generality that $y = 0$. Let R be the convex closure of the image of C in Y, R_1 the convex closure of the image of $h^{-1}(h(\text{cl } (G)) \cap R)$. Then R is a compact absolute retract and so is R_1, and in particular both are compact absolute neighborhood retracts. If we apply the last conclusion of Proposition (12.4), we see that

$$\deg_{LS} (I - Ch^{-1}, h(G), 0) = \text{ind } (Ch^{-1}, R \cap h(G), R),$$

and

$$\deg_{LS} (I - h^{-1}C, C^{-1}h(G), 0) = \text{ind } (h^{-1}C, C^{-1}(h(G)) \cap R_1, R_1).$$

Hence, it suffices to prove that

$$\text{ind } (Ch^{-1}, R \cap h(G), R) = \text{ind } (h^{-1}C, R_1 \cap C^{-1}(h(G)), R_1).$$

We note first that Ch^{-1} does indeed map $R \cap h(G)$ into R, and $h^{-1}C$ does indeed map $R_1 \cap C^{-1}(h(G))$ into R_1. By the definition of R_1, h^{-1} maps $R \cap h(G)$ into R_1 while C maps all of cl (G) into R. Hence, we may apply property (5) of Proposition (12.4) for the local fixed point index, and we find that

$$\text{ind } (Ch^{-1}, R \cap h(G), R) = \text{ind } (Ch^{-1}, R \cap (h^{-1})^{-1}(G), R)$$
$$= \text{ind } (h^{-1}C, R_1 \cap C^{-1}(h(G) \cap R), R_1)$$
$$= \text{ind } (h^{-1}C, R_1 \cap C^{-1}(h(G)), R_1).$$

Hence the proof of the desired equality is complete.
q.e.d.

In the earlier part of the discussion of the present section, we have defined the degree function for a class of mappings f of cl (G) into Y which were limits of mappings represented in terms of permissible homeomorphisms and compacts, but only in the case when all the homeomorphisms involved belong to a convex family M. It is very natural to ask whether a definition of a degree function can be carried through for such limit mappings f without assuming that the basic class M is convex. A partial affirmative answer to this question is given by the following:

DEFINITION (12.8). *Let X and Y be Banach spaces. Then:*

(I) *H_0 is the class of mappings S of $X \times X$ into Y such that for each v in X, the mapping S_v of X into Y, given by $S_v(u) = S(u, v)$, is a homeomorphism of X onto Y, while the mapping $v \to S_v$ is a continuous mapping of X into the space of homeomorphisms of X onto Y with the topology of uniform convergence on bounded subsets of X, and the image of X under the mapping $v \to S_v$ is relatively compact in the space of homeomorphisms of X onto Y.*

(II) *H is the class of mappings S of $X \times X$ into Y such that there exists a sequence $\{S_k\}$ from H_0 which converges to S uniformly on each bounded subset of $X \times X$, while for each y in Y and each bounded subset B of X, $\{S_{k,v}^{-1}(y) : k \geq 1, v \in B\}$ is a bounded subset of X.*

THEOREM (12.10). *Let X and Y be Banach spaces, G a bounded open subset of X, f a continuous mapping of cl (G) into Y such that for a mapping S of class H, $f(x) = S(x, x)$ for all x in cl (G). Let y be a point of $y -$ cl $(f(\text{bdry }(G)))$. Then:*

(a) *If $\{S_k\}$ is a sequence in the class H_0 converging to S in the sense of Definition (12.8)(II), then for f_k given by $f_k(x) = S_k(x, x)$ for x in cl (G), it follows that for some $K > 0$ and all $k > K$, $\deg_2 ([f_k, S_k], G, y)$ is well defined and independent of k. This common value is independent of the choice of the sequence $\{S_k\}$, and we denote it by $\deg_3 ([f, S], G, y)$.*

(b) *The degree function $\deg_3$ defined in part (a) has the following properties:*

(1) *If $\deg_3 ([f, S], G, y)$ is well defined and $\neq 0$, then y lies in cl $(f(g))$. In particular, if $f(\text{cl }(G))$ is closed in Y, then y lies in $f(G)$.*

(2) *The degree function $\deg_3$ is additive in the domain G in the following sense: If G contains two open subsets G_1 and G_2 and if y lies in $Y -$ cl $(f(G'))$, where $G' = [G - (G_1 \cup G_2)] \cup \text{cl }(G_1 \cap G_2)$, then*

$$\deg_3 ([f, S], G, y) = \deg_3 ([f, S], G_1, y) + \deg_3 ([f, S], G_2, y).$$

(3) *The degree function $\deg_3$ is invariant under homotopies in the following sense: Let $\{f_t : 0 \leq t \leq 1\}$ be a continuous curve of mappings of cl (G) into Y such that there exists a continuous curve $\{S_t : 0 \leq t \leq 1\}$ in H such that for each t in $[0, 1]$, $f_t(x) = S_t(x, x)$ for all x in cl (G). Let $\{y(t) : 0 \leq t \leq 1\}$ be a continuous curve in Y such that for each t in $[0, 1]$, $y(t)$ lies in $Y -$ cl $(f_t(\text{bdry }(G)))$. Then for each t in $[0, 1]$, $\deg_3 ([f_t, S_t], G, y(t))$ is well defined, and this degree function is independent of t in $[0, 1]$.*

PROOF OF THEOREM (12.10). We begin with some general remarks which are relevant to the arguments for both part (a) and part (b), and then give the proofs for part (a) and the three properties stated in part (b). We begin with several lemmas.

LEMMA (12.1). *Let X and Y be Banach spaces, S a mapping of $X \times X$ into Y, G a bounded open set in X. Suppose that S lies in the class H_0 defined in Definition (12.8). We define a mapping C of cl (G) into X by setting $C(v) = S_v^{-1}(y)$, where S_v is the homeomorphism of X onto Y given by*

$$S_v(u) = S(u, v) \qquad (u \in X).$$

Then:

(a) *C is a continuous mapping of cl (G) into a relatively compact subset of X.*

(b) *If for the mapping g of X into Y given by $g(x) = S(x, x)$ ($x \in X$), $g^{-1}(y)$ does not intersect the boundary of G, then C has no fixed points on the boundary of G. More generally, the fixed point set of C coincides with $g^{-1}(y)$ and is contained in the set G_y defined by*

$$G_y = \{v \mid v \in G, S_v(G) \text{ contains } y\}$$

(the latter under the hypothesis that $g^{-1}(y)$ does not intersect bdry (G)).

PROOF OF LEMMA (9.1). Since S_v is a homeomorphism of X onto Y for each v in X, it follows that C is a well-defined mapping of cl (G) into X (and indeed that the definition of C can be extended so that C is defined on the whole space X). To show that C is continuous, suppose that v_k converges to v in cl (G). By the definition of the class H_0, the homeomorphisms S_{v_k} converge to the homeomorphism S_v as $k \to \infty$, uniformly on each bounded subset of X. Let $u_k = S_{v_k}^{-1}(y) = C(v_k)$, $u = S_v^{-1}(y) = C(v)$. Each of the homeomorphisms S_v for any v in X is a permissible homeomorphism in the sense of Definition (12.1) above since S_v maps X homeomorphically onto Y, and hence maps each closed set in X onto a closed set in Y. If N is a connected neighborhood of u in X, then y lies in $S_v(N)$ and there exists a closed ball $B_d(y)$ contained in $S_v(N)$. For all v' in a suitable neighborhood of v, $S_{v'}(N)$ contains a point of $B_d(y)$ and $S_{v'}(\text{bdry }(N))$ is disjoint from $B_d(y)$. Applying the repeated argument used in the proof of Proposition (12.1) and elsewhere above that for such v', $S_{v'}(N)$ contains $B_d(y)$, it follows that for such a choice of the neighborhood N of u and all $k > k(N)$, we have u_k in N. Since X is locally connected, it follows that $u_k \to u$ in X as $k \to \infty$. Hence C is a continuous mapping.

To prove that $C(\text{cl }(G))$ is relatively compact in X, we consider another sequence $\{v_k\}$ in cl (G), without assuming that $\{v_k\}$ converges. By hypothesis, the mapping $v \to S_v$ maps bounded subsets of X into relatively compact subsets of the spaces of homeomorphisms of X onto Y. Hence, by passing to an infinite subsequence, we may assume that S_{v_k} converges to a homeomorphism h of X onto Y, uniformly on each bounded subset of X, as $k \to \infty$. Let $u = h^{-1}(y)$. Then by the same argument as in the preceding paragraph, $u_k = C(v_k) = S_{v_k}^{-1}(y)$ must converge to u as $k \to \infty$, i.e., $C(v_k)$ is strongly convergent in X. Hence, C is a compact mapping of cl (G) into X.

Let $g(x) = S(x, x)$ for all x in cl (G). Then we see that x is a point of $g^{-1}(y)$ if and only if $S(x, x) = y$, i.e., $x = S_x^{-1}(y) = C(x)$. Hence $g^{-1}(y)$ consists of the fixed point set of the mapping C as defined in the statement of the lemma. If $g^{-1}(y)$ does not meet bdry (G), $F(C)$, the fixed point set of C, does not meet bdry (G).

Suppose that $g^{-1}(y)$ does not meet bdry (G). Then the fixed point set of $C = g^{-1}(y)$ is a subset of G. If $x \in g^{-1}(y)$, we have $S(x, x) = y$, i.e., $S_x(G)$ contains y. Hence x lies in the subset G_y of G given by $G_y = \{v \mid v \in G, y \in S_v(G)\}$. q.e.d.

LEMMA (12.2). *Let X and Y be two Banach spaces, S_0 and S_1 two mappings of $X \times X$ into Y of the class H_0 described in Definition (12.8). Let g_0 and g_1 be the two mappings of X into Y given by*

$$g_0(x) = S_0(x, x), \qquad g_1(x) = S_1(x, x) \qquad (x \in X).$$

Suppose that there exist $\epsilon > 0$ and a bounded open set G of X such that for x in bdry (G), the following inequality is valid:

$$\|y - g_0(x)\| > \epsilon.$$

Let C_0 and C_1 be the mappings of cl (G) into X defined in Lemma (12.1) for the mappings S_0 and S_1, respectively, and suppose that for each v in cl (G) and each u in $C_1(\text{cl}(G))$, we have $\|S_0(u, v) - S_1(u, v)\| \leq \epsilon$.

Then there exists a homotopy of C_0 to C_1 through compact mappings C_t of cl (G) into X whose ranges are contained in a fixed compact subset K_0 of X and such that for all x in bdry (G), $C_t(x) \neq x$. In particular,

$$\deg_{\text{LS}}(I - C_0, G, 0) = \deg_{\text{LS}}(I - C_1, G, 0).$$

PROOF OF LEMMA (12.2). We may assume $y = 0$. We construct the homotopy $\{C_t : 0 \leq t \leq 1\}$ by setting

$$C_t(v) = S_{0,v}^{-1}(tS_0(C_1(v), v)) \qquad (t \in [0, 1], v \in \text{cl}(G)).$$

We remark that by Lemma (12.1), the mapping $v \to C_1(v)$ of cl (G) into X is continuous and has relatively compact image in X. Hence $C_t(v)$ is continuous in $[v, t]$ on cl $(G) \times [0, 1]$. We shall now show that the image of cl $(G) \times [0, 1]$ under the mapping C of cl $(G) \times [0, 1]$ into X given by $C(v, t) = C_t(v)$ is relatively compact in X.

Let $\{[v_k, t_k]\}$ be an infinite sequence in cl $(G) \times [0, 1]$. By passing to an infinite subsequence, we may assume that $t_k \to t$ and that $S_{0,v_k} \to h$ as $k \to \infty$, where t lies in $[0, 1]$ and h is a homeomorphism of X onto Y (the convergence of the S_{0,v_k} being uniform on bounded subsets of X). Hence since we may also ensure that $C_1(v_k)$ converges strongly to some u_0 in X by the compactness of the mapping C_1, we have

$$t_k S_0(C_1(v_k), v_k) = t_k S_{0,v_k}(C_1(v_k)) \to th(u_0).$$

On the other hand, it follows by the argument applied in Lemma (12.1) that if $\{w_k\}$ converges strongly to w_0 in Y and if S_{0,v_k} converges uniformly to the homeomorphism h of X onto Y on bounded subsets of X as $k \to \infty$, then

$$S_{0,v_k}^{-1}(w_k) \to h^{-1}(w_0) \qquad (k \to \infty).$$

Thus, finally,
$$C(v_k, t_k) = S_{0,v_k}^{-1}(t_k S_0(C_1(v_k), v_k)) \to h^{-1}(th(u_0)).$$

It follows that the mapping C as defined above is indeed compact.

To complete the proof of the lemma, it suffices to show that for v on the boundary of G and for t in $[0, 1]$, $C_t(v)$ never coincides with v. Suppose it did, however. Then for such an element v, we have $S_0(v, v) = tS_0(C_1(v), v)$ while $S_1(C_1(v), v) = 0$, by definition. Since $S_0(v, v) = g_0(v)$, it follows that $\|S_0(v, v)\| > \epsilon$. Since $u = C_1(v)$ lies in $C_1(\text{cl }(G))$, we have moreover $\|S_1(C_1 v, v) - S_0(C_1 v, v)\| \le \epsilon$. Hence

$$\|S_0(v, v)\| \le \|S_0(C_1(v), v)\| \le \|S_1(C_1(v), v) - S_0(C_1(v), v)\| \le \epsilon$$

contradicting the reverse inequality on $\|S_0(v, v)\|$. q.e.d.

PROOF OF THEOREM (12.10) COMPLETED. PROOF OF (a). Let $\{S_k\}$ be a sequence of mappings from the class H_0 converging uniformly on each bounded subset of X to the mapping S of $X \times X$ into Y. For the given bounded open subset G of X, we can construct the corresponding sequence $\{C_k\}$ of compact mappings of $\text{cl }(G)$ into X, where

$$C_k(v) = S_{k,v}^{-1}(y) \qquad (v \in \text{cl }(G)).$$

Let
$$B_{y,k} = \{v \mid v \in G, S_{k,v}(G) \text{ contains } y\}.$$

Then if $f_k(x) = S_k(x, x)$, $f_k(x) \to f(x)$ uniformly on $\text{cl }(G)$,

$$\deg_2 ([f_k, S_k], G, y) = \deg_{\text{LS}} (I - C_k, G_{y,k}, 0) = \deg_{\text{LS}} (I - C_k, G, 0)$$

provided that f_k does not take on the value y for x on the boundary of G. This will be true for k sufficiently large, since the assumption that y lies in $Y - \text{cl }(f(\text{bdry }(G)))$ implies that for some $d > 0$, $B_d(y)$ does intersect $f(\text{bdry }(G))$ while for k sufficiently large, $\|f_k(x) - f(x)\| < d$ for all x in bdry (G). Hence to show that $\deg_2 ([f_k, S_k], G, y)$ is the same for all sufficiently large k, it suffices to prove the same fact for $\deg_{\text{LS}} (I - C_k, G, 0)$.

We may assume by the remark of the preceding paragraph that for all $k > K$ and some $d_0 > 0$, $\|f_k(v) - y\| > d_0$ for all v on the boundary of G. By hypothesis we can make $\|S_k(u, v) - S(u, v)\| < \epsilon$ for a prescribed $\epsilon > 0$ and all $[u, v]$ in a given bounded subset B of $X \times X$ by choosing $k > K(\epsilon, B)$ with $K(\epsilon, B)$ sufficiently large. If this inequality is valid, it follows moreover that for $j, k > K(\epsilon, B)$, $\|S_j(u, v) - S_k(u, v)\| < 2\epsilon$.

By the hypothesis on the sense in which the mappings S_k approximate the mapping S, the mappings C_k are uniformly bounded, i.e., for a given bounded set B, there exists a constant $M > 0$ independent of k such that for all v in B and all $k \ge 1$, $\|C_k(v)\| \le M$. Since S_k converges to S uniformly on bounded subsets of $X \times X$, we may choose $K > 0$ such that for $j, k > K$, we have

$$\|S_k(u, v) - S_j(u, v)\| < 2\epsilon$$

for v in $\text{cl }(G)$ and u in $\bigcup_r C_r(\text{cl }(G))$.

We let $2\epsilon < d_0$, and apply Lemma (12.2) to the corresponding mappings C_j and C_k for $j, k > K$. For this purpose, we set $S_0 = S_j$ and $S_1 = S_k$. It follows from Lemma (12.2) and the inequalities we have verified in the immediately preceding remarks that for such j and k,

$$\deg_{LS} (I - C_j, G, 0) = \deg_{LS} (I - C_k, G, 0).$$

Hence the proof of part (a) of Theorem (12.10) is complete. q.e.d.

PROOF OF PART (b). PROOF OF PROPERTY (1). Since

$$\deg_3 ([f, S], G, y) = \deg_2 ([f_k, S_k], G, y)$$

for k sufficiently large, it follows from the properties of the degree function $\deg_2$ that for $k > K$, there exists u_k in G such that $f_k(u_k) = y$. Since f_k converges uniformly to f on cl (G), we find that $f(u_k) \to y$, i.e., y lies in cl $(f(G))$. In particular, if $f(\text{cl } (G))$ is closed in Y, it contains cl $(f(G))$ and y lies in $f(\text{cl } (G))$. Since y does not lie in $f(\text{bdry } (G))$, it follows in that case that y lies in $f(G)$. q.e.d.

PROOF OF PROPERTY (2). Under the hypotheses of this property, it follows from the properties of $\deg_2$ that for k sufficiently large,

$$\deg_2 ([f_k, S_k], G, y) = \deg_2 ([f_k, S_k], G_1, y) + \deg_2 ([f_k, S_k], G_2, y).$$

Taking the limit as $k \to \infty$, we obtain the desired result for $\deg_3$. q.e.d.

PROOF OF PROPERTY (3). If we are given a continuous curve $\{S_t : t \in [0, 1]\}$ in the function space H, in order to verify that $\deg_3 ([f_t, S_t], G, y(t))$ is independent of t in $[0, 1]$, it suffices to show that this degree is locally constant in $[0, 1]$. Hence, we may assume without loss of generality that for a prescribed $\epsilon > 0$, we have $\|y(1) - y(0)\| < \epsilon$ and $\|S_0(u, v) - S_1(u, v)\| < \epsilon$ for all u and v in X with $\|u\| \leq \epsilon^{-1}$, $\|v\| \leq \epsilon^{-1}$.

By hypothesis, there exists $d > 0$ such that $B_{3d}(y)$ does not intersect $f_0(\text{bdry } (G))$. We may choose approximations S_0' and S_1' for S_0 and S_1, respectively, in the class of mappings H_0 such that for some constant $M_0 > 0$, and all v in cl (G), $\|(S_{1,v}')^{-1}(y)\| \leq M_0$, while for v in cl (G) and u in $B_{M_0}(0) \cup \text{cl } (G)$, we have

$$\|S_0'(u, v) - S_0(u, v)\| < \epsilon, \qquad \|S_1'(u, v) - S_1(u, v)\| < \epsilon.$$

We choose $\epsilon > 0$ so that

$$0 < \epsilon < d/3, \quad M_0 \leq \epsilon^{-1}, \quad \sup \|v\|, \quad v \in \text{cl } (G) \leq \epsilon^{-1}.$$

Then it follows that if we set $g_0(x) = S_0'(x, x)$, $g_1(x) = S_1'(x, x)$, we have for x in bdry (G),

$$\|g_0(x) - y\| \geq \|f_0(x) - y\| - \|g_0(x) - f_0(x)\| > 3d - \epsilon > d.$$

On the other hand, if C_0 and C_1 are the mappings corresponding to S_0' and S_1'

$$\|S_0'(C_1(v), v) - S_1'(C_1(u), v)\| < 3\epsilon < d.$$

Hence by Lemma (12.2), $\deg_{LS} (I - C_0, G, 0) = \deg_{LS} (I - C_1, G, 0)$. q.e.d.

13. Compact perturbations of nonexpansive, monotone, and accretive mappings.

The discussion of the present section is devoted to the study of nonlinear mappings in Banach spaces which are obtained from mappings of various noncompact types (nonexpansive, monotone, and accretive in particular) by applying compact perturbations, either additive or more generally in an intertwined form. We shall apply to these questions the general theory of the extended topological degree as developed in §12 and compare the results thus obtained to other more special results obtained in particular cases by *ad hoc* methods.

PROPOSITION (13.1). *Let X be a Banach space, and let M_0 denote the class of mappings T, with each T defined on the closure of an open subset G of X, mapping cl (G) into X, and with T of the form $T = I - U$, where I is the identity mapping of cl (G) into X and U is a mapping of cl (G) into X such that for some constant $k < 1$, $\|U(u) - U(v)\| \leq k \|u - v\|$ for all u and v in cl (G).*
Then:

(a) *M_0 is a convex class of permissible homeomorphisms from X to X.*

(b) *The degree function $\deg_{2, M_0}(f, G, v)$ is defined for any mapping f of cl (G) into X having an intertwined representation with respect to M_0, and any v in $X - f(\text{bdry}(G))$.*

PROOF OF PROPOSITION (13.1). PROOF OF (a). Let u and v be points of cl (G), and let $w = T(u)$, $z = T(v)$. Then, we have

$$\|w - z\| = \|(u - v) - (U(u) - U(v))\|$$
$$\geq \|u - v\| - \|U(u) - U(v)\| \geq (1 - k) \|u - v\|.$$

Hence $\|u - v\| \leq (1 - k)^{-1} \|T(u) - T(v)\|$ and T is a one-to-one bicontinuous mapping of cl (G) onto $T(\text{cl}(G))$.

Let u_0 be a point of G, which by hypothesis is an open subset of X, and let $w_0 = T(u_0)$. There exists $d > 0$ such that the closed ball $B_d(u_0)$ is contained in G. Let w be a point of X in the closed ball $B_{d_1}(w_0)$. We shall show that for $d_1 > 0$ and sufficiently small, each such w lies in $T(B_d(u_0))$. To show that $T(u) = w$ for some u in $B_d(u_0)$, we set $u = u_0 + v$, $\|v\| < d$. Then $T(u) = u - U(u) = u_0 + v - U(u_0 + v)$. We set $w = w_0 + z$, $\|z\| < d_1$. Then $T(u) = w$ is equivalent to the equation $v - U(u_0 + v) + U(u_0) = z$, i.e., $v = S(v) = z - U(u_0) + U(u_0 + v)$. Since

$$\|S(v) - S(v_1)\| = \|U(u_0 + v_1) - U(u_0 + v)\| \leq k \|v - v_1\|,$$

it follows from the Picard contraction principle that S will have a fixed point in the closed ball $B_d(0)$ if S maps this closed ball into itself. However, $S(0) = z$, and $\|S(v) - S(0)\| \leq k \|v\| \leq kd$. Hence for $\|v\| \leq d$,

$$\|S(v)\| \leq \|S(0)\| + \|S(v) - S(0)\| \leq \|z\| + kd \leq d_1 + kd.$$

In particular, if we take $d_1 = (1 - k) d$, we find that S has a fixed point for each z in X with $\|z\| \leq (1 - k) d$, i.e., $T(B_d(u_0))$ contains $B_{(1-k)d}(T(u_0))$. Hence $T(G)$ is an open subset of X.

Since T^{-1} is Lipschitzian and hence uniformly continuous on $T(\text{cl}(G))$, it follows that $T(\text{cl}(G))$ is closed in X. Indeed, suppose that $\{w_k\}$ is a sequence in $T(\text{cl}(G))$ with $w_k = T(u_k)$, and such that w_k converges strongly to w in X. Then $\|u_j - u_k\| \le (1-k)^{-1} \|w_j - w_k\| \to 0$, so that u_k converges strongly to an element u of $\text{cl}(G)$. Since T is continuous, $w = T(u)$, so that $T(\text{cl}(G))$ is closed in X. Since $T(\text{cl}(G))$ is closed in X, it contains $\text{cl}(T(G))$. However, for each point u in $\text{cl}(G)$, there exists a sequence $\{u_k\}$ in G converging strongly to u. Hence $T(u_k)$ converges strongly in $T(u)$, so that $T(u)$ lies in $\text{cl}(T(G))$. Hence $T(\text{cl}(G)) = \text{cl}(T(G))$.

We have thereby verified the fact that each mapping T in M_0 is indeed a permissible homeomorphism of $\text{cl}(G)$ into X. To complete the proof of part (a) of our proposition, we need only to note that M_0 is convex since the class of mappings U of $\text{cl}(G)$ into X which satisfy the inequality $\|U(u) - U(v)\| \le k\|u - v\|$ for some $k < 1$ and all u and v in $\text{cl}(G)$ is obviously convex. q.e.d.

PROOF OF PART (b). This is simply an application of the result of part (a) together with Theorem (12.6) of §12. q.e.d.

DEFINITION (13.1). *Let X be a Banach space, G an open subset of X, g a continuous mapping of $\text{cl}(G)$ into X. Then g is said to be weakly semicontractive from $\text{cl}(G)$ to X if there exists a mapping S of $\text{cl}(G) \times \text{cl}(G)$ into X for which the following properties are valid:*

(a) *For each u in $\text{cl}(G)$, $g(u) = S(u, u)$.*

(b) *For each v in $\text{cl}(G)$, the mapping S_v of $\text{cl}(G)$ into X given by $S_v(u) = S(u, v)$ for all u in $\text{cl}(G)$ is a nonexpansive mapping of $\text{cl}(G)$ into X.*

(c) *The mapping $v \to S_v$ of $\text{cl}(G)$ into the space of continuous mappings of $\text{cl}(G)$ into X is compact.*

THEOREM (13.1). *Let X be a Banach space, G an open subset of X, g a weakly semicontractive mapping of $\text{cl}(G)$ into X such that $g(\text{cl}(G))$ is bounded in X. Let M_0 be the class of permissible homeomorphisms from X to X described in Proposition (13.1). Then:*

(a) *The generalized topological degree $\deg_{2,M_0}(I - g, G, v)$ is well defined by Theorem (12.7) for any point v in $X - \text{cl}(((I - g)(\text{bdry}(G))))$ and has the properties described in Theorem (12.7).*

(b) *Suppose that G is a convex open subset of X and that g maps the boundary of G into $\text{cl}(G)$. Then if $(I - g)(\text{cl}(G))$ is closed in X, g must have a fixed point in $\text{cl}(G)$.*

(c) *If in addition g has no fixed points on $\text{bdry}(G)$ and $(I - g)(\text{bdry}(G))$ is closed in X, then $\deg_{2,M_0}((I - g), G, 0) = +1$.*

PROOF OF THEOREM (13.1). PROOF OF (a). Since M_0 is a convex class of permissible homeomorphisms from X to X by Proposition (13.1), it follows from Theorem (12.7) that we need only show that there exists a sequence $\{f_k\}$ converging uniformly to $I - g$ on $\text{cl}(G)$, where each f_k has an intertwined representation (in the restricted sense) with respect to the class M_0 as described in Definition (12.7). We obtain such a sequence as follows: For each $k \ge 1$, we set $\lambda_k = 1 - k^{-1}$, and let $f_k = I - \lambda_k g$, where $f_k(u) = R_k(u, u)$ for the mapping R_k of cl $(G) \times$ cl (G) into X given by $R_k(u, v) = u - \lambda_k S(u, v)$. For each given element v of cl (G), $R_{kv} = I - \lambda_k S_v$ is of the form $I - U$, with U a strict contraction. By condition (c) of Definition (13.1), the mapping $v \to R_{k,v}$ is a compact mapping of cl (G) into the class M_0 for each fixed k. Finally, $\|(I - g)(u) - f_k(u)\| = \|(1 - \lambda_k)g(u)\| \to 0$ as $k \to \infty$, uniformly for u on cl (G) since $g(\text{cl}(G))$ is assumed to be a bounded subset of X.

q.e.d.

PROOF OF PART (b). Suppose that g has no fixed points on cl (G) and that $(I - g)(\text{cl}(G))$ is a closed subset of X. Then there exists a ball $B_d(0)$ for some $d > 0$ such that $B_d(0)$ has no points in common with $(I - g)(\text{bdry}(G))$. We shall apply the homotopy invariance property of the generalized degree function $\deg_{2, M_0}$ by constructing the homotopy

$$f_t(u) = u - [(1-t)g(u) + tx_0] \qquad (u \in \text{cl}(G), t \in [0, 1])$$

for a given interior point x_0 of G. It follows immediately from this definition that we obtain thereby a homotopy of $(I - g)$ to the homeomorphism $I - x_0$ through mappings of the form $I - g_1$, with each g_1 semicontractive.

To show that for all t in $[0, 1]$, $\deg_{2, M_0}(f_t, G, 0)$ is well defined and that this degree is independent of t in $[0, 1]$, it suffices to show that there exists $d > 0$ such that $\|f_t(x)\| \geq d$ for all x in bdry (G) and all t in $[0, 1]$. If this were not true, however, there would exist a sequence $\{t_k\}$ in $[0, 1]$ and a corresponding sequence $\{x_k\}$ in bdry (G) such that $\|f_{t_k}(x_k)\| \to 0$ as $k \to \infty$. If the sequence $\{t_k\}$ has an infinite subsequence converging to 0, we may assume that $t_k \to 0$ for the whole sequence and we obtain

$$\|(I - g)(x_k)\| \leq \|f_{t_k}(x_k)\| + t_k \|x_0 - g(x_k)\| \to 0,$$

contradicting the fact that $(I - g)(\text{bdry}(G))$ stays a positive distance away from 0. In the other case, we may assume that $t_k \to d_1 > 0$. Since x_k lies in bdry (G) for each k, it follows from the hypothesis of part (b) that $g(x_k)$ lies in cl (G) for each k. Since $t_k \to d_1$, we obtain

$$\|f_{d_1}(x_k)\| \to 0 \qquad (k \to \infty).$$

On the other hand, f_{d_1} has a representation of the form $f_{d_1}(u) = u - S_{d_1}(u, u)$, where S_{d_1} is obtained from the mapping S representing g by

$$S_{d_1}(u, v) = (1 - d_1)S(u, v) + d_1 x_0.$$

Hence for each v in cl (G), $S_{d_1, v} = (1 - d_1)S_v + d_1 x_0$ is a strict contraction, and f_{d_1} has a restricted intertwined representation with respect to the class M_0 in the sense of Definition (12.7). It follows from part (b) of Theorem (12.6) that f_{d_1} maps bdry (G) into a closed subset of X. Hence there must exist x in bdry (G) such that $f_{d_1}(x) = 0$, i.e., $x = (1 - d_1)g(x) + d_1 x_0$. However, x lies in bdry (G) while the convex linear combination $(1 - d_1)g(x) + d_1 x_0$ with $d_1 > 0$ of the point $g(x)$ of cl (G) and the point x_0 of G must lie in G. This contradiction shows that the sequences $\{t_k\}$ and $\{x_k\}$ as described above cannot exist and therefore that $\deg_{2, M_0}(f_t, G, 0)$ is well defined for all t in $[0, 1]$ and independent of t.

For $t = 1$, however, $f_1 = I - x_0$ lies in the class M_0. Hence $\deg_{2, M_0}(f_1, G, 0) = +1$ since 0 lies in $(I - x_0)(G)$, because x_0 is assumed to be a point of G. Thus $\deg_{2, M_0}((I - g), G, 0) = +1$ and by Theorem (12.7), there exists a point x in cl (G) such that $x = g(x)$ since $(I - g)(\text{cl}(G))$ is closed in X. q.e.d.

PROOF OF PART (c). Suppose that $(I - g)(\text{bdry}(G))$ is closed in X and that g has no fixed points on bdry (G), where G is a convex open subset of X. Then it follows that $\deg_{2, M_0}((I - g), G, 0)$ is well defined. We may apply the argument of part (b) under the present hypotheses and verify that for the homotopy f_t constructed in the proof of part (b), $\deg_{2, M_0}(f_t, G, 0)$ is well defined and independent of t in $[0, 1]$. Hence $\deg_{2, M_0}((I - g), G, 0) = \deg_{2, M_0}(f_1, G, 0) = +1$, as before. q.e.d.

DEFINITION (13.2). *Let X be a Banach space, G a subset of X, g a continuous mapping of cl (G) into X. Then g is said to be strongly semicontractive from cl (G) to X if there exists a continuous mapping S of cl $(G) \times$ cl (G) into X which satisfies the following conditions:*

(a) *For each u in cl (G), $g(u) = S(u, u)$.*

(b) *For each v in cl (G), the mapping S_v of cl (G) into X given by $S_v(u) = S(u, v)$ lies in the class of strict contractions of cl (G) into X.*

(c) *The mapping $v \to I - S_v$ is a compact mapping of cl (G) into the class of mappings of the form $I - U$, U a strict contraction.*

THEOREM (13.2). *Let X be a Banach space, G a convex open subset of X, g a strongly semicontractive mapping of cl (G) into X with $g(\text{cl}(G))$ bounded in X. Suppose that g maps the boundary of G into cl (G). Then*

(a) *g has a fixed point in cl (G).*

(b) *If g has no fixed points on bdry (G), we have $\deg_{2, M_0}(I - g, G, 0) = +1$.*

PROOF OF THEOREM (13.2). PROOF OF (a). By the results of part (b) of Theorem (13.1), it suffices to show that $(I - g)(\text{cl}(G))$ is closed in X. However, $I - g$ has a restricted intertwined representation with respect to the class M_0 in the sense of Definition (12.7). Hence, it follows from the result of part (b) of Theorem (12.7) that $(I - g)$ is a closed mapping on cl (G). q.e.d.

PROOF OF (b). By part (c) of Theorem (13.1), it suffices to show that $(I - g)$ $\cdot (\text{bdry}(G))$ is closed in X. This follows from the fact that $(I - g)$ is a closed mapping on cl (G). q.e.d.

THEOREM (13.3). *Let X be a Banach space, G a convex open subset of X, U and C two mappings of cl (G) into X with C a compact mapping of cl (G) into X and U a strict contraction of cl (G) into X with $U(\text{cl}(G))$ bounded in X. Suppose that $(U + C)$ maps the boundary of G into cl (G).*

Then $(U + C)$ has a fixed point in cl (G).

PROOF OF THEOREM (13.3.) We need only to verify that the mapping $g = U + C$ is strongly semicontractive in the sense of Definition (13.2). For this purpose, we set $S(u, v) = U(u) + C(v)$, and verify immediately that $S_v = (I - U) - C(v)$ is an element of M_0 varying compactly with v in cl (G). q.e.d.

DEFINITION (13.3). *Let X be a Banach space, G an open subset of X, g a continuous mapping of cl (G) into X. Then g is said to be a semicontractive mapping of cl (G) into X if there exists a mapping S of cl $(G) \times$ cl (G) into X for which the following conditions hold:*

(a) *For each u in cl (G), $g(u) = S(u, u)$.*

(b) *For each v in cl (G), the mapping S_v of cl (G) into X given by $S_v(u) = S(u,v)$ is a nonexpansive mapping of cl (G) into X.*

(c) *If $\{v_k\}$ is a sequence in cl (G) such that v_k converges weakly to v in cl (G) as $k \to \infty$, then S_{v_k} converges uniformly to S_v on cl (G) as $k \to \infty$.*

THEOREM (13.4). *Let X be a uniformly convex Banach space, G a bounded convex open subset of X, g a semicontractive mapping of cl (G) into X. Suppose that g maps bdry (G) into cl (G).*

Then g has a fixed point in cl (G).

PROOF OF THEOREM (13.4). Let $\lambda_k = 1 - k^{-1}$ for each integer $k \geq 1$. Then if x_0 is a point of cl (G), each of the mappings $g_k = (1 - \lambda_k)x_0 + \lambda_k g$ is strongly semicontractive from cl (G) to X. Moreover, g maps cl (G) into a bounded subset of X. (Indeed, suppose that for a sequence $\{v_k\}$ in cl (G), $\|g(v_k)\| \to \infty$. We may assume that v_k converges weakly to an element of the weakly sequentially compact set cl (G) as $k \to \infty$. Then $\|S_v(v_k) - g(v_k)\| \to 0$ by condition (c) of Definition (13.3). It follows that S_v is an unbounded mapping of cl (G) into X. This contradicts the fact that G is bounded and S_v is nonexpansive.) Hence each g_k maps cl (G) into a bounded subset of G.

We may apply Theorem (13.2) to each g_k since $g_k(\text{bdry}(G))$ is contained in cl (G) for each $k \geq 1$. We find that for each k, we may choose a fixed point v_k of g_k in cl (G). Since cl (G) is closed, convex, and bounded in the reflexive space X, we may extract a weakly convergent subsequence of the original sequence $\{v_k\}$, which we can identify with the original sequence for simplicity of notation, i.e., v_k converges weakly to v in cl (G) as $k \to \infty$.

Since

$$\|g_k(v) - g(v)\| \leq (1 - \lambda_k) \|x_0 - g(v)\| \to 0 \qquad (k \to \infty)$$

uniformly for v in cl (G), it follows that $\|v_k - g(v_k)\| = \|g_k(v_k) - g(v_k)\| \to 0$ as $k \to \infty$. However, $\|S_v(u) - S_{v_k}(u)\| \to 0$ $(k \to \infty)$ uniformly for u in cl (G). Hence $\|S(v_k, v) - g(v_k)\| \to 0$. Finally, we see that $\|(I - S_v)(v_k)\| \to 0$ $(k \to \infty)$.

Since the Banach space X is uniformly convex and the mapping S is assumed to be nonexpansive, we may apply Theorem (8.4)(a) and conclude that $v - S_v(v) = 0$, i.e., $v = g(v)$.   q.e.d.

THEOREM (13.5). *Let X be a uniformly convex Banach space, G a bounded open convex subset of X, U and C two continuous mappings of cl (G) into X with U nonexpansive and C completely continuous (i.e., if v_k converges weakly to v in cl (G), the $C(v_k)$ converges strongly to $C(v)$ in X). Suppose that $(U + C)$ maps the boundary of G into cl (G).*

Then $(U + C)$ has a fixed point in cl (G).

PROOF OF THEOREM (13.5). The result of Theorem (13.5) is a special case of that of Theorem (13.4) if we note that the mapping $g = U + C$ is semicontractive under the representation $S(u, v) = U(u) + C(v)$. q.e.d.

DEFINITION (13.4). *Let M be a metric space. Then:*

(a) *For any subset A of M, the measure of noncompactness $\gamma(A)$ of A is defined to be*

$$\gamma(A) = \inf \{\epsilon \mid \epsilon > 0, A \text{ can be covered by a finite family of subsets of diameter} < \epsilon\}.$$

(b) *A mapping V of A into M is said to be a condensing mapping if for each $A_1 \subset A$ with $0 < \gamma(A_1) < \infty$, we have $\gamma(V(A_1)) < \gamma(A_1)$.*

PROPOSITION (13.2). *Let X be a Banach space, G a bounded subset of X, g a continuous mapping of $\mathrm{cl}\,(G)$ into X such that g is strongly semicontractive (i.e., such that there exists a mapping S of $\mathrm{cl}\,(G) \times \mathrm{cl}\,(G)$ into X such that the conditions of Definition (13.2) hold).*

Then g is a condensing mapping of $\mathrm{cl}\,(G)$ into X.

PROOF OF PROPOSITION (13.2). We must show that for any subset A of $\mathrm{cl}\,(G)$ with $\gamma(A) > 0$, we have $\gamma(g(A)) < \gamma(A)$.

For each v in $\mathrm{cl}\,(G)$, S_v is a strict contraction of $\mathrm{cl}\,(G)$ into X, i.e., there exists a constant $k_v < 1$ such that for all u and u_1 in $\mathrm{cl}\,(G)$, $\|S_v(u) - S_v(u_1)\| \leq k_v \|u - u_1\|$. The constant k of strict contractivity is lower-semicontinuous on the space of strict contractions of $\mathrm{cl}\,(G)$ into X with the topology of uniform convergence. Since the image of $\mathrm{cl}\,(G)$ under the mapping $v \to S_v$ is relatively compact in the class of strict contractions, there exists a constant k independent of v in $\mathrm{cl}\,(G)$ such that $\|S_v(u) - S_v(u_1)\| \leq k \|u - u_1\|$ $(u, u_1 \in \mathrm{cl}\,(G))$.

We may find a finite open covering $\{N_k\}$ of $\mathrm{cl}\,(G)$ such that for any two elements v and v_1 of a given N_k and any u in $\mathrm{cl}\,(G)$,

$$2 \|S_v(u) - S_{v_1}(u)\| < (1 - k)\gamma(A) - \epsilon$$

for $\epsilon > 0$ with $\epsilon < (1 - k)\gamma(A)$. Let $r > 0$ be any constant with $\gamma(A) < r < \gamma(A) + \epsilon$. Then there exists a finite family $\{A_1, \ldots, A_s\}$ of subsets of A with diameter $< r$, which covers A. Then $\{g(A_1), \ldots, g(A_s)\}$ is a covering of $g(A)$. It follows moreover that

$$g(A) = \bigcup_{j=1}^{s} \bigcup_{k=1}^{r} g(A_j \cap N_k).$$

Hence, it suffices to show for each j and k that $\gamma(g(A_j \cap N_k)) < \gamma(A)$.

For the latter purpose, we observe that there exists a constant $d > 0$ with $2d \leq (1 - k)\gamma(A) - \epsilon$ such that for all u in $\mathrm{cl}\,(G)$ and for a given point v_k in N_k, we have $\|S_{v_k}(u) - S_v(u)\| \leq d$ for all v in N_k. Since $g(v) = S_v(v)$, it follows that $2 \|g(v) - S_{v_k}(v)\| \leq d$. If we can decompose $S_{v_k}(A_j \cap N_k)$ into a finite family of subsets of diameter less than r_1 for a given $r_1 > 0$, then taking the d_1-neighborhood of each of these sets, we have a corresponding decomposition of $g(A_j \cap N_k)$ into a finite family of subsets of diameter less than $r_1 + d$. Hence,

$$\gamma(g(A_j \cap N_k)) \leq \mathrm{diam}\,(S_{v_k}(A_j \cap N_k)) + d.$$

On the other hand, S_{v_k} is strongly contractive with constant k. Hence

$$\mathrm{diam}\,(S_{v_k}(A_j \cap N_k)) \leq k\,\mathrm{diam}\,(A_j \cap N_k) \leq kr.$$

Thus

$$\gamma(g(A_j \cap N_k)) \leq kr + d < kr + (1-k)\gamma(A) - \epsilon < \gamma(A). \qquad \text{q.e.d.}$$

A sharpening of Proposition (13.2) is given in the following:

THEOREM (13.6). *Let M be a bounded metric space, g a mapping of a subset A_0 of M into M. Suppose that g is semicondensive in the following sense: There exists a continuous mapping S of $A_0 \times A_0$ into M such that $g(u) = S(u, u)$ for all u in A_0, the mapping S_v of A_0 into M is condensive for each v in A_0 (where $S_v(u) = S(u, v)$ for u and v in A_0), and the mapping $v \to S_v$ is a compact mapping of A_0 into the space of condensive mappings from A_0 to M with the topology of uniform convergence on A_0.*

Then g is a condensive mapping from A_0 to M.

We note that the effective gist of Theorem (13.6) is that the class of condensive mappings is already saturated with respect to possible compact extensions in the form of intertwined representations in the class of condensive mappings.

PROOF OF THEOREM (13.6). Let A be a subset of A_0 with $\gamma(A) > 0$. We must show that $\gamma(g(A)) < \gamma(A)$.

Let f be a condensive mapping of A_0 into M. Then $\gamma(f(A)) < \gamma(A)$. If $\epsilon > 0$ is a constant such that $2\epsilon + \gamma(f(A)) < \gamma(A)$, then for any mapping f_1 of A_0 into M such that $\|f_1(u) - f(u)\| < \epsilon$ ($u \in A_0$), it follows that for any subset B of A_0, $f_1(B)$ lies in the $\epsilon/2$-neighborhood of $f(B)$.

By hypothesis, the mapping $v \to S_v$ is a compact mapping of A_0 into the space of condensive mappings of A_0 into M, i.e., it is continuous in the topology of uniform convergence on the condensive mappings and has relatively compact image in that family. For each mapping S_v, we have a neighborhood N_v in the family of condensive mappings such that for f_1 in that neighborhood

$$\|f_1(u) - S_v(u)\| < \epsilon \qquad (u \in A_0),$$

where ϵ_v is the positive constant defined above for the mapping $f = S_v$ and the given subset A_0. Since the family $\{S_v : v \in A_0\}$ is relatively compact in the space of condensive mappings of A_0 into M, we may find a finite family $\{S_{v_j}, j = 1, \ldots, r\}$ which has the property that the corresponding neighborhoods N_{v_j} form a covering of the set $\{S_v : v \in A_0\}$. Since for v near v_j, S_v lies in N_{v_j}, if we let ψ denote the mapping of A_0 into the space of condensive maps given by $\psi(v) = S_v$, the family $\psi^{-1}(N_j)$ forms a covering of A_0.

Since

$$A_0 = \bigcup_{j=1}^{r} (A \cap \psi^{-1}(N_j)),$$

we know that
$$g(A_0) = \bigcup_{j=1}^{r} (g(A_0) \cap N_j).$$

Hence, $(g(A_0)) \leq \max_{1 \leq j \leq r} (g(A_0) \cap N_j)$ and it suffices to show that for each j, $\gamma(g(A_0) \cap N_j) < \gamma(A_0)$.

On the other hand, $\gamma(S_{v_j}(A_0)) < \gamma(A_0) - 2\epsilon_{v_j}$. For any subset A_1 of A_0 with diam $(S_{v_j}(A_1)) < r$, we know that for u in $A_1 \cap \psi^{-1}(N_j)$, $g(u)$ lies within the ϵ_{v_j}-neighborhood of $S_{v_j}(A_1)$. Hence diam $(g(A_1 \cap N_j)) < r + 2\epsilon_{v_j}$. It follows in particular that $\gamma(g(A_0) \cap N_j) \leq \gamma(S_{v_j}(A_0)) + 2\epsilon_{v_j} < \gamma(A_0)$. q.e.d.

A simple and useful fixed point theorem for condensive mappings is the following:

THEOREM (13.7). *Let X be a Banach space, G a closed bounded convex subset of X, g a condensive mapping of G into G. Then g has a fixed point in G.*

It is convenient to preface the proof of Theorem (13.7) with the following auxiliary result giving various detailed conclusions about the structure of condensive mappings:

PROPOSITION (13.3). *Let X be a Banach space, G a bounded closed convex subset of X, V a condensive mapping of G into G. Then:*

(a) *For two subsets A and B of G,*
$$\gamma(A \cup B) = \max(\gamma(A), \gamma(B)), \quad \gamma(\text{cl}(A)) = \gamma(A).$$

(b) *If x_0 is a point of G, then* cl $(\bigcup_{j \geq 0} V^j(x_0))$ *is compact subset of G.*

(c) $(I - V)$ *is a proper mapping (i.e., $(I - V)^{-1}(K)$ is compact if K is compact). Hence $I - V$ maps closed subsets of G into closed subsets of G.*

(d) *If A is a subset of G, then for the convex closure $C(A)$ of A, $\gamma(C(A)) = \gamma(A)$.*

(e) *There exists a minimal closed convex subset of G invariant under V and intersecting* cl $(\bigcup_{j \geq 0} \{V^j(x_0)\})$ *for a given point x_0 of G.*

PROOF OF PROPOSITION (13.3). PROOF OF (a). If A is a subset of A_1, then it follows trivially from the definition of the measure of noncompactness that $\gamma(A) \leq \gamma(A_1)$. Hence $\gamma(A) \leq \gamma(A \cup B)$, $\gamma(B) \leq \gamma(A \cup B)$, so that
$$\max(\gamma(A), \gamma(B)) \leq \gamma(A \cup B).$$

Suppose that, on the other hand, $r > \gamma(A)$, $r > \gamma(B)$. Then we can find two finite families of subsets $\{A_1, \ldots, A_s\}$ of A and $\{B_1, \ldots, B_t\}$ of B such that each A_j and each B_k have diameter less than r, while
$$A = \bigcup_{j=1}^{s} A_j, \quad B = \bigcup_{k=1}^{t} B_k.$$

Then
$$A \cup B = \bigcup_{j=1}^{s} A_j \cup \bigcup_{k=1}^{t} B_k,$$

so that $\gamma(A \cup B) \leq r$. Hence $\gamma(A \cup B) \leq \max(\gamma(A), \gamma(B))$, and finally

$$\gamma(A \cup B) = \max(\gamma(A), \gamma(B)).$$

Since A is a subset of cl (A), $\gamma(A) \leq \gamma(\text{cl}(A))$. On the other hand, suppose that $r > \gamma(A)$. Then A can be written as the union of a finite family of subsets $\{A_1, \ldots, A_s\}$ with diam $(A_j) < r$ for each j. Then

$$\text{cl}(A) = \bigcup_{j=1}^{s} \text{cl}(A_j),$$

and for each j, diam $(\text{cl}(A_j)) \leq r$. Hence $\gamma(\text{cl}(A)) \leq r$. Since r was any constant greater than $\gamma(A)$, it follows that $\gamma(\text{cl}(A)) \leq \gamma(A)$. Finally, we have $\gamma(\text{cl}(A)) = \gamma(A)$. q.e.d.

PROOF of (b). Let A be the set $\bigcup_{j=0}^{\infty} \{V^j(x_0)\}$. To show that cl (A) is compact, it suffices to prove that $\gamma(A) = 0$. Suppose on the contrary that $\gamma(A) > 0$. By hypothesis, V is condensive so that $\gamma(V(A)) < \gamma(A)$. On the other hand, $A = \{x_0\} \cup V(A)$, so that $\gamma(A) \leq \max(\gamma\{x_0\}, \gamma(V(A)))$. For the single-point set, $\gamma(\{x_0\}) = 0$. Hence $\gamma(A) \leq \gamma(V(A)) < \gamma(A)$, which is a contradiction following from the assumption that $\gamma(A) > 0$. Thus $\gamma(A) = 0$, and cl (A) is compact. q.e.d.

PROOF of (c). Let K be a compact subset of G, and let $A = (I - V)^{-1}(K)$. Then $A \subset V(A) + K$. We note that for two subsets A and B of X, $\gamma(A + B) \leq \gamma(A) + \gamma(B)$. Indeed, suppose that $r > \gamma(A)$, $r_1 > \gamma(B)$. Then A can be written as the union of a finite family $\{A_1, \ldots, A_s\}$ of subsets of diameter less than r and B can be written as the union of a finite family of subsets $\{B_1, \ldots, B_t\}$ of diameter less than r_1. Then

$$A + B = \bigcup_{j=1}^{s} \bigcup_{k=1}^{t} (A_j + B_k).$$

On the other hand, for any pair of elements a and a' of A_j, $\|a - a'\| < r$, while for any pair of elements b and b' of B_k, $\|b - b'\| < r_1$. Hence

$$\|(a + b) - (a' + b')\| < r + r_1,$$

i.e., diam $(A_j + B) < r + r_1$. Hence $\gamma(A + B) \leq r + r_1$, i.e., $\gamma(A + B) \leq \gamma(A) + \gamma(B)$.

Applying this conclusion to A, we find that $\gamma(A) \leq \gamma(V(A)) + \gamma(K)$, while $\gamma(K) = 0$. Hence $\gamma(A) \leq \gamma(V(A))$. Since V is condensive, it follows that $\gamma(A) = 0$ and A is relatively compact in G. Since A is closed in G and G is closed in the complete space X, A is compact.

Since every proper continuous mapping of metric spaces is a closed mapping, it follows that $(I - V)$ maps closed sets of G onto closed sets of G. q.e.d.

PROOF of (d). Let A be a subset of G, $C(A)$ its convex closure. Then $\gamma(A) \leq \gamma(C(A))$. To show the reverse inequality, let $\epsilon > 0$ be given and choose $r > \gamma(A)$. Then there exists a finite family $\{A_1, \ldots, A_s\}$ of subsets of A which covers A and such that diam $(A_j) \leq r$ for each j.

We note first that for each j, $C(A_j)$ is of diameter $\leq r$. Indeed,
$$A_j \subset \bigcap_{x \in A_j} B_r(x)$$
where this intersection is a closed convex subset of G. (We note that $B_d(x)$ denotes the closed ball of radius d about x in G.) Hence
$$C(A_j) \subset \bigcap_{x \in A_j} B_r(x).$$
In particular, for each u in $C(A_j)$, $A_j \subset B_r(u)$. Hence $A_j \subset \bigcap_{u \in C(A_j)} B_r(u)$, and the convexity of the latter intersection implies that $C(A_j) \subset \bigcap_{u \in C(A_j)} B_r(u)$. Hence diam $(C(A_j)) \leq r$.

To show that $\gamma(C(A)) \leq r$, it suffices to show that if A and B are convex subsets of G, then for the convex closure C of $A \cup B$, $\gamma(C) \leq \max(\gamma(A), \gamma(B))$. (Indeed, we need only apply this principle to the union of the first k subsets $((A_1), \ldots, (A_k))$.) On the other hand, it follows from the convexity of A and B that
$$C = \{u \mid u = (1-t)x + ty, x \in A, y \in B, 0 \leq t \leq 1\}.$$
For a given $\epsilon > 0$, we can find a finite subset $\{t_1, \ldots, t_n\}$ of $[0, 1]$ such that for each t in $[0, 1]$, there exists an element t_j in the subset such that
$$2|t - t_j| \text{ diam } (G) < \epsilon.$$
We consider the family of subsets of C given by $C_j = (1-t_j)A + t_j B$. Then
$$\gamma(C_j) \leq \gamma((1-t_j)A) + \gamma(t_j B) \leq (1-t_j)\gamma(A) + t_j \gamma(B) \leq \max(\gamma(A), \gamma(B)).$$
On the other hand, for each u in C,
$$u = (1-t)x + ty \qquad (x \in A, y \in B, t \in [0,1]).$$
There exists an index j such that t is close to t_j as above. Hence for $u_j = (1-t_j)x + t_j y \in C_j$, we have $\|u - u_j\| = \|(t_j - t)(x - y)\| \leq |t - t_j|$ diam (G) $< \epsilon/2$. Hence C is contained in the $(\epsilon/2)$-neighborhood of the set $\bigcup_j^2 C_j$, so that
$$\gamma(C) \leq \epsilon + \gamma\left(\bigcup_j C_j\right) \leq \epsilon + \max_j \gamma(C_j) \leq r + \epsilon.$$
Since $\epsilon > 0$ is arbitrary, it follows that $\gamma(C) \leq r$, i.e., $\gamma(C) \leq \max(\gamma(A), \gamma(B))$.
q.e.d.

PROOF OF (e). Let x_0 be a given point of G, and let
$$A = \text{cl}\left(\bigcup_{j \geq 0} \{V^j(x_0)\}\right).$$
Then A is a compact subset of G by the result of part (b).

We consider the family Λ of subsets G_1 of G, where
$$\Lambda = \{G_1 \mid G_1 \text{ is a closed convex subset of } G, V(G_1) \subset G_1, G_1 \cap A \neq \emptyset\}.$$
The family Λ is nonempty since G itself is an element of Λ. For any linearly

ordered subfamily $\{G_\alpha\}$ of Λ, $\bigcap_\alpha G_\alpha$ is a closed convex subset of G which is invariant under V. Moreover, since A is compact,

$$A \cap \bigcap_\alpha G_\alpha \neq \varnothing.$$

Hence $\bigcap_\alpha G_\alpha$ is an element of Λ. Therefore, we may apply Zorn's lemma to Λ where the latter is ordered by inclusion, and we obtain the existence of a minimal element G_1 of Λ. q.e.d.

With the proof of part (e), the proof of Proposition (13.3) is complete. q.e.d.

PROOF OF THEOREM (13.7). Let x_0 be a given point of G, and let $A = \text{cl}\,(\bigcup_{j \geq 0} \{V^j(x_0)\})$. Then A is a compact subset of G, with A invariant under V. By part (e) of Proposition (13.3), there exists a minimal closed convex subset G_1 of G which is invariant under V and has nonempty intersection with A.

We assert that G_1 is compact. If it were not, then $\gamma(G_1) > 0$, and $V(G_1)$ would be a subset of G_1 such that $\gamma(V(G_1)) < \gamma(G_1)$. If G_2 is the convex closure of $V(G_1)$, then by the conclusion of part (d) of Proposition (13.3), $\gamma(G_2) = \gamma(V(G_1)) < \gamma(G_1)$. Hence, G_2 is a proper subset of G_1, with G_2 closed, convex, and invariant under the mapping V.

There exists a point x in $G_1 \cap A$. Then $V(x)$ lies in $A \cap V(G_1)$, and therefore in $A \cap G_2$. Hence G_2 is an element of Λ which is properly contained in G_1, contradicting the minimality of G_1 in the family Λ.

Thus G_1 must be compact. V is a continuous self-mapping of the compact convex subset G_1 of G and hence V has a fixed point in G_1 by the Schauder fixed point theorem. q.e.d.

THEOREM (13.8). *Let X be a Banach space, G a closed bounded convex subset of X, f a continuous mapping of G into G such that for each subset A of G, $\gamma(f(A)) \leq \gamma(A)$. Suppose that $(I - f)(G)$ is closed in X.*

Then f has a fixed point in G.

PROOF OF THEOREM (13.8). For each $\epsilon > 0$, let $V_\epsilon(u) = (1 - \epsilon)f(u) + \epsilon x_0$, for a given point x_0 of G. Then for each subset A of G,

$$\gamma(V_\epsilon(A)) \leq \gamma((1 - \epsilon)f(A)) + \gamma(\epsilon x_0) = (1 - \epsilon)\gamma(f(A)) \leq (1 - \epsilon)\gamma(A).$$

Hence if $\gamma(A) > 0$, $\gamma(V_\epsilon(A)) < \gamma(A)$.

V_ϵ is therefore a condensive self-mapping of G, and if we apply the result of Theorem (13.7), it follows that V_ϵ has a fixed point u_ϵ in G. For this fixed point, we have

$$u_\epsilon - f(u_\epsilon) = [u_\epsilon - V_\epsilon(u_\epsilon)] + [V_\epsilon(u_\epsilon) - f(u_\epsilon)] = \epsilon[x_0 - f(u_\epsilon)],$$

so that $\|u_\epsilon - f(u_\epsilon)\| \leq \epsilon \,\text{diam}\,(G) \to 0$ ($\epsilon \to 0$). It follows that 0 lies in $\text{cl}\,((I - f)(G))$. Since $(I - f)(G)$ is assumed to be closed in X, 0 lies in $(I - f)(G)$. Hence there exists a point x in G such that $x = f(x)$. q.e.d.

THEOREM (13.9). *Let X be a Banach space, G a closed bounded convex subset of X, g a strongly semicontractive continuous mapping of G into G. Then g has a fixed point in G.*

PROOF OF THEOREM (13.9). By Proposition (13.2), each strongly semicontractive mapping of G into X is a condensive mapping. Hence, the conclusion of Theorem (13.9) follows from Theorem (13.7). q.e.d.

THEOREM (13.10). *Let X be a Banach space, G a closed bounded convex subset of X, U and C two continuous mappings of G into X with C compact and U a strict contraction. Suppose that $(U + C)$ maps G into G.*
Then $(U + C)$ has a fixed point in G.

PROOF OF THEOREM (13.10). The mapping $U + C$ is strongly semicontractive. Hence, the desired result follows from Theorem (13.10). q.e.d.

THEOREM (13.11). *Let X be a Banach space, G a closed bounded convex subset of X, g a continuous mapping of G into G such that g is weakly semicontractive. Suppose that $(I - g)(G)$ is closed in X.*
Then g has a fixed point in G.

COROLLARY TO THEOREM (13.11). *Let X be a Banach space, G a closed bounded convex subset of X, U and C two continuous mappings of G into X with C compact and U nonexpansive. Suppose that $(U + C)$ maps G into G and that $(I - U - C)(G)$ is closed in X.*
Then $(U + C)$ has a fixed point in G.

PROOF OF THEOREM (13.11). It follows easily that for each λ with $0 < \lambda < 1$, $g_\lambda(u) = \lambda g(u) + (1 - \lambda)x_0$ for a given point x_0 in G yields a strongly semicontractive self-mapping g_λ of G. By Theorem (13.9), for each such λ, g_λ has a fixed point u_λ in G. Since

$$\|u_\lambda - g(u_\lambda)\| = |1 - \lambda| \, \text{diam}\, (G) \to 0 \qquad (\lambda \to 1-),$$

it follows that 0 lies in cl $((I - g)(G))$. By hypothesis, $(I - g)(G)$ is closed in X. Hence there exists x in G such that $(I - g)(x) = 0$, i.e., $x = g(x)$. q.e.d.

PROOF OF THE COROLLARY TO THEOREM (13.11). If $T = U + C$ with U nonexpansive and C compact, then T is weakly semicontractive. Hence the result for $(U + C)$ follows from Theorem (13.11). q.e.d.

THEOREM (13.12). *Let X be a uniformly convex Banach space, G a closed bounded convex subset of X, g a continuous semicontractive mapping of G into G. Then g has a fixed point in G.*

PROOF OF THEOREM (13.12). For each λ with $0 < \lambda < 1$, the mapping g_λ of G into G given by $g_\lambda(u) = (1 - \lambda)x_0 + \lambda g(u)$ for u in G yields a strongly semicontractive self-mapping of G. Hence each g_λ has a fixed point u_λ in G for $0 < \lambda < 1$. As $\lambda \to 1-$, it follows as before that $\|(I - g)(u_\lambda)\| \to 0$.

It follows as in the proof of Theorem (13.4) that $(I - g)(G)$ is closed in X. Hence g has a fixed point in G. q.e.d.

THEOREM (13.13). *Let X be a uniformly convex Banach space, G a closed bounded convex subset of X, U and C two continuous mappings of G into X with U*

nonexpansive and C completely continuous (i.e., continuous from sequential weak convergence in G to the strong topology on X). Suppose that $(U + C)$ maps G into G. Then $(U + C)$ has a fixed point in G.

PROOF OF THEOREM (13.13). The mapping $g = U + C$ is semicontractive from G to G. Hence the result of Theorem (13.13) follows from that of Theorem (13.12).   q.e.d.

THEOREM (13.14). *Let G be the unit ball of a separable Hilbert space H. Then there exist mappings U and C of G into G with U nonexpansive and C compact such that $(U + C)(G) \subset G$ which $(U + C)$ does not have a fixed point in G.*

For a given $\epsilon > 0$, C may be chosen Lipschitzian with Lipschitz constant less than ϵ, as well as of finite rank.

PROOF OF THEOREM (13.14). We may choose the Hilbert space H to be the sequence space of square summable sequences $x = (x_1, x_2, \ldots)$ with $\|x\|^2 = \sum_{j=1}^{\infty} x_j^2$ (and all the components x_j real numbers). The nonexpansive mapping U is taken to be the shift mapping $U(x) = (0, x_1, x_2, \ldots)$, i.e.,

$$(U(x))_j = 0, \qquad j = 1,$$
$$= x_{j-1}, \qquad j \geq 2.$$

U is an isometric linear mapping and therefore certainly nonexpansive.

For the mapping C, we take $C(x) = (\tfrac{1}{2}\epsilon(1 - \|x\|^2), 0, 0, \ldots)$ where all the components of $C(x)$ are taken to be zero except the first. Then C is continuous and maps the unit ball G into a one-dimensional subset. Therefore C is obviously compact. Moreover,

$$\|C(x) - C(y)\| < \frac{\epsilon}{2}|(1 - \|x\|^2) - (1 - \|y\|^2)| = \frac{\epsilon}{2}|\|x\|^2 - \|h\|^2|$$
$$\leq \epsilon |\|x\| - \|y\||$$
$$\leq \epsilon \|x - y\|.$$

We assert that for $\epsilon \leq 1$, $(U + C)$ maps G into G. Indeed, suppose that $\|x\| \leq 1$. Then if $y = (U + C)(x)$, we have

$$\|y\|^2 \leq \|U(x)\|^2 + (1 - \|x\|^2)^2 = \|x\|^2 + 1 - 2\|x\|^2 + \|x\|^4$$
$$= 1 + [\|x\|^4 - \|x\|^2] \leq 1.$$

Finally, we assert that $(U + C)$ does not have a fixed point in G. Indeed, suppose that $(U + C)(x) = x$. Then by an inspection of the components of $(U + C)(x)$, we see that for $j \geq 2$, $x_j = x_{j-1}$. Hence all components of x must vanish, i.e., $x = 0$. However $(U + C)(0) = (\epsilon/2, 0, 0, \ldots) \neq 0$.   q.e.d.

We now turn to the consideration of the compact perturbation theory of accretive and pseudo-contractive mappings.

DEFINITION (13.5). *Let X be a Banach space. Then we denote by M_1 the class of mappings h of $\mathrm{cl}\,(G)$ into X, where G is a bounded open subset of X, h is locally* uniformly continuous on cl (G), and for a given duality mapping J of X into X^* there exists a continuous strictly increasing function $c\colon R^+ \to R^+$ with $c(0) = 0$ such that for all u and v of cl (G),

$$(J(u - v), h(u) - h(v)) \geq c(\|u - v\|).$$

DEFINITION (13.6). *Let X be a Banach space with X^* uniformly convex. Then we denote by M_2 the class of mappings h of cl (G) into X, where G is a bounded open subset of X, h is a continuous mapping of cl (G) into X, and there exists a continuous, strictly increasing function $c\colon R^+ \to R^+$ with $c(0) = 0$ such that for all u and v of cl (G),*

$$(J(u - v), h(u) - h(v)) \geq c(\|u - v\|).$$

PROPOSITION (13.4). *Let M_1 and M_2 denote the classes of mappings described in Definitions (13.5) and (13.6). Then:*

(a) M_1 and M_2 are convex classes of permissible homeomorphisms from X to X for the appropriate Banach spaces X.

(b) The degree functions $\deg_{2, M_1}(f, G, v)$, $\deg_{2, M_2}(f, G, v)$ are defined for any mapping f of cl (G) into X having an intertwined representation with respect to the class M_1 or the class M_2, respectively, and for any v in $X - f(\mathrm{bdry}\,(G))$.

PROOF OF PROPOSITION (13.4). PROOF OF (a). We note first that if h satisfies the condition that $(J(u - v), h(u) - h(v)) \geq c(\|u - v\|)$ and if $h(u) = h(v)$, then $c(\|u - v\|) = 0$. Since $c(r) > 0$ for $r > 0$, it follows that $u = v$. Hence h is an injective mapping of cl (G) into X.

Let $M =$ the diameter of G. Then it follows from the assumed inequality on the mapping h that for all u and v in cl (G), $c(\|u - v\|) \leq M \|h(u) - h(v)\|$. Hence h^{-1} is uniformly continuous from $h(\mathrm{cl}\,(G))$ to cl (G) and $h(\mathrm{cl}\,(G))$ is a closed subset of X. Since h is continuous, $h(G)$ is dense in $h(\mathrm{cl}\,(G))$, while $h(\mathrm{cl}\,(G))$ contains cl $(h(G))$. Hence cl $(h(G)) = h(\mathrm{cl}\,(G))$.

To prove that the homeomorphisms h in our two classes are permissible (in the sense of §12), it therefore suffices to prove that for each case, $h(G)$ is an open subset of X. Suppose then that u_0 is a point of G and that for a given constant $d > 0$, the closed ball $B_d(u_0)$ is contained in G. Let $S_d(u_0)$ be the bounding sphere of this ball. By our assumed inequality, we know that for u in $S_d(u_0)$, we have

$$(J(u - u_0), h(u) - h(u_0)) \geq c(d) > 0.$$

Let v be any element of X with $\|v\| < d^{-1}c(d)$. To complete our argument, we shall show that for any such v, $h(u_0) + v$ lies in $h(B_d(u_0))$, i.e., for any $d_1 > 0$ with $d_1 < d^{-1}c(d)$, $B_{d_1}(h(u_0))$ is contained in $h(G)$. It will then follow that $h(G)$ is an open subset of X.

To show that $h(u_0) + v$ lies in $h(B_d(u_0))$, we note first that we have already shown that $h(B_d(u_0))$ is a closed subset of X. Hence, it suffices to show that $h(u_0) + v$ can be approximated arbitrarily closely in X by elements of $h(B_d(u_0))$. To establish this last fact, we choose a constant $\epsilon > 0$, and consider the differential equation

$$du(t)/dt = -h(u(t)) - \epsilon u(t) + h(u_0) + v \qquad (t \geq 0)$$

with the initial condition $u(0) = u_0$. The existence of solutions $u(t)$ for this initial value problem follows from Theorems (9.7) and (9.9), respectively, and the solutions $u(t)$ can be continued as long as they lie in $B_d(u_0)$. We shall show that for sufficiently small $\epsilon > 0$, they will lie in $B_d(u_0)$ for all t in R^+.

To establish this last fact, suppose that $u(t)$ lies in $B_d(u_0)$ for $0 \le t < t_0$ and that $u(t_0)$ lies in $S_d(u_0)$. Let $u_0(t)$ be the constant function $u_0(t) \equiv u_0$ ($t \ge 0$). Then $u_0(t)$ satisfies the differential equation

$$du_0(t)/dt = -h(u_0(t)) + h(u_0).$$

For $t < t_0$ and for $w(t) = u(t) - u_0(t)$, we have

$$\|w(t)\|^2 \ge \|w(t_0)\|^2 + 2(J(w(t_0)), w(t) - w(t_0)).$$

Since $\|w(t)\| \le d$ and $\|w(t_0)\| = d$, we find that $2(J(w(t_0)), w(t) - w(t_0)) \le 0$. Dividing by the negative quantity $(t - t_0)$ and taking the limit as $t \to t_0-$, we obtain the inequality $(J(w(t_0)), dw(t_0)/dt) \ge 0$, i.e.,

$$(J(u(t_0) - u_0), -h(u(t_0)) + h(u_0) - \epsilon u(t_0) + v) \ge 0.$$

If we use the fact that $(J(u(t_0) - u_0), h(u(t_0)) - h(u_0)) \ge c(d)$, we find that

$$c(d) \le (J(u(t_0) - u_0), v - \epsilon u(t_0)) \le d(\|v\| + \epsilon R_0)$$

where $R_0 = \|u_0\| + d$. Since $\|v\| d < c(d)$, it follows that if we choose $\epsilon > 0$ so small that $\epsilon R_0 < c(d) - \|v\| d$, the above inequality becomes $c(d) \le \|v\| d + \epsilon R_0 < c(d)$, which is contradictory. Hence for such ϵ, the solutions $u(t)$ described above will exist for all t in R^+ and lie in $B_d(u_0)$.

It follows from Theorem (9.10) that for the solutions $u(t)$ thus obtained on R^+, $\|du(t)/dt\| \le M_0 \exp(-\epsilon t)$, i.e.,

$$\|h(u(t)) - [h(u_0) + v] + \epsilon u(t)\| \le M_0 \exp(-\epsilon t).$$

Hence for large t, we have $\|h(u(t)) - [h(u_0) + v]\| \le 2\epsilon R_0$. Since $\epsilon > 0$ can be made arbitrarily small in this argument, it follows that $h(u_0) + v$ lies in $\text{cl}(h(B_d(u_0)))$ and therefore in $B_d(u_0)$.

q.e.d.

PROOF OF (b). This follows from the result of part (a) together with Theorem (12.6) of §12.

q.e.d.

DEFINITION (13.7). *Let X be a Banach space, G a bounded open subset of X, g a continuous mapping of $\text{cl}(G)$ into X. Then g is said to be strongly semiaccretive on $\text{cl}(G)$ if there exists a continuous mapping S of $\text{cl}(G) \times \text{cl}(G)$ into X for which the following conditions are all satisfied:*

(a) *For each u in $\text{cl}(G)$, $g(u) = S(u, u)$.*

(b) *For each v in $\text{cl}(G)$, the mapping S_v of $\text{cl}(G)$ into X given by $S_v(u) = S(u, v)$ ($u \in \text{cl}(G)$) is a permissible homeomorphism of $\text{cl}(G)$ into X, lying in the class M_1 or M_2, respectively.*

(c) *The mapping $v \to S_v$ is a continuous mapping of $\text{cl}(G)$ into a relatively compact subset of M_1 or M_2, respectively.*

We shall speak of a g being strongly semiaccretive with respect to M_1 or with respect to M_2 according as to whether the class M_1 or the class M_2 is applied in the above definition.

THEOREM (13.15). *Let X be a Banach space, G a bounded open subset of X, g a continuous strongly semiaccretive mapping of* cl (G) *into X. Then:*

(a) *For each v in $X - g(\text{bdry }(G))$, the generalized degree function* $\deg_{2, M_1}(g, G, v)$ *or* $\deg_{2, M_2}(g, G, v)$ *is defined according as to whether g is strongly semiaccretive with respect to M_1 or with respect to M_2. This degree function has the properties described in Theorem (12.7).*

(b) *Suppose that G is convex and that $f = I - g$ maps the boundary of G into* cl (G). *Then there exists a point x of* cl (G) *such that $g(x) = 0$.*

(c) *Suppose that G is convex, that $f = I - g$ maps the boundary of G into* cl (G), *and that g has no zeros on* bdry (G). *Then* $\deg_{2, M_j}(g, G, 0) = +1$ *(according to which class M_j, g is semiaccretive with respect to).*

(d) *Suppose that $G = B_R(0)$ for some $R > 0$, and that for each u in* bdry (G), $(J(u), g(u)) \geq 0$. *Then there exists a point x in* cl (G) *such that $g(x) = 0$.*

PROOF OF THEOREM (13.15). PROOF OF (a). The conclusion of part (a) follows from Theorem (12.6) and the observation that each strongly semiaccretive mapping g is thereby defined in such a way as to have a restricted intertwined representation with respect to one of the classes M_1 and M_2 of permissible homeomorphisms from X to X.

PROOF of (b). We define a homotopy $\{g_t : 0 \leq t \leq 1\}$ of $g = g_1$ to the mapping $g_0 = I - x_0$ where x_0 is an interior point of G, by setting $g_t = tg_1 + (1-t)g_0$. This is, by its construction, a permissible deformation of strongly semiaccretive mappings with respect to the convex class M_1 or M_2 of permissible homeomorphisms.

If g has a zero on bdry (G), then the conclusion of part (b) is obviously verified. In the other case, g has no zeros on bdry (G). We then assert that for each t in $[0, 1]$, $g_t(x) \neq 0$ for each x in bdry (G). Indeed, let $f = I - g$, and for each t in $[0, 1]$, let $f_t(x) = (1-t)x_0 + tf(x)$. Then $g_t(x) = x - f_t(x)$ $(0 \leq t \leq 1, x \in \text{cl }(G))$. Suppose that $g_t(x) = 0$ for some x in bdry (G) and some t in $[0, 1]$. Then $f_t(x) = x$. Since $g = g_1$ has no zero points on bdry (G), we know that $t < 1$. For such t, $f_t(x)$ is a proper convex linear combination of the point x_0 of G and the point $f(x)$ of cl (G). Hence $f_t(x)$ is itself a point of G, and this is impossible since $x = f_t(x)$ lies in bdry (G). This contradiction shows that $g_t(x) \neq 0$ for any t in $[0, 1]$ and any x in bdry (G).

Applying the homotopy invariance property of the generalized degree function, we see that for the appropriate choice of $j = 1$ or 2, $\deg_{2, M_j}(g_t, G, 0)$ is well defined and independent of t in $[0, 1]$. For $t = 0$, g_0 lies in the corresponding class of homeomorphisms M_j, since $g_0 = I - x_0$. It follows for $t = 0$ that

$$\deg_{2, M_j}(g_0, G, 0) = +1.$$

It follows that $\deg_{2, M_j}(g, G, 0) = +1$, and hence there exists a point x in G such that $g(x) = 0$. q.e.d.

PROOF of (c). The proof of the assertion of part (c) was actually established in the second case of the proof of part (b). q.e.d.

PROOF of (d). We construct the homotopy $\{g_t : 0 \leq t \leq 1\}$ by setting $g_t = tg + (1-t)I$. Then this is a permissible homotopy of strongly semiaccretive mappings. For u in bdry (G), we have

$$(J(u), g_t(u)) = t(J(u), g(u)) + (1-t)(J(u), u) \geq (1-t) \|u\|^2.$$

If for $t = 1$, $g_1 = g$ has a zero point on bdry (G), the conclusion of part (d) follows. In the other case, g_1 has no zeros on bdry (G). By the above inequality, $g_t(u)$ is never zero for u on bdry (G) and $(1-t) > 0$. Thus, $g_t(u)$ is never zero on bdry (G), and $\deg_{2,M_j}(g_t, G, 0)$ is well defined and independent of t in $[0, 1]$.

For $t = 0$, $g_0 = I$, which is a homeomorphism lying in the appropriate class M_j of permissible homeomorphisms. Hence $\deg_{2,M_j}(g_0, G, 0) = +1$. Therefore, $\deg_{2,M_j}(g, G, 0) = +1$, and as a consequence, there exists a point x in G such that $g(x) = 0$. q.e.d.

DEFINITION (13.8). *Let X be a Banach space, G a bounded open subset of X, g a continuous mapping of* cl (G) *into X. Then g is said to be weakly semiaccretive on* cl (G) *if for each $\epsilon > 0$, the mapping $g_\epsilon = g + \epsilon I$ is strongly semiaccretive from* cl (G) *to X.*

In particular, aside from the continuity conditions on the intertwined representation of the mapping g or g_ϵ, this will be true if $g(u) = S(u, u)$ where S is a mapping of cl $(G) \times$ cl (G) *into X such that for each v in* cl (G), S_v *is an accretive mapping from* cl (G) *to X while the mapping $v \to S_v$ carries* cl (G) *continuously into a relatively compact subset of the space of accretive mappings of* cl (G) *into X (where this last space is taken with the topology of uniform convergence on* cl (G)).

THEOREM (3.16). *Let X be a Banach space, G a bounded open subset of X, g a continuous mapping of* cl (G) *into X which is weakly semiaccretive on* cl (G). *Then:*

(a) *For the appropriate class M_j ($j = 1$, or 2), the generalized degree function* $\deg_{2,M_j}(g, G, v)$ *is defined by Definition (12.6) for any point $v \in X -$ cl $(g(\text{bdry}(G)))$, and the conclusions of Theorem (12.7) are valid for these degree functions.*

(b) *Suppose that G is convex, that $f = I - g$ maps the boundary of G into* cl (G), *and that g (cl (G)) is closed in X. Then there exists a point x of* cl (G) *such that $g(x) = 0$.*

(c) *Suppose that G is convex, that $f = I - g$ maps the boundary of G into* cl (G), *and that 0 does not lie in the closure of $g(\text{bdry}(G))$. Then $\deg_{2,M_j}(g, G, 0) = +1$.*

(d) *Suppose that $G = B_R(0)$ for some $R > 0$, and that for each u in bdry (G), $(J(u), g(u)) \geq 0$. Suppose further that $g(\text{cl}(G))$ is closed in X. Then there exists a point x in* cl (G) *such that $g(x) = 0$.*

PROOF OF THEOREM (3.16). PROOF OF (a). The result of (a) follows from part (a) of Theorem (3.15) together with Theorem (12.7) if we note that g is the limit of $g_\epsilon = g + \epsilon I$, uniformly on cl (G), with each g_ϵ for $\epsilon > 0$ a strongly semiaccretive mapping. q.e.d.

PROOF of (b). We may assume without loss of generality that 0 is an interior point of G, since translation of the independent variable for a mapping does not affect its accretiveness and hence its semiaccretiveness. Hence for $\epsilon > 0$, $(1-\epsilon)f$ also maps bdry (G) into cl (G) for $\epsilon < 1$. Since g is semiaccretive, the mapping $g + \epsilon(1-\epsilon)^{-1}I$ is strongly semiaccretive, and therefore, so is

$$(1-\epsilon)[g + \epsilon(1-\epsilon)^{-1}I] = (1-\epsilon)g + \epsilon I.$$

If we set $g'_\epsilon = (1-\epsilon)g + \epsilon I$, then

$$(I - g'_\epsilon) = (1-\epsilon)(I-g) = (1-\epsilon)f$$

which by our previous remark also maps bdry (G) into cl (G). Applying the result of part (b) of Theorem (13.15), we see that there must exist a point x_ϵ in cl (G) for each ϵ with $0 < \epsilon < 1$ such that $g'_\epsilon(x_\epsilon) = 0$, i.e., $g(x_\epsilon) = -\epsilon(1-\epsilon)^{-1}x_\epsilon \to 0$ ($\epsilon \to 0$). Hence 0 lies in cl $(g(\text{cl}\,(G)))$, and since $g(\text{cl}\,(G))$ is assumed to be closed in X, there exists a point x in cl (G) such that $g(x) = 0$. q.e.d.

PROOF of (c). For each ϵ with $0 < \epsilon < 1$, let $g'_\epsilon = (1-\epsilon)g + \epsilon I$. Since 0 does not lie in cl $(g(\text{bdry}\,(G)))$. Then there exists $d > 0$ such that for all u in bdry (G), $\|g(u)\| \geq d$. Hence, for all u in bdry (G), $\|g'_\epsilon(u)\| \geq (1-\epsilon)d - \epsilon\|u\| > 0$ for ϵ sufficiently small, independent of u in the bounded set bdry (G). Therefore, by the conclusion of part (c) of Theorem (13.15), we have $\deg_{2,M_j}(g'_\epsilon, G, 0) = +1$ for all $\epsilon > 0$ sufficiently small. Passing to the limit as $\epsilon \to 0$, the desired result follows from Theorem (12.7) and the definition of the generalized degree for limit mappings. q.e.d.

PROOF of (d). If for each $\epsilon > 0$, $g_\epsilon = g + \epsilon I$, then for each u in bdry (G),

$$(J(u), g_\epsilon(u)) = (J(u), g(u)) + \epsilon\|u\|^2 > 0.$$

Therefore, by the result of part (d) of Theorem (13.15), there must exist a point x_ϵ in cl (G) for each $\epsilon > 0$ such that $g_\epsilon(x_\epsilon) = 0$. For this point, we have $g(x_\epsilon) = -\epsilon x_\epsilon \to 0$ ($\epsilon \to 0$). Thus, 0 lies in the closure of $g(\text{cl}\,(G))$. Since $g(\text{cl}\,(G))$ is assumed to be closed, there exists a point in cl (G) such that $g(x) = 0$. q.e.d.

With the proof of part (d), the proof of Theorem (13.16) is complete. q.e.d.

DEFINITION (13.9). *Let X be a Banach space, G a bounded open subset of X, g a continuous mapping of cl (G) into X. Then g is said to be a semiaccretive mapping of cl (G) into X if there exists a continuous mapping S of cl $(G) \times$ cl (G) into X such that all of the following conditions are satisfied:*

(a) *For each u in cl (G), $g(u) = S(u, u)$.*

(b) *For each v in cl (G) and the mapping S_v of cl (G) into X given by $S_v(u) = S(u, v)$ ($u \in \text{cl}\,(G)$), $S_v + \epsilon I$ is a permissible strongly accretive homeomorphism of cl (G) into X which lies in one of the two classes M_1 or M_2 (the same class for all v in cl (G)) for each $\epsilon > 0$.*

(c) *The mapping $v \to S_v$ is continuous from the weak topology on cl (G) to the topology of uniform convergence on the mapping classes M_1 or M_2, respectively.*

THEOREM (13.17). *Let X be a uniformly convex Banach space, G a bounded open subset of X, g a semiaccretive mapping of cl (G) into X. Then:*

(a) *For any v in $X - \text{cl}\,(g(\text{bdry}\,(G)))$, $\deg_{2,M_j}(g, G, v)$ is well defined and has all the properties described in the conclusions of Theorem (12.7).*

(b) *If G_1 is a convex open subset of G with cl (G_1) at positive distance from $X - G$, and $(I - g)(\text{bdry } G_1) \subset \text{cl } (G_1)$, then there exists a point x in cl (G_1) such that $g(x) = 0$.*

(c) *Suppose G_1 as in (b) and that $(I - g)$ maps the boundary of G_1 into cl (G_1), and 0 does not lie in the closure of $g(\text{bdry }(G_1))$. Then $\deg_{2, M_j}(g, G_1, 0) = +1$.*

(d) *Suppose that $G_1 = B_R(0)$, and for each u in bdry (G_1), $(J(u), g(u)) \geq 0$. Then there exists a point x in cl (G_1) such that $g(x) = 0$.*

PROOF OF THEOREM (13.17). PROOF OF (a). By Theorem (13.16), it suffices to prove that each semiaccretive mapping g in a uniformly convex space X is also weakly semiaccretive, i.e., that for each $\epsilon > 0$, $g + \epsilon I$ is strongly semiaccretive in the sense of Definition (13.7). Let S be the mapping of cl $(G) \times$ cl (G) into X given for g by the semiaccretiveness described in Definition (13.9). For the given mapping $g_\epsilon = g + \epsilon I$, we consider the mapping S_ϵ of cl $(G) \times$ cl (G) into X defined by $S_\epsilon(u, v) = S(u, v) + \epsilon u$. Then by property (b) of Definition (13.9), $S_{\epsilon, v} = S_v + \epsilon I$ lies in the given class M_j. Hence, it suffices to show that the mapping $v \to S_{\epsilon, v}$ is a compact mapping of cl (G) into the space M_j. Since cl (G) is bounded, from any sequence $\{x_j\}$ in cl (G), we may extract a weakly convergent subsequence. Identifying this with the original sequence, we see that $S_{x_j, \epsilon}$ converges uniformly on cl (G) to a mapping h_ϵ, and $h_\epsilon = T + \epsilon I$ with T accretive. Hence h_ϵ lies in the appropriate class M_j, and we have shown that $g + \epsilon I$ is strongly semiaccretive. Hence g is weakly semiaccretive. q.e.d.

PROOF of (b). To derive the conclusion of part (b) from the results of Theorem (13.16), it suffices to show that for the convex bounded open subset G_1 of G, $g(\text{cl }(G_1))$ is closed in X. For a given sequence $\{x_j\}$ in cl (G_1), let $w_j = g(x_j)$ converge strongly to w in X. We may assume without loss of generality that x_j converges weakly to an element x of cl (G_1) as $j \to \infty$. Then $S(x_j, x) - S(x_j, x_j) \to 0$ as $j \to \infty$ by property (c) of Definition (13.9), i.e., $S_x(x_j) \to w$. Hence by Theorems (10.7) and (10.8), we have $S_x(x) = w$, i.e., $g(x) = w$. Thus w lies in $g(\text{cl }(G_1))$, and the latter set is closed in X. q.e.d.

The proofs of parts (c) and (d) follow from the above proofs and the corresponding results in Theorem (13.16). q.e.d.

Let us now note the following specializations of the arguments and results for the theory of semiaccretive mappings:

THEOREM (13.18). *Let X be a Banach space, G a bounded convex open subset of X, T and C two continuous mappings of cl (G) into X with $T \in M_j$ ($j = 1$, or 2) and C compact. Suppose further that $(I - T - C)$ maps the boundary of G into cl (G). Then there exists a point x in cl (G) such that $(T + C)(x) = 0$.*

PROOF OF THEOREM (13.18). The present result follows from Theorem (13.15) if we note that the mapping $g = T + C$ is strongly semiaccretive with the mapping S of cl $(G) \times$ cl (G) into X given by $S(u, v) = T(u) + C(v)$. q.e.d.

THEOREM (13.19). *Let X be a Banach space, G a bounded convex open subset of X, U and C two mappings of cl (G) into X with C compact and U strongly pseudo-*

contractive in the following sense

$$(J(u - v), U(u) - U(v)) \leq k \|u - v\|^2$$

for a fixed constant $k < 1$ and all u and v in cl (G). Suppose that either U is locally uniformly continuous or that X^* is uniformly convex and U is merely continuous. Suppose finally that $(U + C)$ maps the boundary of G into cl (G).

Then $(U + C)$ has a fixed point in cl (G).

PROOF OF THEOREM (13.19). If we set $T = I - U$, T is a strongly accretive mapping lying in one of the two classes M_j, $j = 1$ or 2. Hence the present result follows from Theorem (13.18). q.e.d.

THEOREM (13.20). *Let X be a Banach space, G a bounded convex open subset of X, T and C two mappings of cl (G) into X with T accretive and C compact. Suppose that either T is locally uniformly continuous on cl (G) or that T is continuous and X^* is uniformly convex. Suppose that $(T + C)(\text{cl}\,(G))$ is closed in X and that $(I - T - C)$ maps bdry (G) into cl (G).*

Then there exists x in cl (G) such that $(T + C)(x) = 0$.

PROOF OF THEOREM (13.20). For each $\epsilon > 0$, $(T + C + \epsilon I)$ is strongly semiaccretive with $S(u, v) = T(u) + \epsilon u + C(v)$. q.e.d.

THEOREM (13.21). *Let X be a Banach space, $G = B_R(0)$ for some $R > 0$ and let T and C be two mappings of cl (G) into X with C compact and T satisfying one of the two following conditions:*

(i) *T is strongly accretive or*

(ii) *T is accretive and $(T + C)(\text{cl}\,(G))$ is closed in X. Suppose that either T is locally uniformly continuous or that T is merely continuous and X^* is uniformly convex.*

Suppose finally that for each u on the boundary sphere of G, $(J(u), (T + C)(u)) \geq 0$. Then there exists a point x in $B_R(0)$ such that $(T + C)(x) = 0$.

PROOF OF THEOREM (13.21). If T satisfies condition (i), $T + C$ is strongly semiaccretive. If T and C satisfy condition (ii), $T + C$ is weakly semiaccretive. In the first instance, our result follows from Theorem (13.15)(d). In the second, it follows from Theorem (13.16)(d). q.e.d.

THEOREM (13.22). *Let X be a Banach space, G a bounded convex open subset of X, U and C two continuous mappings of cl (G) into X with C compact and U pseudo-contractive (i.e., $(J(u - v), U(u) - U(v)) \leq \|u - v\|^2$ for all u and v of cl (G)). Suppose that $(I - U - C)(\text{cl}\,(G))$ is closed in X, and that either U is locally uniformly continuous or X^* is uniformly convex.*

(a) *If $(U + C)$ maps the boundary of G into cl (G), then $(U + C)$ has a fixed point in cl (G).*

(b) *If $G = B_R(0)$ for some $R > 0$, and if for all u in bdry (G),*

$$(J(u), (U + C)(u)) \leq \|u\|^2$$

then $U + C$ has a fixed point in $B_R(0)$.

PROOF OF THEOREM (13.22). $T = I - U$ is accretive. Hence the result follows from the preceding results if one interprets $T + C$ as a semiaccretive mapping.   q.e.d.

THEOREM (13.23). *Let X be a uniformly convex Banach space, G a bounded open subset of X, G_1 a convex open subset of G such that* cl (G_1) *has positive distance from* $X - G$. *Let T and C be two continuous mappings of* cl (G) *into X with T accretive and C completely continuous (i.e., C is continuous from the weak topology of G to the strong topology of X). Suppose that either T is locally uniformly continuous or X^* is uniformly convex.*

(a) *If $(I - T - C)$ maps* bdry (G_1) *into* cl (G_1), *then there exists an element x of* cl (G_1) *such that* $(T + C)(x) = 0$.

(b) *If $G_1 = B_R(0)$ for some $R > 0$ and if for all u in* bdry (G),

$$(J(u), (T + C)(u)) \geq 0$$

then there exists an element x in cl (G_1) *such that* $(T + C)(x) = 0$.

PROOF OF THEOREM (13.23). The mapping $(T + C)$ is semiaccretive with the mapping S given by $S(u, v) = T(u) + C(v)$. Hence the result of Theorem (13.23) follows from Theorem (13.17) above.   q.e.d.

THEOREM (13.24). *Let X be a uniformly convex Banach space, G a bounded open subset of X, U and C two continuous mappings of* cl (G) *into X with C completely continuous and U pseudo-contractive. Let G_1 be a convex open subset of G such that* cl (G_1) *has positive distance from $X - G$, and suppose that either U is locally uniformly continuous or that X^* is uniformly convex.*

Then if $(U + C)$ maps the boundary of G_1 into cl (G_1), *it follows that $(U + C)$ must have a fixed point in* cl (G_1).

PROOF OF THEOREM (13.24). If $T = I - U$, T is accretive. Hence the present result follows from Theorem (13.23).   q.e.d.

We now turn to the corresponding theory for monotone mappings from a reflexive Banach space X to its conjugate space X^*.

DEFINITION (13.10). *Let X be a reflexive Banach space, X^* its conjugate space. Then we denote by M_3 the class of continuous mappings T from the closure* cl (G) *of bounded open subsets G of X to X^* for which there exist continuous, strictly increasing functions $c: R^+ \to R^+$ with $c(0) = 0$ such that for all u and v in* cl (G),

$$(T(u) - T(v), u - v) \geq c(\|u - v\|).$$

PROPOSITION (13.35). *The class M_3 is a convex class of permissible homeomorphisms from X to X^*.*

PROOF OF PROPOSITION (13.35). If T is a mapping in the class M_3, then by Definition (13.10), T is a continuous mapping of cl (G) into X^*, while

$$\|u - v\| \cdot \|T(u) - T(v)\| \geq c(\|u - v\|)$$

for all u and v in cl (G). Since G is assumed to be bounded, there exists a constant

R_0 such that $\|u - v\| \leq R_0$ for all u and v in cl (G). Hence $c(\|u - v\|) \leq R_0 \|T(u) - T(v)\|$, and since the function $c(r)$ has a continuous inverse $\zeta(s)$, it follows that $\|u - v\| \leq \zeta(\|T(u) - T(v)\| R_0)$. Hence, T^{-1} is a continuous mapping from $T(\text{cl }(G))$ into cl (G), and $T(\text{cl }(G))$ is a closed subset of X^*. Since $T(G)$ is dense in $T(\text{cl }(G))$, it follows that $T(\text{cl }(G)) = \text{cl }(T(G))$. Hence, in order to show that T is a permissible homeomorphism, it suffices to show that $T(G)$ is an open subset of X^*.

Let x_0 be a point of G, and suppose that for a constant $d > 0$, $B_d(x_0)$ is contained in G. For each point u in bdry $(B_d(x_0))$, we have $(T(u) - T(x_0), u - x_0) \geq c(d) > 0$. We define a new mapping S of $B_d(0)$ into X^* by setting:

$$S(v) = T(x_0 + v) - T(x_0) - w$$

for any w in X^* with $\|w\| d < c(d)$. Then for v in X with $\|v\| = d$, we have

$$(S(v), v) = (T(v + x_0) - T(x_0), v) - (w, v)$$
$$= (T(u) - T(x_0), u - x_0) - (w, u - x_0),$$

where $u = x_0 + v \in$ bdry $(B_d(x_0))$. Hence

$$(S(v), v) \geq c(d) - \|w\| d > 0 \qquad (\|v\| = d).$$

Since T is continuous and monotone on G, S is continuous and monotone on an open set containing $B_d(0)$. Hence by Theorem (7.2), since S is continuous and coercive on $B_d(0)$ with respect to 0, there exists a point x in $B_d(0)$ such that $S(x) = 0$, i.e., $T(x_0 + x) = T(x_0) + w$. Hence $T(B_d(x_0))$ contains $B_{d_1}(T(x_0))$ where d_1 is any positive constant with $d_1 < d^{-1} c(d)$. Hence $T(G)$ is an open subset of X^*.

Finally, it follows immediately from Definition (13.10) that M_3 is a convex class of mappings. q.e.d.

DEFINITION (13.11). *Let X be a reflexive Banach space, G a bounded open subset of X, g a continuous mapping of* cl (G) *into X^*. Then:*

(a) *g is said to be strongly semimonotone if there exists a continuous mapping S of* cl $(G) \times$ cl (G) *into X^* such that $g(u) = S(u, u)$ for all u in* cl (G) *while each S_v for v in* cl (G) *is a homeomorphism of* cl (G) *into X^* lying in the class M_3 of strongly monotone mappings, while the mapping $v \to S_v$ yields a compact mapping of* cl (G) *into the space of mappings M_3 (with the latter space being given the topology of uniform convergence on* cl (G)*).*

(b) *g is said to be weakly semimonotone if there exists a sequence $\{g_k\}$ of strongly semimonotone mappings converging uniformly to g on* cl (G).

THEOREM (13.25). *Let X be a reflexive Banach space, G a bounded open subset of X, g a continuous mapping of* cl (G) *into X^*. Then if g is weakly semimonotone from* cl (G) *to X^* and w is any point of $X^* -$ cl $(g(\text{bdry }(G)))$, $\deg_{2, M_3}(g, G, w)$ is well defined by Theorem (12.7) and satisfies the conclusions of that theorem.*

PROOF OF THEOREM (13.25). This follows from the above definitions as compared with the definition of restricted intertwined representation as given in Definition (12.6). q.e.d.

To avoid a further expansion of the length of this section, we shall not give the further details of the results for semimonotone mappings which are analogous to the previous results for semiaccretive mappings. Since each bounded strongly semimonotone mapping is *pseudo-monotone* in the sense of §7, sharper conclusions can be obtained from the latter theory in this case. Moreover, the theory of semimonotone mappings follows as a special case of the theory of A-proper mappings treated below in §17, and will be considered in that section.

14. Nonlinear Fredholm mappings. Let X and Y be Banach spaces, G an open subset of X, f a continuous mapping of G into Y which is once continuously Fréchet differentiable at each point x of G. Let f'_x denote the Fréchet derivative of f at x in G, so that f'_x is an element of the space $L(X, Y)$ of bounded linear mappings of X into Y and we have

$$f(x + \epsilon u) = f(x) + \epsilon f'_x(u) + R_x(\epsilon, u),$$

where
$$\|R_x(\epsilon, u)\| \leq \epsilon g(\epsilon) \qquad (0 < \epsilon < \epsilon_0, \|u\| \leq 1)$$

with $g(\epsilon) \to 0$ $(\epsilon \to 0)$.

DEFINITION (14.1). *The mapping f is said to be a (nonlinear) Fredholm mapping of G into Y if for each x in G, f'_x is a Fredholm (linear) mapping from X to Y (i.e., the null space of f'_x is a finite-dimensional subspace $N(f'_x)$ of X and the range $R(f'_x)$ of f'_x is closed and has finite codimension in Y).*

DEFINITION (14.2). *Let G be connected, f a nonlinear Fredholm mapping of G into Y. Then the index of f (written ind (f)) is defined to be the index of the linear Fredholm mapping f'_x for any x in G, i.e.,*

$$\begin{aligned} \text{ind } (f) &= \text{dimension } N(f'_x) - \text{codimension } R(f'_x) \\ &= \dim (N(f'_x)) - \dim (Y/R(f'_x)). \end{aligned}$$

Since the mapping of G into $L(X, Y)$ given by $x \to f'_x$ is assumed to be continuous from the strong topology on G to the norm topology on $L(X, Y)$, the index of f'_x is locally constant on G. Since G is assumed to be connected, this index is constant on G and this fact makes the definition of ind (f) given in Definition (14.2) meaningful.

It is our purpose in the present section to apply the theory of the generalized degree developed in §12 to obtain a proof of the following theorem:

THEOREM (14.1). *Let X and Y be Banach spaces, G a convex open subset of X, f a once continuously differentiable mapping of G into Y which is a nonlinear Fredholm mapping of index zero. Suppose that f is proper (i.e., the inverse image under f of each compact subset K of Y is a compact subset of G), and suppose further that for a given point y_0 of Y, $f^{-1}(y_0)$ consists of a finite subset $\{x_1, \ldots, x_s\}$ of G with an odd number of elements such that for each x_j in the set, $N(f'_{x_j})$ is trivial.*

Then f maps G onto Y.

We shall obtain the proof of Theorem (14.1) by applying a number of technical results which together imply the representability of the mapping f with respect to a suitable class M of permissible homeomorphisms from X to Y.

PROPOSITION (14.1). *Let $\{R_1, \ldots, R_s\}$ be a finite family of closed subspaces of a Banach space X, with each R_j having a finite-dimensional complementary subspace in X. Then*

$$R = \bigcap_{j=1}^{s} R_j$$

is a closed subspace of X which has a finite-dimensional complementary subspace in X.

PROOF OF PROPOSITION (14.1). For each j, let M_j be the finite-dimensional subspace of X^*, the conjugate space of X, which is the annihilator of R_j. Then if M is the span of the finite family $\{M_1, \ldots, M_s\}$, M is of finite dimension and M is the annihilator of the subspace R of X. Hence R has a finite-dimensional complement in X. q.e.d.

PROPOSITION (14.2). *Let K be a compact subset of G, f a Fredholm mapping of G into Y. Then there exists a continuous mapping C of K into the space of linear mappings $L_f(X, Y)$ of X into Y with finite rank such that, for all x in K, the null space of the linear mapping $(f'_x + C_x)$ is independent of x in K.*

PROOF OF PROPOSITION (14.2). For each x in K, let N_x be the null space of f'_x and choose a closed complementary subspace R_x for the finite-dimensional subspace N_x in X. Since $X = N_x + R_x$, f'_x is an injective mapping of R_x into Y and the range of f'_x restricted to R_x is the whole of the range of f'_x and hence a closed subspace of Y. Hence there exists a positive constant $c_x > 0$ such that for u in R_x, $\|f'_x(u)\| \geq c_x \|u\|$.

We may choose a neighborhood V_x of x in K such that for all v in V,

$$\|f'_x - f'_v\| < c_x.$$

Then for all such v, and all u in R_x,

$$\|f'_v(u)\| \geq \|f'_x(u)\| - \|(f'_v - f'_x)(u)\| \geq c_x \|u\| - \|f'_v - f'_x\| \cdot \|u\| \geq c_{x,v} \|u\|$$

where $c_{x,v} = c_x - \|f'_v - f'_x\| > 0$ for all v in V.

Since K is assumed to be compact, we may cover K by a finite number of the neighborhoods $\{V_1, \ldots, V_m\}$ of the above form, corresponding to the points $\{x_1, \ldots, x_m\}$ of K. For $1 \leq j \leq m$, let R_j be the corresponding closed subspace as above with finite-dimensional complementary subspace such that for all v in V_j and all u in R_j, $\|f'_v(u)\| \geq c_{j,v} \|u\|$, with $c_{j,v} > 0$ for all v in V_j. Let

$$R = \bigcap_{j=1}^{m} R_j.$$

By Proposition (14.1), R is a closed subspace of X with a finite-dimensional complementary subspace N, $X = R + N$. Let P be a bounded linear projection mapping of X onto N such that $R(I - P) = R$.

We note that for all u in R and for all v in K, we have $\|f'_v(u)\| \geq c_v \|u\|$ for some $c_v > 0$ since v lies in at least one of the neighborhoods V_j and u lies in the intersection of all the spaces R_j.

We now define the mapping C_x for each x in K by setting $C_x(u) = -f'_x(P(u))$. Since P is of finite rank, C_x is of finite rank, and indeed of rank at most the rank of P. Moreover, since f'_x is continuous in x on G, C_x is continuous in x on K. Finally, if we set

$$L_x = f'_x + C_x = f'_x - f'_x P = f'_x(I - P)$$

we see that for u in N, $L_x(u) = f'_x(u) - f'_x(P(u)) = 0$, so that N is contained in the null space of each L_x. On the other hand, if $0 = L_x(u) = f'_x((I - P)(u))$, we note that $(I - P)(u)$ lies in R and since f'_x annihilates only the null element of R, $(I - P)(u) = 0$. Hence u lies in N.

Thus, with this choice for C_x, $N(f'_x + C_x) = N$ for all x in K. q.e.d.

PROPOSITION (14.3). *Let K be a compact convex subset of the open set G of X, f a Fredholm mapping of index zero of G into Y. Then there exists a continuous mapping C of K into the space of linear mappings $L_f(X, y)$ of finite rank from X to Y such that for each x in K, $(f'_x + C_x)$ is a linear isomorphism of X onto Y.*

PROOF OF PROPOSITION (14.3). By Proposition (14.2), we can find a continuous function C^1 from K to the space of linear operators of finite rank from X to Y such that for each x in K, $f'_x + C^1_x$ has the same null space N. If we set $L_x = f'_x + C^1_x$, it suffices to find another continuous function C from K to $L_f(X, Y)$ such that for each x in K, $L_x + C_x$ is an isomorphism of X onto Y.

For each x in K, f'_x is a linear Fredholm mapping of index zero. Since C^1_x is of finite rank and hence compact, $L_x = f'_x + C^1_x$ is a Fredholm mapping of index zero. By the choice of C^1_x, the null space of L_x is the fixed subspace N and hence $\dim(N(L_x))$ is the same for all x in K. Since

$$0 = \operatorname{ind}(f'_x) = \operatorname{ind}(L_x) = \dim(N(L_x)) - \dim(Y/R(L_x)),$$

it follows that $\dim(Y/R(L_x)) = \dim(N)$ is independent of x in K.

To construct the mapping C_x, we split up X as the direct sum of N and a closed subspace R on which each L_x is injective, and if we let P be the corresponding bounded linear mapping which is a projection of X on N and which annihilates R, we shall construct C_x by letting $C_x = Q_x P$, where Q_x is a linear mapping of the finite-dimensional space N into Y. To make sure that $L_x + C_x$ is an isomorphism of X onto Y, it suffices to ensure that the null space of $L_x + C_x$ is trivial since $L_x + C_x$ like L_x itself will be a linear Fredholm mapping of index zero for each x in K. We shall choose Q_x to be an injective mapping of N into Y. Since L_x is an injective mapping of R into Y for each x in K, in order that $L_x + C_x$ be injective on the whole of X, it suffices to ensure that $R(Q_x) \cap R(L_x) = \{0\}$.

To construct such a mapping Q_x for each x in K, with Q_x varying continuously with x in K, we note that for each such x, $R(L_x)$ is a closed subspace of a fixed finite codimension n in Y. Suppose that $\{u_1, \ldots, u_n\}$ is a basis for N, and that we can find a continuous function Φ from K to Y such that for each x in K, $\Phi(x) \in Y - R(L_x)$. Then we may define $L_x^{(1)}(u + cu_n) = L_x(u) + c\Phi(x)$ for each u in the closed subspace $R^{(1)} = \text{span}\,(R, \{u_n\})$ and the codimension in Y of $R(L)_x^{(1)}$ will be $(n-1)$ for each x in K. If we can proceed step-by-step in this fashion, we can extend the mappings $L_x^{(j)}$ with decreasing j in such a fashion that on the increasing sequence $R_x^{(j)}$ of closed subspaces with $R^{(n)} = X$, $L_x^{(j)}$ is an injective mapping while $L_x^{(j)}$ differs from L_x on $R_x^{(j)}$ by a mapping of finite rank. Thus, it suffices under our original conditions to show that there exists a mapping Φ as described above such that for each x in K, $\Phi(x) \in Y - R(L_x)$ while Φ is continuous from K to Y.

To demonstrate the existence of such a mapping Φ, we consider two auxiliary spaces, $(K \times Y)$ and the space Z given by

$$Z = \bigcup_{x \in K} \{x\} \times (Y/R(L_x))$$

(i.e., Z is the disjoint union indexed by x in K of the quotient spaces $Y/R(L_x)$). We introduce mappings π_1 and π_2 of $K \times Y$ into Z and of Z into K, by setting $\pi_1(x, y) = [[x, y] + R(L_x)]$ and $\pi_2(x, y + R(L_x)) = x$.

We shall show that with a suitable topology on Z, $K \times Y$ is a C^0 vector bundle over Z with fibre isomorphic to $R(L_x)$ for each x in K, and similarly that Z is a vector bundle over K with π_2 as fibre mapping and with fibre an n-dimensional Euclidean space.

We introduce the desired topology on Z by showing that for each x_0 in K, there is a neighborhood V in K such that $\pi_2^{-1}(V)$ can be represented in the form $V \times R^n$ in such a way that for each point x in V, $\pi^{-1}(x)$ is of the form $\{x\} \times R^n$ in this representation. We identify the topology of the open subset $\pi_2^{-1}(V)$ of Z with $V \times R^n$ by this representation which is obtained as follows: Let N_0 be a given n-dimensional subspace of Y which is a complementary subspace of $R(L_{x_0})$. Then N_0 is the representative n-dimensional Euclidean space, and for a suitably small neighborhood V of x_0 in K, for each x in V, the quotient mapping π_x of Y onto $Y/R(L_x)$ maps N_0 isomorphically onto $Y/R(L_x)$. Indeed, let R be the given complementary subspace to the common null space N of L_x for each x in K, and define a mapping ψ_x of $N_0 \times R(L_{x_0})$ into Y by setting

$$\psi_x(w, v) = w + L_x(L_{x_0}^{-1}(y)),$$

where we restrict all the mappings L_x to act only upon R. Since ψ_{x_0} is the identity and ψ_x is a family of bounded linear mappings depending continuously upon x in the norm topology on ψ_x, there is a neighborhood V of X_0 such that for x in V, ψ_x is an isomorphism onto Y. For such x, N_0 is a complementary subspace to $R(L_x)$ and π_x maps N_0 isomorphically onto $Y/R(L_x)$. Finally, we identify $\pi_2^{-1}(V)$ with $V \times N_0$ by the mapping $\xi: V \times N_0 \to \pi_2^{-1}(V)$ by setting

$$\xi(x, w) = [x, \pi_x(w)] \in Z.$$

By the above remarks, ξ is a one-to-one mapping of $V \times N_0$ onto $\pi_2^{-1}(V)$ with the property that ξ is a mapping of $\{x\} \times N_0$ onto $\pi^{-1}(\{x\})$. It follows in addition that for another n-dimensional complementary subspace N_1 to $R(L_{x_0})$ in Y, the transition mapping from N_0 to N_1 given by $(\pi_x \mid N_1)^{-1}(\pi_x \mid N_0)$ is an isomorphism of N_0 on N_1 for x near x_0 varying continuously with x in a neighborhood V of x_0 in K.

By the remarks above, we have exhibited a topology on Z and shown that Z is a real vector bundle over K with π_2 as fibre map with a n-dimensional Euclidean space as fibre. Let us now consider $K \times Y$ as a potential fibre space over Z with π_1 as a possible fibre mapping. Let V be a neighborhood in K of the type considered in the preceding paragraph with $\pi_2^{-1}(V)$ represented as above in the form $V \times N_0$ by the homeomorphism $\xi \colon V \times N_0 \to \pi_2^{-1}(V)$. We take $\pi_2^{-1}(V)$ as a neighborhood U in Z over which we shall show that $\pi_1^{-1}(U)$ has a suitable product representation. We shall identify $\pi_1^{-1}(U)$ with $U \times R(L_{x_0})$ by the mapping $\zeta \colon (\pi_2^{-1}(V)) \times R(L_{x_0}) \to (K \times Y)$ where for x in V, y in Y, and v in $R(L_{x_0})$,

$$\zeta(x, y + R(L_x), v) = [x, (\pi_x \mid N_0)^{-1}(\pi_x(y)) + L_x(L_{x_0}^{-1}(v))]$$

i.e., $\zeta(\xi(x, w), v) = [x, w + L_x(L_{x_0}^{-1}(v))]$. We see immediately that

$$\pi_1(\zeta(\xi(x, w), v)) = [x, \pi_x(w)] = \xi(x, w)$$

so that ζ does give us a product representation of $\pi_1^{-1}(U)$ with the property that for each point z in U, $\pi_1^{-1}(z)$ coincides with $\zeta(z, R(L_{x_0}))$. It follows as before that π_1 is a fibre mapping of the vector bundle $K \times Y$ over Z.

We now apply the classification theory of fibre bundles. By a result of Steenrod [431] (Corollary (11.6), p. 53), if X is compact and contractible over itself to a point, then any bundle over X is equivalent to a product bundle. In particular, our given compact convex subset K of G satisfies the conditions on X given in this last result, so that Z is homeomorphic with $K \times N_0$ by a mapping $h \colon Z \to K \times N_0$ such that $h(z) = [\pi_1(z), \alpha(z)]$. Hence, we can find a continuous section Φ_1 of the bundle Z over K such that for each x in K, $\Phi_1(x)$ is never equal to the zero element of $(Y/R(L_x))$. Similarly, since Z is homeomorphic to $K \times N_0$, Z also satisfies the conditions on X in the result stated at the beginning of this paragraph. Hence $K \times Y$ is equivalent to a product bundle over Z, i.e., there exists a homeomorphism h_1 of $R(L_{x_0}) \times Z$ into $K \times Y$ for a fixed element x_0 of K such that

$$h_1(v, z) = [\pi_1(z), \beta(y, z)],$$
$$\pi_2 h_1(v, z) = z.$$

In particular, there exists a continuous mapping Φ_2 of Z into Y such that

$$\pi_2(\pi_1(Z), \Phi_2(Z)) = Z.$$

In conclusion, we set $\Phi(x) = \Phi_2(\Phi_1(x))$, $x \in K$. Then by the definition of Φ_1 and Φ_2, $\Phi(x)$ never lies in $R(L_x)$ for any x in K. q.e.d.

If we use the result of Proposition (14.3), the conclusion of Theorem (14.1) follows immediately from the following theorem:

THEOREM (14.2). *Let X and Y be Banach spaces, G a convex open subset of X, f a nonlinear Fredholm mapping of index zero from G to Y. Suppose that f is proper* and that for each compact convex subset K of G, there exists a continuous mapping C of K into the space $L_f(X, Y)$ of bounded linear mappings of finite rank from X to Y such that for each x in K, $f'_x + C_x$ is an isomorphism of X onto Y.

Suppose further that the inverse image $f^{-1}(y_0)$ of some given point y_0 of Y consists of a finite odd number of points at each of which $N(f'_x)$ is trivial.

Then f maps G onto Y.

PROOF OF THEOREM (14.2). Let y_1 be any point of Y different from y_0 and consider the closed line segment K_0 joining y_0 to y_1. Since f is a proper mapping from G to Y, $f^{-1}(K_0)$ is a compact subset K_1 of G. Since X is a Banach space, the convex closure K of K_1 is a compact convex subset of X.

We now assert that K is actually a compact subset of G. Indeed for each point x_0 of K_1, there exists a convex neighborhood V of x_0 whose closure in X is contained in G. Since K_1 is compact, we may choose a finite family $\{V_1, \ldots, V_m\}$ of such neighborhoods which covers K_1. If for each j with $1 \leq j \leq m$, we denote by Q_j the convex closure of $K_1 \cap \text{cl}\,(V_j)$, then each Q_j is a compact convex subset of $\text{cl}\,(V_j)$ and hence a compact convex subset of G. On the other hand, it follows immediately from the compactness and convexity of the various Q_j that

$$K = \left\{u \,\bigg|\, u = \sum_{j=1}^m \lambda_j u_j, \, u_j \in Q_j, \, 0 \leq \lambda \leq 1, \sum_{j=1}^m \lambda_j = 1\right\}$$

since the set on the right is a convex compact subset of X which contains K_1. In particular, K_1 is a subset of G.

Since K is a compact convex subset of the Banach space X, K is an absolute retract and there exists a continuous retraction r of X on K, i.e., a continuous mapping $r: X \to K$ such that $r(x) = x$ for x in K. We now define a mapping S of $G \times G$ into Y by setting

$$S(u, v) = f(u) + C_{r(v)}(u - v) \qquad (u, v \in G).$$

Let M be the class of all permissible homeomorphisms from the closures of open subsets of X to Y. For each element v on G, the mapping f_v of G into Y given by $f_v(u) = f(u) + C_{r(v)}(u - v)$ is once continuously Fréchet differentiable at v and its differential is given by $(f'_v)_x = f'_x + C_{r(v)}$ at each point x of G. In particular, since $f'_{r(v)} + C_{r(v)}$ is an isomorphism of X onto Y, $(f'_v)_x$ is an isomorphism of X onto Y for all v in some neighborhood V of $r(v)$ in G. Thus there exists an open neighborhood U of each v_0 in K such that for all v in U, $f_v = f + C_{r(v)} - v$ is a permissible homeomorphism of $\text{cl}\,(U)$ into Y.

It follows from the result of the last paragraph that if we consider the subset W of $G \times G$ given by $W = \{[u, v] \mid u \in U(v)\} \subset G \times G$ where $U(v)$ is a suitably chosen neighborhood of v in G on which f_v is a homeomorphism with the neighborhoods $U(v)$ chosen from a finite open covering of K in G, then W is an open subset of $G \times G$ and S is a representation of the mapping $S(u, u) = f(u)$ on W in the sense of Definition (12.6) with respect to the broadest class M of all permissible homeomorphisms. Hence, we may apply Theorem (12.4) to obtain the fact that for each

t in $[0, 1]$ and $y_t = (1 - t)y_0 + ty_1$, $\deg_2([f, S], G_1, y_t)$ is well defined for a suitable open neighborhood G_1 of K in G such that $S(u, u) = f(u)$ for all u in cl (G_1). Since $f^{-1}(y_t)$ is contained in the fixed compact subset K of G_1 for all t in $[0, 1]$, it follows from Theorem (12.5), moreover, that $\deg_2([f, S], G_1, y_t)$ is independent of t in $[0, 1]$. If we can show that $\deg_2([f, S], G_1, y_1)$ is nonnull, then by Theorem (12.6) once more, $f^{-1}(y_1)$ is nonempty. In this case, the conclusion of our theorem follows since y_1 was chosen to be an arbitrary point of Y. Hence, to complete the proof of the present theorem, it suffices to show that $\deg_2([f, S], G_1, y_0) \neq 0$.

To prove this last fact, we may replace G_1 by a finite family of pairwise disjoint small balls B_j about the various points x_j of $f^{-1}(y_0)$. If we can show for each such ball B of sufficiently small radius that $\deg_2([f, S], B, y_0) = \pm 1$, then

$$\deg_2([f, S], G_1, y_0) = \sum_{j=1}^{s} \deg_2([f, S], B_j, y_0)$$

being the sum of an odd number s of summands, each must be different from zero.

To calculate $\deg_2([f, S], B, y_0)$, we first deform the representation S for f through a family of permissible representations for f of the same type by setting $S_t(u, v) = f(u) + C_{tr(v)+(1-t)x_0}(u - v)$, with x_0 the point of $f^{-1}(y_0)$ which is the center of the ball B. By the homotopy invariance property of the generalized degree $\deg_2([f, S], B, y_0) = \deg_2([f, S_0] B, y_0)$, where

$$S_0(u, v) = f(u) + C_{x_0}(u - v) = [f(u) + C_{x_0}(u)] - C_{x_0}(v).$$

On the other hand, the ball B being taken sufficiently small, the mapping $f + C_{x_0}$ is a homeomorphism of B into Y. Hence the representation S_0 is of the form $S(u, v) = h(u) - C(v)$ with h a permissible homeomorphism from cl (B) to Y and C a compact mapping. Theorefore, we may apply Theorem (12.9) to conclude that $\deg_2([f, S_0], B, y_0) = \deg_1([f, h], B, y_0)$ where h is the homcomorphism of B into Y given by $h(u) = f(u) + C_{x_0}(u)$ ($u \in B$).

Finally, we may apply the properties of the generalized degree function $\deg_1$ as given in Theorem (12.1) and by property (4) of that degree function, it follows from the fact that f is injective that $\deg_1([f, h], B, y_0) = \pm 1$. Hence $\deg_2([f, S], B, y_0) = \pm 1$. q.e.d.

15. Orientation-preserving and complex analytic mappings.

In the present section, we study the general class of nonlinear mappings which preserve orientation in some generalized sense and, in particular, the important subclass of complex analytic mappings of complex Banach spaces. For these classes of nonlinear mappings, we have the general phenomenon that the generalized degree represents a more precise count of the number of solutions of functional equations than in the general case, a phenomenon which corresponds to the general principle that there is no cancellation of degrees for different solutions since all the degrees have the same sign.

We begin with the simplest and most basic case, that of finite-dimensional mappings.

DEFINITION (15.1). *Let X_n be a real Banach space of finite dimension n, L a linear mapping of X_n into X_n. Then L is said to be orientation-preserving if for any given basis of X_n and the corresponding representation of L as a matrix with respect to this basis, $\det(L) \geq 0$. (We denote by $\det(L)$ the determinant of the corresponding matrix.)*

DEFINITION (15.2). *Let X_n be a real Banach space of finite dimension n, G an open subset of X_n, f a mapping of G into X_n. Then f is said to be strictly orientation-preserving if all of the following conditions are valid:*

(a) f is a mapping of class C^1. For each x in G, the linear mapping f'_x of X_n into X_n is orientation-preserving in the sense of Definition (15.1).

(b) The subset of G given by

$$S_f = \{x \mid x \in G, f'_x \text{ is singular, ie., } \det(f'_x) = 0\}$$

is nowhere-dense in G.

DEFINITION (15.3). *Let X_n be a real Banach space of finite dimension n, G an open subset of X_n, f a continuous mapping of G into X_n. Then f is said to be orientation-preserving in the general sense (or more briefly, orientation-preserving, without modifiers) if for each open subset G_1 of G with $\mathrm{cl}(G_1)$ compact in G, there exists a sequence $\{f_k\}$ converging uniformly to f on G_1 with each f_k strictly orientation-preserving on G_1 in the sense of Definition (15.2).*

THEOREM (15.1). *Let X_n be a real Banach space of finite dimension n, G an open subset of X_n, f a continuous mapping of G into X_n which is orientation-preserving in the sense of Definition (15.3). Let V be an open subset of G such that $\mathrm{cl}(V)$ is a compact subset of U, and let y be a point of $X_n - f(\mathrm{bdry}(V))$. Then:*

(a) $\deg_{LS}(f, V, y) \geq 0$.

(b) More precisely, $\deg_{LS}(f, G, y) > 0$ if and only if y lies in $f(V)$.

(c) $\deg_{LS}(f, V, y) = +1$ implies that $f^{-1}(y) \cap V$ is connected.

(d) Let C be a component of $X_n - f(\mathrm{bdry}(V))$ which contains a point of $f(V)$. Then C is contained in $f(V)$.

We begin the proof of Theorem (15.1) with the proof of the following special case:

PROPOSITION (15.1). *The conclusions of Theorem (15.1) hold for a C^1 mapping f which is strictly orientation-preserving in the sense of Definition (15.2).*

PROOF OF PROPOSITION (15.1). Let f be a C^1 mapping of G into X_n which is strictly orientation-preserving and let V be an open subset of G with $\mathrm{cl}(V)$ a compact subset of G.

Since the mapping f is of class C^1, we may apply Sard's theorem [334], [426] to obtain the conclusion that the image under f of the critical set S_f of f is of Lebesgue n-measure zero in X_n. Since $\mathrm{cl}(V)$ is compact and S_f is closed in G, $S_f \cap \mathrm{cl}(V)$ is also compact. Hence $f(\mathrm{cl}(V) \cap S_f)$ is a closed set of measure zero in X_n and therefore nowhere dense in X_n.

Let C be any component of $X_n - f(\mathrm{bdry}(V))$, y any point of C. Since $X_n - f(\mathrm{bdry}(V))$ is open in X_n and X_n is locally connected, C is a connected open subset of X_n and hence pathwise connected. Since $\deg_{LS}(f, V, y_t)$ is constant for any continuous curve $\{y_t, 0 \leq t \leq 1\}$ in C, it follows that $\deg(f, V, y)$ is independent of the choice of y in C. For y in $C - f(S_f \cap \text{cl}(V))$ we know that each point x which lies in $f^{-1}(y) \cap \text{cl}(V)$ has the property that f'_x is nonsingular. By condition (a) in Definition (15.2) for strict orientation-preserving mappings, $\det(f'_x) \geq 0$. Since f'_x is nonsingular, we know in addition that $\det(f'_x) \neq 0$. Hence $\det(f'_x) > 0$. Hence, if we consider a sufficiently small ball B about x in V, the homotopy

$$f_\lambda(u) = (1 - \lambda)f(u) + \lambda[y + (u - x)] \qquad (u \in B, 0 \leq \lambda \leq 1)$$

deforms $f = f_0$ into the mapping f_1 with $\deg_{LS}(f_1, B, y) = +1$ and $f_\lambda(u) \neq y$ on bdry (B). Thus $\deg_{LS}(f, B, y) = +1$ for a sufficiently small ball B about each point x of $f^{-1}(y) \cap V$, where the latter set is finite. Hence for a point y in $C - f(S_f \cap \text{cl}(V))$, the degree function $\deg_{LS}(f, V, y) = m_y$ where m_y is the number of points x in V such that $f(x) = y$.

PROOF OF ASSERTION (a). Let C be any component of $X_n - f(\text{bdry}(V))$. If C does not intersect $f(V)$, then for each point y in C, $\deg_{LS}(f, V, y) = 0$. Since C is open in X_n and $f(S_f \cap \text{cl}(V))$ is nowhere dense in X_n, there exists a point y in $C - f(S_f \cap \text{cl}(V))$. By the discussion of the preceding paragraph, $\deg_{LS}(f, V, y) = m_y$ where m_y is the number of points in $f^{-1}(y) \cap V$.

Hence, in particular, $\deg_{LS}(f, V, y) \geq 0$. For any other point y_1 in C, and each y in $C - f(S_f \cap \text{cl}(V))$, $\deg_{LS}(f, V, y) = \deg_{LS}(f, V, y_1)$. Hence $\deg_{LS}(f, V, y_1) \geq 0$ for any point y_1 in $X_n - f(\text{bdry}(V))$. q.e.d.

PROOF OF (b). If $\deg_{LS}(f, V, y) > 0$ for a given point y in $X_n - f(\text{bdry}(V))$, then it follows from the basic properties of the degree function that y must lie in $f(V)$.

Suppose on the other hand that for a given point y of a given component C of $X_n - f(\text{bdry}(V))$, y lies in $f(V)$. Let x be any point in $f^{-1}(y) \cap V$. Since C is open in X_n, we may choose a neighborhood N of x in V such that $f(N) \subset C$. By condition (b) for f to be strictly orientation-preserving in Definition (15.2), S_f is nowhere dense in U. Hence there exists a point x_1 in $N - S_f$ such that $y_1 = f(x_1)$ lies in the component C. We choose a neighborhood N_1 of x_1 in V such that $f^{-1}(y_1) \cap \text{cl}(N_1)$ consists of the single point x_1. In particular, $f^{-1}(y_1) \cap \text{cl}(N_1)$ does not intersect S_f. Hence by our preceding discussion, $\deg_{LS}(f, N_1, y_1) = +1$. Since $[f^{-1}(y_1) - N_1] \cap V$ is compact and disjoint from cl (N_1) we may find a neighborhood N_2 of this set in V from N_1. By the additivity of the degree,

$$\deg_{LS}(f, V, y_1) = \deg_{LS}(f, N_1, y_1) + \deg_{LS}(f, N_2, y_1) \geq \deg_{LS}(f, N_1, y_1) = +1$$

since by our preceding proof for part (a), $\deg_{LS}(f, N_2, y_1) \geq 0$. Therefore, $\deg_{LS}(f, V, y) = \deg_{LS}(f, V, y_1) \geq 1$ for any point y in C, and the conclusion of part (b) follows. q.e.d.

PROOF OF (c). Suppose that for a given point y in $X_n - f(\text{bdry}(V))$ that $\deg_{LS}(f, V, y) = +1$ while $f^{-1}(y) \cap V$ is not connected. Since $f^{-1}(y) \cap V = f^{-1}(y) \cap \text{cl}(V)$ is compact, there exist two disjoint open subsets N_1 and N_2 of V

such that $f^{-1}(y) \cap V \subset N_1 \cap N_2$ while both sets contain points of $f^{-1}(y)$. By the conclusion for part (b) which we have just proved,

$$\deg_{\text{LS}}(f, N_1, y) \geq 1, \quad \deg_{\text{LS}}(f, N_2, y) \geq 1.$$

Hence by the additivity of the degree function,

$$\deg_{\text{LS}}(f, V, y) = \deg_{\text{LS}}(f, N_1, y) + \deg_{\text{LS}}(f, N_2, y) \geq 2,$$

which contradicts the assumption that $\deg_{\text{LS}}(f, V, y) = +1$. This contradiction proves that $f^{-1}(y) \cap V$ is indeed connected. q.e.d.

PROOF OF (d). Suppose that C is a component of $X_n - f(\text{bdry}(V))$ which contains a point y_1 of $f(V)$, and let y be an arbitrary point of C. By the conclusion of part (b), $\deg_{\text{LS}}(f, V, y_1) \geq 1$. By the fact that C is pathwise connected, $\deg_{\text{LS}}(f, V, y) = \deg_{\text{LS}}(f, V, y_1) > 0$. Hence y lies in $f(V)$, i.e., $C \subset f(V)$. q.e.d.

With the proof of part (d), the proof of Proposition (15.1) is complete. q.e.d.

PROOF OF THEOREM (15.1). Suppose that f is an orientation-preserving mapping of V, where V is an open subset of G with cl (V) a compact subset of G. We choose a second open subset G_1 of G with compact closure in G such that cl $(V) \subset G_1$. By Definition (15.3), there exists a sequence $\{f_k\}$ of strict orientation-preserving mappings of class C^1 from G_1 to X_n which converges uniformly to f on G_1 as $k \to \infty$. The conclusions of Theorem (15.1) are valid for each of the mappings f_k by Proposition (15.1). We shall now prove that they are valid for the limit mapping f.

PROOF OF (a). Let y be a point in $X_n - f(\text{bdry}(V))$. Let d_0 be the distance from y to $f(\text{bdry}(V))$, so that $d_0 > 0$. There exists an index k_0 such that $k > k_0$ and all x in bdry (V), $\|f_k(x) - f(x)\| \leq d_1 < d_0$. Hence for each x in bdry (V),

$$\|f_k(x) - y\| \geq \|f(x) - y\| - \|f_k(x) - f(x)\| \geq d_0 - d_1 > 0.$$

In particular, y lies in $X_n - f_k(\text{bdry}(V))$ for $k > k_0$ and hence $\deg_{\text{LS}}(f_k, V, y)$ is well defined for all such k, with $\deg_{\text{LS}}(f_k, V, y) \geq 0$.

For any t in $[0, 1]$, we let

$$g_t(x) = tf(x) + (1-t)f_k(x) \quad (x \in \text{cl}(V))$$

for each $k > k_0$. For any point x in bdry (V), we know that

$$g_t(x) - y = t(f(x) - y) + (1-t)(f_k(x) - y)$$
$$= [f(x) - y] + (1-t)[f_k(x) - f(x)].$$

Therefore, we have $\|g_t(x) - y\| \geq d_0 - (1-t)d_1 > 0$ for each x in bdry (V). Hence $\{g_t : 0 \leq t \leq 1\}$ is a permissible homotopy of f to f_k and by the familiar properties of the degree function, $\deg_{\text{LS}}(g_t, V, y)$ is defined for all t in $[0, 1]$ and independent of t in $[0, 1]$. Thus $\deg_{\text{LS}}(f, V, y) = \deg_{\text{LS}}(f_k, V, y) \geq 0$. q.e.d.

PROOF OF (b). If $y \in X_n - f(\text{bdry}(V))$ and $\deg_{\text{LS}}(f, V, y) > 0$, then it follows from the basic properties of the degree function that y must lie in $f(V)$.

Suppose on the other hand that y lies in $f(V) \cap (X_n - f(\text{bdry}(V)))$. Let C be the component of $X_n - f(\text{bdry}(V))$ which contains the point y. Let $d_0 > 0$ be the distance from y to $f(\text{bdry}(V))$, and choose a point x_0 of V such that $f(x_0) = y$. For a sufficiently large index k_0 and all $k > k_0$, we may ensure that $\|f_k(x) - f(x)\| \leq d_1/2$, $(d_1 < d_0)$, for all points x in cl (V). In particular, for each point x in bdry (V), we have

$$\|f_k(x) - y\| \geq \|f(x) - y\| - \|f_k(x) - f(x)\| > d_0 - d_0/2 = d_0/2.$$

Hence for $k > k_0$, each point y_1 in the ball of radius $(d_0/2)$ about y lies in the same component of $X_n - f_k(\text{bdry}(V))$ as does y itself.

However, we know that $\|f_k(x_0) - y\| \leq \|f_k(x_0) - f(x_0)\| < d_0/2$. Therefore, $f_k(x_0)$ lies in the same component of $X_n - f_k(\text{bdry}(V))$ as does y itself. If we now apply conclusion (b) of Theorem (15.1) to the mapping f_k (for which the results of Theorem (15.1) have been proven valid in Proposition (15.1)), we see that $\deg_{\text{LS}}(f_k, V, y) > 0$ for all $k > k_0$. On the other hand, it follows by the argument of part (a) that for $k > k_0$, $\deg_{\text{LS}}(f, V, y) = \deg_{\text{LS}}(f_k, V, y)$. Hence $\deg_{\text{LS}}(f, V, y) > 0$ for y in $f(V) - f(\text{bdry}(V))$. q.e.d.

PROOF OF (c). Suppose that for a given point y of $X_n - f(\text{bdry}(V))$, we have $\deg_{\text{LS}}(f, V, y) = +1$. We wish to show that $f^{-1}(y) \cap V$ is connected.

Suppose on the contrary that $f^{-1}(y) \cap V$ is disconnected. Since $f^{-1}(y) \cap V = f^{-1}(y) \cap \text{cl}(V)$ is compact, there exist two open subsets N_1 and N_2 whose closures are disjoint such that each set intersects $f^{-1}(y)$ while $f^{-1}(y) \cap V \subset N_1 \cup N_2$.

By the conclusion of part (b) which has already been proved,

$$\deg_{\text{LS}}(f, N_1, y) \geq 1, \qquad \deg_{\text{LS}}(f, N_2, y) \geq 1.$$

By the additivity of the degree function,

$$\deg_{\text{LS}}(f, V, y) = \deg_{\text{LS}}(f, N_1, t) + \deg_{\text{LS}}(f, N_2, y) \geq 2,$$

which contradicts the assumption that $\deg_{\text{LS}}(f, V, y) = +1$. This contradiction proves that $f^{-1}(y) \cap V$ is connected. q.e.d.

PROOF OF (d). Suppose that for a given component C of $X_n - f(\text{bdry}(V))$, we have a point y_1 in $f(V) \cap C$. By the conclusion of part (b), $\deg_{\text{LS}}(f, V, y_1) \geq 1$. Since C is pathwise connected, we know by the basic properties of the degree function that for all y in C, $\deg_{\text{LS}}(f, V, y) = \deg_{\text{LS}}(f, V, y_1) \geq 1$. Hence, each y in C lies in $f(V)$, i.e., $C \subset f(V)$. q.e.d.

With the completion of the proofs of parts (a), (b), (c), and (d), the proof of Theorem (15.1) is thereby complete. q.e.d.

PROPOSITION (15.2). *Let Y_m be a complex Banach space of complex dimension $m \geq 1$, X_n the corresponding real Banach space of real dimension $n = 2m$ obtained from Y_m forgetting its complex structure. Let G be an open subset of Y_m, f a complex analytic mapping of G into Y_m.*

Then, considered as a mapping of the subset G of X_n, f is an orientation-preserving mapping in the sense of Definition (15.3).

PROOF OF PROPOSITION (15.2). We may assume without loss of generality that G is connected. For each $\epsilon > 0$, let f_ϵ be the mapping of G into Y_m given by $f_\epsilon(z) = f(z) + \epsilon z$. Then each f_ϵ is also complex analytic and for any sequence $\epsilon_j \to 0$, the corresponding sequence $\{f_{\epsilon_j}\}$ converges uniformly to f on any bounded subset of G. Hence it suffices to prove that for some such sequence, we can ensure that each f_{ϵ_j} is strict orientation-preserving in the sense of Definition (15.2).

We know by the complex analyticity of f and the f_ϵ that all of these mappings are of class C^∞ and *a fortiori* of class C_1. For any complex analytic mapping g of G into Y_m and any point x of G, if we choose a fixed complex basis $\{z_1, \ldots, z_m\}$ of Y_m, then with respect to the real basis of the corresponding real space X_n given by $\{x_1, \ldots, x_m, y_1, \ldots, y_m\}$ with $x_j + iy_j = z_j$ $(1 \leq j \leq m)$, it follows by an easy computation that $\det(g'_z) = |\det(J(g))(z)|^2$ where $J(g)(z)$ is the complex Jacobian matrix for g at Z, i.e.,

$$\{J(g)(z)\}_{j,k} = \partial g_j / \partial z_k(z).$$

In particular $\det(g'_z) \geq 0$ $(z \in G)$ for any such basis. Setting $g = f_\epsilon$, it therefore suffices to prove that for a sequence $\epsilon_j \to 0$, $S(f_\epsilon)$ is nowhere dense for each j.

By the formula for $\det(f_\epsilon)'_x$ derived in the preceding paragraph, we know that a point x lies in $S(f_\epsilon)$ if and only if $J(f_\epsilon)(x) = 0$ where $J(f_\epsilon)(x)$ is the complex Jacobian matrix of f_ϵ with respect to a fixed complex basis. Since $S(f_\epsilon)$ is a closed subset of G, it follows that $S(f_\epsilon)$ is somewhere dense if and only if $J(f_\epsilon)(x)$ vanishes for all x in an open subset of G. Since $J(f_\epsilon)$ is complex analytic and G is connected, this will happen if and only if $J(f_\epsilon)$ is identically null on G.

Choose a point z_0 in G. Then $(Jf_\epsilon)(z_0)$ is a polynomial in ϵ whose top-order coefficient is one. Hence for some $d_0 > 0$, it follows that for all ϵ with $0 < \epsilon < d_0$, $(Jf_\epsilon)(z_0) \neq 0$. For such ϵ, it follows from the discussion immediately above that $S(f_\epsilon)$ is nowhere dense in G. q.e.d.

PROPOSITION (15.3). *Let X_n be a finite-dimensional real Banach space, G an open subset of X_n, f a continuous accretive mapping of G into X_n. Then f is an orientation-preserving mapping in the sense of Definition (15.3).*

PROOF OF PROPOSITION (15.3). By hypothesis, f is an accretive mapping of G into X_n, so that there exists a single-valued section $J: X_n \to X_n^*$ of the duality mapping such that for all u and v in G, $(J(u - v), f(u) - f(v)) \geq 0$. If we consider an open subset G_1 of G with $\mathrm{cl}\,(G_1)$ a compact subset of G, then $\mathrm{cl}\,(G_1)$ is bounded in X_n and we may approximate f uniformly on $\mathrm{cl}\,(G_1)$ by the mappings f_ϵ with small $\epsilon > 0$ given by $f_\epsilon = f + \epsilon I$. Hence, if we replace f by f_ϵ, we may assume that for a given positive constant $c > 0$, $(J(u - v), f(v)) \geq c \|u - v\|^2$.

Since X_n is of finite dimension n and $\mathrm{cl}\,(G_1)$ is compact, f is uniformly continuous on some uniform neighborhood of $\mathrm{cl}\,(G_1)$ contained in G. Let ψ be a C^∞ scalar function with compact support in X_n such that $0 \leq \psi(x)$ for all x in X_n and $\int_{X_n} \psi(x)\,dx = 1$ (where the integration is taken with respect to Lebesgue n-measure on X_n). For each $\epsilon > 0$, we set $\psi_\epsilon(x) = \epsilon^{-n}\psi(x/\epsilon)$, and define the sequence of functions $g_\epsilon(x) = \int_G \psi_\epsilon(x - y) f(y)\,dy$. It follows by familiar arguments that the functions $g_\epsilon(x)$ converge to $f(x)$ uniformly for x in $\mathrm{cl}\,(G_1)$, while for all u and v in cl (G_1), we have

$$(J(u - v), g_\epsilon(u) - g_\epsilon(v)) = \int_G \psi_\epsilon(y)(J([u - y] - [v - y]), f(u - y) - f(v - y)) \, dy$$

$$\geq c \int_G \psi_\epsilon(y) \|u - v\|^2 \, dy = c \|u - v\|^2$$

as soon as the ϵ-neighborhood of cl (G_1) is contained in G. Moreover, each g_ϵ is a C^1-mapping of G_1 into X_n.

Let $h_{\epsilon,t} = (1 - t)g_\epsilon + tI$ $(0 \leq t \leq 1)$. For each t in $[0, 1]$, we have $h_{\epsilon,t}$ a C^1-mapping of G_1 into X_n with the property that for a fixed constant $c_1 > 0$,

$$(J(u - v), h_{\epsilon,t}(u) - h_{\epsilon,t}(v)) \geq c \|u - v\|^2$$

for all u and v in G_1. It follows that

$$c \|u - v\|^2 \leq \|J(u - v)\| \cdot \|h_{\epsilon,t}(u) - h_{\epsilon,t}(v)\| = \|u - v\| \cdot \|h_{\epsilon,t}(u) - h_{\epsilon,t}(v)\|.$$

In particular, we obtain $\|h_{\epsilon,t}(u) - h_{\epsilon,t}(v)\| \geq c \|u - v\|$ so that each $h_{\epsilon,t}$ is a homeomorphism of G_1 into X_n. Moreover, if we set $u = v + \lambda w$ in the above inequality for $\lambda > 0$, we find that

$$\|\lambda^{-1}[h_{\epsilon,t}(u + \lambda w) - h_{\epsilon,t}(u)]\| \geq c \|w\|.$$

Letting $\lambda \to 0+$, we see that $c \|w\| \leq \|(h_{\epsilon,t})'(w)\|$ for each u in G_1 and each w in X_n. Hence each $h_{\epsilon,t}$ is a diffeomorphism from G into X_n. It follows that since we have a continuous family of diffeomorphisms with det $(h'_{\epsilon,t})$ nonnull and varying continuously in t, each $h_{\epsilon,t}$ is a strict orientation-preserving mapping. Hence so is g_ϵ and thereby f is an orientation-preserving mapping. q.e.d.

DEFINITION (15.4). *Let X be a real Banach space (of finite or infinite dimension), G an open subset of X, f a continuous mapping of G into X. Let $\{X_\alpha, P_\alpha : \alpha \in \Omega\}$ be a family of finite-dimensional subspaces X_α of X and of projection mappings P_α of X onto X_α, such that $\bigcup_\alpha X_\alpha$ is dense in X.*

Then f is said to be an orientation-preserving mapping with respect to the family $\{X_\alpha, P_\alpha\}$ if for each α in Ω, the mapping

$$f_\alpha = P_\alpha f|_{G \cap X_\alpha} : G \cap X_\alpha \to X_\alpha$$

is an orientation-preserving mapping in the sense of Definition (15.3) of $G \cap X_\alpha$ into the finite-dimensional Banach space X_α.

As a consequence of Proposition (15.2), we have the following result:

PROPOSITION (15.4). *Let Y be a complex Banach space, G an open subset of Y, f a complex analytic mapping of G into Y. Let $\{Y_\alpha, P_\alpha\}$ be a family of finite-dimensional complex subspaces of Y and of projections P_α of Y onto Y_α (where each P_α is a bounded, complex linear mapping). Let X be the real Banach space obtained from Y by ignoring the complex structure of Y, let X_α be the corresponding real subspace of X obtained from Y_α, and let P_α denote the corresponding projection of X on X_α. Suppose that $\bigcup_\alpha Y_\alpha$ is dense in Y.*

Then f is an orientation-preserving mapping of G into X with respect to the family $\{X_\alpha, P_\alpha\}$ in the sense of Definition (15.4).

PROOF OF PROPOSITION (15.4). For each α, $f_\alpha = P_\alpha f$ is a complex analytic mapping of $G \cap Y_\alpha$ into Y_α. The result then follows from Proposition (15.2).
q.e.d.

A corresponding result for accretive mappings is the following:

PROPOSITION (15.5). *Let X be a real Banach space with X^* strictly convex, G an open subset of X, f a continuous mapping of G into X. Suppose that f is accretive and that we are given a family $\{X_\alpha, P_\alpha\}$ of finite-dimensional subspaces X_α of X and of projection mappings P_α of X on X such that $\bigcup_\alpha X_\alpha$ is dense in X. Suppose further that for each α, $\|P_\alpha\| = 1$.*

Then f is orientation-preserving in the sense of Definition (15.4) with respect to the family $\{X_\alpha, P_\alpha\}$.

PROOF OF PROPOSITION (15.5). By Proposition (15.3), it suffices to prove that the mapping $f_\alpha = P_\alpha f$ is accretive from $G \cap X_\alpha$ into X_α. For u and v in $G \cap X_\alpha$, however (if we let J_α be the duality mapping of X_α into X^*) we have

$$(J_\alpha(u-v), f_\alpha(u) - f_\alpha(v)) = (J(u-v), f_\alpha(u) - f_\alpha(v))$$
$$= (J(u-v), P_\alpha(f(u) - f(v)))$$
$$= (P_\alpha^* J(u-v), f(u) - f(v)).$$

On the other hand,

$$\|P_\alpha^* J(u-v)\| \le \|P_\alpha^*\| \cdot \|J(u-v)\| = \|P_\alpha\| \cdot \|u-v\| \le \|u-v\|$$

while

$$(P_\alpha^* J(u-v), u-v) = (J(u-v), P_\alpha(u-v)) = (J(u-v), u-v) = \|u-v\|^2.$$

Thus by the strict convexity of X^*, it follows that $P_\alpha^* J(u-v) = J(u-v)$ ($u, v \in X_\alpha$). Finally

$$(J_\alpha(u-v), f_\alpha(u) - f_\alpha(v)) = (P_\alpha^* J(u-v), f(u) - f(v))$$
$$= (J(u-v), f(u) - f(v)) \ge 0,$$

so that f_α is indeed accretive on $G \cap X_\alpha$.
q.e.d.

We now prove the first and simplest of our general results on orientation-preserving mappings in the infinite-dimensional case:

THEOREM (15.2). *Let X be a real Banach space, G an open subset of X, f a continuous mapping of G into X of the form $f = I + C$, with I the identity mapping and C a compact mapping of G into X. Suppose that f is an orientation-preserving mapping with respect to a family $\{X_\alpha, P_\alpha\}$ of finite-dimensional subspaces X_α of X and linear projection mappings P_α of X onto X_α. Suppose further that for each compact subset K of X, each space X_β in the family, and each $\epsilon > 0$, there exists a space X_α in the family which contains X_β and such that for all x in K,*

$$\|P_\alpha(x) - x\| < \epsilon.$$

Let V be a bounded open subset of G with cl (V) *contained in G. Let y be a point of $X - f(\text{bdry}(V))$. Then:*

(a) $\deg_{LS}(f, V, y) \geq 0$.
(b) $\deg_{LS}(f, V, y) > 0$ *if and only if y lies in $f(V)$.*
(c) *If* $\deg_{LS}(f, V, y) = +1$, *then* $f^{-1}(y) \cap V$ *is connected.*
(d) *If for a component C of $X - f(\text{bdry}(V))$, C intersects $f(V)$, then C is contained in $f(V)$.*

PROOF OF THEOREM (15.2). PROOF OF (a). Let y be a point of $X - f(\text{bdry}(V))$ for a given open subset V of G with cl (V) bounded and contained in G. By hypothesis, $\bigcup_\beta X_\beta$ is dense in X. Since $X - f(\text{bdry}(V))$ is an open subset of X, there exists a ball $B_d(y)$ for some $d > 0$ which is contained in $X - f(\text{bdry}(V))$. In particular, there exists a point y_β of some X_β in this ball. For this point y_β, it follows by the invariance of the degree under deformations during which the degree remains defined that $\deg_{LS}(f, V, y) = \deg_{LS}(f, V, y_\beta)$. Hence, we may assume without loss of generality that the point y itself lies in a given subspace X_β.

Let K be the compact subset cl $(C(V))$ of X. By the definition of the Leray-Schauder degree, there exists a constant $\epsilon > 0$ such that for any subspace X_α of our given family of finite-dimensional spaces such that K is at distance at most ϵ from X_α with $X_\beta \subset X_\alpha$ and for any continuous mapping C_α of cl $(V) \cap X_\alpha$ into X_α such that

$$\|C_\alpha(x) - C(x)\| < \epsilon \qquad (x \in \text{cl}(V) \cap X_\alpha),$$

it follows that $\deg_{LS}(f, V, y) = \deg_{LS}(I - C_\alpha, V \cap X_\alpha, y)$ where the second degree is taken in the space X_α. However, by our hypothesis such a space X_α exists with $C_\alpha = P_\alpha C$. Hence $\deg_{LS}(f, V, y) = \deg_{LS}(f_\alpha, V \cap X_\alpha, y) \geq 0$ since f_α is assumed to be an orientation-preserving mapping in the finite-dimensional space X_α to which the conclusions of Theorem (15.1) apply. q.e.d.

PROOF OF (b). If $\deg_{LS}(f, V, y) > 0$, then it follows from the general properties of the Leray-Schauder degree that y must lie in $f(V)$.

Suppose on the other hand that y lies in $f(V)$ while y does not lie in $f(\text{bdry}(V))$. Let $d_0 > 0$ be the distance from y to $f(\text{bdry}(V))$. We may choose a point y_1 in one of the spaces X_β such that $\|y - y_1\| < d_0/3$. By the hypothesis of Theorem (15.2), we can find a space X_α containing X_β while for each point u in $C(\text{cl}(V))$, $\|P_\alpha(u) - u\| < d_0/3$. Hence for each x in cl $(V) \cap X_\alpha$, we have for $f_\alpha = P_\alpha f$,

$$\|f(x) - f_\alpha(x)\| = \|x - C(x) - P_\alpha x + P_\alpha C(x)\| = \|C(x) - P_\alpha C(x)\| < d_0/3.$$

By a still stricter choice of the space X_α, we may also ensure as in the proof of part (a) that $\deg_{LS}(f, V, y_1) = \deg_{LS}(f_\alpha, V \cap X_\alpha, y_1)$, and in addition that for a given point x in V such that $f(x) = y$, there exists a point x_1 in $X_\alpha \cap V$ such that $\|f(x_1) - f(x)\| < d_0/3$.

Combining the above inequalities, we see that $\|f_\alpha(x_1) - y\| < 2d_0/3$. On the other hand, for each x in bdry $(V) \cap X_\alpha$,

$$\|y - f_\alpha(v)\| \geq \|y - f(v)\| - \|f(v) - f_\alpha(v)\| \geq d_0 - d_0/3 = 2d_0/3.$$

Hence the open ball of radius $2d_0/3$ about y intersects X_α in a set which is contained in a single component of $X_\alpha - f_\alpha(\text{bdry}\,(V \cap X_\alpha))$ (since $\text{bdry}\,(V \cap X_\alpha) \subset \text{bdry}\,(V) \cap X_\alpha$). Since the point $f_\alpha(x_1)$ lies in the intersection of this ball with $f_\alpha(V)$ and since f_α is assumed to be orientation-preserving, it follows from Theorem (15.1) that for y_1 in this component, $\deg_{LS}(f_\alpha, V \cap X_\alpha, y_1) > 0$. Hence, we have finally

$$\deg_{LS}(f, V, y) = \deg_{LS}(f_\alpha, V \cap X_\alpha, y_1) > 0. \quad \text{q.e.d.}$$

PROOF OF (c). Suppose that $f^{-1}(y) \cap V$ is not connected for a point y in $X - f(\text{bdry}\,(V))$ for which $\deg_{LS}(f, V, y) = +1$. Since $f^{-1}(y) \cap V = f^{-1}(y) \cap \text{cl}\,(V)$ and since for each point x in $f^{-1}(y) \cap V$, we have $x = y + C(x)$ lies in the relatively compact subset $y + C(V)$ of X, it follows that $f^{-1}(y) \cap V$ is compact. If $f^{-1}(y) \cap V$ is not connected, there exist two disjoint open subsets V_1 and V_2 of Y with $f^{-1}(y) \cap V \subset V_1 \cup V_2$, and both of the sets containing points of $f^{-1}(y)$. Hence by the conclusion of part (b),

$$\deg_{LS}(f, V_1, y) \geq 1, \quad \deg_{LS}(f, V_2, y) \geq 1.$$

On the other hand, by the additivity property of the degree function,

$$\deg_{LS}(f, V, y) = \deg_{LS}(f, V_1, y) + \deg_{LS}(f, V_2, y) \geq 2,$$

which contradicts the assumption that $\deg_{LS}(f, V, y) = +1$. q.e.d.

PROOF OF (d). Let y be a point of a component C of $X - f(\text{bdry}\,(V))$ which contains a point y_1 of $f(V)$. Then by part (b), $\deg_{LS}(f, V, y_1) > 0$. Hence $\deg_{LS}(f, V, y) = \deg_{LS}(f, V, y_1) > 0$. Hence y lies in $f(V)$, i.e., C is contained in $f(V)$. q.e.d.

With the proof of parts (a), (b), (c), and (d) completed, the proof of Theorem (15.2) is thereby complete. q.e.d.

We combine the result of Theorem (15.2) with the following simple observation:

PROPOSITION (15.6). *Let X and Y be topological spaces, f a mapping of X into Y. Suppose that for a point y of Y, $f^{-1}(y)$ is both connected and totally disconnected. Then $f^{-1}(y)$ is either empty or a single point.*

The proof of Proposition (15.6) is obvious.

The particular context in which we shall obtain conclusions like those of Proposition (15.6) is that contained in the following theorem:

THEOREM (15.3). *Let X and Y be complex Banach spaces, G an open subset of X, f a continuous mapping of G into Y which is complex analytic and also a Fredholm mapping. Suppose that $f^{-1}(y_0)$ is compact for a given point y_0 of Y. Then $f^{-1}(y_0)$ is totally disconnected.*

To carry through the proof of Theorem (15.3) using a number of auxiliary results, we recall a number of basic definitions and results from the theory of analytic functions of several complex variables, in the form expounded in the book by Gunning and Rossi [234]. We note first that if G_0 is an open subset of C^n (the complex n-dimensional space), then a subset V of G_0 is said to be a subvariety if for every z in G_0 there exist a neighborhood G_z of z in U_0 and a finite family $\{f_1, \ldots, f_t\}$ of functions holomorphic on G_z such that

$$V \cap G_z = \{x \mid x \in G_z, f_j(x) = 0, 1 \leq j \leq t\}.$$

By a ringed space, one means a pair (X, S) where X is a Hausdorff space and S is a sheaf of subrings with unity of the sheaf of germs of continuous complex-valued functions on X. If we have two ringed spaces (X, S) and (Y, S_1), a continuous mapping f of X into Y is called a mapping of ringed spaces if for each x in X and each local section of S_1 near $f(x)$ the composition $h \circ f$ is a local section of S over some neighborhood of x in X. If V is a subvariety of a domain G_0, then V can be identified with the ringed space (V, S_0) where S_0 is the sheaf of germs of functions holomorphic on V.

If (X, S) and (Y, S_1) are two ringed spaces, a mapping f of X into Y is said to be an isomorphism of ringed spaces if f is a homeomorphism of X onto Y and if f is an injection (i.e., for each x in X, the mapping f^* which sends each local section h of S_1 on a neighborhood of $f(x)$ into the local section $f^*(h) = h \circ f$ on some neighborhood of x maps the stalk of S_1 over $f(x)$ onto the stalk of S over x).

A ringed space (X, S) is said to be an *analytic space* if each x in X has a neighborhood G in X such that $(G, S \mid G)$ is isomorphic to a ringed space (V, S) where V is a subvariety of a domain in C^n and S_0 is the sheaf of holomorphic functions on V.

Let X be a Hausdorff space. Suppose that $X = \bigcup_\alpha X_\alpha$ and that there are mappings $\Phi_\alpha : X_\alpha \to D_\alpha$ such that (i) Φ_α is surjective, (ii) D_α is a subvariety of a domain in C^n (for some n), (iii) for all α and β such that $X_\alpha \cap X_\beta \neq \varnothing$ the mapping $\Phi_\beta \Phi_\alpha^{-1} : D_\alpha \cap \Phi_\alpha(X_\alpha \cap X_\beta) \to D_\beta \cap \Phi(X_\alpha \cap X_\beta)$ is a mapping of ringed spaces. Suppose further that the Φ_α are homeomorphisms. Then there exists a unique sheaf S on X such that (X, S) is an analytic space and the mappings Φ_α are isomorphisms.

The sheaf S for an analytic space (X, S) is called the sheaf of germs of holomorphic functions on X, and a section of this sheaf is called a holomorphic function on X. In particular, we have the following result:

PROPOSITION (15.7). *If (X, S) is an analytic space with X compact and connected, then S is the sheaf of constants.*

PROOF OF PROPOSITION (15.7). This is simply Theorem (V, 6) of Gunning and Rossi [234, p. 15]. q.e.d.

PROOF OF THEOREM (15.3). For each point x_0 in the compact set $f^{-1}(y_0)$, f'_{x_0} is a linear Fredholm operator. Hence, its null space N_x is a finite-dimensional complex subspace of X, and we can find a closed complex subspace R_{x_0} of X which is a closed complement to N_{x_0}. For h in R_{x_0}, f'_{x_0} satisfies an inequality of the form $\|f'_{x_0}(h)\| \geq c_0 \|h\|$ for some $c_0 > 0$ since the range of the linear mapping f'_{x_0} is a closed subspace of Y. It follows in particular that there exists a neighborhood G_{x_0}

in G such that for all x in $G_{x_0} \cap f^{-1}(y_0)$, f'_x satisfies a similar inequality on the space R_{x_0} with c_0 replaced by $\frac{1}{2}c_0$. Since $f^{-1}(y_0)$ is compact, it can be covered by a finite number of such neighborhoods $\{G_1, G_2, \ldots, G_r\}$ corresponding to a finite family of points $\{x_1, \ldots, x_r\}$ and with the corresponding family of closed subspaces $\{R_1, \ldots, R_r\}$. If we set $R = \bigcap_{j=1}^r R_j$, then R is a closed complex subspace of X with a finite-dimensional complementary complex subspace N, and there exists a constant $c > 0$ such that for any x in $f^{-1}(y_0)$ and any h in R, $\|f'_x(h)\| \geq c\|h\|$.

Consider x_0 once more in $f^{-1}(y_0)$, and let R and N be the two subspaces of X constructed in the preceding paragraph (which do not depend upon the choice of x_0), with N of finite dimension. Then $Y_0 = f'_{x_0}(R)$ is a closed subspace of Y of finite codimension, and there exists a complex linear mapping P_0 projecting Y on Y_0. Let x be any point in X near x_0. Then x may be written in one and only one way in the form
$$x = x_0 + u + h \qquad (u \in N, h \in R)$$
with both u and h small in norm. By the complex analyticity of f, we have $f(x) = f(x_0) + f'_{x_0}(h) + f'_{x_0}(u) + s(u, h)$ where $s(u, h)$ is analytic in u and h with a power series beginning with quadratic terms. If x is to lie in $f^{-1}(y_0)$ (since $f(x_0) = y_0$), it is necessary and sufficient that $f'_{x_0}(h) = -f'_{x_0}(u) - s(u, h)$, which in turn is equivalent to the pair of equations:

(1) $$f'_{x_0}(h) + P_0 s(u, h) = -P_0 f'_{x_0}(u),$$
(2) $$(I - P_0)[f'_{x_0}(u) + s(u, h)] = 0.$$

We note that f'_{x_0} is an isomorphism of the space R onto the space Y_0 while for each fixed u in a suitable neighborhood V of 0 in N, the mapping $P_0 s(u, \cdot): V_1 \to Y_0$ satisfies a Lipschitz condition with small Lipschitz constant for a suitable small neighborhood V_1 of 0 in R. Hence the equation (1) has one and only one solution h in V_1 for each u in a still smaller neighborhood V_2 of 0 in N, i.e., the equation (1) is equivalent to $h = \Phi(u)$, where by the complex analyticity of the mappings involved in equation (1), Φ is a holomorphic mapping of V_2 into V_1. If we now substitute $h = \Phi(u)$ into equation (2), the necessary and sufficient condition that $x = x_0 + u + h$ should satisfy the condition that $f(x) = y_0$ becomes the new equation

(3) $$(I - P_0)[f'_{x_0}(u) + s(u, \Phi(u))] = 0.$$

If we set
$$\psi(u) = (I - P_0)[f'_{x_0}(u) + s(u, \Phi(u))],$$
ψ is a holomorphic mapping of V_2, an open subset of N, into the range of $(I - P_0)$, which is a finite-dimensional complex subspace of Y. Hence the subset of u in V_2 for which the equation (3) holds is a subvariety of the neighborhood V_2.

We know, however, that for $x = x_0 + h + u$ with u in V_2 and h in V_1 to lie in $f^{-1}(y_0)$ it is necessary and sufficient that $h = \Phi(u)$ and that u lies in this analytic subvariety of V_2, i.e., if W is the subset of $f^{-1}(y_0)$ lying in this neighborhood and if

P is the projection of X on N with $(I - P)$ projecting on R, we have P mapping $W - x_0$ homeomorphically on the subvariety F of V_2 with $P(x) = P(x_0) + f$ ($f \in F$) and $x = x_0 + P(x - x_0) + \Phi(P(x - x_0))$. It follows that W is an analytic space with respect to the family of mappings ψ_α of neighborhoods of points of W into subvarieties of neighborhoods in N, where each ψ_α is defined on a neighborhood of the point x_α and maps all x in the neighborhood of x_α into $\psi_\alpha(x) = P(x)$.

By Proposition (15.7), all holomorphic functions on W are constants on components of W. In particular, the holomorphic mapping P maps each component of W into a single point. Since P is a local homeomorphism of W into N, W is totally disconnected. q.e.d.

We now apply the result of Theorem (15.3) to establish the following extension of Theorem (15.2):

THEOREM (15.4). *Let X be a complex Banach space, G an open subset of X, f a continuous mapping of G into X which is holomorphic and such that f is of the form $f = I - C$, where C maps each bounded subset of G into a relatively compact subset of X. Let V be a bounded open subset of G with cl (V) contained in G and let y be a point of $X - f(\text{bdry } (V))$. Then:*

(a) $f^{-1}(y) \cap V$ *is a finite set.*

(b) $\deg_{LS} (f, V, y) \geq 0$.

(c) $\deg_{LS} (f, V, y) > 0$ *if and only if $f^{-1}(y) \cap V$ is nonempty.*

(d) $\deg_{LS} (f, V, y) = +1$ *implies that there exists exactly one point x in V such that $f(x) = y$.*

(e) *For any component C of $X - f(\text{bdry } (V))$, C intersects $f(V)$ if and only if C is contained in $f(V)$.*

PROOF OF THEOREM (15.4). Since f is a holomorphic mapping, f is of class C^∞ and in particular of class C^1. The mapping C is thereby also holomorphic and of class C^1. For each x in G, C'_x, the derivative of C at x, is defined by

$$C'_x(h) = \lim_{\epsilon \to 0} \epsilon^{-1}[C(x + \epsilon h) - C(x)] \quad (h \in X)$$

and is a bounded linear mapping of X into X. The above limit exists uniformly for h in any given bounded subset of X, and therefore each C'_x is a compact linear mapping of X into X. In particular, $f'_x = I - C'_x$ is a Fredholm mapping (in the linear sense) of X into X. Thus f itself is a (nonlinear) Fredholm mapping of G into X.

Suppose y does not lie in $f(\text{bdry } (V))$. Then $f^{-1}(y) \cap V = f^{-1}(y) \cap \text{cl } (V)$ is a closed subset of X since cl (V) is contained in G. For each x in $f^{-1}(y) \cap V$, $x - C(x) = y$, so that $x \in y + C(V)$. Since V is a bounded subset of G and since C maps each bounded subset of G into a relatively compact subset of X, $C(V)$ is relatively compact in X. Hence $f^{-1}(y) \cap V = K$ is contained in a relatively compact subset of X, and since $f^{-1}(y) \cap V$ is a closed subset of X, $K = f^{-1}(y) \cap V$ is compact. Hence by Theorem (15.3), $f^{-1}(y) \cap V$ is totally disconnected.

Let x_0 be a point of K. Since f'_{x_0} is a Fredholm operator of index zero from X to X, there exists a finite-dimensional subspace N of X with complex dimension n, for some finite n, which is the null space of f'_x and a complex linear subspace N_1 of the same dimension n which is a closed complement of $R(f'_{x_0})$ in X. Let P be a complex linear projection mapping of X on N, and let S be a nonsingular complex linear mapping of N on N_1. Then $f'_{x_0} + SP$ is an isomorphism of X onto X, so that the mapping g is given by

$$g(u) = f(u) + SP(u - x_0) \qquad (u \in V)$$

is a holomorphic diffeomorphism of some neighborhood V_0 of x_0 in V on a neighborhood of $f(x_0) = y$. In particular, $f(u) = y$ for any u in V_0 if and only if $g(u) = y + SP(u - x_0)$, i.e., if and only if $v = g(u)$ satisfies the equation $v - SP(g^{-1}(v) - x_0) = y$. The last equation involves the new holomorphic mapping h defined by $h(v) = v - SP(g^{-1}(v) - x_0)$ which is of the form $I - C_1$ with C_1 of finite-dimensional range in the neighborhood of y.

PROOF OF (a). If we wish to prove that $f^{-1}(y) \cap V$ is finite, it suffices to show that this is true in the neighborhood of each point x_0 in the compact set $K = f^{-1}(y) \cap V$. If we apply the remarks of the preceding paragraph, we may assume without loss of generality that C maps cl (V_0) into a finite-dimensional subspace F of X. By taking the complex linear span of F and the point y, we may assume that y lies in F. If we replace V_0 by a possibly smaller neighborhood, we may apply the fact that $f^{-1}(y) \cap V$ is totally disconnected to choose the open neighborhood V_0 of x_0 so that $f^{-1}(y) \cap V_0$ itself is compact, i.e., $f^{-1}(y)$ does not intersect the boundary of V_0.

For a point x of V_0, we know that $f(x) = x - C(x) = y$ only if x lies in F, since both y and $C(x)$ lie in the finite-dimensional subspace F. Hence $f^{-1}(y) \cap V_0$ is a subset of F. Moreover,

$$\deg_{\mathrm{LS}} (f, V_0, y) = \deg_{\mathrm{LS}} (f|_{V_0 \cap F}, V_0 \cap F, y) \geq 0$$

by Theorem (15.2) and Proposition (15.2). Let $m = \deg_{\mathrm{LS}} (f, V_0, y)$. We assert that $f^{-1}(y) \cap V_0$ does not have more than m points. If it did, we could apply the total disconnectedness of $f^{-1}(y) \cap V_0$ and obtain a family of $(m + 1)$ disjoint open subsets $\{G_1, \ldots, G_{m+1}\}$ of F, each containing at least one point of $f^{-1}(y)$ such that $f^{-1}(y)$ is a compact subset of the union of the set G_j. Then

$$\deg_{\mathrm{LS}} (f, V_0, y) = \sum_{j=1}^{m+1} \deg_{\mathrm{LS}} (f, G_j, y) \geq m + 1$$

since by Theorem (15.2), it follows that, for each j, $\deg_{\mathrm{LS}} (f, G_j, y) \geq 1$.

Since $f^{-1}(y) \cap V_0$ is a finite set for a neighborhood of each point x_0 of the compact set $f^{-1}(y) \cap V$, it follows that $f^{-1}(y) \cap V$ is a finite set. q.e.d.

PROOF OF (b). Since $f^{-1}(y) \cap V$ is finite, we may replace V by the union of a finite family of small balls about the various points in $f^{-1}(y) \cap V$, i.e., $V = \bigcup_{j=1}^{r} G_j$, where each G_j contains exactly one point x_j such that $f(x_j) = y$. Since

$$\deg_{\mathrm{LS}} (f, V, y) = \sum_{j=1}^{r} \deg_{\mathrm{LS}} (f, G_j, y),$$

it suffices to prove that $\deg_{\mathrm{LS}} (f, G_j, y) \geq 1$.

If we choose the ball G_j about x_j of sufficiently small radius, we may apply the argument of the paragraph preceding the proof of part (a) and obtain the existence of a complex analytic diffeomorphism g of G_j on an open neighborhood of y and a finite-dimensional mapping C_1 such that $f = g + C_1$, where g is also of the form $g = I - C_2$ with C_2 a compact mapping of G_j into X. Hence $f = (I + C_1 g^{-1})g = (I + C_1 g^{-1})(I - C_2)$. Since g is a homeomorphism of G_j on an open set G'_j of X, it follows from the product formula for the degree that $\deg_{\mathrm{LS}}(f, G_j, y) = \deg_{\mathrm{LS}}(I + C_1 g^{-1}, G'_j, y)$. Since we may assume that y lies in the finite-dimensional subspace F which contains the range of the holomorphic mapping $C_1 g^{-1}$, it follows that $\deg_{\mathrm{LS}}(I + C_1 g^{-1}, G'_j, y) = \deg_{\mathrm{LS}}((I + C_1 g^{-1})|_{U'_j \cap F}, U'_j \cap F, y) \geq 1$ by Theorem (15.2) and Proposition (15.2). q.e.d.

PROOF OF (c). It follows from the argument of part (b) that if r is the cardinality of $f^{-1}(y) \cap V$, then $\deg_{\mathrm{LS}}(f, V, y) \geq r$. In particular, if $r \geq 1$, $\deg_{\mathrm{LS}}(f, V, y) > 0$. q.e.d.

PROOF OF (d). If $\deg_{\mathrm{LS}}(f, V, y) = +1$, then $f^{-1}(y) \cap V$ is nonempty while its cardinality is at most 1. Hence $f^{-1}(y) \cap V$ consists exactly of one point.

PROOF OF (e). If y is a point of the component C of $X - f(\mathrm{bdry}(V))$ lying in $f(V)$, then by part (c) above, $\deg_{\mathrm{LS}}(f, V, y) > 0$. On the other hand, for any y_1 in C, $0 < \deg_{\mathrm{LS}}(f, V, y) = \deg_{\mathrm{LS}}(f, V, y_1)$. Hence C lies in $f(V)$. q.e.d.

With the completion of the proof of parts (a) through (e) of Theorem (15.4), the proof of that theorem is complete. q.e.d.

We may now extend the conclusions of Theorem (15.4) in the following way to holomorphic Fredholm mappings of index zero.

THEOREM (15.5). *Let X and Y be complex Banach spaces, G an open subset of X, f a continuous (nonlinear) Fredholm mapping of index zero of G into Y which is holomorphic. Then:*

(a) *For any y in Y such that $f^{-1}(y)$ is compact, $f^{-1}(y)$ is a finite subset of G.*

(b) *If $f^{-1}(y)$ is compact for a given point y of Y, then for each x in $f^{-1}(y)$, we may define the multiplicity of x as a solution of $f(x) = y$ by finding a neighborhood V_1 of x containing no other elements of $f^{-1}(y)$ on which $f = (I - C)g$ with C a compact holomorphic mapping and g a holomorphic diffeomorphism of V_1 into X, and then setting*

$$m_y(x) = \deg_{\mathrm{LS}}(I - C, g(V_1), y) \geq 1.$$

(c) *Suppose that f is a proper mapping of G into the subset G_1 of Y. Then for each y in Y, the sum of the multiplicities of the points of $f^{-1}(y)$ is constant on each component of G_1. In particular, each component of G_1 which contains a point of $f(G)$ must lie entirely in $f(G)$.*

(d) *If f is a proper mapping of G into Y, then f maps G onto Y.*

PROOF OF THEOREM (15.5). PROOF OF (a). By hypothesis $f^{-1}(y)$ is compact. To show that $f^{-1}(y)$ is finite, it suffices to prove that each point x_0 of $f^{-1}(y)$ has a neighborhood V_1 such that $f^{-1}(y) \cap V_1$ is finite. By Theorem (15.3), $f^{-1}(y)$ is totally disconnected. Hence, we may choose an arbitrarily small neighborhood V_1 of x_0 such that $f^{-1}(y) \cap V_1$ is compact. By hypothesis, f'_{x_0} is a Fredholm mapping of index zero of X into Y. Hence the null space N of f'_{x_0} is a finite-dimensional complex subspace of X of dimension n for some finite n, and there exists a finite-dimensional subspace N_1 of Y of dimension n which is a closed complement in Y to the range of f'_{x_0}. Let P be a projection of X on N and let S be an isomorphism of N on N_1. Then the holomorphic mapping g of G into Y given by $g(u) = f(u) + SP(u - x_0)$ is a diffeomorphism of a suitably small neighborhood V_1 of x_0 onto an open neighborhood of y. In particular, it follows that $f(u) = y$ for any u in V_1 if and only if the corresponding point $v = g(u)$ of $g(V_1)$ satisfies the equation $v - SP(g^{-1}(v) - x_0) = y$. This last equation is of the type treated in Theorem (15.4) and since the set of solutions is precisely $g(f^{-1}(y) \cap V_1)$ which is a compact subset of $g(V_1)$, it follows from Theorem (15.4) that $g(f^{-1}(y) \cap V_1)$ is a finite set. Hence $f^{-1}(y) \cap V_1$ is a finite set.

Finally, $f^{-1}(y)$ must be a finite set. q.e.d.

PROOF OF (b). By the result of part (a), $f^{-1}(y)$ is a finite subset of G. For each point x in $f^{-1}(y)$, it follows from the argument of the proof of part (a) that there exists a neighborhood V_1 of x in G such that on V_1, f is of the form

$$f = g - C_1 = (I - C_1 g^{-1})g = (I - C)g$$

with C compact (or even finite-dimensional), g a holomorphic diffeomorphism of V_1 on an open neighborhood $g(V_1)$ of y, and C holomorphic. In particular, it follows from Theorem (15.4) that for each such x (if we choose V_1 to be so small that no other points of $f^{-1}(y)$ lie in V_1), $\deg_{LS}(I - C, g(V_1), y) \geq 1$.

To complete the proof of part (b), we must show that $\deg_{LS}(I - C, g(V_1), y)$ is independent of the choice of the representation of f in the form $f = (I - C)g$. Suppose $f = (I - C_0)g_0$ is another such representation of f on another neighborhood of x_0. We may assume without loss of generality that the two representations are defined on the same neighborhood V_1 of x in G. Since $(I - C_0)g_0 = (I - C)g$, we have $(I - C) = (I - C_0)(g_0 g^{-1})$, and it suffices to assume that f is of the form $f = I - C$ with C compact and has the representation $f = (I - C_0)g_0$. Then $g_0 - C_0 g_0 = I - C$, so that $g = I - C_2$ with C_2 compact. Hence, by the product formula for the degree,

$$\deg_{LS}(I - C, V_1, y) = \deg_{LS}((I - C_0)g_0, V_1, y) = \deg_{LS}(I - C_0, g_0(V_1), y). \quad \text{q.e.d.}$$

PROOF OF (c). It suffices to show that the sum of the multiplicities of the solutions of $f(x) = y$ as defined in part (b) is locally constant in y on each line segment in G_1. Let y_0 be a given element of G_1 and let $\{V_1, \ldots, V_r\}$ be a finite family of open subsets of G on each of which f is representable in the form $f = (I - C_j)g_j$ with g_j a holomorphic diffeomorphism of V_j into Y and C_j holomorphic compact mappings of V_j into Y. Consider the points of a line segment $y_t = (1 - t)y_0 + ty_1$ lying in G_1 ($0 \leq t \leq 1$). Since f is a proper mapping of G into G_1, there exists a compact subset K of G such that $f^{-1}(y_t)$ is contained in K for each t in [0, 1]. We assert that for some $\epsilon > 0$ and all t in the interval $[0, \epsilon]$, $f^{-1}(y_t)$ must be contained in $\bigcup_{j=1}^{r} V_j$; since otherwise

$$\bigcup_{0 \leq t \leq \epsilon} \left\{ \left(K - \bigcup_{j=1}^{r} V_j \right) \cap f^{-1}(y_t) \right\}$$

would be a family of compact sets with an empty intersection which has the finite intersection property.

It follows that for $0 \leq t < \epsilon$, $\sum_{j=1}^{r} \deg_{\mathrm{LS}}(I - C_j, g_j(V_j), y_t)$ is independent of t. However, it follows easily from the additivity property of the Leray-Schauder degree that this sum is equal for all t in $[0, \epsilon)$ to the sum of the multiplicities in the sense of part (b) of the solutions of the equation $f(x) = y_t$. q.e.d.

PROOF OF (d). Here $Y = G_1$ is connected and the result follows from that of part (c). q.e.d.

As an extension of Theorem (15.5), we establish the following:

THEOREM (15.6). *Let X and Y be complex Banach spaces, G an open subset of X, $\{f_t, 0 \leq t \leq 1\}$ a family of holomorphic Fredholm mappings of index zero from G to Y. Suppose that f_t varies continuously in t on $[0, 1]$ in the sense of local uniform convergence. Suppose further that for a given point y of G_1, there exists a fixed compact subset K of G such that $f_t^{-1}(y)$ is contained in K.*

Then the multiplicity of solutions of $f_t(x) = y$ is independent of t in $[0, 1]$.

PROOF OF THEOREM (15.6). It follows from an application of the Cauchy integral formula in finite-dimensional subspaces of X that $(f_t)_x'$ varies continuously with t in the sense of local uniform convergence in G. It suffices to prove that the sum of the multiplicities is locally constant for t in $[0, 1]$, and in particular constant for $0 \leq t < \epsilon$. Let $\{V_1, \ldots, V_j\}$ be a family of disjoint neighborhoods of the distinct points of $f_0^{-1}(y)$. It follows as in the proof of part (c) of Theorem (15.5) that for $\epsilon < 0$ sufficiently small and all t in $[0, \epsilon)$, $f_t^{-1}(y)$ must be contained in $\bigcup_{j=1}^{r} V_j$.

It suffices to choose the V_j so small that on each V_j, $f_t = (I - C_t)g_t$ with g_t a family of permissible homeomorphisms of V_j into Y varying continuously in t in the sense of uniform convergence on V_j and C_t a family of compact mappings of $g_t(V_j)$ into Y varying continuously in t in the sense of local uniform convergence. We can do this for $0 \leq t \leq d_0$ since $h_j = (f_t)_{x_j}'u + S_j P_j(u - x_j)$ will be an isomorphism for t near 0 if it is an isomorphism for $t = 0$. If this is true however, $g_t = f_t + S_j P_j(I - x_j)$ will be a continuous family of diffeomorphisms on a fixed neighborhood V_j of x_j for $t \leq d_0$.

Finally, the sum of the multiplicities of solutions of the equation $f(x_t) = y$ will be equal to $\sum_{j=1}^{r} \deg_{\mathrm{LS}}(I - C_t, g_t(V_j), y)$ which is independent of t in $[0, \epsilon)$ by the arguments of §12. q.e.d.

As an application of Theorem (15.6), we have the following:

THEOREM (15.7). *Let X be a complex Banach space, G a bounded open subset of X, Φ a holomorphic mapping of G into X of the form $\Phi = g - C$ with C a compact* *holomorphic mapping of G into X and g a mapping of G into X such that for all x and u in G and a fixed constant $k < 1$, $\|g(x) - g(u)\| \leq k\, \|x - u\|$. Let G_0 be a nonempty open convex subset of G with $\operatorname{cl}(G_0)$ contained in G and suppose that Φ maps the boundary of G_0 into $\operatorname{cl}(G_0)$ but has no fixed points on the boundary of G_0.*

Then Φ has exactly one fixed point in G_0.

PROOF OF THEOREM (15.7). We may assume without loss of generality that 0 lies in the interior of G_0. For each t in $[0, 1]$, we let $f_t(u) = I - tf(u)$, $u \in G$. Then $\{f_t\}$ is a continuous family of mappings in the sense of uniform convergence. For each t in $[0, 1]$, $f_t = (I - tg) - tC$, where $h_t = (I - tg)$ is a local diffeomorphism of G into X since $(h_t)'_x = I - tg'_x$ is an isomorphism for each x in G. (Indeed, it follows from the fact that g is a strict contraction with constant $k < 1$ that for each x in G, g'_x is one also by a straightforward limit process.)

Suppose that $f_t(u) = 0$. Then $u - tg(u) = tC(u)$ where $tC(u)$ lies in the convex closure of $C(G)$ and 0, which is compact since $C(G)$ is relatively compact in X. Let $\{u_j\}$ and $\{t_j\}$ be a pair of sequences such that $f_{t_j}(u_j) = 0$. We may assume that $t_j \to t$ as $j \to \infty$ and that $t_j C(u_j)$ converges to an element v of X. Hence $u_j - tg(u_j) = (I - tg)(u_j)$ converges to v in X. Since g is a strict contraction and $0 \leq t \leq 1$, it follows that u_j must converge in X as $j \to \infty$. Hence, there exists a compact set K in $\operatorname{cl}(G_0)$ such that for all t in $[0, 1]$, $f_t^{-1}(0) \cap \operatorname{cl}(G_0) \subset K$.

The family $\{f_t : t \in [0, 1]\}$ is a continuous family of holomorphic Fredholm mappings of G into X. To show that the sum of multiplicities of the solutions of the equation $f_t(u) = 0$ in G_0 is independent of t and equal to $+1$, it suffices to prove that $f_t(u) = 0$ has no solution on the boundary of G_0. If it did, however, we would have $u = t\Phi(u)$ ($t \in [0, 1]$), so that either $t = 1$, in which case u would be a fixed point of Φ on bdry (G_0), or $t < 1$ in which case $t\Phi(u)$ would be in G_0 since 0 is an interior point of G_0. Both cases are excluded, the first by hypothesis, and the second since a boundary point of G_0 cannot also be an interior point.

For $t = 0$, $f_0 = I$, and the sum of multiplicities of solutions of $f_t(u) = 0$ is equal to $+1$ for all t in $[0, 1]$. It follows immediately that there exists one point u_0 in G_0 such that $f_1(u_0) = 0$, i.e., $\Phi(u_0) = u_0$.   q.e.d.

16. **Asymptotic fixed point theorems.** Let X be a metric space, f a continuous mapping of X into X. By an asymptotic fixed point theorem for such a mapping, we mean a theorem which derives the existence of fixed points of f from hypotheses on the behavior of iterates f^n for large n. In the present section, we derive such theorems for self-mappings of convex subsets of a Banach space.

An important part of the basic data for our results is embodied in the following concept:

DEFINITION (16.1). *Let f be a self-mapping of X. Then for each positive integer n, we let $C_n = f^n(X)$, and we set $C_\infty = \bigcup_{n=1}^{\infty} f^n(X)$. The set C_∞ is called the core of the mapping f.*

Obviously, each fixed point of f must be a point of the core C_∞ of f, though the opposite need not be true. (For example, in a rotation of a circle, there are no fixed points though the core is the whole circle.) It is the object of our results to establish the existence of fixed points in the core of a mapping under suitable hypotheses.

DEFINITION (16.2). *If A is a subset of X, the orbit of A under f is the invariant subset $O(A)$ given by $O(A) = \bigcup_{n \geq 0} f^n(A)$.*

PROPOSITION (16.1). *Let f be a continuous self-mapping of the metric space X, and suppose that a compact set A has a relatively compact orbit under f. Let G be an open subset of X which contains the core C_∞ of f. Then there exists a positive integer N such that for $n \geq N$, $f^n(A) \subset G$.*

PROOF OF PROPOSITION (16.1). By hypothesis, the orbit $O(A)$ of A is relatively compact in X, and by its construction, $O(A)$ is invariant under the mapping f. Hence the closure K of $O(A)$ in X is a compact invariant subset of X.

Suppose the result of Proposition (16.1) were false. Then there would exist an infinite sequence $\{x_j\}$ in A and a corresponding sequence $\{n_j\}$ of positive integers with $n_j \to \infty$ as $j \to \infty$, such that for each j, $f^{n_j}(x_j) \in K - G$. Since K is compact and G is open in X, $K - G$ is a compact subset of X. Hence, by passing to an infinite subsequence, we may assume that for the original sequence, $f^{n_j}(x_j) \to y$ $(j \to \infty)$ for some point y of $K - G$. Since $n_j \to \infty$ as $j \to \infty$, for a given positive integer k, there exists an index j_k such that for $j > j_k$, we have $n_j > k$. Hence, for such j, $f^{n_j}(x_j) = f^k(f^{n_j-k}(x_j))$. On the other hand, x_j lies in A so that for each $j > j_k$, $f^{n_j-k}(x_j) \in K$. Since K is compact, we may consider an infinite subsequence (which we again denote by the original index) such that $f^{n_j-k}(x_j)$ converges as $j \to \infty$ to a point u of K. Hence $f^k(f^{n_j-k}(x_j)) \to f^k(u)$, i.e., $y = f^k(u)$. Hence y lies in $f^k(K) \subset f^k(X)$.

Since this last conclusion is true for each positive integer k, it follows that $y \in C_\infty$, which contradicts the fact that y lies outside of the set G which contains C_∞.
q.e.d.

Another useful result in a similar direction is the following:

PROPOSITION (16.2). *Let f be a continuous self-mapping of the metric space X and suppose that each compact subset A of X has a relatively compact orbit under f. Let G be an open neighborhood of C_∞.*

Then there exists an open subset G_1 of G which also contains C_∞, with G_1 invariant under f. In particular, G_1 can be chosen so that it contains all subsets of G which are invariant under f.

PROOF OF PROPOSITION (16.2). We set $G_1 = \bigcap_{n=0}^\infty f^{-n}(G)$, where $f^{-n}(G)$ denotes as usual the subset of x in X such that $f^n(x)$ lies in G. If A is an invariant subset of G, then for each x in A, $f^n(x)$ lies in A for each $n \geq 1$ and hence $f^n(x)$ lies in G. Thus each invariant subset of G lies in G_1 and, in particular, C_∞ is contained in G_1.

Obviously G_1 is itself invariant under the mapping f, since for each integer n and each x in G_1, $f^{n+1}(x) = f^n(f(x))$ lies in G. Hence $f^n(f(x))$ lies in G for all $n \geq 0$, so that $f(x)$ lies in G_1. We need only prove therefore that G_1 is an open subset of G.

Suppose that G_1 were not open in G. Then there would exist a point u in G_1 and a sequence $\{x_j\}$ in $G - G_1$ converging to u. For each integer n, $f^n(u)$ lies in G, and G being an open set and f continuous, there exists a neighborhood V_n of u such that $f^n(V_n) \subset G$. All the points x_j eventually lie in the neighborhood V_n since $x_j \to u$. Hence, for all except a finite number of indices for each n, $f^n(x_j) \in G$. Since each x_j does not lie in G_1, however, there exists an index n_j such that $f^{n_j}(x_j) \in G$, $f^{n_j+1}(x_j) \in X - G$. By thinning out the sequence, we may assume that n_j is strictly increasing in j.

The sequence $\{x_j\}$ and its limit point u together form a compact subset A of X. By hypothesis, each compact subset A of X has relatively compact orbit in X. Hence, the closure K of the orbit $O(A)$ in X is a compact invariant subset of X containing all the points x_j. In particular, it contains the points $f^{n_j+1}(x_j)$, so that replacing the given sequence by an infinite subsequence, we may assume that $f^{n_j+1}(x_j) \to y$ for some point y of $K - G$.

Let k be a fixed positive integer. Then for $j > j_k$, $n_j + 1 > k$. Hence for such j,
$$f^{n_j+1}(x_j) = f^k(f^{n_j+1-k}(x_j)).$$

By passing to an infinite subsequence, we may assume that the sequence $f^{n_j+1-k}(x_j)$ in the compact set K converges as $j \to \infty$ to an element u_k of K. Hence $f^k(f^{n_j+1-k}(x_j)) \to f^k(u_k)$, i.e., $y = f^k(u_k)$.

Thus y lies in $f^k(X)$ for all $k \geq 1$, and therefore y lies in C_∞. On the other hand, y lies in $K - G$ and G contains C_∞. This contradiction proves the proposition.

q.e.d.

The first of our asymptotic fixed point theorems is the following:

THEOREM (16.1). *Let X be a Banach space, G a closed convex subset of X, f a continuous mapping of G into G such that its core, $C_\infty(f) = \bigcap_{j=0}^{\infty} f^j(G)$ is a nonempty relatively compact subset of G. Suppose further that each point x of G has a relatively compact orbit under f and that there exists an open neighborhood V_0 of* cl (C_∞) *in G such that $f(V_0)$ is relatively compact in G.*

Then f has a fixed point in G.

PROOF OF THEOREM (16.1). By hypothesis, each point x in G has a relatively compact orbit under f. We assert first that for each point x in G, there exists an integer $m(x)$ such that for $m \geq m(x)$, $f^m(x)$ lies in V_0. Indeed, we may apply Proposition (16.1) to the set A consisting of the single point x to obtain this fact.

We may assume without loss of generality that $f(V_0)$ is a subset of V_0. Indeed, if we take any open subset V_1 of V_0 with cl $(C_\infty) \subset V_1$, it will still be true that $f(V_1)$ is relatively compact in G, while by Proposition (16.2), we can choose such an open neighborhood of the invariant subset cl (C_∞) such that V_1 is invariant under f. Hence V_0 itself can be assumed invariant under f.

For each point x in G, there exists an integer $m(x)$ such that $f^{m(x)}(x)$ lies in V_0. Since $f^{m(x)}$ is continuous and V_0 is an open subset of G, there exists a neighborhood U_x of x in G such that for all u in U_x, $f^{m(x)}(u)$ lies in V_0. For $m > m(x)$, we note that

$f^m(u) = f^{m-m(x)}(f^{m(x)}(u)) \in V_0$ since V_0 is invariant under f, i.e., $f^m(U_x) \subset V_0$, $m \geq m(x)$.

We now apply the following proposition:

PROPOSITION (16.3). *Let G be a metric space, V_0 an open subset of G, f a continuous self-mapping of G such that $f(V_0) \subset V_0$ and cl $(f(V_0))$ is a compact subset of G. Let K be a compact invariant subset of V_0 which contains the core of the mapping f on G. Then there exists an open neighborhood V_1 of K in V_0 such that $f(V_1)$ is a compact subset of V_1.*

PROOF OF PROPOSITION (16.3). We may replace G by the closure of V_0, and since $f(\text{cl}(V_0)) \subset \text{cl}(f(V_0))$ is compact, we may assume without loss of generality that $f(G)$ is compact. With this normalization of the situation, we know by Proposition (16.1) that if we take a small neighborhood V_2 of K in V, there exists an integer m such that $f^m(G) \subset V_2$. We now construct a decreasing sequence of open subsets G_j of V_0 with

$$\text{cl}(V_2) \subset G_{m-1} \subset \text{cl}(G_{m-1}) \subset G_{m-2} \subset \cdots \subset \text{cl}(G_j) \subset G_{j-1} \subset \cdots \subset \text{cl}(G_1) \subset V_0$$

(as we can do by the normality of the metric space G). We then set

$$V_1 = V_0 \cap \bigcap_{j=1}^{m-1} f^{-1}(G_j).$$

Since V_1 is an intersection of a finite family of open subsets of V_0, V_1 itself is an open subset of V_0. Since K is invariant under f and lies in each G_j, K is contained in V_1. Finally,

$$f(V_1) = \bigcap_{j=1}^{m-1} f(f^{-j}(G_j)) = \bigcap_{j=1}^{m-1} f^{-(j-1)}(G_j) \cap V_0,$$

so that

$$\text{cl}(f(V_1)) \subset \bigcap_{j=1}^{m-1} \text{cl}(f^{-(j-1)}(G_j)).$$

Since f is continuous on G, as are its iterates,

$$\text{cl}(f^{-(j-1)}(G_j)) \subset f^{-(j-1)}(\text{cl}(G_j)) \subset f^{-(j-1)}(G_{j-1}).$$

Hence

$$\text{cl}(f(V_1)) \subset \bigcap_{j=2}^{m-1} f^{-(j-1)}(G_{j-1}).$$

On the other hand

$$f^{m-1}(\text{cl}(f(V_1))) \subset \text{cl}(f^m(V_1)) \subset \text{cl}(f^m(G)) \subset \text{cl}(V_2) \subset G_{m-1}.$$

Thus

$$\text{cl}(f(V_1)) \subset \bigcap_{j=2}^{m} f^{-(j-1)}(G_{j-1}) \subset V_1.$$

Thus $\text{cl}(f(V_1))$ is a subset of V_1, and being a closed subset of the compact set $\text{cl}(f(V_0))$, $\text{cl}(f(V_1))$ is compact. q.e.d.

PROOF OF THEOREM (16.1) CONTINUED. If we replace V_0 by the smaller neighborhood V_1 of cl (C_∞) constructed in Proposition (16.3), we may assume without loss of generality that cl $(f(V_0))$ is a compact subset K_0 of V_0.

Let K_1 be the convex closure of K_0 in G. Then since G is a closed convex subset of the Banach space X, K_1 is compact. For each point x in K_1 there exist an integer $m(x)$ and a neighborhood U_x of x in K_1 such that for all u in U_x, $f^m(u)$ lies in V_0 for all $m \geq m(x)$. Since K_1 is compact, it can be covered by a finite family of such neighborhoods U_x, i.e., there exists a single integer m such that for all u in K_1, $f^m(u)$ lies in V_0. Since $f(V_0) \subset K_0$, it follows that $f^{m+1}(K_1) \subset K_0$.

Since K_1 lies in $f^{-m}(V_0)$, $f(K_1)$ is a compact subset of the open set $f^{-(m-1)}(V_0)$ in G. Hence there exists a finite family of closed balls contained in $f^{-(m-1)}(V_0)$ and containing $f(K_1)$. We denote this family by $\{B_1, \ldots, B_r\}$. We let $K_2 = \bigcup_{k=1}^{r}$ convex closure $(f(K_1) \cap B_r)$. Then K_2 is a compact subset of $f^{-(m-1)}(V_0)$ which is the finite union of compact convex subsets of G. We now proceed by an iterative argument. Suppose that for $2 \leq s \leq m - 1$, we have constructed compact subsets K_t for $t \leq s - 1$, with K_t contained in $f^{-(m-t+1)}(V_0)$ and each consisting of a finite union of compact convex subsets of G. Then we continue the construction by noting that since K_{s-1} is a subset of $f^{-(m-s+2)}(V_0)$, $f(K_{s-1})$ is a compact subset of $f^{-(m-s+1)}(V_0)$. Hence, it can be covered by a finite family of balls in the open set $f^{-(m-s+1)}(V_0)$, which we denote by $\{B_{1_r}, \ldots, B_{r_s}\}$. We then set $K_s = \bigcup_{k=1}^{s}$ convex closure $(f(K_{s-1}) \cap B_k)$. Then K_s is a compact subset of $f^{-(m-(s-1))}(V_0)$ which is the union of a finite family of compact convex subsets. Thus the construction can be carried through for the whole range of s, $1 \leq s \leq m - 1$.

We thus obtain a family $\{K_1, K_2, \ldots, K_{m-1}\}$ of compact subsets of G with the following properties:

$$f(K_j) \subset K_{j+1} \qquad (1 \leq j \leq m - 1),$$
$$K_j \subset f^{-(m-(j-1))}(V_0) \qquad (1 \leq j \leq m - 1).$$

In particular, K_{m-1} lies in $f^{-1}(V_0)$ so that $f(K_{m-1})$ is a compact subset of V_0. We repeat the construction as above by constructing a compact subset K_m of V_0 which contains $f(K_{m-1})$ and is the union of a finite family of compact convex subsets of V_0. Since K_m is a subset of V_0, $f(K_m) \subset f(V_0) \subset K_0 \subset K_1$.

Let $R = \bigcup_{j=1}^{m} K_j$. Then R is a compact subset of G, and by our construction f is a continuous mapping of R into itself. R is the finite union of compact convex subsets of G, and hence (Borsuk [38]) R is a compact absolute neighborhood retract.

We now apply the following variant of the Lefschetz fixed point theorem for compact absolute neighborhood retracts.

PROPOSITION (16.4). *Let R be a compact absolute neighborhood retract, f a continuous self-mapping of R. Suppose that for some positive integer m, $f^m(R)$ is contained in a contractible subset R_0 of R.*

Then f has a fixed point in R.

PROOF OF PROPOSITION (16.4). By the Lefschetz fixed point theorem for compact absolute neighborhood retracts (Lefschetz [**301**]), a sufficient condition for f to have a fixed point is that $\Lambda(f) \neq 0$, where the Lefschetz number $\Lambda(f)$ of the mapping f is the homology invariant of the mapping defined as follows: For each $n \geq 0$, let $H_n(R)$ be the n-dimensional homology group of R with rational coefficients, and f_{*n} the endomorphism of $H_n(R)$ induced by the mapping f. For each n, $H_n(R)$ is a vector space of finite dimension over the rationals, and the trace tr (f_{*n}) of f_{*n} is well defined. Only a finite number of the groups $H_n(R)$ are nontrivial, and hence we may define

$$\Lambda(f) = \sum_{n \geq 0} (-1)^n \text{ tr } (f_{*n}).$$

By the functorial character of f_{*n}, $(f_{*n})^j = (f^j)_{*n}$ for each positive integer j. In particular, if $f^m(R)$ is contained in a contractible subset of R, $(f_{*n})^m = 0$ for $n > 0$. Since f_{*n} is a nilpotent endomorphism, tr $(f_{*n}) = 0$. Finally, $(f_{*0})^m$ projects $H_0(R)$ on the subgroup spanned by $H_0(R_0)$. Hence $(f_{*0})^m$ has a single nonzero eigenvalue, and it follows immediately that tr $(f_{*,0}) = +1$. Finally, we have

$$\Lambda(f) = \sum_{n \geq 0} (-1)^n \text{ tr } (f_{*n}) = \text{ tr } (f_{*,0}) = 1,$$

and the existence of a fixed point for f follows from the Lefschetz theorem.   q.e.d.

PROOF OF THEOREM (16.1) CONCLUDED. f maps the compact absolute neighborhood retract R constructed above into itself while $f^{m+1}(R)$ is contained in the contractible set K_1. If we apply Proposition (16.4), we see that f must have a fixed point in R.   q.e.d.

THEOREM (16.2). *Let G be a closed convex subset of a Banach space X, G_0 a closed subset of G, f a continuous mapping of G into G. Suppose that there exists a mapping g of G_0 into a compact absolute neighborhood retract R contained in G_0 such that f and g are homotopic as mappings of G_0 into G_0. Suppose further that for a given positive number $\epsilon > 0$, $\|f(x) - g(x)\| < \epsilon$ for all x in G_0. Suppose further that for each compact subset A of G, there exists $m \geq 1$ such that $f^m(A) \subset G_0$.*

Then there exists a point x_0 in G_0 such that $\|x_0 - f(x_0)\| \leq \epsilon$.

PROOF OF THEOREM (16.2). It suffices to prove that the mapping g has a fixed point in G_0 since for any such fixed point x_0, $\|x_0 - f(x_0)\| = \|g(x_0) - f(x_0)\| \leq \epsilon$.

To show that g has a fixed point, it suffices to prove that if we consider Čech homology with compact support, the Lefschetz number $\Lambda(g)$ is nonnull. Indeed, let h be the injection mapping of R into G_0. Then if we let g_0 denote the mapping of G_0 into R induced by g, then $g = hg_0$, and hence for each $j \geq 0$, $g_{*j} = h_{*j}(g_0)_{*j}$. Since the trace of a product of two linear endomorphs is independent of the order in which the factors are taken,

$$\text{tr } (g_{*,j}) = \text{tr } ((g_0)_{*j} h_{*j}) = \text{tr } (k_{*j})$$

where k is the restriction of the mapping g_0 to R (i.e., $k: R \to R$ is given by $k(x) = g(x)$ for all x in R). In particular, $\Lambda(g) = \Lambda(k)$, and if $\Lambda(k) \neq 0$, it follows from the

Lefschetz fixed point theorem for compact absolute neighborhood retracts that k has a fixed point in R. Such a fixed point x_0 is automatically a fixed point of the extended mapping g.

Since g and f are homotopic as mappings of G_0 into G_0, it follows that $\Lambda(g) = \Lambda(f_0)$, where f_0 is the self-mapping of G_0 induced by f. Hence, it suffices to prove that $\Lambda(f_0) \neq 0$.

Since G is convex, all its reduced Čech homology groups with compact supports vanish. Hence, the homology endomorphisms $(f_0)_{*j}$ coincide with the homology endomorphisms induced by f on $H_{j-1}(G, G_0)$. By hypothesis, for each compact subset A of G, there exists an integer m (depending upon A) such that $f^m(A) \subset G_0$. In particular, the endomorphism f_{*j} on each $H_j(G, G_0)$ is nilpotent, and therefore tr $((f_0)_{*j}) = 0, j \geq 1$. Moreover, tr $(f_0)_{*0} = +1$. Hence $\Lambda(f_0) = +1 \neq 0$. q.e.d.

THEOREM (16.3). *Let H be a Hilbert space, G a closed convex subset of H, f a continuous self-mapping of G. Suppose that there exist a compact subset K of G and two sequences of positive numbers $\{\xi_k\}$ and $\{\epsilon_k\}$ with $\xi_k \to 0$, $\epsilon_k \to 0$ as $k \to \infty$ and $\epsilon_k < \xi_k$ for each k, such that the following two conditions hold:*

(a) *For each neighborhood G_0 of K and each compact subset A of G there exists an integer m such that $f^m(A) \subset G_0$.*

(b) *The image under f of the ξ_k neighborhood of K is contained in the ϵ_k-neighborhood of K for each $k \geq 1$.*

Then f has a fixed point in K.

PROOF OF THEOREM (16.3). Suppose that f does not have a fixed point in K. Then by the compactness of K, there exists a positive number $d > 0$ such that for all x in K, $\|f(x) - x\| \geq d$. We choose ξ_k in our given sequence such that $2\xi_k < d$, and consider the corresponding pair of neighborhoods of K,

$$N_{\xi_k}(K) = \{u \mid u \text{ is at distance at most } \xi_k \text{ from } K\},$$
$$N_{\epsilon_k}(K) = \{u \mid u \text{ is at distance at most } \epsilon_k \text{ from } K\}.$$

By Theorem (16.2), it suffices to take $G_0 = N_{\xi_k}(K)$ and to construct a mapping g of G_0 into a finite simplicial polytope R contained in G_0 with g homotopic to f on G_0 and differing from f by at most $2\xi_k$ on G_0.

We construct such a mapping g by composing the mapping f with a second mapping q, where q is a projection mapping of a set containing the image of $f(G_0)$ into a finite simplicial complex contained in G_0. It suffices to ensure that the mapping q is defined and continuous on $N_{\epsilon_k}(K)$, that its image set $q(N_{\epsilon_k}(K)) \subset N_{\xi_k}(K)$, and that for each point x in $N_{\epsilon_k}(K)$, the line segment joining x to $q(x)$ lies entirely in $N_{\xi_k}(K)$ and is of length at most $2\xi_k$.

To construct the mapping q, we note that since K is compact and $\lambda_k = \xi_k - \epsilon_k > 0$, we can cover $N_{\epsilon_k}(K)$ by a finite family of open balls of radius $d_k = \epsilon_k + \frac{1}{2}\lambda_k$ with centers at various points of K, such that the union of all these closed balls is contained in N_{ϵ_k}. We consider the finite simplicial polytope R which is the nerve of this covering and let the mapping q be a canonical mapping of the space $N_{\epsilon_k}(K)$

into the nerve of a finite open covering of the space. Thus, each point x of $N_{\epsilon_k}(K)$ is mapped into a point $q(x) = \sum_{j=1}^{r} \lambda_j(x) x_j$, where each x_j is a point of K and $\lambda_j(x)$ is a continuous function of x which vanishes if $\|x - x_j\| \geq d_k$. To verify the properties of the mapping q, we use the following result:

PROPOSITION (16.5). *Let H be a Hilbert space, $\{x_1, \ldots, x_r\}$ a finite subset of H such that for a given $d > 0$, $\bigcap_{j=1}^{r} B_d(x_j) \neq \emptyset$, where $B_d(x_j)$ is the closed ball of radius d about the point x_j. Then each point x in the convex spanned by the set $\{x_1, \ldots, x_r\}$ in H is at a distance at most d from at least one of the vertices $\{x_1, \ldots, x_r\}$.*

PROOF OF PROPOSITION (16.5). We carry through the proof by induction on r for $r \geq 2$.

For $r = 2$, the result is obvious since the fact that two balls $B_d(x_1)$ and $B_d(x_2)$ have a nonempty intersection implies that $\|x_1 - x_2\| \leq 2d$. Hence, any point on the line segment joining x_1 to x_2 is at distance at most d from one or the other endpoint.

Suppose the proposition true for $(r - 1)$ for a given $r \geq 3$. We need to prove it true for r. Let K_r be the convex set spanned by the r points $\{x_1, \ldots, x_r\}$. By hypothesis, there exists a point y in H such that $\|y - x_j\| \leq d$ for $1 \leq j \leq r$. Let y_0 be the point of K_r nearest to y. (Such a point exists and is unique by the uniform convexity of the Hilbert space H.) Since the closest-point mapping of the Hilbert space H on the closed convex subset K_r does not increase distances, $\|y_0 - x_j\| \leq d$ for all j, i.e., we can assume without loss of generality that the original point y lies in the convex set K_r itself. Let u be any point of K_r. If $u = y$ or if u lies in one of the faces of K_r (i.e., one of the convex subsets of K_r spanned by fewer than r of the points $\{x_1, \ldots, x_r\}$), then it follows by the choice of y and the induction hypothesis that u is at distance at most d from at least one of the vertices x_j of the set K_r. If u differs from y and lies in the relative interior of the set K_r, the line segment from y to u lies in K_r and when extended intersects one of the faces of K_r in a point v. By the induction hypothesis, there exists a vertex x_j such that $\|v - x_j\| \leq d$. By the choice of y, $\|y - x_j\| \leq d$ for all j. By construction, u is a convex linear combination of y and v. Hence $\|u - x_j\| \leq d$, and the induction step has been proved for r. q.e.d.

PROOF OF THEOREM (16.3) COMPLETED. We construct the finite simplicial polytope R which is the geometrical realization of the nerve of the given finite covering of $N_{\epsilon_k}(K)$ by assigning to the open set $B_{d_k}(x_k)$ the corresponding vertex x_k, which by the construction of the covering is a point of the compact set K. The corresponding vertices $\{x_1, \ldots, x_s\}$ span a simplex of R (possibly degenerate), if and only if the corresponding balls $B_{d_k}(x_j)$ intersect. By Proposition (16.5), it follows that each point u of any such simplex must be at distance $\leq d_k$ from at least one of the vertices x_j of the simplex, i.e., u must lie at distance at most d_k from the set K. Hence each simplex of R and thereby R itself is contained in $N_{\epsilon_k}(K)$, since $d_k < \xi_k$.

We may construct a partition of unity on the metric space $N_{\epsilon_k}(K)$ corresponding to the finite covering by the balls $B_{d_k}(x_j)$ for various points x_j. Let $\lambda_j(x)$ be a function in this partition of unity so that $0 \leq \lambda_j(x) \leq 1$, $\lambda_j(x) = 0$ for x outside of $B_{d_k}(x_j)$, and $1 = \sum_j \lambda_j(x)$ for all x in $N_{\epsilon_k}(K)$. Then we set $q(x) = \sum_j \lambda_j(x)x_j$. By the definition of the nerve R of the covering, q is a continuous mapping of $N_{\epsilon_k}(K)$ into R. To complete the proof, we must verify that q is homotopic to the identity mapping of $N_{\epsilon_k}(K)$ through $N_{\xi_k}(x)$ with a homotopy path of length $\leq 2\xi_k$ for each x in $N_{\epsilon_k}(K)$.

To prove this last fact, however, we note that if x is a point of $N_{\epsilon_k}(K)$ and $q(x) = \sum_j \lambda_j(x)x_j$, with each $\lambda_j(x)$ which is given explicitly being positive, then x lies in $\bigcap_j B_d(x_j)$, the intersection being taken over the corresponding set of indices j. By Proposition (16.5), we can choose a vertex x_j in this family such that $\|q(x) - x_j\| \leq d_k$. Hence, the line segment joining x to $q(x)$ must lie in the ball $B_{d_k}(x_j)$ for this particular choice of j, and hence this line segment is contained in $N_{\xi_k}(K)$. Moreover, the length of this line segment is at most $2d_k \leq 2\xi_k$.   q.e.d.

17. A-proper mappings, approximation methods, and related generalizations of the topological degree. In §12, we considered extensions of the concept of the topological degree based upon the decomposition of mappings in terms of homeomorphisms and compact mappings. In the present section, we turn to another extension of the topological degree based upon replacing the uniform approximation of compact mappings by finite-dimensional ones by the more general use of approximation methods of Galerkin types to provide finite-dimensional approximations for general classes of noncompact mappings.

The general class of mappings for which this theory makes sense, the *A-proper mappings* defined below, is defined by means and in terms of some given approximation scheme of the following type:

DEFINITION (17.1). *Let X and Y be two Banach spaces. By an oriented approximation scheme for mappings from X to Y, we mean: two sequences $\{X_n\}$ and $\{Y_n\}$ of oriented finite-dimensional spaces with dimension $(X_n) =$ dimension (Y_n) for each n, as well as two sequences $\{P_n\}$ and $\{Q_n\}$ of continuous (possibly nonlinear) mappings with P_n mapping X_n into X and Q_n mapping Y into Y_n for each n.*

Using the above definition, we can formulate the following:

DEFINITION (17.2). *Let X and Y be two Banach spaces, $\Gamma = (\{X_n\}, \{Y_n\}, \{P_n\}, \{Q_n\})$ an approximation scheme in the sense of Definition (17.1) for mappings from X to Y. Let G be an open subset of X, T a continuous mapping of $\mathrm{cl}\,(G)$ into Y. Then T is said to be A-proper on $\mathrm{cl}\,(G)$ with respect to the approximation scheme Γ if for any sequence $\{n_j\}$ of integers with $n_j \to \infty$ as $j \to \infty$ and a corresponding sequence $\{x_{n_j}\}$ with x_{n_j} in X_{n_j} and $P_{n_j}(x_{n_j}) \in \mathrm{cl}\,(G)$ for all j, for which for some y in Y,*

$$\|Q_{n_j}TP_{n_j}(x_{n_j}) - Q_{n_j}(y)\|_{Y_{n_j}} \to 0 \quad (j \to +\infty),$$

there exist an infinite subsequence $\{x_{n_{j(k)}}\}$ and an element x of X such that $P_{n_{j(k)}}(x_{n_{j(k)}}) \to x$ as $k \to \infty$, and $T(x) = y$.

PROPOSITION (17.1). *Suppose that T is an A-proper mapping of $\mathrm{cl}\,(G)$ into Y for a given approximation scheme Γ as in Definition (17.2), and for each n, let*

$T_n = Q_n T P_n$ on $G_n = P_n^{-1}(G)$. Let y be a point in $Y - T(\text{bdry}(G))$. Suppose that G_n is bounded for each n.

Then there exists an integer n_0 such that for $n \geq n_0$, $T_n(\text{bdry}(G_n))$ does not contain $Q_n(y)$ and $\deg(T_n, G_n, Q_n y)$ is well defined. Moreover, for a suitable choice of n_0, there exists a constant $d > 0$ such that for all $n \geq n_0$ and all x_n in bdry (G_n), $\|T_n(x_n) - Q_n(y)\| \geq d$.

REMARK. By deg, we denote the Brouwer degree of the mapping of an open subset G of the oriented finite-dimensional space X_n into the oriented finite-dimensional space Y_n of the same dimension. We write norms with the usual norm symbol in all of the various spaces, X, Y, X_n, Y_n in which they are taken, and trust to the context and the fact that we do not identify elements of distinct spaces to make it clear in each case which norm is being applied. (Only one can be applied in any given case.)

PROOF OF PROPOSITION (17.1). The last assertion of the conclusion of Proposition (17.1) implies the other assertions. To prove this last assertion, we suppose that it is false. Then there exists a sequence $\{x_{n_j}\}$ with $x_{n_j} \in \text{bdry}(G_{n_j})$ for each j such that $\|T_{n_j}(x_{n_j}) - Q_{n_j}(y)\| \to 0$ as $j \to \infty$. It follows from the assumption that T is A-proper with respect to the given approximation scheme that if we replace this sequence by an infinite subsequence, we may assume that $P_{n_j}(x_{n_j}) \to x$ as $T(x) = y$. However, the fact that x_{n_j} lies in bdry (G_{n_j}) implies that $P_{n_j}(x_{n_j}) \in$ bdry (G). Hence x lies in bdry (G), which contradicts the fact that $y \in Y - T(\text{bdry}(G))$.
q.e.d.

DEFINITION (17.3). *Let X and Y be Banach spaces, Γ an approximation scheme for mappings from X to Y in the sense of Definition (17.1), G an open subset of X, T a continuous mapping of cl (G) into Y. Suppose that T is an A-proper mapping of cl (G) into Y with respect to Γ in the sense of Definition (17.2), and for each n, suppose that $G_n = P_n^{-1}(G)$ is bounded.*

Let $T_n = Q_n T P_n$ mapping cl (G_n) into Y_n. Let y be a point of $Y - T(\text{bdry}(G))$. Then:

(a) *Z' is the extended set of integers (positive and negative) together with $\{+\infty\}$ and $\{-\infty\}$.*

(b) *Deg (T, G, y) is the subset of Z' which is defined as follows: An integer m lies in Deg (T, G, y) if there exists an infinite sequence of integers $\{n_j\}$ with $n_j \to \infty$ as $j \to \infty$ such that $\deg(T_{n_j}, G_{n_j}, Q_{n_j}(y)) = m$. $\pm\infty$ lies in Deg (T, G, y) if there exists an infinite sequence of integers $\{n_j\}$ with $n_j \to \infty$ as $j \to \infty$ such that $\deg(T_{n_j}, G_{n_j}, Q_{n_j}(y)) \to \pm\infty$.*

THEOREM (17.1). *Let X and Y be Banach spaces, Γ an approximation scheme for mappings from X to Y, G an open subset of X, T a continuous mapping of cl (G) into Y which is A-proper with respect to Γ. Let y be a point of $Y - T(\text{bdry}(G))$ and suppose that $G_n = P_n^{-1}(G)$ is bounded for each n. Let $T_n = Q_n T P_n$: cl $(G_n) \to Y_n$. Then:*

(a) *There exists n_0 such that for $n \geq n_0$, $\deg(T_n, G_n, Q_n y)$ is well defined. Hence the subset Deg (T, G, y) of Z' is always nonempty.*

(b) *If* Deg (T, G, y) *does not consist of* $\{0\}$, *then there exists an element* x *in* G *such that* $T(x) = y$.

(c) *Let* T *be a continuous mapping of* cl $(G) \times [0, 1]$ *into* Y, *and for each* t *in* $[0, 1]$, *let* T_t *be the mapping of* cl (G) *into* Y *given by* $T_t(x) = T(x, t)$. *Suppose that* $T_t(x)$ *is continuous in* t, *uniformly for* x *in* cl (G), *while for each* t *in* $[0, 1]$, T_t *is* A-*proper with respect to the given scheme* Γ. *Suppose that the mappings* Q_n *satisfy the following equicontinuity condition: For any given* $\epsilon > 0$ *and any bounded subset* B *of* Y, *there exists* $\xi > 0$ *such that for any* y *and* y' *in* B *with* $\|y - y'\| < \xi$, $\|Q_n(y) - Q_n(y')\| < \epsilon$ *for all* n.

Then if y *is a point of* $Y - T(\text{bdry } (G) \times [0, 1])$, *it follows that* Deg (T_t, G, y) *is well defined and independent of* t *in* $[0, 1]$.

(d) *Let* $G = G_1 \cup G_2$, *and for*

$$G' = (G_1 \cap G_2) \cup \text{bdry } (G_1) \cup \text{bdry } (G_2),$$

suppose that y *does not lie in* $T(G')$. *Then*

$$\text{Deg } (T, G, y) \subseteq \text{Deg } (T, G_1, y) + \text{Deg } (T, G_2, y),$$

with the convention that $(+\infty) + (-\infty) = Z'$. *Equality holds if either summand is a singleton.*

PROOF OF THEOREM (17.1). PROOF OF (a). This follows from Proposition (17.1). q.e.d.

PROOF OF (b). If deg $(T, G, y) \neq \{0\}$, then there exists an infinite sequence $n_j \to \infty$ such that deg $(T_{n_j}, G_{n_j}, Q_{n_j}(y)) \neq 0$. For each such n_j, there exists a point x_{n_j} in G_{n_j} such that $T_{n_j}(x_{n_j}) = Q_{n_j}(y)$. Hence, we may apply the assumption that T is A-proper with respect to Γ and by passing to an infinite subsequence, we may assume that $P_{n_j}(x_{n_j}) \to x$ with $T(x) = y$. Hence there exists a point x in cl (G) such that $T(x) = y$. Since x does not lie in bdry (G) because $T(\text{bdry } (G))$ excludes y, it follows that x lies in G. q.e.d.

PROOF OF (c). By the conclusion of part (a), we know that for each t in $[0, 1]$, Deg (T_t, G, y) is well defined since y never lies in $T_t(\text{bdry } (G))$ and each T_t is A-proper with respect to the approximation scheme Γ. To show that Deg (T_t, G, y) is independent of t in $[0, 1]$, it suffices to show that there exists an integer n_1 such that for $n \geq n_1$, deg $((T_t)_n, G_n, Q_n y)$ is independent of t in $[0, 1]$. Suppose, however, that this last assertion is false. Then there must exist sequences $\{n_j\}$ of integers with $n_j \to \infty$ as $j \to \infty$, $\{t_j\}$ of elements of $[0, 1]$ with $t_j \to t$ as $j \to \infty$ for some t in $[0, 1]$, and $\{x_{n_j}\}$ with each x_{n_j} in bdry (G_{n_j}) such that $(T_{t_j})_{n_j}(x_{n_j}) = Q_{n_j}(y)$, i.e., $Q_{n_j} T P_{n_j}(x_{n_j}) = Q_{n_j}(y)$ $(j \geq 1)$. By our hypothesis upon the homotopy of mappings T_t, T_{t_j} converges to T_t uniformly on cl (G) as $t_j \to t$. Hence

$$\|T_t P_{n_j}(x_{n_j}) - T_{t_j} P_{n_j}(x_{n_j})\| \to 0 \quad (j \to \infty).$$

By the equicontinuity of the mappings $\{Q_n\}$ which we have assumed,

$$\|Q_{n_j} T_t P_{n_j}(x_{n_j}) - (T_{t_j})_{n_j}(x_{n_j})\| \to 0,$$

i.e., $\|(T_t)_{n_j}(x_{n_j}) - Q_{n_j}(y)\| \to 0$. Since T_t is A-proper with respect to the given approximation scheme, we may pass to an infinite subsequence and assume that $P_{n_j}(x_{n_j}) \to x$ in X and $T_t(x) = y$. Since each $P_{n_j}(x_{n_j})$ lies in bdry (G), the point x must lie in bdry (G), which contradicts the hypothesis that $T_t(\text{bdry }(G))$ never contains y for any t in $[0, 1]$. This contradiction proves the assertion of (c). q.e.d.

PROOF OF (d). To prove (d), it suffices to show that for a suitable n_2 and all $n \geq n_2$,

$$\deg(T_n, G_n, Q_n y) = \deg(T_n, (G_1)_n, Q_n y) + \deg(T_n, (G_2)_n, Q_n y).$$

However,

$$\text{bdry }((G_1)_n) \subset P_n^{-1}(\text{bdry }(G_1)),$$

$$\text{bdry }(G_2)_n \subset P_n^{-1}(\text{bdry }(G_2)),$$

$$(G_1)_n \cap (G_2)_n \subset P_n^{-1}(G_1 \cap G_2).$$

If we apply the additivity of the degree in the finite-dimensional case, then the equality of degrees written above can only be violated for large n if there exist a sequence of integers $\{n_j\}$ with $n_j \to +\infty$ as $j \to \infty$ and a corresponding sequence of points $\{x_{n_j}\}$ in $P_{n_j}^{-1}(G')$ such that $T_{n_j}(x_{n_j}) = Q_{n_j}(y)$. If we apply the fact that T is A-proper and pass to an infinite subsequence, we may assume that $P_{n_j}(x_{n_j}) \to x$ as $j \to \infty$ for some point x of cl (G) for which $T(x) = y$. By its definition, however, G' is a closed subset of cl (G) since the boundary of $(G_1 \cap G_2)$ is contained in the union bdry $(G_1) \cup$ bdry (G_2). Hence there must exist a point x in G' for which $T(x) = y$, a fact excluded by the hypothesis. This contradiction establishes the conclusion of (d). q.e.d.

With the proof of part (d), the proof of Theorem (17.1) is complete. q.e.d.

THEOREM (17.2). *Let X and Y be Banach spaces with a given approximation scheme Γ for mappings from X to Y with Q_n equicontinuous on bounded sets, and let T be a continuous mapping of cl (G) into Y where G is an open subset of X. Suppose that T is A-proper with respect to Γ on cl (G), that G is invariant under the involution $\pi(x) = -x$, and that 0 does not lie in $T(\text{bdry }(G))$. Suppose that each G_n is bounded and invariant under π, that each T_n is an odd mapping (i.e., $T_n(-x) = -T_n(x)$) on bdry (G_n), and that $Q_n(0) = 0$ for all n.*

Then:

(a) Deg $(T, G, 0)$ contains no even integers, and, in particular, does not coincide with $\{0\}$.

(b) If for a given $d > 0$, $B_d(0)$ does not meet $T(\text{bdry }(G))$, then $T(G)$ contains the entire ball $B_d(0)$.

PROOF OF THEOREM (17.2). PROOF OF (a). For each $n \geq n_0$,

$$\deg(T_n, G_n, Q_n(0)) = \deg(T_n, G_n, 0)$$

is well defined. Since T_n is odd on bdry (G_n), it follows from the classical Borsuk-Ulam theorem [37] that deg $(T_n, G_n, 0)$ is an odd integer for each $n \geq n_0$. The conclusion of (a) follows from the definition of Deg $(T, G, 0)$.

PROOF OF (b). Let y_1 be a point in $B_d(0)$ and let $T_t = T - y_t$ where $y_t = ty_1$. Then for each t in $[0, 1]$, $T_t(\text{bdry } (G))$ does not contain 0. By the homotopy property of the multi-valued degree as established in part (c) of Theorem (17.1), Deg $(T_t, G, 0)$ does not depend on t. By part (a), Deg $(T, G, 0) = $ Deg $(T_0, G, 0)$ is not $\{0\}$. Hence Deg $(T_t, G, 0) \neq \{0\}$. By property (b) of the multi-valued degree, there exists x_1 in G such that $T_1(x_1) = 0$, i.e., $T(x_1) = y_1$. q.e.d.

A number of classes of nonlinear mappings of the types treated in the earlier sections (strongly monotone, strongly accretive, $T = I - U$ with U a strict contraction) can be shown to be A-proper with respect to suitable approximation schemes (cf. [**131**] for a more detailed study). In the present section, we shall concentrate our attention upon the following class:

DEFINITION (17.4). *Let X be a Banach space, X^* its conjugate space, G an open subset of X, T a continuous mapping of* cl (G) *into X^*. Then T is said to be of type* (S) *if it satisfies the following condition:*

(S) *For any sequence $\{u_j\}$ in* cl (G) *for which for some u, $u_j \rightharpoonup u$ and*

$$(T(u_j) - T(u), u_j - u) \to 0,$$

u_j *must converge strongly to u in X.*

DEFINITION (17.5). *Let X be a separable Banach space. Then by an injective approximation scheme for mappings from X to the conjugate space X^* of X, we mean one of the following form: The sequence $\{X_n\}$ is an increasing sequence of finite-dimensional subspaces of X with dense union in X, P_n is the injection mapping of X_n into X, $Y_n = X_n^*$ for each n, and $Q_n = P_n^*$ mapping X^* onto X_n^*.*

THEOREM (17.3). *Let X be a separable, reflexive Banach space with X^* its conjugate space, Γ a given injective approximation scheme for mappings from X to X^*. Let G be a bounded open subset of X, T a bounded continuous mapping of* cl (G) *into X^* which is of type* (S) *on* cl (G) *in the sense of Definition (17.4). Then:*

(a) *T maps closed subsets of* cl (G) *into closed subsets of Y.*

(b) *T is A-proper on* cl (G) *with respect to the given injective approximation scheme.*

PROOF OF THEOREM (17.3). PROOF OF (a). Let $y_j = T(x_j)$ for an infinite sequence of points $\{x_j\}$ in cl (G) lying in a closed subset A of cl (G). Suppose that y_j converges strongly to y in X^*. Since X is reflexive and cl (G) is bounded, we may assume without loss of generality (after passing to an infinite subsequence) that x_j converges weakly in X to some element x. Then

$$(T(x_j) - T(x), x_j - x) = (y_j - T(x), x_j - x) \to 0$$

as $j \to \infty$, since $y_j - T(x)$ converges strongly to $y - T(x)$ while $x_j - x$ converges weakly to 0. If we apply the condition (S), it follows that an infinite subsequence of the given sequence $\{x_j\}$ converges strongly to x. Hence x lies in A, and by

the continuity of T, $T(x_j) \to T(x) = y$. Thus y lies in $T(A)$ and $T(A)$ is closed in Y.
q.e.d.

PROOF OF (b). To prove that T is A-proper, we consider an infinite sequence of integers $\{n_j\}$ with $n_j \to \infty$, and a corresponding sequence of points $\{x_{n_j}\}$ in X_{n_j}, $G = G_{n_j}$ such that for some point y in X^*,

$$\|P^*_{n_j} T(x_{n_j}) - P^*_{n_j}(y)\|_{X_{n_j}^*} \to 0.$$

Since X is reflexive and cl (G) is bounded, we may assume that x_{n_j} converges weakly in X to some point x as $j \to \infty$.

Let v be a point in some one of the spaces X_n, i.e., $v \in \bigcup_n X_n$ where the latter subset is dense in X. Then for j sufficiently large, $x_{n_j} - v$ lies in X_{n_j}. Hence

$$|(T(x_{n_j}) - y, x_{n_j} - v)| = |(P^*_{n_j} T(x_{n_j}) - P^*_{n_j}(y), x_{n_j} - v)|$$
$$\leq \|P^*_{n_j} T(x_{n_j}) - P^*_{n_j}(y)\|_{X_{n_j}^*} \|x_{n_j} - v\| \to 0$$

since the first factor on the right tends to zero, and the second factor is bounded. Hence

$$(T(x_{n_j}), x_{n_j} - v) \to (y, x - v)$$

for all such v, since

$$(y, x_{n_j} - v) \to (y, x - v)$$

by the weak convergence of x_{n_j} to x. Since T is a bounded mapping and $\|x_{n_j}\|$ is uniformly bounded, $\|T(x_{n_j})\|$ is uniformly bounded. Hence

$$(T(x_{n_j}), x_{n_j} - v) \to (y, x - v)$$

for all v in X, since the convergence already holds on a dense subset of X.

If we apply this last result with v replaced by x, we obtain

$$(T(x_{n_j}) - T(x), x_{n_j} - x) \to (y, x - x) - (T(x), x - x) = 0.$$

Hence, we apply the hypothesis (S) on the mapping T to conclude that an infinite subsequence $\{x_{n_{j(k)}}\}$ converges strongly to x. Hence $T(x_{n_{j(k)}})$ converges strongly to $T(x)$ by the continuity of T. On the other hand

$$(T(x_{n_{j(k)}}), x_{n_{j(k)}} - v) \to (T(x), x - v) \qquad (v \in X).$$

In particular,

$$(y, x - v) = (T(x), x - v) \qquad (v \in X)$$

which implies that $y = T(x)$.
q.e.d.

THEOREM (17.4). *Let X be a separable reflexive Banach space, G an open subset of X symmetric about the origin with $0 \in G$. Let T be a mapping of cl (G) into X^* which is continuous, bounded, and satisfies condition (S) on cl (G). Suppose that T is odd on the boundary of G, and 0 does not lie in $T(\mathrm{bdry}\,(G))$.*

Then 0 lies in $T(G)$ and $T(G)$ contains the whole component of $X^* - T(\mathrm{bdry}\,(G))$ which contains 0. In particular, if $B_d(0) \cap T(\mathrm{bdry}\,(G)) = \varnothing$, then $B_d(0)$ is contained in $T(G)$.

PROOF OF THEOREM (17.4). This is simply a special case of Theorem (17.2) after the details of the hypothesis of Theorem (17.2) have been verified in this case. The fact that T is A-proper with respect to an arbitrarily chosen injective scheme follows from part (b) of Theorem (17.3). q.e.d.

A useful restriction of the class of mappings satisfying the condition (S) is the following:

DEFINITION (17.5). *Let X be a Banach space, G an open subset of X, T a continuous mapping of $\operatorname{cl}(G)$ into X^*. Then T is said to be a mapping of type $(S)_+$ if it satisfies the following condition:*

$(S)_+$ *From any sequence $\{u_j\}$ in $\operatorname{cl}(G)$ for which $u_j \rightharpoonup u$ in X, and*

$$\overline{\lim}\,(T(u_j) - T(u), u_j - u) \leq 0,$$

u_j converges strongly to u in X.

THEOREM (17.5). *The class of mappings satisfying condition $(S)_+$ on $\operatorname{cl}(G)$ for a given open set of a reflexive space X is a subclass of the mappings of type (S) and is convex.*

PROOF OF THEOREM (17.5). The fact that mappings of type $(S)_+$ are automatically of type (S) follows trivially from the definitions. We need only prove the convexity of the class.

Let T_0 and T_1 be mappings of type $(S)_+$ and for a fixed t with $0 < t < 1$, let $T_t = (1-t)T_0 + tT_1$. Suppose that we are given a sequence $\{u_j\}$ in $\operatorname{cl}(G)$ and an element u in X such that $u_j \rightharpoonup u$, and

$$\overline{\lim}\,(T_t(u_j) - T_t(u), u_j - u) \leq 0.$$

We note that

$$(T_t(u_j) - T_t(u), u_j - u) = (1-t)(T_0(u_j) - T_0(u), u_j - u)$$
$$+ t(T_1(u_j) - T_1(u), u_j - u).$$

Since both t and $(1-t)$ are positive, it follows that for an infinite subsequence $\{u_{j(k)}\}$ of $\{u_j\}$, we have either

$$\overline{\lim}\,(T_0(u_{j(k)}) - T_0(u), u_{j(k)} - u) \leq 0$$

or

$$\overline{\lim}\,(T_1(u_{j(k)}) - T_1(u), u_{j(k)} - u) \leq 0.$$

In either case, we can extract a further infinite subsequence which is strongly convergent to u. Hence T_t satisfies condition $(S)_+$. q.e.d.

THEOREM (17.6). *Let X be a separable reflexive Banach space, G an open subset of X with $0 \in G$ and G symmetric about the origin in X. Let T be a mapping of $\operatorname{cl}(G)$ into X^* which is continuous and bounded on $\operatorname{cl}(G)$ and satisfies condition $(S)_+$ on $\operatorname{cl}(G)$. Suppose that $T(\operatorname{bdry}(G))$ does not contain 0 and that for all x in $\operatorname{bdry}(G)$, $T(x) \neq cT(-x)$ $(0 \leq c \leq 1)$.*

Then 0 lies in $T(G)$ and $T(G)$ contains the entire component of $X^* - T(\mathrm{bdry}\,(G))$ which contains 0.

PROOF OF THEOREM (17.6). We note first that if T satisfies condition $(S)_+$, then so does the mapping T_1 of cl (G) into X^* given by $T_1(x) = -T(-x)$. By Theorem (17.5), it follows that for each t with $0 \le t \le 1$, the mapping $T_t = (1-t)T + tT_1$ also satisfies condition $(S)_+$ and therefore is A-proper with respect to a fixed injective approximation scheme Γ for mappings from X to X^* by Theorem (17.3). For $t = \frac{1}{2}$, $T_{1/2}$ is an odd mapping on bdry (G) and indeed on all of cl (G).

For x in bdry (G), and for any t with $0 \le t \le \frac{1}{2}$, $T_t(x) = (1-t)T(x) - tT(-x) \ne 0$ since otherwise, we should have

$$T(x) = t(1-t)^{-1}T(-x) = cT(-x) \qquad (0 \le c \le 1).$$

Hence, we may apply the multi-valued degree theory for A-proper mappings with respect to the fixed injective approximation scheme. We find by the homotopy invariance property for the degree function, Deg $(T_t, G, 0)$ is independent of t in $[0, \frac{1}{2}]$. For $t = \frac{1}{2}$, Deg $(T_{1/2}, G, 0) \ne \{0\}$. Hence Deg $(T, G, 0) \ne \{0\}$. In particular, 0 lies in $T(G)$ and any point connectible to 0 by a path in the open set $X^* - T(\mathrm{bdry}\,(G))$ must lie in $T(G)$. This last set coincides with the component of $X^* - T(\mathrm{bdry}\,(G))$.   q.e.d.

We now apply these results to obtain a sharper result for the much broader class of pseudo-monotone mappings T, which includes both bounded monotone mappings and the mappings of type $(S)_+$.

THEOREM (17.7). *Let X be a separable reflexive Banach space, G a bounded convex open subset of X containing 0 and symmetric about 0. Let T be a continuous mapping of cl (G) into X^* which is bounded and pseudo-monotone, i.e., if $\{u_j\}$ is a weakly convergent sequence in cl (G) with weak limit u and if*

$$\overline{\lim}\,(T(u_j) - T(u), u_j - u) \le 0,$$

then for each v in X,

$$(T(u), u - v) \le \underline{\lim}\,(T(u_j), u_j - v).$$

Then:

(a) If 0 does not lie in $T(\mathrm{bdry}\,(G))$ and T is an odd mapping on bdry (G), then 0 lies in $T(G)$.

(b) If T is an odd mapping on bdry (G) and if there exists $d > 0$ such that $B_d(0)$ does not meet $T(\mathrm{bdry}\,(G))$, then $T(G)$ contains the interior of the ball $B_d(0)$.

(c) Suppose that $T(\mathrm{bdry}\,(G))$ does not contain 0 and that there exists $d > 0$ such that for all x in bdry (G) and all c with $0 \le c \le 1$, $\|T(x) - c(T(-x))\| \ge d$. Then 0 lies in $T(G)$ and $T(G)$ contains any closed ball about 0 in X^* which does not meet $T(\mathrm{bdry}\,(G))$.

The proof of Theorem (17.7) is based upon the following relation between the class of pseudo-monotone mappings and the class of mappings satisfying condition $(S)_+$:

THEOREM (17.8). *Let X be a reflexive Banach space which is locally uniformly convex, G a bounded open set in G, T a continuous mapping of cl (G) into X^* with T pseudo-monotone. Suppose that X^* is locally uniformly convex, and let J be the duality mapping of X into X^*.*

Then for each $\epsilon > 0$, $T_\epsilon = T + \epsilon J$ is a mapping of cl (G) into X^* which satisfies condition $(S)_+$.

REMARK. Essentially the force of Theorem (17.8) is based upon the fact that it enables us to carry through arguments for pseudo-monotone mappings by using the more structured class of "nonsingular" mappings satisfying condition $(S)_+$ in the same way as we earlier used the approximation of monotone mappings by strongly monotone maps which are homeomorphisms.

PROOF OF THEOREM (17.8). Let $\{u_j\}$ be a sequence in cl (G), u an element in X such that for a given $\epsilon > 0$, $u_j \rightharpoonup u$ and $(T_\epsilon(u_j) - T_\epsilon(u), u_j - u) \to 0$. By its definition,

$$(T_\epsilon(u_j) - T_\epsilon(u), u_j - u) = (T(u_j) - T(u), u_j - u) + \epsilon(J(u_j) - J(u), u_j - u),$$

where $(J(u_j) - J(u), u_j - u) \geq 0$ by the monotonicity of J. Since $\epsilon > 0$ implies that the second summand is nonnegative for each j, it follows that

$$\overline{\lim}\, (T(u_j) - T(u), u_j - u) \leq 0.$$

Since T is pseudo-monotone, we know that for each v in X,

$$(T(u), u - v) \leq \underline{\lim}\, (T(u_j), u_j - u),$$

while by the weak convergence of u_j to u, $(T(u), u_j - u) \to 0$. Hence, setting $v = u$ above, we have

$$\underline{\lim}\, (T(u_j) - T(u), u_j - u) = \underline{\lim}\, (T(u_j), u_j - u) \geq (T(u), u - u) = 0.$$

Therefore, $\lim (T(u_j) - T(u), u_j - u) = 0$.

Returning to the inequality involving the mappings T_ϵ, we obtain

$$(J(u_j) - J(u), u_j - u) \to 0.$$

Since the Banach space X is locally uniformly convex, it follows that u_j must converge strongly to u in X.

Since X^* is locally uniformly convex, the mapping J is continuous from the strong topology of X to the strong topology of X^*, and J is automatically bounded. Hence by the result of the preceding discussion, each $T_\epsilon = T + \epsilon J$ is a mapping satisfying condition $(S)_+$ for $\epsilon > 0$. q.e.d.

PROOF OF THEOREM (17.7). Since X is separable and reflexive as in X^*, each of the spaces X and X^* by a result of Kadec [**262**] has an equivalent norm which is locally uniformly convex. Hence by the result of Asplund [**10**], there exists an equivalent norm on X such that in the new norm both X and X^* are locally uniformly convex. Since both the hypotheses and conclusions of Theorem (17.7) are independent of any passage to an equivalent norm, it follows that we may assume without loss of generality that both X and X^* are locally uniformly convex.

Thus, we may now apply Theorem (17.8) which we have just proved above and conclude that if J is the duality mapping of X into X^*, then for each $\epsilon > 0$, the mapping $T_\epsilon = T + \epsilon J$ satisfies condition $(S)_+$ for "nonsingularity." In particular, we may apply the results of Theorem (17.4) to the mapping T_ϵ for each $\epsilon > 0$ since J is an odd mapping by its definition, while by Theorem (17.5) each mapping of type $(S)_+$ is certainly of type (S). We now consider the separate assertions in the conclusions for Theorem (17.7).

PROOF OF (a). For each $\epsilon > 0$, either 0 belongs to $T_\epsilon(\text{bdry}\,(G))$ or 0 lies in $T_\epsilon(G)$. Thus, there exists an element u_ϵ of cl (G) such that $T_\epsilon(u_\epsilon) = 0$. The set G is bounded and J maps bounded sets into bounded sets. Therefore,

$$T(u_\epsilon) = T_\epsilon(u_\epsilon) - \epsilon J(u_\epsilon) = -\epsilon J(u_\epsilon) \to 0 \qquad (\epsilon \to 0+).$$

By the reflexivity of the Banach space X and the convexity of the set G, cl (G) is a sequentially weakly compact space. Hence, we may find a sequence $\epsilon_j \to 0$ such that the corresponding elements u_{ϵ_j} converge weakly in X to an element u of cl (G). By hypothesis, T is pseudo-monotone. Hence by Theorem (7.7), $T(\text{cl}\,(G))$ is a closed subset of X^*. In particular, there exists u_0 in cl (G) such that $T(u_0) = 0$. Since u_0 with this property does not lie in bdry (G), it follows that $u_0 \in G$. q.e.d.

PROOF OF (b). If for the given constant d, $B_d(0)$ does not meet $T(\text{bdry}\,(G))$, then if $M = \sup\{\|x\|, x \in G\}$, it follows by a simple calculation that for $\epsilon M < d$ and $d_1 = d - \epsilon M > 0$, $T_\epsilon(\text{bdry}\,(G))$ does not meet $B_{d_1}(0)$. Applying Theorem (17.4) to the mapping T_ϵ, it follows that for such ϵ, $T_\epsilon(G)$ contains $B_{d_1}(0)$.

Let y be a point of X^* lying in $B_{d_0}(0)$ for $d_0 < d$. By the result of the previous paragraph, for $0 < \epsilon < M^{-1}(d - d_0)$, y lies in $T_\epsilon(\text{cl}\,(G))$. Passing to the limit as $\epsilon \to 0+$ as in the proof of part (a), it follows that y lies in $T(\text{cl}\,(G))$. Since y is excluded from $T(\text{bdry}\,(G))$, it follows that $y \in T(G)$. Hence the interior of $B_d(0)$ is contained in $T(G)$. q.e.d.

PROOF OF (c). For the proof of part (c), we again consider the mapping $T_\epsilon = T + \epsilon J$ for $\epsilon > 0$. For each x in bdry (G), we have for $0 \leq c \leq 1$,

$$\|T_\epsilon(x) - cT_\epsilon(-x)\| \geq \|T(x) - cT(-x))\| - \|\epsilon J(x)\| - \epsilon \|cJ(-x)\| \geq d - 2\epsilon M.$$

If $\epsilon > 0$ is chosen so small that $2\epsilon M < d$, then T_ϵ satisfies the conditions imposed upon T and we may apply Theorem (17.6) to the mapping T_ϵ for each sufficiently small $\epsilon > 0$. In particular, it follows that for a suitably small $B_{d_1}(0)$, the ball lies in $T_\epsilon(\text{cl}\,(G))$ for all sufficiently small $\epsilon > 0$. Passing to the limit as $\epsilon \to 0+$, as in the proof of part (a), we find that $B_{d_1}(0)$ lies in $T(\text{cl}\,(G))$. Since the hypothesis excludes the possibility that for d_1 sufficiently small, $B_{d_1}(0)$ intersects $T(\text{bdry}\,(G))$, it follows that $B_{d_1}(0)$ is contained in $T(G)$.

Let y be a point of X^* which can be joined to 0 by a curve in X^* at positive distance from $T(\text{bdry}\,(G))$. Then for $\epsilon > 0$ sufficiently small, this curve does not meet $T_\epsilon(\text{bdry}(G))$. Hence for $0 < \epsilon < \epsilon_0$, it follows from Theorem (17.4) that this curve C_0 is actually contained in $T_\epsilon(G)$. Passing to the limit as $\epsilon \to 0+$, we find that C_0 is contained in $T(\text{cl}(G))$ and hence in $T(G)$. Since each point of the component of $X^* - \text{cl}(T(\text{bdry}(G)))$ which contains 0 (if indeed, 0 does not lie in $T(\text{bdry}(G))$) can be connected to 0 by such a curve C_0, it follows that any closed ball about 0 in X^* which does not intersect $T(\text{bdry}(G))$ has its interior in $T(G)$. Hence the ball itself lies in $T(G)$ since $T(\text{cl}(G))$ is closed in X^*. q.e.d.

As a consequence of Theorem (17.7), we obtain the following:

THEOREM (17.9). *Let X be a separable, reflexive Banach space, T a continuous, bounded, pseudo-monotone mapping of X into X^*. Suppose that T is odd outside of some ball in X (i.e., there exists $R > 0$ such that for $\|x\| \geq R$, $T(-x) = -T(x)$). Suppose further that the inverse image under T of any bounded set is bounded.*

Then the range of T is all of X^*.

PROOF OF THEOREM (17.9). If $d > 0$ is any given positive constant, there exists $R_d > R$ such that $T^{-1}(B_d(0))$ is contained in the interior of the ball $B_{R_d}(0)$ in X. We apply Theorem (17.7) to the convex set G which is the interior of $B_{R_d}(0)$. Then $T(\text{bdry}(G))$ misses the ball $B_d(0)$ and hence $B_d(0)$ lies in the image of $B_{R_d}(0)$ under T. Since $d > 0$ is arbitrary, it follows that $T(X) = X^*$. q.e.d.

As a special case of the situation encountered in Theorem (17.9), we have the following:

THEOREM (17.10). *Let X be a separable, reflexive Banach space, and let T be a continuous, bounded, pseudo-monotone mapping of X into X^*. Suppose that for some $R > 0$, T is an odd mapping outside of $B_R(0)$ and is homogeneous of some positive order outside of that ball (i.e., $T(\lambda x) = \lambda^a T(x)$ for $\|x\| \geq R$, $\lambda \geq 1$ ($a > 0$)). Suppose that 0 does not lie in the closure of the image under T of $\text{bdry}(B_R(0))$.*

Then the range of T is all of X^*.

PROOF OF THEOREM (17.10). By hypothesis, there exists $d > 0$ such that $T(\text{bdry}(B_R(0))) \cap B_d(0) = \emptyset$. Let $S_R(0)$ be the sphere of radius R about 0. Then by the homogeneity of the mapping T, we have

$$T(S_{\lambda R}(0)) \cap B_{\lambda^a d}(0) = \emptyset \qquad (\lambda > 1).$$

It follows from this fact that for $r \geq d$, $\lambda = (r/d)^{1/a}$, $T^{-1}(B_r(0)) \subset B_{R\lambda}(0)$. q.e.d.

We close this section with the following simple result which characterizes A-proper mappings on the whole of a Banach space X:

THEOREM (17.11). *Let X and Y be Banach spaces, T a continuous mapping of X into Y, Γ an approximation scheme with equicontinuous $\{Q_n\}$ and $\{P_n\}$, for mappings from X to Y in the sense of Definition (17.1) with $P_n(X_n)$ increasing with n and with dense union in X. Let $T_n = Q_n T P_n$, and suppose that T is uniformly stable with respect to this approximation scheme, i.e., there exists a function α from R^+ to R^+ such that $\alpha(r_j) \to 0$ implies that $r_j \to 0$, for which*

(i) $\qquad \|T_n(u) - T_n(v)\| \geq \alpha(\|u - v\|) \qquad (u, v \in X_n)$.

Then T is A-proper from X to Y with respect to Γ if and only if the range of T is the whole of the Banach space Y.

PROOF OF THEOREM (17.11). Since the spaces X_n and Y_n are of the same finite dimension, it follows from the Brouwer theorem on invariance of domain that each T_n is a homeomorphism of X_n onto Y_n, i.e., for each y_n in Y_n, there is exactly one x_n in X_n such that $T_n(x_n) = y_n$.

Suppose first that T is A-proper, and let y be an element of Y. For each n, we can find a solution x_n in X_n of the approximating equation $T_n(x_n) = Q_n(y)$. If we apply the A-properness of T, we can find an element x of X which the limit of a sequence of $P_{n_j}(x_{n_j})$ such that $T(x) = y$.

Suppose conversely that $R(T) = Y$, and suppose that we are given a sequence $\{u_{n_j}\}$ in X such that $\|T_{n_j}(u_{n_j}) - Q_{n_j}(y)\| \to 0$. By hypothesis, there exists an element x in X such that $T(x) = y$. By the assumption that $P_n(X_n)$ increase with n and have dense union in X, we can find a sequence $\{x_{n_j}\}$ with $x_{n_j} \in X_{n_j}$ and $n_j \to \infty$ such that $P_{n_j}(x_{n_j}) \to x$. By the continuity of T, $TP_{n_j}(x_{n_j}) \to T(x) = y$. Since the mappings Q_n are equicontinuous by assumption,

$$\|Q_{n_j} T P_{n_j}(x_{n_j}) - Q_{n_j}(y)\| = \epsilon_j \to 0 \qquad (j \to +\infty).$$

If we apply the inequality (i) to the elements $u = x_{n_j}$ and $v = u_{n_j}$, we find that

$$\alpha(\|u_{n_j} - x_{n_j}\|) \leq \|T_{n_j}(u_{n_j}) - T_{n_j}(x_{n_j})\| \to 0.$$

Hence $\|u_{n_j} - x_{n_j}\| \to 0 \, (j \to \infty)$. By the assumed equicontinuity of the mappings P_n,

$$\|P_{n_j}(u_{n_j}) - P_{n_j}(x_{n_j})\| \to 0.$$

However, by their construction, $P_{n_j}(x_{n_j}) \to x$. Hence $P_{n_j}(u_{n_j}) \to x$. q.e.d.

BIBLIOGRAPHY

1. R. Abraham, A. Kelly and J. Robbin, *Transversal mappings and flows*, Benjamin, New York, 1967. MR **39** #2181.

2. P. G. Aĭzengendler and M. M. Vaĭnberg, *Branching of periodic solutions of autonomous systems and of differential equations in Banach spaces*, Dokl. Akad. Nauk SSSR **176** (1967), 9–12 = Soviet Math. Dokl. **8** (1967), 1013–1017. MR **36** #4106.

3. P. Alexandroff and H. Hopf, *Topologie*, Springer, Berlin, 1935; reprint, Band 1, Chelsea, New York, 1965. MR **32** #3023.

4. M. Altman, *A fixed point theorem in Hilbert space*, Bull. Acad. Polon. Sci. Cl. III. **5** (1957), 19–22. (Russian) MR **19**, 297.

5. ———, *An extension to locally convex spaces of Borsuk's theorem on antipodes*, Bull. Acad. Polon. Sci. Sér. Sci. Math. Astr. Phys. **6** (1958), 293–295. MR **20** #3531.

6. H. Amann, *Ein Existenz und Eindeutigkeitssatz für die Hammesteinsche Gleichung im Banachräum*, Math. Z. **111** (1969), 175–190. MR **40** #7894.

7. A. Ambrosetti, *Un teorema di esistenza per le equazioni differenziali negli spazi di Banach*, Rend. Sem. Mat. Univ. Padova **39** (1967), 349–361. MR **36** #5478.

8. N. Aronszajn, *La correspondant topologique de l'unicité dans la théorie des équations différentielles*, Ann. of Math. (2) **43** (1942), 730–738. MR **4**, 100.

9. E. Asplund, *Positivity of duality mappings*, Bull. Amer. Math. Soc. **73** (1967), 200–203. MR **34** #6481.

10. ———, *Averaged norms*, Israel J. Math. **5** (1967), 227–233. MR **36** #5660.

11. ———, *Fréchet differentiability of convex functions*, Acta Math. **121** (1968), 31–47. MR **37** #6754.

12. ———, *A monotone convergence theorem for sequences of nonlinear mappings*, Proc. Sympos. Pure Math., vol. 18, part 1, Amer. Math. Soc., Providence, R.I., 1970, pp. 1–9. MR **43** #997.

13. E. Asplund and R. T. Rockafellar, *Gradients of convex functions*, Trans. Amer. Math. Soc. **139** (1969), 443–467. MR **39** #1968.

14. J. P. Aubin, *Un théorème de compacité*, C.R. Acad. Sci. Paris **256** (1963), 5042–5044. MR **27** #2832.

15. ———, *Approximation of variational inequations*, Functional Analysis and Optimization, Academic Press, New York, 1966, pp. 7–14. MR **35** #4743.

16. S. Banach, *Sur les opérations dans les ensembles abstraits et leur application aux équations intégrales*, Fund. Math. **3** (1922), 133–181.

17. ———, *Théorie des opérations linéaires*, Monografie Mat., PWN, Warsaw, 1932.

18. S. Banach and S. Mazur, *Über mehrdeutige stetige Abbildungen*, Studia Math. **5** (1934), 174–178.

19. C. Bardos and H. Brezis, *Sur une classe de problèmes d'évolution non linéaires*, C.R. Acad. Sci. Paris Sér. AB **266** (1968), A56–A59. MR **38** #436.

20. ———, *Sur une classe de problèmes d'évolution non linéaires*, J. Differential Equations **6** (1969), 345–394. Mr **39** #3355.

21. R. G. Bartle, *Implicit functions and solutions of equations in groups*, Math. Z. **62** (1955), 335–346. MR **17**, 62.

22. ———, *On the openness and inversion of differentiable mappings*, Ann. Acad. Sci. Fenn. Ser. AI No. 257 (1958). MR **21** #2932.

23. R. G. Bartle and L. M. Graves, *Mappings between function spaces*, Trans. Amer. Math. Soc. **72** (1952), 400–413. MR **13**, 951.

24. L. P. Belluce and W. A. Kirk, *Fixed-point theorems for families of contraction mappings*, Pacific J. Math. **18** (1966), 213–217. MR **33** #7846.

25. ———, *Nonexpansive mappings and fixed points in Banach spaces*, Illinois J. Math. **11** (1967), 474–479. MR **35** #5988.

26. L. P. Belluce, W. A. Kirk and E. F. Steiner, *Normal structure in Banach spaces*, Pacific J. Math. **26** (1968), 433–440. MR **38** #1501.

27. C. Berge, *Espaces topologiques: Fonctions multivoques*, Coll. Univ. Math., vol. III, Dunod, Paris, 1959. MR **21** #4401.

28. M. S. Berger, *A Sturm-Liouville theorem for nonlinear elliptic partial differential equations*, Ann. Scuola Norm. Sup. Pisa (3) **20** (1966), 543–582; Correction, ibid. (3) **22** (1968), 351–354. MR **35** #2181; **37** #5726.

29. G. M. Bergman and B. R. Halpern, *A fixed point theorem for inward and outward maps*, Trans. Amer. Math. Soc. **130** (1968), 353–358. MR **36** #4397.

30. C. Bessaga, *Topological equivalence of unseparable reflexive Banach spaces. Ordinal resolutions of identity and monotone bases*, Bull. Acad. Polon. Sci. Sér. Sci. Math. Astronom. Phys. **15** (1967), 397–399. MR **36** #4321.

31. A. Beurling and A. E. Livingston, *A theorem on duality mappings in Banach spaces*, Ark. Math. **4** (1962), 405–411. MR **26** #2851.

32. G. D. Birkhoff and O. D. Kellogg, *Invariant points in function spaces*, Trans. Amer. Math. Soc. **23** (1922), 96–115.

33. E. Bishop and R. R. Phelps, *The support functionals of a convex set*, Proc. Sympos. Pure Math., vol. 7, Amer. Math. Soc., Providence, R.I., 1963, pp. 27–35. MR **27** #4051.

34. H. F. Bohnenblust and S. Karlin, *On a theorem of Ville*, Contributions to the Theory of Games, Ann. of Math. Studies, no. 24, Princeton Univ. Press, Princeton, N.J., 1950, pp. 155–160. MR **12**, 844.

35. R. Bonic and J. Frampton, *Smooth functions on Banach manifolds*, J. Math. Mech. **15** (1966), 877–898. MR **33** #6647.

36. F. F. Bonsall, *Lectures on some fixed point theorems of functional analysis*, Tata Institute Lecture Notes, no. 26, Bombay, 1962.

37. K. Borsuk, *Drei Sätze über die n-dimensionale Euklidische Sphäre*, Fund. Math. **21** (1933), 177–190.

38. ———, *Theory of retracts*, Monografie Math., Tom 44, PWN, Warsaw, 1967. MR **35** #7306.

39. D. G. Bourgin, *Fixed points on neighborhood retracts*, Rev. Math. Pures Appl. **2** (1957), 371–374. MR **20** #1297.

40. ———, *Un indice dei punti uniti*. I, II, III, Atti Accad. Naz. Lincei Rend. Cl. Sci. Fis. Mat. Nat. (8) **19** (1955), 435–440; **20** (1956), 43–48; **21** (1956), 395–400. MR **17**, 1120; **19**, 571.

41. ———, *Modern algebraic topology*, Macmillan, New York, 1963. MR **28** #3415.

42. N. Bourbaki, *Éléments de mathématique*. Fasc. XV. Livre V: *Espaces vectoriels topologiques*, Actualités Sci. Indust., no. 1189, Hermann, Paris, 1953. MR **14**, 880.

———, *Éléments de mathématique*. XIX. Part 1. *Les structures fondamentales de l'analyse*. Livre V: *Espaces vectoriels topologiques (Fascicule de résultats)*, Actualités Sci. Indust., no. 1230, Hermann, Paris, 1955. MR **17**, 1109.

43. F. Brauer and S. Sternberg, *Local uniqueness, existence in the large, and the convergence of successive approximations*, Amer. J. Math. **80** (1958), 421–430; ibid. **81** (1959), 797. MR **20** #1806; **21** #5038.

44. L. M. Brègman, *The method of successive projection for finding a common point of convex sets*, Dokl. Akad. Nauk SSSR **162** (1965), 487–490 = Soviet Math. Dokl. **6** (1965), 688–692. MR **33** #6499.

45. H. Brézis, *Une généralisation des opérateurs monotones*, C.R. Acad. Sci. Paris Sér. A-B **264** (1967), A683–A686. MR **35** #5983.

46. ———, *Inéquations d'évolution abstraits*, C.R. Acad. Sci. Paris Sér. A-B **264** (1967), A732–A735. MR **42** #2320.

47. ———, *Équations et inéquations non linéaires dans les espaces vectoriels en dualité*, Ann. Inst. Fourier (Grenoble) **18** (1968), fasc. 1, 115–175. MR **42** #5113.

48. ———, *On some degenerate nonlinear parabolic equations*, Proc. Sympos. Pure Math., vol. 18, part 1, Amer. Math. Soc., Providence, R.I., 1970, pp. 28–38.

49. H. Brézis and J.-L. Lions, *Sur certains problèmes unilatéraux hyperboliques*, C.R. Acad. Sci. Paris Sér. A-B **264** (1967), A928–A931. MR **38** #2633.

50. H. Brézis and M. Sibony, *Méthodes d'approximation et d'itération pour les opérateurs monotones*, Arch. Rational Mech. Anal. **28** (1967/68), 59–82. MR **36** #3177.

51. H. Brézis and G. Stampacchia, *Sur la régularité de la solution d'inéquations elliptiques*, Bull. Soc. Math. France **96** (1968), 153–180. MR **39** #659.

52. M. S. Brodskiĭ and D. P. Mil'man, *On the center of a convex set*, Dokl. Akad. Nauk SSSR **59** (1948), 837–840. (Russian) MR **9**, 448.

53. A. Brøndsted and R. T. Rockafellar, *On the subdifferentiability of convex functions*, Proc. Amer. Math. Soc. **16** (1965), 605–611. MR **31** #2361.

54. R. B. S. Brooks, *Coincidence, roots, and fixed points*, Ph.D. Thesis, UCLA, 1967.

55. L. E. J. Brouwer, *Über Abbildungen vom Mannigfaltigkeiten*, Math. Ann. **71** (1912), 97–115.

56. F. E. Browder, *The topological fixed point theory and its applications to functional analysis*, Ph.D. Thesis, Princeton University, Princeton, N.J., 1948.

57. ———, *Covering spaces, fibre spaces, and local homeomorphisms*, Duke Math. J. **21** (1954), 329–336. MR **15**, 978.

58. ———, *Nonlinear functional equations in locally convex spaces*, Duke Math. J. **24** (1957) 579–589. MR **19**, 1184.

59. ———, *On a generalization of the Schauder fixed point theorem*, Duke Math. J. **26** (1959) 291–303. MR **21** #4368.

60. ———, *On the fixed point index for continuous mappings of locally connected spaces*, Summa Brasil. Math. **4** (1960), 253–293. MR **26** #4354.

61. ———, *On continuity of fixed points under deformations of continuous mappings*, Summa Brasil. Math. **4** (1960), 183–191. MR **24** #A543.

62. ———, *The solvability of nonlinear functional equations*, Duke Math. J. **30** (1963), 557–566. MR **27** #6133.

63. ———, *Variational boundary value problems for quasilinear elliptic equations of arbitrary order*, Proc. Nat. Acad. Sci. U.S.A. **50** (1963), 31–37. MR **29** #3750.

64. ———, *Variational boundary value problems for quasilinear elliptic equations*. II, Proc. Nat. Acad. Sci. U.S.A. **50** (1963), 592–598. MR **29** #3751.

65. ———, *Variational boundary value problems for quasilinear elliptic equations*. III, Proc. Nat. Acad. Sci. U.S.A. **50** (1963), 794–798. MR **29** #3752.

66. ———, *Nonlinear elliptic boundary value problems*, Bull. Amer. Math. Soc. **69** (1963), 862–874. MR **27** #6048.

67. ———, *Non-linear parabolic boundary value problems of arbitrary order*, Bull. Amer. Math. Soc. **69** (1963), 858–861. MR **27** #6049.

68. ———, *Non-linear equations of evolution*, Ann. of Math. (2) **80** (1964), 485–523. MR **30** #4167.

69. ———, *Nonlinear elliptic problems*. II, Bull. Amer. Math. Soc. **70** (1964), 299–302. MR **28** #3247.

70. ———, *Continuity properties of monotone nonlinear operators in Banach spaces*, Bull. Amer. Math. Soc. **70** (1964), 551–553. MR **29** #502.

71. ———, *Strongly non-linear parabolic boundary value problems*, Amer. J. Math. **86** (1964), 339–357. MR **29** #3764.

72. ———, *Remarks on non-linear functional equations*, Proc. Nat. Acad. Sci. U.S.A. **51** (1964), 985–989. MR **29** #503.

73. ———, *Remarks on non-linear functional equations*. II, Illinois J. Math. **9** (1965), 608–616. MR **32** #2941.

74. ———, *Remarks on non-linear functional equations*. III, Illinois J. Math. **9** (1965), 617–622. MR **32** #2941.

75. ———, *Non-linear elliptic boundary value problems*. II, Trans. Amer. Math. Soc. **117** (1965), 530–550. MR **30** #4054.

76. ———, *On a theorem of Beurling and Livingston*, Canad. J. Math. **17** (1965), 367–372. MR **31** #595.

77. ———, *Multivalued monotone non-linear mappings and duality mappings in Banach spaces*, Trans. Amer. Math. Soc. **118** (1965), 338–351. MR **31** #5114.

78. ———, *Nonlinear initial value problems*, Ann. of Math. (2) **82** (1965), 51–87. MR **34** #7923.

79. ———, *Another generalization of the Schauder fixed point theorem*, Duke Math. J. **32** (1965), 399–406. MR **34** #3567.

80. ———, *A further generalization of the Schauder fixed point theorem*, Duke Math. J. **32** (1965), 575–578. MR **34** #3568.

81. ———, *Fixed point theorems on infinite dimensional manifolds*, Trans. Amer. Math. Soc. **119** (1965), 179–194. MR **33** #3287.

82. ———, *Non-linear monotone operators and convex sets in Banach spaces*, Bull. Amer. Math. Soc. **71** (1965), 780–785. MR **31** #5112.

83. F. E. Browder, *Variational methods for non-linear elliptic eigenvalue problems*, Bull. Amer. Math. Soc. **71** (1965), 176–183. MR **31** #3707.

84. ——, *Lusternik-Schnirelman category and non-linear elliptic eigenvalue problems*, Bull. Amer. Math. Soc. **71** (1965), 644–648. MR **31** #2506.

85. ——, *Infinite dimensional manifolds and nonlinear elliptic eigenvalue problems*, Ann. of Math. (2) **82** (1965), 459–477. MR **34** #3102.

86. ——, *Remarks on the direct method of the calculus of variations*, Arch. Rational Mech. Anal. **20** (1965), 251–258. MR **32** #4576.

87. ——, *Existence of periodic solutions for non-linear equations of evolution*, Proc. Nat. Acad. Sci. U.S.A. **53** (1965), 1100–1103. MR **31** #1558.

88. ——, *Fixed point theorems for non-compact mappings in Hilbert space*, Proc. Nat. Acad. Sci. U.S.A. **53** (1965), 1272–1276. MR **31** #2582.

89. ——, *Mapping theorems for non-compact non-linear operators in Banach spaces*, Proc. Nat. Acad. Sci. U.S.A. **54** (1965), 337–342. MR **31** #5113.

90. ——, *Non-expansive nonlinear operators in a Banach space*, Proc. Nat. Acad. Sci. U.S.A. **54** (1965), 1041–1044. MR **32** #4574.

91. ——, *Nonlinear operators in Banach spaces*, Math. Ann. **162** (1965/66), 280–283. MR **32** #8187.

92. ——, *Nonlinear elliptic functional equations in a non-reflexive Banach space*, Bull. Amer. Math. Soc. **72** (1966), 89–95. MR **32** #2755.

93. ——, *Fixed point theorems for non-linear semi-contractive mappings in Banach spaces*, Arch. Rational Mech. Anal. **21** (1965/66), 259–269. MR **34** #641.

94. ——, *Further remarks on nonlinear functional equations*, Illinois J. Math. **10** (1966), 275–286. MR **34** #8243.

95. ——, *Existence and uniqueness theorems for solutions of nonlinear boundary value problems*, Proc. Sympos. Appl. Math., vol. 17, Amer. Math. Soc., Providence, R.I., 1965, pp. 24–49. MR **33** #6092.

96. ——, *Problèmes non linéaires*, Séminaire de Mathématiques Supérieures, no. 15 (Été, 1965), Les Presses de l'Université de Montréal, Montréal, Qué., 1966. MR **40** #3380.

97. ——, *On the unification of the calculus of variations and the theory of monotone nonlinear operators in Banach spaces*, Proc. Nat. Acad. Sci. U.S.A. **56** (1966), 419–425. MR **34** #3383.

98. ——, *Existence and approximation of solutions of non-linear variational inequalities*, Proc. Nat. Acad. Sci. U.S.A. **56** (1966), 1080–1086. MR **34** #3384.

99. ——, *Convergence of approximants to fixed points of non-expansive nonlinear mappings in Banach spaces*, Arch. Rational Mech. Anal. **24** (1967), 82–90. MR **34** #6582.

100. ——, *Periodic solutions of nonlinear equations of evolution in infinite dimensional spaces*, Lectures in Differential Equations, vol. 1, Van Nostrand, Princeton, N.J., 1969, pp. 71–96. MR **42** #8049.

101. ——, *Convergence theorems for sequences of nonlinear operators in Banach spaces*, Math. Z. **100** (1967), 201–225. MR **35** #5984.

102. ——, *Existence and perturbation theorems for nonlinear maximal monotone operators in Banach spaces*, Bull. Amer. Math. Soc. **73** (1967), 322–327. MR **35** #3495.

103. ——, *Nonlinear accretive operators in Banach spaces*, Bull. Amer. Math. Soc. **73** (1967), 470–476. MR **35** #3496.

104. ——, *Approximation-solvability of nonlinear functional equations in normed linear spaces*, Arch. Rational Mech. Anal. **26** (1967), 33–42. MR **36** #3185.

105. ——, *Topological methods for non-linear elliptic equations of arbitrary order*, Pacific J. Math. **17** (1966), 17–31. MR **34** #3098.

106. ——, *Non-local elliptic boundary value problems*, Amer. J. Math. **86** (1964), 735–750. MR **30** #2215.

107. ——, *Families of linear operators depending upon a parameter*, Amer. J. Math. **87** (1965), 752–758. MR **32** #2917.

108. ——, *A new generalization of the Schauder fixed point theorem*, Math. Ann. **174** (1967), 285–290. MR **36** #6991.

109. ——, *Nonlinear maximal monotone operators in Banach space*, Math. Ann. **175** (1968), 89–113. MR **36** #6989.

110. ——, *Nonlinear equations of evolution and nonlinear accretive operators in Banach spaces*, Bull. Amer. Math. Soc. **73** (1967), 867–874. MR **38** #580.

111. F. E. Browder, *Nonlinear mappings of non-expansive and accretive type in Banach spaces*, Bull. Amer. Math. Soc. **73** (1967), 875–882. MR **38** #581.

112. ———, *The fixed point theory of multi-valued mappings in topological vector spaces*, Math. Ann. **177** (1968), 283–301. MR **37** #4679.

113. ———, *Nonlinear eigenvalue problems and Galerkin approximations*, Bull. Amer. Math. Soc. **74** (1968), 651–656. MR **37** #2043.

114. ———, *Semicontractive and semiaccretive nonlinear mappings in Banach spaces*, Bull. Amer. Math. Soc. **74** (1968), 660–665. MR **37** #5742.

115. ———, *On the convergence of successive approximations for nonlinear functional equations*, Nederl. Akad. Wetensch. Proc. Ser. A **71** = Indag. Math. **30** (1968), 27–35. MR **37** #5743.

116. ———, *Nonlinear variational inequalities and maximal monotone mappings in Banach spaces*, Math. Ann. **183** (1969), 213–231. MR **42** #6661.

117. ———, *Asymptotic fixed point theorems*, Math. Ann. **185** (1970), 38–60. MR **43** #1165.

118. ———, *Nonlinear mappings of analytic type in Banach spaces*, Math. Ann. **185** (1970), 259–278. MR **41** #4318.

119. ———, *Local and global properties of nonlinear mappings in Banach spaces*, Symposia Mathematica, vol. II (INDAM, Rome, 1968), Academic Press, London, 1969, pp. 13–35. MR **41** #2476.

120. ———, *Topology and non-linear functional equations*, Studia Math. **31** (1968), 189–204. MR **38** #6410.

121. ———, *Remarks on nonlinear interpolation in Banach spaces*, J. Functional Analysis **4** (1969), 390–403.

122. ———, *Existence theorems for nonlinear partial differential equations*, Proc. Sympos. Pure Math., vol. 16, Amer. Math. Soc., Providence, R. I., 1970, pp. 1–60. MR **42** #4855.

123. ———, *Nonlinear eigenvalue problems and group invariance*, Functional Analysis and Related Fields (Proc. Conf. for M. Stone, Univ. of Chicago, Chicago, Ill., 1968), Springer, New York, 1970, pp. 1–58. MR **42** #6662.

124. ———, *Nonlinear monotone and accretive operators in Banach spaces*, Proc. Nat. Acad. Sci. U.S.A. **61** (1968), 388–393. MR **44** #7389.

125. F. E. Browder and D. J. de Figueiredo, *J-monotone nonlinear operators in Banach spaces*, Nederl. Akad. Wetensch. Proc. Ser. A **69** = Indag. Math. **28** (1966), 412–420. MR **34** #4957.

126. F. E. Browder and C. P. Gupta, *Topological degree and non-linear mappings of analytic type in Banach spaces*, J. Math. Anal. Appl. **26** (1969), 390–402. MR **41** #2475.

127. F. E. Browder and R. D. Nussbaum, *The topological degree for non-compact nonlinear mappings in Banach spaces*, Bull. Amer. Math. Soc. **74** (1968), 671–676. MR **38** #583.

128. F. E. Browder and W. V. Petryshyn, *The solution by iteration of nonlinear functional equations in Banach spaces*, Bull. Amer. Math. Soc. **72** (1966), 571–575. MR **32** #8155b.

129. ———, *Construction of fixed points of nonlinear mappings in Hilbert space*, J. Math. Anal. Appl. **20** (1967), 197–228. MR **36** #747.

130. ———, *The topological degree and Galerkin approximations for noncompact operators in Banach spaces*, Bull. Amer. Math. Soc. **74** (1968), 641–646. MR **37** #4678.

131. ———, *Approximation methods and the generalized topological degree for nonlinear mappings in Banach spaces*, J. Functional Analysis **3** (1969), 217–245. MR **39** #6126.

132. F. E. Browder and B. A. Ton, *Nonlinear functional equations in Banach spaces and elliptic super-regularization*, Math. Z. **105** (1968), 177–195. MR **38** #582.

133. ———, *Convergence of approximants by regularization for solutions of nonlinear functional equations in Banach spaces*, Math. Z. **106** (1968), 1–16. MR **38** #829.

134. R. F. Brown, *On the Nielsen fixed point theorem for compact maps*, Duke Math. J. **36** (1969), 699–708. MR **40** #3529.

135. H. D. Brunk, *On an extension of the concept of conditional expectation*, Proc. Amer. Math. Soc. **14** (1963), 298–304. MR **26** #5599.

136. R. Cacciopoli, *Un teorema generale sull' esistenza di elementi uniti in una transformazione funzionale*, Rend. Accad. Naz. Lincei **11** (1930), 794–799.

137. ———, *Sugli elementi uniti delle transformazioni funzionale; un osservazione sui problemi di valeri ai limiti*, Rend. Accad. Naz. Lincei **13** (1931), 498–502.

138. ———, *Sugli elementi uniti delle trasformazioni funzionali; un teorema di esistenza e di unicita od alcunne sue applicazioni*, Rend. Sem. Mat. Univ. Padova **3** (1932), 1–15.

139. R. Cacciopoli, *Un principio di inversione per le corrispondenze funzionali e sue applicazioni alle equazioni a derivate parziali*, Rend. Accad. Naz. Lincei **16** (1932), 390–395, 484–489.

140. ———, *Problemi non lineari in analisis funzionale*, Rend. Sem. Nat. Roma (3) **1** (1931/32), 13–22.

141. ———, *Sulle equazioni ellitiche non lineari a derivate parziali*, Rend. Accad. Naz. Lincei (6) **1** (1933), 103–106.

142. ———, *Sulle corrispondenze funzionali inverse diramate; teoria generale e applicazioni ad alcune equazioni funzionali non lineari e al problema di Plateau*. I, II, Rend. Accad. Naz. Lincei (6) **24** (1936), 258–263, 416–421.

143. C. Carathéodory and H. Rademacher, *Über die Eineindeutigkeit in Kleinen und in grossen Stetiger Abbildungen von Gebieten*, Arch. Math. Phys. **26** (1917), 1–9.

144. H. Cartan, *Sur les transformations localement topologiques*, Acta Litt. Sci. Szeged **6** (1933), 85–104.

145. L. Cesari, *Functional analysis and periodic solutions of nonlinear differential equations*, Contributions to Differential Equations **1** (1963), 149–187. MR **27** #1662.

146. ———, *Functional analysis and Galerkin's method*, Michigan Math. J. **11** (1964), 385–414. MR **30** #4047.

147. W. Cheney and A. A. Goldstein, *Proximity maps for convex sets*, Proc. Amer. Math. Soc. **10** (1959), 448–450. MR **21** #3755.

148. M. Cotsaftis, *Conditions nécessaires et suffisantes de stabilité globale d'une classe de mouvements non linéaires nondissipatifs*, C. R. Acad. Sci. Paris Sér. A-B **265** (1967), A911–A914. MR **36** #4084.

149. M. Crandall, *Differential equations in convex sets*, J. Math. Soc. Japan **22** (1970), 443–455. MR **42** #3388.

150. M. Crandall and A. Pazy, *Semigroups of nonlinear contractions and dissipative sets*, J. Functional Analysis **3** (1969), 376–418. MR **39** #4705.

151. ———, *On accretive sets in Banach spaces*, J. Functional Analysis **5** (1970), 204–217. MR **41** #4201.

152. J. Cronin, *The existence of multiple solutions of elliptic differential equations*, Trans. Amer. Math. Soc. **68** (1950), 105–131. MR **11**, 361.

153. ———, *Branch points of solutions of equations in Banach spaces*, Trans. Amer. Math. Soc. **69** (1950), 208–231. MR **12**, 716.

154. ———, *Branch points of solutions of equations in Banach space*. II, Trans. Amer. Math. Soc. **76** (1954), 207–222. MR **16**, 47.

155. ———, *Topological degree of some mappings*, Proc. Amer. Math. Soc. **5** (1954), 175–178. MR **16**, 60.

156. ———, *Analytic functional mappings*, Ann. of Math. (2) **58** (1953), 175–181. MR **15**, 234.

157. ———, *The Dirichlet problem for nonlinear elliptic equations*, Pacific J. Math. **5** (1955), 335–344. MR **17**, 494.

158. ———, *Fixed points and topological degree in nonlinear analysis*, Math. Surveys, no. 11, Amer. Math. Soc., Providence, R.I., 1964. MR **29** #1400.

159. ———, *Using Leray-Schauder degree*, J. Math. Anal. Appl. **25** (1969), 414–424. MR **40** #3389.

160. ———, *Upper and lower bounds for the number of solutions of nonlinear equations*, Proc. Sympos. Pure Math., vol. 18, part 1, Amer. Math. Soc., Providence, R.I., 1970, pp. 50–61. MR **43** #5374.

161. D. F. Cudia, *Rotundity*, Proc. Sympos. Pure Math., vol. 7, Amer. Math. Soc., Providence, R.I., 1963, pp. 73–97. MR **27** #5106.

162. ———, *The geometry of Banach spaces, smoothness*, Trans. Amer. Math. Soc. **110** (1964), 284–314. MR **29** #446.

163. G. Darbo, *Punti uniti in trasformazioni a codeminio non compatto*, Rend. Sem. Mat. Univ. Padova **24** (1955), 84–92. MR **16**, 1140.

164. M. M. Day, *Linear normed spaces*, 2nd rev. ed., Academic Press, New York; Springer-Verlag, Berlin, 1962. MR **26** #2847.

165. H. Debrunner and P. Flor, *Ein Erweiterungssatz für monotone Mengen*, Arch. Math. **15** (1964), 445–447. MR **30** #428.

166. A. Deleanu, *Théorie des points fixes: Sur les rétractes de voisinage des espaces convexoides*, Bull. Soc. Math. France **87** (1959), 235–243. MR **26** #763.

167. A. Deleanu, *Une généralization du théorème de point fixe de Schauder*, Bull. Soc. Math. France **89** (1961), 223–226. MR **26** #769.

168. R. De Marr, *Common fixed points for commuting contraction mappings*, Pacific J. Math. **13** (1963), 1139–1141. MR **28** #2446.

169. J. B. Diaz and F. T. Metcalf, *On the structure of the set of sub-sequential limits points of successive approximations*, Bull. Amer. Math. Soc. **73** (1967), 516–519. MR **35** #2268.

170. A. Dold, *Partitions of unity in the theory of fibration*, Ann. of Math. (2) **78** (1963), 223–255. MR **27** #5264.

171. ———, *Fixed point index and fixed point theorem for Euclidean neighborhood retracts*, Topology **4** (1965), 1–8. MR **33** #1850.

172. C. L. Dolph, *Nonlinear integral equations of the Hammerstein type*, Trans. Amer. Math. Soc. **66** (1949), 289–307. MR **11**, 367.

173. C. L. Dolph and G. J. Minty, *On nonlinear integral equations of the Hammerstein type*, Nonlinear Integral Equations (Proc. Advanced Seminar Conducted by Math. Res. Center, U.S. Army, Univ. of Wisconsin, Madison, Wis., 1963), Univ. of Wisconsin Press, Madison, Wis., 1964, pp. 99–154. MR **28** #4322.

174. J. R. Dorroh, *Some classes of semigroups of nonlinear transformations and their generators*, J. Math. Soc. Japan **20** (1968), 437–455. MR **37** #6796.

175. A. Douady, *Le problème des modules pour les sousespaces analytiques compacts d'un espace analytique donné*, Ann. Inst Fourier (Grenoble) **16** (1966), fasc 1, 1–95. MR **34** #2940.

176. A. S. Downs and D. E. Edmunds, *Sur les opérateurs nonlinéaires monotones*, C.R. Acad. Sci. Paris **261** (1965), 1157–1159. MR **32** #2942.

177. Ju. A. Dubinskiĭ, *Some integral inequalities and the solvability of degenerate quasilinear elliptic systems of differential equations*, Mat. Sb. **64** (**106**) (1964), 458–480; English transl., Amer. Math. Soc. Transl. (2) **53** (1966), 167–191. MR **29** #6160.

178. ———, *Weak convergence in nonlinear elliptic and parabolic equations*, Mat. Sb. **67** (**109**) (1965), 609–642; English transl., Amer. Math. Soc. Transl. (2) **67** (1968), 226–258. MR **32** #7958.

179. ———, *Nonlinear parabolic equations on the plane*, Mat. Sb. **69** (**111**) (1966), 470–496; English transl., Amer. Math. Soc. Transl. (2) **67** (1968), 259–288. MR **33** #4475.

180. ———, *Quasilinear elliptic and parabolic equations of arbitrary order*, Uspehi Mat. Nauk **23** (1968), no. 1 (139), 45–90 = Russian Math. Surveys **23** (1968), no. 1, 45–91. MR **37** #4405.

181. Ju. A. Dubinskiĭ and S. I. Pohožaev, *On a certain class of operators and the solvability of quasilinear elliptic equations*, Mat. Sb. **72** (**114**) (1967), 226–236 = Math. USSR Sb. **1** (1967), 199–208. MR **34** #6569.

182. J. Dugundji, *An extension of Tietze's theorem*, Pacific J. Math. **1** (1951), 353–367. MR **13**, 373.

183. N. Dunford and J. T. Schwartz, *Linear operators.* I: *General theory*, Pure and Appl. Math., vol. 7, Interscience, New York, 1958. MR **22** #8302.

184. C. J. Earle and J. Eells, *On the differential geometry of Teichmüller spaces*, J. Analyse Math. **19** (1967), 35–52. MR **36** #3975.

185. ———, *Foliations and fibrations*, J. Differential Geometry **1** (1967), no. 1, 33–41. MR **35** #6161.

186. C. J. Earle and R. Hamilton, *A fixed point theorem for holomorphic mappings*, Proc. Sympos. Pure Math., vol. 16, Amer. Math. Soc., Providence, R.I., 1970, pp. 61–65. MR **42** #918.

187. M. Edelstein, *An extension of Banach's contraction principle*, Proc. Amer. Math. Soc. **12** (1961), 7–10. MR **22** #11375.

188. ———, *On fixed and periodic points under contractive mappings*, J. London Math. Soc. **37** (1962), 74–79. MR **24** #A2936.

189. ———, *On predominantly contractive mappings*, J. London Math. Soc. **38** (1963), 81–86. MR **26** #6726.

190. ———, *On nonexpansive mappings of Banach spaces*, Proc. Cambridge Philos. Soc. **60** (1964), 439–447. MR **29** #1521.

191. ———, *A remark on a theorem of M. A. Krasnoselski*, Amer. Math. Monthly **73** (1966), 509–510. MR **33** #3072.

192. M. Edelstein and A. C. Thompson, *Contractions, isometrics, and some properties of inner-product spaces*, Nederl. Akad. Wetensch. Proc. Ser. A **70** = Indag. Math. **29** (1967), 326–331. MR **36** #5672.

193. D. E. Edmunds, *Remarks on nonlinear functional equations*, Math. Ann. **174** (1967), 233–239. MR **36** #3180.

194. ———, *Nonlinear functional equations in locally convex spaces*, J. London Math. Soc. **41** (1966), 750–754. MR **34** #643.

195. J. Eells, *A setting for global analysis*, Bull. Amer. Math. Soc. **72** (1966), 751–807. MR **34** #3590.

196. ———, *Alexander-Pontryagin duality in function spaces*, Proc. Sympos. Pure Math., vol. 3, Amer. Math. Soc., Providence, R.I., 1961, pp. 109–129. MR **23** #A2879.

197. ———, *Fibring spaces of maps*, Sympos. on Infinite Dimensional Topology, Ann. of Math. Studies, no. 69, Princeton Univ. Press, Princeton, N.J., 1972, pp. 43–58.

198. ———, *Fredholm structures*, Proc. Sympos. Pure Math., vol. 18, part 1, Amer. Math. Soc., Providence, R.I., 1970, pp. 62–85. MR **45** #2753.

199. J. Eells and J. H. Sampson, *Harmonic mappings of Riemannian manifolds*, Amer. J. Math. **86** (1964), 109–160. MR **29** #1603.

200. S. Eilenberg, *Sur quelques propriétés des transformations localement homeomorphes*, Fund. Math. **24** (1935), 35–42.

201. K. D. Elworthy, *Fredholm maps and $GL_c(E)$-structures*, Bull. Amer. Math. Soc. **74** (1968), 582–586. MR **36** #7160.

203. K. D. Elworthy and A. J. Tromba, *Degree theory on Banach manifolds*, Proc. Sympos. Pure Math., vol. 18, part 1, Amer. Math. Soc., Providence, R.I., 1970, pp. 86–94. MR **43** #2746.

204. K. Fan, *Fixed point and minimax theorems in locally convex linear spaces*, Proc. Nat. Acad. Sci. U.S.A. **38** (1964), 121–126.

205. ———, *Existence theorems and extreme solutions for inequalities concerning convex functions or linear transformations*, Math. Z. **68** (1957), 205–216. MR **19**, 1183.

206. ———, *Invariant cross sections and invariant linear subspaces*, Israel J. Math. **2** (1964), 19–26. MR **30** #1382.

207. ———, *A generalization of Tychonoff's fixed point theorem*, Math. Ann. **142** (1960/61), 305–310. MR **24** #A1120.

208. ———, *Sur un théorème minimax*, C.R. Acad. Sci. Paris **259** (1964), 3925–3928. MR **30** #5145.

209. ———, *Applications of a theorem concerning sets with convex sections*, Math. Ann. **163** (1966), 189–203. MR **32** #8101.

210. K. Fan and I. Glicksberg, *Some geometric properties of the sphere in a normed linear space*, Duke Math. J. **25** (1958), 553–568. MR **20** #5421.

211. C. Fenske, *Lokales Fixpunktverhalten bei stetigen Abbildungen in kompakten konvexen Mengen*, Forschungsberichte des Landes Nordrhein-Westfalen, no. 1931, Westdeutscher Verlag, Cologne, 1968, pp. 17–52. MR **38** #5081.

212. G. Fichera, *Problemi elastostatici com vincoli unilatori: il problema di signorini con ambigue condizioni al contorno*, Atti Accad. Naz. Lincei Mem. Cl. Sci. Fis. Mat. Natur. Sez. I (8) **7** (1963/64), 91–140. MR **31** #2888.

213. ———, *Electrostatics problems with unilateral constraints*, Séminaire sur les équations aux dérivées partielles, Collège de France, 1966/67.

214. F. A. Ficken, *The continuation method for functional equations*, Comm. Pure Appl. Math. **4** (1951), 435–456. MR **13**, 562.

215. D. G. de Figueiredo, *Fixed point theorems for weakly continuous mappings*, Report #638, Mathematics Research Center, Madison, Wis., 1966.

216. ———, *Fixed point theorems for nonlinear operators and Galerkin approximations*, J. Differential Equations **3** (1967), 271–281. MR **34** #6578.

217. ———, *Topics on nonlinear functional analysis*, Lecture Notes, Inst. for Fluid Dynamics and Appl. Math., University of Maryland, College Park, Md., 1967.

218. ———, *Some remarks on fixed point theorems for nonlinear operators in Banach spaces*, Technical Note BN-489, University of Maryland, College Park, Md., 1967.

219. D. G. de Figueiredo and C. P. Gupta, *Solvability of non-linear integral equations of Hammerstein type* (unpublished).

220. D. G. de Figueiredo and L. A. Karlovitz, *On the radial projection in normed spaces*, Bull. Amer. Math. Soc. **73** (1967), 364–368. MR **35** #2130.

221. ———, *On the extension of contractions in normed spaces*, Proc. Sympos. Pure Math., vol. 18, part 1, Amer. Math. Soc., Providence, R.I., 1970, pp. 95–104. MR **43** #877.

222. R. L. Frum-Ketkov, *Mapping into a Banach space sphere*, Dokl. Akad. Nauk SSSR **175** (1967), 1229–1231 = Soviet Math. Dokl. **8** (1967), 1004–1006. MR **36** #3181.

223. K. Geba, *Algebraic topology methods in the theory of compact fields in Banach spaces*, Fund. Math. **54** (1964), 179–209. MR **28** #5317.

224. K. Geba and A. Granas, *Algebraic topology in linear normed spaces*. I. *Basic categories*, Bull. Acad. Polon. Sci. Sér. Sci. Math. Astronom. Phys. **13** (1965), 287–290. MR **32** #1708.

———, *Algebraic topology in linear normed spaces*. II. *The functor* $\Pi(X, U)$, Bull. Acad. Polon. Sci. Sér. Sci. Math. Astronom. Phys. **13** (1965), 341–345. MR **32** #4694.

———, *Algebraic topology in linear normed spaces*. III. *The cohomology functors* H^{oo-n}, Bull. Acad. Polon. Sci. Sér. Sci. Math. Astronom. Phys. **15** (1967), 137–143. MR **35** #6141.

———, *Algebraic topology in linear normed spaces*. IV. *The Alexander-Pontriagin invariance theorem in E*, Bull. Acad. Polon. Sci. Sér. Sci. Math. Astronom. Phys. **15** (1967), 145–152. MR **35** #6142.

225. M. D. George, *The spectrum of an operator in a Banach space*, Proc. Amer. Math. Soc. **16** (1965), 980–982. MR **32** #4542.

226. I. J. Glicksberg, *A further generalization of the Kakutani fixed point theorem with applications to Nash equilibrium points*, Proc. Amer. Math. Soc. **3** (1952), 170–174. MR **13**, 764.

227. D. Göhde, *Über Fixpunktsatze und die Theorie des Abbildungsgrades in Funktionalräumen*, Math. Nachr. **20** (1959), 356–371. MR **22** #11285.

228. ———, *Über Fixpunkte bei stetigen Selbstabbildungen mit kompakten Iterierten*, Math. Nachr. **28** (1964), 45–55. MR **31** #4020.

229. ———, *Zum Prinzip der kontraktiven Abbildung*, Math. Nachr. **30** (1965), 251–258. MR **32** #8129.

230. A. Granas, *Sur la notion du degré topologique pour une certaine classe de transformations multivalentes dans les espaces de Banach*, Bull. Acad. Polon. Sci. Sér. Sci. Math. Astronom. Phys. **7** (1959), 191–194. MR **21** #7457.

231. ———, *The theory of compact fields and some of its applications to the topology of functional spaces*. I, Rozprawy Mat. **30** (1962), 93 pp. MR **26** #6743.

232. ———, *Introduction à la topologie des espaces de Banach*, Lecture Notes, Institut Poincaré, Paris, 1966.

233. L. M. Graves, *Some mapping theorems*, Duke Math. J. **17** (1950), 111–114. MR **11**, 729.

234. R. C. Gunning and H. Rossi, *Analytic functions of several complex variables*, Prentice-Hall Series in Modern Analysis, Prentice-Hall, Englewood Cliffs, N.J., 1965. MR **31** #4927.

235. K. Gustafson, *Stability inequalities for semimonotonically perturbed nonhomogeneous boundary problems*, SIAM J. Appl. Math. **15** (1967), 368–391. MR **35** #6960.

236. J. Hadamard, *Sur les transformations punctuelles*, Bull. Soc. Math. France **34** (1960), 71–84.

237. B. Halpern, *Fixed points of nonexpanding mappings*, Bull. Amer. Math. Soc. **73** (1967), 957–961. MR **36** #2022.

238. ———, *Fixed points for iterates*, Pacific J. Math. **25** (1968), 255–275. MR **41** #1039.

239. ———, *A general fixed-point theorem*, Proc. Sympos. Pure Math., vol. 18, part 1, Amer. Math. Soc., Providence, R.I., 1970, pp. 114–131. MR **42** #8356.

240. H. Hanani, E. Netanyaku and M. Reichaw-Reichbach, *The sphere in the image*, Isarel J. Math. **1** (1963), 188–195. MR **29** #453.

241. O. Hanner, *Some theorems on absolute neighborhood retracts*, Arch. Mat. **1** (1951), 389–408. MR **13**, 266.

242. P. Hartman, *On stability in the large for systems of ordinary differential equations*, Canad. J. Math. **13** (1961), 480–492. MR **23** #A1113.

243. ———, *Ordinary differential equations*, Wiley, New York, 1964. MR **30** #1270.

244. ———, *Generalized Lyapunov functions and functional equations*, Ann. Mat. Pura Appl. (4) **69** (1965), 305–320. MR **33** #595.

245. ———, *The existence and stability of stationary points*, Duke Math. J. **33** (1966), 281–290. MR **35** #480.

246. ———, *On homotopic harmonic maps*, Canad. J. Math. **19** (1967), 673–687. MR **35** #4856.

247. P. Hartman and L. Nirenberg, *On spherical image maps whose Jacobians do not change sign*, Amer. J. Math. **81** (1959), 901–920. MR **23** #A4106.

248. P. Hartman and G. Stampacchia, *On some nonlinear elliptic functional differential equations*, Acta Math. **115** (1966), 271–310. MR **34** #6355.

249. E. Hille and R. S. Phillips, *Functional analysis and semi-groups*, rev. ed., Amer. Math. Soc. Colloq. Publ., vol. 31, Amer. Math. Soc., Providence, R.I., 1957. MR **19**, 664.

250. S. T. Hu, *Theory of retracts*, Wayne State Univ. Press, Detroit, Mich., 1965. MR **31** #6202.

251. R. C. James, *Reflexivity and the supremum of linear functionals*, Ann. of Math.(2) **66** (1957), 159–169. MR **19**, 755.

252. ———, *Characterizations of reflexivity*, Studia Math. **23** (1963/64), 205–216. MR **30** #431.

253. F. John, *On quasi-isometric mappings*. I, Comm. Pure Appl. Math. **21** (1968), 77–110. MR **36** #5716.

254. G. S. Jones, *Asymptotic fixed point theorems and periodic systems of functional differential equations*, Contributions to Differential Equations **2** (1963), 385–405. MR **28** #1361.

255. ———, *Periodic motions in Banach spaces and applications to functional-differential equations*, Contributions to Differential Equations **3** (1964), 75–106. MR **29** #342.

256. ———, *Stability and asymptotic fixed point theory*, Proc. Nat. Acad. Sci. U.S.A. **53** (1965), 1262–1264. MR **31** #4959.

257. R. I. Kačurovskiĭ, *On monotone operators and convex functionals*, Uspehi Mat. Nauk **15** (1960), 213–215. (Russian)

258. ———, *Monotone nonlinear operators in Banach spaces*, Dokl. Akad. Nauk SSSR **163** (1965), 559–562 = Soviet Math. Dokl. **6** (1965), 953–956. MR **34** #644.

259. ———, *Approximation methods in the solution of nonlinear operator equations*, Izv. Vysš. Učebn. Zaved. Matematika **1967**, no. 12 (67), 27–37. (Russian) MR **37** #2044.

260. ———, *Nonlinear equations with monotonic and other operators*, Dokl. Akad. Nauk SSSR **173** (1967), 515–518 = Soviet Math. Dokl. **8** (1967), 427–430. MR **35** #3498.

261. ———, *Nonlinear monotone operators in Banach spaces*, Uspehi Mat. Nauk **23** (1968), no. 2 (140), 121–168 = Russian Math. Surveys **23** (1968), no. 2, 117–165. MR **37** #2045.

262. M. I. Kadec, *Spaces isomorphic to a locally uniformly convex space*, Izv. Vysš. Učebn. Zaved. Matematika **1959**, no. 6 (13), 51–57. (Russian) MR **23** #A3987.

263. ———, *Topological equivalence of all separable Banach spaces*, Dokl. Akad. Nauk SSSR **167** (1966), 23–25 = Soviet Math. Dokl. **7** (1966), 319–322. MR **34** #1828.

264. S. Kakutani, *A generalization of Brouwer's fixed point theorem*, Duke Math. J. **8** (1941), 457–459. MR **3**, 60.

265. S. Kaniel, *Quasicompact nonlinear operators in Banach space and applications*, Arch. Rational Mech. Anal. **20** (1965), 259–278. MR **32** #4575.

266. L. V. Kantorovič, *The method of successive approximations for functional equations*, Acta Math. **71** (1939), 63–97. MR **1**, 18.

267. T. Kato, *Demicontinuity, hemicontinuity, and monotonicity*, Bull. Amer. Math. Soc. **70** (1964), 548–550. MR **29** #501.

268. ———, *Nonlinear evolution equations in Banach spaces*, Proc. Sympos. Appl. Math., vol. 17, Amer. Math. Soc., Providence, R.I., 1965, pp. 50–67. MR **32** #1573.

269. ———, *Demicontinuity, hemicontinuity, and monotonicity*. II, Bull. Amer. Math. Soc. **73** (1967), 886–889. MR **38** #6411.

270. ———, *Nonlinear semigroups and evolution equations*, J. Math. Soc. Japan **19** (1967), 508–520. MR **37** #1820.

271. ———, *Accretive operators and nonlinear evolution equations in Banach spaces*, Proc. Sympos. Pure Math., vol. 18, part 1, Amer. Math. Soc., Providence, R.I., 1970, pp. 138–161. MR **42** #6663.

272. ———, *Note on the differentiability of nonlinear semigroups*, Proc. Sympos. Pure Math., vol. 16, Amer. Math. Soc., Providence, R.I., 1970, pp. 91–94. MR **42** #5100.

273. B. von Kerekjarto, *Zur Theorie der mehrdeutigen stetigen Abbildungen*, Math. Z. **8** (1920), 310–319.

274. Khoan Vo-Khac, *Q-solutions d'un système différentiel*, C.R. Acad. Sci. Paris **258** (1964), 3430–3433. MR **29** #500.

275. A. V. Kibenko, M. A. Krasnosel'skiĭ and Ja. D. Mamedov, *One-sided estimates and conditions for existence of solutions of differential equations in function spaces*, Azerbaĭdžan. Gos. Univ. Učen. Zap. Ser. Fiz.-Mat. Nauk **1961**, no. 3, 13–19. (Russian) MR **36** #4108.

276. W. A. Kirk, *A fixed point theorem for mappings which do not increase distance*, Amer. Math. Monthly **72** (1965), 1004–1006. MR **32** #6436.

277. ——, *Non expansive mappings and the weak closure of iterates*, Duke Math. J. **36** (1969), 639–645. MR **40** #1831.

278. ——, *Fixed point theorems for non-expansive mappings*, Proc. Sympos. Pure Math., vol. 18, part 1, Amer. Math. Soc., Providence, R.I., 1970, pp. 162–168. MR **42** #6677.

279. M. D. Kirszbraun, *Über die zussamenziehende und Lipschitzsche Transformationen*, Fund. Math. **22** (1934), 77–108.

280. V. Klee, *Convex bodies and periodic homeomorphisms in Hilbert space*, Trans. Amer. Math. Soc. **74** (1953), 10–43. MR **14**, 989.

281. ——, *Some topological properties of convex sets*, Trans. Amer. Math. Soc. **78** (1955), 30–45. MR **16**, 1030.

282. ——, *Leray-Schauder theory without local convexity*, Math. Ann. **141** (1960), 286–296. MR **24** #A1004.

283. K. Klingelhöfer, *Über nichtlineare Randwertaufgaben der Potential-theorie*, Mitt. Math. Sem. Giessen **76** (1967), 1–70. MR **41** #3793.

284. G. Köthe, *Topologische lineare Räume*. I, Die Grundlehren der math. Wissenschaften, Band 107, Springer-Verlag, Berlin, 1960. MR **24** #A411.

285. I. Kolodner, *Equations of Hammerstein type in Hilbert space*, J. Math. Mech. **13** (1964), 701–750. MR **30** #1415.

286. J. Kolomy, *The solvability of non-linear integral equations*, Comment. Math. Univ. Carolinae **8** (1967), 273–289. MR **35** #5878.

287. Y. Kômura, *Nonlinear semigroups in Hilbert space*, J. Math. Soc. Japan **19** (1967), 493–507. MR **35** #7176.

288. ——, *Differentiability of nonlinear semigroups*, J. Math. Soc. Japan **21** (1969), 375–402. MR **40** #3358.

289. M. A. Krasnosel'skiĭ, *On the theory of completely continuous fields*, Ukrain. Mat. Ž. **3** (1951), 174–183. (Russian) MR **14**, 1109.

290. ——, *Topological methods in the theory of nonlinear integral equations*, GITTL, Moscow, 1956; English transl., Macmillan, New York, 1964. MR **20** #3464; **28** #2414.

291. ——, *Two notes on the method of successive approximations*, Uspehi Mat. Nauk **10** (1955), no. 1 (63), 123–127. (Russian) MR **16**, 833.

292. ——, *Positive solutions of operator equations*, Fizmatgiz, Moscow, 1962; English transl., Noordhoff, Groningen, 1964. MR **26** #2862; **31** #6107.

293. ——, *The operator of translation along the trajectories of differential equations*, "Nauka", Moscow, 1966; English transl., Trans. Math. Monographs, vol. 19, Amer. Math. Soc., Providence, R.I., 1968. MR **34** #3012; **36** #6688.

294. M. A. Krasnosel'skiĭ and Ja. D. Mamedov, *Notes on the application of differential and integral inequalities to problems on the correctness of Cauchy's problem for ordinary differential equations in Banach spaces*, Naučn. Dokl. Vysš. Škol. Fiz.-Mat. Nauk **1959**, 32–37. (Russian)

295. M. A. Krasnosel'skiĭ and A. I. Perov, *On the existence of solutions of some nonlinear integral operator equations*, Dokl. Akad. Nauk SSSR **126** (1959), 15–18. (Russian) MR **21** #5153.

296. V. Lakshmikantham, *Differential systems and extensions of Lyapunov's method*, Michigan Math. J. **9** (1962), 311–320. MR **31** #442.

297. M. Landsberg, *Über die Fixpunkte kompakter Abbildungen*, Math. Ann. **154** (1964), 427–431. MR **29** #2629.

298. A. Lasota, *Une généralisation du premier théorème de Fredholm et ses applications à la théorie des équations différentielles ordinaires*, Ann. Polon. Math. **18** (1966), 65–77. MR **33** #2849.

299. A. Lasota and Z. Opial, *On the existence and uniqueness of solutions of nonlinear functional equations*, Bull. Acad. Polon. Sci. Sér. Sci. Math. Astronom. Phys. **15** (1967), 797–800. MR **38** #5078.

300. M. Lees and M. H. Schultz, *A Leray-Schauder principle for A-compact mapping and the numerical solution of non-linear two-point boundary value problems*, Numerical Solutions of Nonlinear Differential Equations (Proc. Adv. Sympos., Madison, Wis., 1966), Wiley, New York, 1966, pp. 167–179. MR **35** #819.

301. S. Lefschetz, *Topics in topology*, Ann. of Math. Studies, no. 10, Princeton Univ. Press, Princeton, N.J., 1942. MR **4**, 86.

302. S. Lefschetz, *Algebraic topology*, Amer. Math. Soc. Colloq. Publ., vol. 27, Amer. Math. Soc., Providence, R.I., 1942. MR **4**, 84.

303. A. Lelek and J. Mycielski, *Some conditions for a mapping to be a covering*, Fund. Math. **49** (1960/61), 295–300. MR **30** #5278.

304. J. Leray, *Topologie des espaces abstraits de M. Banach*, C.R. Acad. Sci. Paris **200** (1935), 1089–1093.

305. ———, *Sur les équations et les transformations*, J. Math. Pures Appl. (9) **24** (1946), 201–248. MR **7**, 468.

306. ———, *La théorie des points fixes et ses applications en analyse*, Proc. Internat. Congress Math. (Cambridge, 1950), vol. 2, Amer. Math. Soc., Providence, R.I., 1952, pp. 202–208. MR **13**, 859.

307. ———, *Théorie des points fixes, indice total, et nombre de Lefschetz*, Bull. Soc. Math. France **87** (1959), 221–233. MR **26** #762.

308. J. Leray and J.-L. Lions, *Quelques résultats de Višik sur les problèmes elliptiques non-linéaires par les méthodes de Minty-Browder*, Bull. Soc. Math. France **93** (1965), 97–107. MR **33** #2939.

309. J. Leray and J. Schauder, *Topologie et équations fonctionnelles*, Ann. Sci. École Norm. Sup. **51** (1934), 45–78.

310. C. Lescarret, *Cas d'addition des applications monotones maximales dans un espace de Hilbert*, C.R. Acad. Sci. Paris **261** (1965), 1160–1163. MR **34** #645.

311. ———, *Sur la sous-différentiabilité d'une somme de fonctionnelles convexes semicontinues inférieurment*, C.R. Acad. Sci. Paris Sér. A-B **262** (1966), A443–A446. MR **33** #4211.

312. P. Levy, *Sur les fonctions des lignes implicites*, Bull. Soc. Math. France **48** (1920), 13–27.

313. D. C. Lewis, *Metric properties of differential equations*, Amer. J. Math. **71** (1949), 294–312. MR **10**, 708.

314. ———, *Differential equations referred to a variable metric*, Amer. J. Math. **73** (1951), 48–58. MR **15**, 873.

315. L. Lichtenstein, *Verlesungen über einige Klassen nichtlinearen Integralgleichungen und Integraodifferentialgleichungen nebst Anwendungen*, Berlin, 1931.

316. J. Lindenstrauss, *On non-separable reflexive Banach spaces*, Bull. Amer. Math. Soc. **72** (1966), 967–970. MR **34** #4875.

317. J.-L. Lions, *Sur certains systèmes hyperboliques nonlinéaires*, C.R. Acad. Sci. Paris **257** (1963), 2057–2060. MR **28** #351.

318. ———, *Sur certaines équations paraboliques non linéaires*, Bull. Soc. Math. France **93** (1965), 155–175. MR **33** #2966.

319. ———, *Remarks on evolution inequalities*, J. Math. Soc. Japan **18** (1966), 331–342. MR **35** #7179.

320. ———, *Sur quelques problèmes de calcul des variations*, Symposia Math. **2** (1968).

321. J.-L. Lions and C. Stampacchia, *Inéquations variationelles non-coercives*, C.R. Acad. Sci. Paris **261** (1965), 25–27. MR **31** #6140.

322. ———, *Variational inequalities*, Comm. Pure Appl. Math. **20** (1967), 493–519. MR **35** #7178.

323. J.-L. Lions and W. A. Strauss, *Some non-linear evolution equations*, Bull. Soc. Math. France **93** (1965), 43–96. MR **33** #7663.

324. L. A. Ljusternik, *The topology of the calculus of variations in the large*, Trudy Mat. Inst. Steklov. **19** (1947); English transl., Transl. Math. Monographs, vol. 16, Amer. Math. Soc., Providence, R.I., 1966. MR **9**, 596; **36** #906.

325. G. G. Lorentz and T. Shimogaki, *Interpolation theorems for operators in function spaces*, J. Functional Analysis **2** (1968), 31–51. MR **41** #2424.

326. A. R. Lovaglia, *Locally uniformly convex Banach spaces*, Trans. Amer. Math. Soc. **78** (1955), 225–238. MR **16**, 596.

327. G. Lumer and R. S. Phillips, *Dissipative operators in Banach spaces*, Pacific J. Math. **11** (1961), 679–698. MR **74** #A2248.

328. A. Lyapunov, *Sur les figures d'equilibre peu différentes des ellipsoides d'une masse liquids homogène donné d'une mouvement de rotation*. I, St. Petersburg, 1906, pp. 1–225.

329. Ja. D. Mamedov, *One-sided estimates in the conditions for existence and uniqueness of solutions of the limit Cauchy problem in Banach spaces*, Sibirsk. Mat. Ž. **6** (1965), 1190–1196. (Russian) MR **32** #7908.

330. Ja. D. Mamedov, *One-sided conditional estimates for investigating the solutions of differential equations of parabolic type in a Banach space*, Azerbaĭdžan. Gos. Univ. Učen. Zap. Ser. Fiz.-Mat. i Him. Nauk **1964**, no. 5, 17–25. (Russian) MR **36** #6750.

331. J. T. Markin, *A fixed point theorem for set valued mappings*, Bull. Amer. Math. Soc. **74** (1968), 639–640. MR **37** #3409.

232. E. Michael, *Continuous selections*. I, Ann. of Math. (2) **63** (1956), 361–382. MR **17**, 990.

333. ———, *A selection theorem*, Proc. Amer. Math. Soc. **17** (1966), 1404–1406. MR **34** #3551.

334. J. W. Milnor, *Topology from the differentiable viewpoint*, Univ. Press of Virginia, Charlottesville, Va., 1965. MR **37** #2239.

335. G. J. Minty, *On the simultaneous solution of a certain system of linear inequalities*, Proc. Amer. Math. Soc. **13** (1962), 11–12. MR **26** #573.

336. ———, *Monotone (nonlinear) operators in a Hilbert space*, Duke Math. J. **29** (1962), 341–346. MR **29** #6319.

337. ———, *On the monotonicity of the gradient of a convex function*, Pacific J. Math. **14** (1964), 243–247. MR **29** #5125a.

338. ———, *On a "monotonicity" method for the solution of nonlinear equations in Banach spaces*, Proc. Nat. Acad. Sci. U.S.A. **50** (1963), 1038–1041. MR **28** #5358.

339. ———, *On the solvability of nonlinear functional equations of monotonic type*, Pacific J. Math. **14** (1964), 249–255. MR **29** #5125b.

340. ———, *A theorem on maximal monotonic sets in Hilbert space*, J. Math. Anal. Appl. **11** (1965), 434–439. MR **33** #6462.

341. ———, *Monotone operators and certain systems of nonlinear ordinary differential equations*, Proc. Sympos. Systems Theory, Brooklyn Polytechnic Institute, 1965, pp. 39–55.

342. ———, *On a generalization of the direct method of the calculus of variations*, Bull. Amer. Math. Soc. **73** (1967), 315–321. MR **35** #3501.

343. C. Miranda, *Problemi di esistenza in analisi funzionale*, Scuola Normale Superiore, Pisa, Quaderni Mat., no. 3, Litografia Tacchi, Pisa, 1950. MR **12**, 265.

344. ———, *Equazioni alle derivate parziali di tipo ellittico*, Ergebnisse der Mathematik und ihrer Grenzgebiete, Heft 2, Springer-Verlag, Berlin, 1955. MR **19**, 421.

345. J. J. Moreau, *Proximité et dualité dans un espace Hilbertien*, Bull. Soc. Math. France **93** (1965), 273–299. MR **34** #1829.

346. ———, *Fonctionelles convexes*, Lecture Notes, Collège de France, 1967.

347. J. Moser, *A rapidly converging iteration method and nonlinear partial differential equations*. I, II, Ann. Scuola Norm. Sup. Pisa (3) **20** (1966), 265–315, 499–535. MR **33** #7667; **34** #6280.

348. U. Mosco, *A remark on a theorem of F. E. Browder*, J. Math. Anal. Appl. **20** (1967), 90–93. MR **36** #3178.

349. ———, *Approximation of the solutions of some variational inequalities*, Ann. Scuola Norm. Sup. Pisa (3) **21** (1967), 373–394. MR **37** #1966.

350. ———, *Convergence of convex sets and of solutions of variational inequalities*, Advances in Math. **3** (1969), 510–585.

351. ———, *Perturbation of variational inequalities*, Proc. Sympos. Pure Math., vol. 18, part 1, Amer. Math. Soc., Providence, R.I., 1970, pp. 182–194. MR **42** #6669.

352. K. K. Mukherjea, *Coincidence theory for infinite dimensional manifolds*, Bull. Amer. Math. Soc. **74** (1968), 493–496. MR **36** #5965.

353. M. Nagumo, *A theory of degree of mapping based on infinitesimal analysis*, Amer. J. Math. **73** (1951), 485–496. MR **13**, 150.

354. ———, *Degree of mapping in convex linear topological spaces*, Amer. J. Math. **73** (1951), 497–511. MR **13**, 150.

355. I. Namioka and E. Asplund, *A geometric proof of Ryll-Nardzewski's fixed point theorem*, Bull. Amer. Math. Soc. **73** (1967), 443–445.

356. J. Nash, *The imbedding problem for Riemannian manifolds*, Ann. of Math. (2) **63** (1956), 20–63. MR **17**, 782.

357. J. W. Neuberger, *An exponential formula for one-parameter semigroups of nonlinear transformations*, J. Math. Soc. Japan **18** (1966), 154–157. MR **34** #622.

358. ———, *Product integral formulae for nonlinear expansive semigroups and non-expansive evolution systems*, J. Math. Mech. **19** (1969/70), 403–409. MR **40** #6301.

359. R. Nevanlinna, *Über die Methode der Sukzessive Approximationen*, Ann. Acad. Fenn. Ser. AI No. 291 (1960). MR **22** #11286.

360. R. D. Nussbaum, *The fixed point index and asymptotic fixed point theorems for k-set contractions*, Bull. Amer. Math. Soc. **75** (1969), 490–495. MR **39** #7589.

361. ———, *Fixed point and mapping theorems for non-linear k-set contraction mappings*, Ph.D. Thesis, University of Chicago, Chicago, Ill., 1969.

362. S. Ôharu, *Note on the representation of semigroups of nonlinear operators*, Proc. Japan Acad. **42** (1966), 1149–1154. MR **36** #3167.

363. ———, *A note on the generation of nonlinear semigroups in locally convex spaces*, Proc. Japan Acad. **43** (1967), 847–851. MR **37** #4670.

364. H. Okamura, *Sur l'unicité de la solution de $dy/dx = f(x, y)$*, Mem. Coll. Sci. Kyoto **17** (1934).

365. C. Olech and A. Plis, *Monotonicity assumptions in uniqueness criteria for differential equations*, Colloq. Math. **18** (1967), 43–58. MR **37** #485.

366. H. Omori, *On the group of diffeomorphisms on a compact manifold*, Proc. Sympos. Pure Math., vol. 15, Amer. Math. Soc., Providence, R.I., 1970, pp. 167–183. MR **42** #6864.

367. Z. Opial, *Weak convergence of the sequence of successive approximations for non-expansive mappings*, Bull. Amer. Math. Soc. **73** (1967), 591–597. MR **35** #2183.

368. ———, *Nonexpansive and monotone mappings in Banach spaces*, Lecture Notes, Division of Applied Mathematics, Brown University, Providence, R.I., 1967.

369. R. S. Palais, *Natural operations on differential forms*, Trans. Amer. Math. Soc. **92** (1959), 125–141. 7R **22** #7140.

370. ———, *Morse theory on Hilbert manifolds*, Topology **2** (1963), 299–340. MR **28** #1633.

371. ———, *Lyusternik-Schnirelman theory on Banach manifolds*, Topology **5** (1966), 115–132. MR **41** #4584.

372. ———, *Homotopy theory of infinite dimensional manifolds*, Topology **5** (1966), 1–16. MR **32** #6455.

373. ———, *Critical point theory and the minimax principle*, Proc. Sympos. Pure Math., vol. 15, Amer. Math. Soc., Providence, R.I., 1970, pp. 185–212. MR **41** #9303.

374. ———, *Foundations of global nonlinear analysis*, Benajmin, New York, 1968. MR **40** #2130.

375. W. V. Petryshyn, *On the extension and the solution of nonlinear operator equations*, Illinois J. Math. **10** (1966), 255–274. MR **34** #8242.

376. ———, *Construction of fixed points of demicompact mappings in Hilbert spaces*, J. Math. Anal. Appl. **14** (1966), 274–284. MR **33** #3147.

377. ———, *On a fixed point theorem for nonlinear p-compact operators in Banach space*, Bull. Amer. Math. Soc. **72** (1966), 329–334. MR **33** #1768.

378. ———, *On nonlinear P-compact operators in Banach space with applications to constructive fixed-point theorems*, J. Math. Anal. Appl. **15** (1966), 228–242. MR **34** #1890.

379. ———, *Further remarks on nonlinear P-compact operators in Banach space*, Proc. Nat. Acad. Sci. U.S.A. **55** (1966), 684–687. MR **33** #3148.

380. ———, *Further remarks on nonlinear P-compact operators in Banach space*, J. Math. Anal. Appl. **16** (1966), 243–253. MR **33** #6458.

381. ———, *Projection methods in nonlinear numerical functional analysis*, J. Math. Mech. **17** (1967), 353–372. MR **36** #2025.

382. ———, *Remarks on the approximation-solvability of nonlinear functional equations*, Arch. Rational Mech. Anal. **26** (1967), 43–49. MR **36** #3186.

383. ———, *On the approximation solvability of nonlinear equations*, Math. Ann. **177** (1968), 156–164. MR **37** #2048.

384. ———, *Fixed point theorems involving P-compact, semicontractive and accretive operators not defined on all of a Banach space*, J. Math. Anal. Appl. **23** (1968), 336–354. MR **38** #588.

385. ———, *On the projectional solvability and the Fredholm alternative for equations involving linear A-proper operators*, Arch. Rational Mech. Anal. **30** (1968), 270–284. MR **37** #6776.

386. ———, *Nonlinear equations involving non-compact operators*, Proc. Sympos. Pure Math., vol. 18, part 1, Amer. Math. Soc., Providence, R.I., 1970, pp. 206–233. MR **42** #6670.

387. W. V. Petryshyn and T. S. Tucker, *On functional equations involving nonlinear generalized P-compact operators*, Trans. Amer. Math. Soc. **135** (1969), 343–373. MR **40** #804.

388. S. I. Pohožaev, *The solvability of nonlinear equations with odd operators*, Funkcional. Anal. i Priložen. **1** (1967), no. 3, 66–73. (Russian) MR **36** #4396.

389. E. Rakotch, *A note on contractive mappings*, Proc. Amer. Math. Soc. **13** (1962), 459–465. MR **26** #5555.

390. R. T. Rockafellar, *Characterization of the subdifferentials of convex functions*, Pacific J. Math. **17** (1966), 497–510. MR **33** #1769.

391. ——, *On the virtual convexity of the domain and the range of a nonlinear maximal monotone operator*, Math. Ann. **185** (1970), 81–90. MR **41** #4330.

392. ——, *Local boundedness of nonlinear maximal monotone operators*, Michigan Math. J. **16** (1969), 397–407. MR **40** #6229.

393. ——, *On the maximality of sums of nonlinear monotone operators*, Trans. Amer. Math. Soc. **149** (1970), 75–88. MR **43** #7984.

394. ——, *On the maximal monotonicity of subdifferential operators*, Pacific J. Math. **33** (1970), 209–216. MR **41** #7432.

395. ——, *Convex functions, monotone operators and variational inequalities*, Theory and Applications of Monotone Operators (Proc. NATO Advanced Study Inst., Venice, 1968), Edizioni Oderisi, Gubbio, 1969, pp. 35–65. MR **41** #6028.

396. ——, *Monotone operators associated with saddle-functions and minimax problems*, Proc. Sympos. Pure Math., vol. 18, part 1, Amer. Math. Soc., Providence, R.I., 1970, pp. 241–250. MR **44** #3159.

397. E. H. Rothe, *Zur Theorie des topologischen Ordnung und der Vektorfelder in Banachschen Raumen*, Compositio Math. **1** (1937), 177–197.

398. ——, *Theory of topological order in some linear topological spaces*, Iowa State J. Sci. **13** (1936), 373–390.

399. ——, *Topological proofs of uniqueness theorems in the theory of differential and integral equations*, Bull. Amer. Math. Soc. **45** (1939), 606–613. MR **1**, 18.

400. ——, *On non-negative functional transformations*, Amer. J. Math. **66** (1944), 245–254. MR **6**, 71.

401. ——, *Critical points and gradient fields of scalars in Hilbert space*, Acta Math. **85** (1951), 73–98. MR **13**, 254.

402. ——, *Critical point theory in Hilbert space under general boundary conditions*, J. Math. Anal. Appl. **11** (1965), 357–409. MR **32** #8361.

403. ——, *Gradient mappings*, Bull. Amer. Math. Soc. **59** (1953), 5–19. MR **14**, 657.

404. ——, *A note on the Banach spaces of Calkin and Morrey*, Pacific J. Math. **3** (1953), 493–499. MR **15**, 39.

405. ——, *Remarks on the application of gradient mappings to the calculus of variations and the connected boundary value problems in partial differential equations*, Comm. Pure Appl. Math. **9** (1956), 551–568. MR **18**, 808.

406. ——, *Some remarks on vector fields in Hilbert space*, Proc. Sympos. Pure Math., vol. 18, part 1, Amer. Math. Soc., Providence, R.I., 1970, pp. 251–269.

407. B. N. Sadovskiĭ, *On a fixed point principle*, Funkcional. Anal. i Priložen. **1** (1967), no. 2, 74–76. (Russian) MR **35** #2184.

408. H. Schaefer, *Zur Theorie nichtlinearen Integralgleichungen*, Math. Nachr. **11** (1954), 193–211. MR **16**, 487.

409. ——, *Über die Methode der sukzessiven Approximationen*, Jber. Deutsch. Math. Verein. **59** (1957), Abt. 1, 131–140. MR **18**, 811.

410. ——, *On nonlinear positive operators*, Pacific J. Math. **9** (1959), 847–860. MR **22** #1827.

411. ——, *Über die Methode der a priori-Schranken*, Math. Ann. **129** (1955), 415–416. MR **17**, 175.

412. J. Schauder, *Zur Theorie stetiger Abbildungen in Funktionalraumen*, Math. Z. **26** (1927), 47–65, 417–431.

413. ——, *Der Fixpunktsatz in Funktionalraumen*, Studia Math. **2** (1930), 171–180.

414. ——, *Invarianz des Gebietes in Funktionalraumen*, Studia Math. **1** (1929), 123–139.

415. ——, *Über den Zusammenhang zwischen der Eindeutigkeit und Losbarkeit partiellen Differentialgleichungen zweiter ordnung vom elliptischen Typus*, Math. Ann. **106** (1932), 661–721.

416. E. Schmidt, *Zur Theorie der linearen und nichtlinearen Integralgleichungen*, Math. Ann. **65** (1907/08), 370–399.

417. J. Schroder, *Das Iterationsverfahren bei allgemeneiren Abstandsbegriff*, Math. Z. **66** (1956), 111–116. MR **18**, 765.

418. ———, *Anwendungen von Fixpunktsätzen bei der numerischen Behandlung nichtlinearen Gleichungen in halbgeordeneton Räumen*, Arch. Rational Mech. Anal. **4** (1959), 177–192. MR **22** #319.

419. J. T. Schwartz, *On Nash's implicit function theorem*, Comm. Pure Appl. Math. **13** (1960), 509–530. MR **22** #4971.

420. ———, *Compact analytic mappings of B-space and a theorem of J. Cronin*, Comm. Pure Appl. Math. **16** (1963), 253–260. MR **29** #481.

421. ———, *Generalizing the Lusternik-Schnirelman theory of critical points*, Comm. Pure Appl. Math. **17** (1964), 307–315. MR **29** #4069.

422. ———, *Intersection-theoretic principles for the existence of critical points and fixed points*, Lecture Series on Differential Equations vol. 1 (1969) van Nostrand, New York, pp. 123–146. MR **46** #906.

423. ———, *Nonlinear functional analysis*, Lecture Notes, New York University, 1964.

424. M. Shinbrot, *A fixed point theorem, and some applications*, Arch. Rational Mech. Anal. **17** (1964), 255–271. MR **29** #6323.

425. S. Smale, *Morse theory and a non-linear generalization of the Dirichlet problem*, Ann. of Math. (2) **80** (1964), 382–396. MR **29** #2820.

426. ———, *An infinite dimensional version of Sard's theorem*, Amer. J. Math. **87** (1965), 861–866. MR **32** #3067.

427. V. L. Smulian, *Sur la dérivabilité de la norme dans l'espace de Banach*, C.R. (Dokl.) Acad. Sci. URSS **27** (1940), 643–648. MR **2**, 102.

428. G. Stampacchia, *Formes bilinéaires coercitives sur les ensembles convexes*, C.R. Acad. Sci. Paris **258** (1964), 4413–4416. MR **29** #3864.

429. ———, *Regularity of solutions of some variational inequalities*, Proc. Sympos. Pure Math., vol. 18, part 1, Amer. Math. Soc., Providence, R.I., 1970, pp. 271–281. MR **43** #2581.

430. ———, *Équations elliptiques du second ordre à coefficients discontinus*, Séminaire de Mathématiques Supérieures, no. 16 (Été, 1965), Les Presses de l'Université de Montréal, Montréal, Qué., 1966. MR **40** #4603.

431. N. E. Steenrod, *The topology of fibre bundles*, Princeton Math. Series, vol. 14, Princeton Univ. Press, Princeton, N.J., 1951. MR **12**, 522.

432. S. Sternberg and R. Swan, *On maps with non-negative Jacobians*, Michigan Math. J. **6** (1959), 339–342. MR **22** #1916.

433. S. Stoilow, *Sur les transformations continus des espaces topologiques*, Bull. Math. Roumaine Sci. **35** (1934), 229–235.

434. W. A. Strauss, *Evolution equations nonlinear in the time derivative*, J. Math. Mech. **15** (1966), 49–82. MR **32** #8217.

435. ———, *The initial value problem for certain nonlinear evolution equations*, Amer. J. Math. **89** (1967), 249–259. MR **36** #553.

436. ———, *On the solution of abstract nonlinear equations*, Proc. Amer. Math. Soc. **18** (1967), 116–119. MR **36** #746.

437. ——— *Further applications of monotone methods to partial differential equations*, Proc. Sympos. Pure Math., vol. 18, part 1, Amer. Math. Soc., Providence R.I., 1970, pp. 282–288. MR **42** #8113.

438. J. G. Taylor, *Topological degree of non-compact mappings*, Proc. Cambridge Philos. Soc. **63** (1967), 335–347. MR **36** #863.

439. R. B. Thompson, *A unified approach to local and global fixed point indices*, Advances in Math. **3** (1969), 1–71. MR **40** #891.

440. ———, *A metatheorem for fixed point theories*, Comment. Math. Univ. Carolinae **11** (1970).

441. ———, *On the semicomplexes of F. Browder*, Bull. Amer. Math. Soc. **73** (1967), 531–536. MR **35** #4902.

442. ———, *Retracts of semicomplexes*, Illinois J. Math. **15** (1971), 258–272. MR **43** #4028.

443. A. J. Tromba, *Degree theory on Banach manifolds*, Ph.D. Thesis, Princeton University, Princeton, N.J., 1968.

444. A. Tychonoff, *Ein Fixpunktsatz*, Math. Ann. **111** (1935), 767–776.

445. M. M. Vainberg, *Variational methods for the study of nonlinear operators*, GITTL, Moscow, 1956; English transl., Holden-Day, San Francisco, Calif., 1964. MR **19**, 567; **31** #638.

446. M. M. Vaĭnberg, *On a new principle in the theory of nonlinear equations*, Uspehi Mat. Nauk **15** (1960), 243–244. (Russian)

447. ———, *On the convergence of the method of steepest descent for nonlinear equations*, Dokl. Akad. Nauk SSSR **130** (1960), 9–12 = Soviet Math. Dokl. **1** (1960), 1–4. MR **25** #751.

448. ———, *On the convergence of the process of steepest descent for nonlinear equations*, Sibirsk. Mat. Ž. **2** (1961), 201–220. (Russian) MR **23** #A4026.

449. M. M. Vaĭnberg and P. G. Aĭzengendler, *Methods of development in the bifurcation theory of solutions*, Progress in Math., vol. 2, Plenum Press, New York, 1968.

450. M. M. Vaĭnberg and R. I. Kačurovskiĭ, *On the variational theory of nonlinear operator equations*, Dokl. Akad. Nauk SSSR **129** (1959), 1199–1202. (Russian) MR **22** #4930; **30**, 1201.

451. M. M. Vaĭnberg and I. V. Šragin, *Nonlinear operators and Hammerstein's equations in Orlicz spaces*, Dokl. Akad. Nauk SSSR **128** (1959), 9–12. (Russian) MR **21** #7414.

452. M. I. Višik, *Solutions of boundary value problems for quasi-linear parabolic equations of arbitrary order*, Mat. Sb. **59** (**101**) (1962), 289–325; English transl., Amer. Math. Soc. Transl. (2) **65** (1967), 1–40. MR **28** #361.

453. ———, *Quasilinear strongly elliptic systems of differential equations in divergence form*, Trudy Moskov. Mat. Obšč. **12** (1963), 125–184 = Trans. Moscow Math. Soc. **1963**, 140–208. MR **27** #6017.

454. T. Wazewski, *Sur la convergence d'approximations successives pour les équations différentielles ordinaires au cas de l'espace de Banach*, Ann. Polon. Math. **16** (1965), 231–235. MR **30** #3285.

455. J. P. Williams, *Spectra of products and numerical ranges*, J. Math. Anal. Appl. **17** (1967), 214–220. MR **34** #3341.

456. S. P. Williams, *A connection between the Cesari and Leray-Schauder methods*, Michigan Math. J. **15** (1968), 441–448.

457. A. Wouk, *Direct iteration, existence, and uniqueness*, Nonlinear Integral Equations, Madison, Wis., 1964, pp. 3–34.

458. S. Yamamuro, *Some fixed point theorems in locally convex linear spaces*, Yokohama Math. J. **11** (1963), 5–12. MR **29** #5095.

459. T. Yoshizawa, *Stability theory by Liapunov's second method*, Publ. Math. Soc. Japan, no. 9, Math. Soc. Japan, Tokyo, 1966. MR **34** #7896.

460. K. Yosida, *Functional analysis*, Die Grundlehren der math. Wissenschaften, Band 123, Academic Press, New York; Springer-Verlag, Berlin, 1965. MR **31** #5054.

461. P. P. Zabreĭko, R. I. Kačurovskiĭ and M. A. Krasnosel'skiĭ, *On a fixed point principle for operators in Hilbert space*, Funkcional. Anal. i Priložen. **1** (1967), no. 2, 93–94. (Russian) MR **35** #3505.

462. E. H. Zarantenello, *Solving functional equations by contractive averaging*, Math. Research Center Report #160, Madison, Wis., 1960.

463. ———, *The closure of the numerical range contains the spectrum*, Bull. Amer. Math. Soc. **70** (1964), 781–787. MR **30** #3389.

464. ———, *The closure of the numerical range contains the spectrum*, Pacific J. Math. **22** (1967), 575–595. MR **37** #4657.

AUTHOR INDEX

Roman numbers refer to pages on which a reference is made to an author or work of an author.

Italic numbers refer to pages on which a complete reference to a work by the author is given.

Abraham, R., *285*
Aĭzengendler, P. G., *285, 301*
Alexandroff, P., *285*
Altman, M., *285*
Amann, H., *285*
Ambrosetti, A., *285*
Aronszjan, N., *285*
Ascoli, G., 8
Asplund, E., 94, 281, *285, 297*
Aubin, J. P., *285*

Banach, S., 1, 5, 20, 101, *285*
Bardos, C., *285*
Bartle, R. G., *285*
Belluce, L. P., *285*
Berge, C., *286*
Berger, M. S., *286*
Bergman, G. M., *286*
Bernstein, S., 1
Bessaga, C., *286*
Beurling, A., *286, 287*
Birkhoff, G. D., *286*
Bishop, E., *286*
Bohnenblust, H. F., *286*
Bonic, R., *286*
Bensall, F. F., *286*
Borsuk, K., 269, 276, *286*
Bourgin, D. G., *286*
Bourbaki, N., *286*
Brauer, F., *286*
Brègman, L. M., *286*
Brézis, H., *285, 286*
Brodskiĭ, M. S., *286*
Brøndsted, A., *287*
Brooks, R. B. S., *287*
Brouwer, L. E. J., 183, *287*
Browder, F. E., 5, 212, 277, *287, 288, 289*
Brown, R. F., *289*
Brunk, H. D., *289*

Cacciopoli, R., 2, 5, *289, 290*
Carathéodory, C., *290*

Cartan, H., *290*
Cesari, L., *290*
Cheney, W., *290*
Cotsaftis, M., *290*
Crandall, M., *290*
Cronin, J., *290*
Cudia, D. F., *290*

Darbo, G., *290*
Day, M. M., 27, *290*
Debrunner, H., *290*
Deleanu, A., *290, 291*
De Marr, R., *291*
Diaz, J. B., *291*
Dold, A., *291*
Dolph, C. L., *291*
Dorroh, J. R., *291*
Douady, A., *291*
Downs, A. S., *291*
Dubinskiĭ, Ju. A., *291*
Dugundji, J., *291*
Dunford, N., *291*

Earle, C. J., *291*
Edelstein, M., *291*
Edmunds, D. E., *291, 292*
Eells, J., *291, 292*
Eilenberg, S., *292*
Elworthy, K. D., *292*

Fan, K., *292*
Fenske, C., *292*
Fichera, G., *292*
Ficken, F. A., *292*
de Figueiredo, D. J., *289, 292*
Flor, P., *290*
Frampton, J., *286*
Frum-Ketkov, R. L., *293*

Geba, K., *293*
George, M. D., *293*
Glicksberg, I. J., *292, 293*

303

Göhde, D., *293*
Goldstein, A. A., *290*
Granas, A., *293*
Graves, L. M., *285, 293*
Gunning, R. C., 258, *293*
Gupta, C. P., *289, 292*
Gustafson, K., *293*

Hadamard, J., *293*
Halpern, B. R., *286, 293*
Hamilton, R., *291*
Hanani, H., *293*
Hanner, O., *293*
Hartman, P., *293, 294*
Hille, E., 145, *294*
Hopf, H., *285*
Hu, S. T., *294*

James, R. C., *294*
John, F., *294*
Jones, G. S., *294*

Kačurovkiĭ, R. I., *294, 301*
Kadec, M. I., 89, 281, *294*
Kakutani, S., *294*
Kaniel, S., *294*
Kantorovič, L. V., *294*
Karlin, S., *286*
Karlowitz, L. A., *292*
Kato, T., *294*
Kellogg, O. D., *286*
Kelly, A., *285*
von Kerekjarto, B., *294*
Khoan Vo-Khac, *294*
Kibenko, A. V., *294*
Kirk, W. A., *285, 295*
Kirszbraun, M. D., *295*
Klee, V., *295*
Klingelhöfer, K., *295*
Kolodner, I., *295*
Kolomy, J., *295*
Kômura, Y., *295*
Köthe, G., *295*
Krasnosel'kiĭ, M. A., *294, 295, 301*

Lakshmikantham, V., *295*
Landsberg, M., *295*
Lasota, A., *295*
Lees, M., *295*
Lefschetz, S., 270, *295, 296*
Lelek, A., *296*
Leray, J., 2, 4, 183, *296*

Lescarret, C., *296*
Levy, P., *296*
Lewis, D. C., *296*
Lichtenstein, L., 1, *296*
Lindenstrauss, J., *296*
Lions, J.-L., *286, 296*
Livingston, A. E., *286, 287*
Ljusternik, L. A., *296*
Lorentz, G. G., *296*
Lovaglia, A. R., *296*
Lumer, G., *296*
Lyapunov, A., 1, *296*

Mamedov, Ja. D., *294, 295, 296, 297*
Markin, J. T., *297*
Mazur, S., *285*
Metcalf, F. T., *291*
Michael, E., *297*
Mil'man, D. P., *286*
Milnor, J. W., 249, *297*
Minty, G. J., *291, 297*
Miranda, C., *297*
Moreau, J. J., *297*
Moser, J., *297*
Mosco, U., *297*
Mukherjea, K. K., *297*
Mycielski, J., *296*

Nagumo, M., 183, *297*
Namioka, I., *297*
Nash, J., *297*
Netanyaku, E., *293*
Neuberger, J. W., *297*
Nevanlinna, R., *298*
Nirenberg, L., *293*
Nussbaum, R. D., *289, 298*

Ôharu, S., *298*
Okamura, H., *298*
Olech, C., *298*
Omori, H., *298*
Opial, Z., *295, 298*

Palais, R. S., *298*
Pazy, A., *290*
Perov, A. I., *295*
Petryshyn, W. V., 277, *289, 298*
Phelps, R. R., *286*
Phillips, R. S., *294, 296*
Picard, E., 1, 5, 101
Plis, A., *298*
Pohožaev, S. I., *291, 299*

Rademacher, H., *290*
Rakotch, E., *299*
Reichaw-Reichbach, M., *293*
Robbin, J., *285*
Rockafellar, R. T., *285, 287, 299*
Rossi, H., 258, *293*
Rothe, E. H., *299*

Sadovskiĭ, B. N., *299*
Sampson, J. H., *292*
Schaefer, H., *299*
Schauder, J., 2, 4, 72, 183, *296, 299*
Schmidt, E., 1, *299*
Schroder, J., *300*
Schultz, M. H., *295*
Schwartz, J. T., *291, 300*
Shimogaki, T., *296*
Shinbrot, M., *300*
Sibony, M., *286*
Smale, S., 249, *300*
Smulian, V. L., 44, 130, *300*
Šragin, I. V., *301*
Stampacchia, G., *286, 294, 296, 300*
Steenrod, N. E., 56, 246, *300*
Steiner, E. F., *285*
Sternberg, S., *286, 300*
Stoilow, S., *300*

Strauss, W. A., *296, 300*
Swan, R., *300*
Taylor, J. G., *300*
Thompson, A. C., *291*
Thompson, R. B., *300*
Ton, B. A., *289*
Tromba, A. J., *292, 300*
Tucker, T. S., *298*
Tychonoff, A., 72, *300*

Ulam, S., 276

Vaĭnberg, M. M., *285, 300, 301*
Višik, M. I., *301*

Wazewski, T., *301*
Williams, J. P., *301*
Williams, S. P., *301*
Wouk, A., *301*

Yamamuro, S., *301*
Yoshizawa, T., *301*
Yosida, K., 145, *301*

Zabreĭko, P. P., *301*
Zarantenello, E. H., *301*

SUBJECT INDEX

accretive, 3, 121
A-proper, 273
asymptotic fixed point theorems, 265

coaccretive, 29
compact perturbations, 220
complex analytic mappings, 248
condensing mapping, 225
continuous curve in the space of Lipschitzian mappings, 24
contractions, strict, 3
contractive mappings, 5
core of mapping, 265
covering mapping, 48
covering space methods, 47

degree,
 generalized, 185
 generalized topological, 4
 topological, 183
 Leray-Schauder, 183
duality mapping J from X to X^*, 28

equation of evolution, 1
 nonlinear, 121

fixed point theory,
 asymptotic fixed point theorems, 265
 for compact multi-valued mappings, 71
 Lefschetz fixed point theorem, 269
 local fixed point index, 211
Fredholm mappping, 241
 nonlinear, 242

Gateaux derivative, 21
generalizations of topological degree of a mapping, 183
generalized degree,
 first definition of, 185
generalized topological degree, 4
global Φ-system, 28
globally Φ-accretive, 28

homeomorphism,
 local, 48
 permissible, 184
hypermaximal accretive, 29
hypermaximal coaccretive, 29

hypermaximal Φ-accretive, 29
hypermaximal Φ-coaccretive, 29
hypermaximal Φ-monotone, 36

interpolation,
 nonlinear, 176
intertwined representation for f with respect to a given class M of permissible homeomorphisms, 196
invertible, 63

Lefschetz fixed point theorem, 269
Leray-Schauder degree, 183
limits of invertible mappings, 63
limits of semi-invertible mappings, 63
Lipschitzian mapping, 3, 23
 continuous curve, 24
Lipschitz norm, 23
local fixed point index for compact mappings of metric absolute neighborhoods, 211
local homeomorphism, 48
locally Lipschitzian mappings, 20, 23

mapping(s),
 accretive, 3, 121
 A-proper, 273
 coaccretive, 29
 compact perturbations of, 220
 complex analytic, 248
 condensing, 225
 contractive, 5
 core of, 265
 covering, 48
 duality, 28
 Fredholm, 241
 hypermaximal accretive, 29
 hypermaximal coaccretive, 29
 hypermaximal Φ-accretive, 29
 hypermaximal Φ-coaccretive, 29
 hypermaximal Φ-monotone, 36
 intertwined representation, 196
 invertible, 63
 limits of, 63
 Lipschitzian, 3, 23
 Lipschitz norm of, 23
 local fixed point index, 211
 locally Lipschitzian, 20, 23
 maximal accretive, 29

maximal coaccretive, 29
maximal monotone, 79
maximal Φ-accretive, 29
maximal Φ-coaccretive, 29
maximal Φ-monotone, 36
monotone, 2, 79
nonexpansive, 101
of type (S), 277
of type (S_+), 279
open, 24
orientation-preserving, 248, 249
Φ-accretive, 4, 28
Φ-coaccretive, 28
pseudo-contractive, 108
pseudo-monotone, 89, 280
semiaccretive, 237
semicontractive, 224
semi-invertible, 63
strictly orientation-preserving, 249
strongly pseudo-contractive, 238
strongly semiaccretive, 234
strongly semicontractive, 223
strongly semimonotone, 241
subgradient, 97
topological degree of, 183
transition, 121
upper-semicontinuous set-valued, 38
weakly semiaccretive, 236
weakly semicontractive, 221
weakly semimonotone, 241
mapping theory for compact multi-valued mappings, 71
maximal accretive, 29
maximal coaccretive, 29
maximal monotone from G to 2^{X^*}, 79
maximal Φ-accretive, 29
maximal Φ-coaccretive, 29
maximal Φ-monotone, 36
monotone mappings in Banach spaces, 2, 79

nonexpansive mappings in Banach spaces, 3, 101
nonlinear equations of evolution, 121
nonlinear Fredholm mappings, 242
nonlinear interpolation, 176

open mapping of X into Y, 24
orientation-preserving mappings, 248, 249
strictly, 249

oriented approximation scheme for mappings, 273

permissible homeomorphism, 184
Φ-accretive, 4, 28
globally, 28
hypermaximal, 29
maximal, 29
with respect to the given Φ-system, 28
Φ-coaccretive, 28
hypermaximal, 29
maximal, 29
Φ-monotone,
hypermaximal, 36
maximal, 36
Φ-system, 28
global, 28
pseudo-contractive, 108
strongly, 238
pseudo-monotone, 89, 280

semiaccretive, 237
strongly, 234
weakly, 236
semicontractive, 224
strongly, 223
weakly, 221
semi-invertible, 63
semimonotone,
strongly, 241
weakly, 241
strict contractions, 3
strictly orientation-preserving, 249
strongly pseudo-contractive, 238
strongly semiaccretive, 234
strongly semicontractive, 223
strongly semimonotone, 241
subgradient mapping, 97

topological degree, 183
transition mapping $U(t)$, 121
type (S), 277
type (S_+), 279

upper-semicontinuous set-valued mapping, 38

weakly semiaccretive, 236
weakly semicontractive, 221
weakly semimonotone, 241

QA
329.8
B76

APR 19 1978